Series in Medical Physics

INTENSITY-MODULATED RADIATION THERAPY

Steve Webb

Professor of Radiological Physics,
Head, Joint Department of Physics,
Institute of Cancer Research and
Royal Marsden NHS Trust,
Sutton, Surrey, UK

IoP
Institute of Physics Publishing
Bristol and Philadelphia

British Library Cataloguing-in-Publication Data

A catalogue record for this book is available from the British Library.

ISBN 0 7503 0699 8

Library of Congress Cataloging-in-Publication Data are available

Cover picture shows a 3D rendering of the volume of the prostate, the adjacent bladder and rectum, which are organs-at-risk when treating the prostate. It is clear that the concave nature of the surface of the prostate and the close proximity to the rectum in particular, requires a concave dose distribution to be delivered. IMRT is required in such circumstances to maximize the dose to the target volume and avoid damage to surrounding healthy tissue.

Commissioning Editor: John Navas
Production Editor: Simon Laurenson
Production Control: Sarah Plenty
Cover Design: Frederique Swist
Marketing Executive: Colin Fenton

Published by Institute of Physics Publishing, wholly owned by The Institute of Physics, London

Institute of Physics Publishing, Dirac House, Temple Back, Bristol BS1 6BE, UK

US Office: Institute of Physics Publishing, The Public Ledger Building, Suite 1035, 150 South Independence Mall West, Philadelphia, PA 19106, USA

Typeset in LaTeX using the IOP Bookmaker Macros
Printed in Great Britain by MPG Books Ltd, Bodmin

The Series in Medical Physics is the official book series of the International Federation for Medical and Biological Engineering (IFMBE) and the International Organization for Medical Physics (IOMP).

IFMBE

The IFMBE was established in 1959 to provide medical and biological engineering with an international presence. The Federation has a long history of encouraging and promoting international cooperation and collaboration in the use of technology for improving the health and life quality of man.
The IFMBE is an organization that is mostly an affiliation of national societies. Transnational organizations can also obtain membership. At present there are 42 national members, and one transnational member with a total membership in excess of 15 000. An observer category is provided to give personal status to groups or organizations considering formal affiliation.

Objectives

- To reflect the interests and initiatives of the affiliated organizations.
- To generate and disseminate information of interest to the medical and biological engineering community and international organizations.
- To provide an international forum for the exchange of ideas and concepts.
- To encourage and foster research and application of medical and biological engineering knowledge and techniques in support of life quality and cost-effective health care.
- To stimulate international cooperation and collaboration on medical and biological engineering matters.
- To encourage educational programmes which develop scientific and technical expertise in medical and biological engineering.

Activities

The IFMBE has published the journal *Medical and Biological Engineering and Computing* for over 34 years. A new journal *Cellular Engineering* was established in 1996 in order to stimulate this emerging field in biomedical engineering. In *IFMBE News* members are kept informed of the developments in the Federation. *Clinical Engineering Update* is a publication of our division of Clinical Engineering. The Federation also has a division for Technology Assessment in Health Care.
Every three years, the IFMBE holds a World Congress on Medical Physics and Biomedical Engineering, organized in cooperation with the IOMP and the IUPESM. In addition, annual, milestone, regional conferences are organized in different regions of the world, such as the Asia Pacific, Baltic, Mediterrancan, African and South American regions.
The administrative council of the IFMBE meets once or twice a year and is the steering body for the IFMBE. The council is subject to the rulings of the General Assembly which meets every three years.

For further information on the activities of the IFMBE, please contact Jos A E Spaan, Professor of Medical Physics, Academic Medical Centre, University of Amsterdam, PO Box 22660, Meibergdreef 9, 1105 AZ, Amsterdam, The Netherlands. Tel: 31 (0) 20 566 5200. Fax: 31 (0) 20 691 7233. Email: IFMBE@amc.uva.nl. WWW: http://vub.vub.ac.be/~ifmbe.

IOMP

The IOMP was founded in 1963. The membership includes 64 national societies, two international organizations and 12 000 individuals. Membership of IOMP consists of individual members of the Adhering National Organizations. Two other forms of membership are available, namely Affiliated Regional Organization and Corporate Members. The IOMP is administered by a Council, which consists of delegates from each of the Adhering National Organization; regular meetings of Council are held every three years at the International Conference on Medical Physics (ICMP). The Officers of the Council are the President, the Vice-President and the Secretary-General. IOMP committees include: developing countries, education and training; nominating; and publications.

Objectives

• To organize international cooperation in medical physics in all its aspects, especially in developing countries.
• To encourage and advise on the formation of national organizations of medical physics in those countries which lack such organizations.

Activities

Official publications of the IOMP are *Physiological Measurement*, *Physics in Medicine and Biology* and the *Series in Medical Physics*, all published by Institute of Physics Publishing. The IOMP publishes a bulletin *Medical Physics World* twice a year.
Two Council meetings and one General Assembly are held every three years at the ICMP. The most recent ICMPs were held in Kyoto, Japan (1991), Rio de Janeiro, Brazil (1994), Nice, France (1997) and Chicago, USA (2000). These conferences are normally held in collaboration with the IFMBE to form the World Congress on Medical Physics and Biomedical Engineering. The IOMP also sponsors occasional international conferences, workshops and courses.

For further information contact: Hans Svensson, PhD, DSc, Professor, Radiation Physics Department, University Hospital, 90185 Umeå, Sweden. Tel: (46) 90 785 3891. Fax: (46) 90 785 1588. Email: Hans.Svensson@radfys.umu.se.

CONTENTS

ACKNOWLEDGMENTS

A book gets written when a number of favourable factors come into conjunction. These synergistic elements are the focus of my acknowledgements. First and foremost for this book was the emergence of the field of IMRT as a primary focus of attention for radiotherapists and physicists. The explosive growth of this field provided the rationale for my study and the scientists who have built the body of topics reviewed have been my main inspiration. It would be invidious to pick out names but all those whose work is cited will recognize their work here and merit acknowledgement. Without their efforts there would be no science of IMRT and without this there would be no book.

Closer to home I want to thank my colleagues at the Institute of Cancer Research (ICR) and Royal Marsden NHS Trust (RMNHST) who have helped me learn so much about radiotherapy. In the Academic Unit of Radiotherapy, these include Alan Horwich, David Dearnaley, John Yarnold, Mike Brada, Gill Ross, Vincent Khoo and Chris Nutting. In the Radiotherapy Physics Teams of the Joint Department of Physics, I am grateful to *inter alia* during the time of writing Liz Adams, James Bedford, Margaret Bidmead, Stephen Breen, David Convery, Viv Cosgrove, Ellen Donovan, Phil Evans, John Fenwick, David Finnigan, Glenn Flux, Vibeke Hansen, Sarah Heisig, Philip Murphy, Alan Nahum, Mark Oldham, Mike Partridge, Mike Rosenbloom, Carl Rowbottom, Bea Sanchez, Richard Symonds-Tayler, Jim Warrington and my students Charlotte Hector, Joao Seco, Sarah Gulliford and Young Lee.

The work of the ICR and RMNHST in intensity-modulated radiotherapy is supported by the Cancer Research Campaign (CRC), the Engineering and Physical Sciences Research Council (EPSRC), Elekta Oncology Systems, Varian, the NOMOS Corporation, the HELAX AB Corporation, and I gratefully acknowledge many insights into IMRT that have resulted from collaborations with these companies. Specifically our membership of the Elekta Oncology International IMRT Consortium has been very fruitful. I was privileged to be co-chairman with Wilfred de Neve (Ghent) in 1998 and 1999.

In the summer of 1996, I became Head of the Radiotherapy Physics Research Team and in the autumn of 1998, Head of the Joint Department of Physics. Performing these duties, attempting to continue my personal research and to write this book have stretched some personal limits, and maybe the tolerance of my colleagues and family. But they have understood that this is how I am, how I need to be and how the Department needs me to be, and they have actively supported me. So, a very big debt of gratitude is extended to them.

I am most grateful for help from Sue Sugden in the Institute of Cancer Research Sutton Branch Library, and to Ray Stuckey and his photographic staff for help in producing illustrations.

I should like to thank all the publishers and authors who have allowed figures to be reproduced. The authors are acknowledged in the figure captions. All publishers were contacted with requests for permission to reproduce copyright material.

I am very grateful to the staff at Institute of Physics Publishing, particularly Nicki Dennis, John Navas and Simon Laurenson for their enthusiasm for this project, and Rachel Lumpkin for production work.

The reference lists are up-to-date as of June 2000. This list concentrates largely, but not exclusively, on the post-1996 literature—earlier work having been reviewed in my previous two books on 3D radiotherapy.

The material reviewed here represents the understanding and personal views of the author offered in good faith. Formulae or statements should not be used in any way concerned with the treatment of patients without checking on the part of the user.

Finally, though they should come first in the list, thank you Linda, Tom and David.

PREFACE

CONFORMAL AND INTENSITY-MODULATED RADIOTHERAPY—WHY AND HOW?

Scientific rationale

Whilst drug development plays an important but expensive role in the treatment of some cancers, it is estimated that, at a time when 2% of patients are cured by drugs, some 30% are cured by a combination of surgery and radiotherapy. Radiotherapy may be the 'sticking plaster of cancer' but it is likely to be necessary for many decades to come and it merits continual development. The development of 3D conformal radiotherapy (CFRT), in which the high-dose volume matches the target volume and avoids normal tissues, has been a major theme for improving the physical basis of radiotherapy. Intensity-modulated radiation therapy (IMRT) is the most advanced form of CFRT. As with medical imaging, its development requires the professional activity of physicists and engineers. In the next ten years, it will be routinely possible to automatically geometrically shape radiation fields, modulate the intensity of such radiation under computer control, verify that radiotherapy is being accurately delivered, predict the clinical outcome via biological models and, if not eliminate uncertainties, quantify them. It may be possible to customize radiotherapy to the individual radiosensitivity of individual patients. Robotic radiotherapy is in its infancy but deserves more attention (Schweikard *et al* 1995, 1996, Gilio *et al* 1998, Delpy *et al* 1998, Shiomi *et al* 1998, Webb 1999a,b,d, 2000a,b, Brahme and Lind 1999). It may solve the problem of accurately irradiating the moving patient and moving target.

Since 'missing the tumour' is the most serious possible CFRT mistake, CFRT is 'driven' by 3D multimodality medical imaging. It is thus evident, given that 3D CFRT is the only way to improve the dose-based estimate of tumour control probability, that a Centre of Excellence in 3D CFRT must have either the same level of capability in medical imaging or at least have telematic access to registered 3D images. The role of ultrasound has, to date, been more diagnostic than therapeutic but may, in future, not only provide a means to monitor tumour regression but also, more importantly, monitor target position fraction-by-fraction. The daily alignment of the high-dose volume with the target, by ultrasound guidance, has been achieved in a research setting, but requires translating to the clinic. This may provide the solution to the ultimate doubts expressed by the critics of CFRT that, whilst the weapon is excellent, it may not hit all of the moving target . Bel *et al* (2000) describe a computer-driven couch which could be used for patient repositioning.

Political dimension

The challenges for Health Services to clinically implement CFRT are evident. 3D CFRT is sadly at present largely a research tool, perhaps less so in the USA than in the UK. There are some national trials in the UK of very limited approaches with conventional equipment. There are few supported programmes of 'translational research' which will bring techniques, well understood by physicists, into the clinic. Moreover, there is very little support for clinical trials of CFRT. In the UK what there is largely comes from research councils and industry. Whilst linacs are widely available, the specialist equipment for CFRT is not. As with medical imaging, questions arise of geographic equality of opportunity. This writer believes that the goal for the next ten years must be to first establish several clinical Centres of Excellence for CFRT, followed by a wider dissemination possibly beyond this timescale. With techniques which are most definitely not 'off the shelf' the requirement for long-term support of professional staff to maintain and grow further expertise is of paramount importance.

There is a 'catch-22'. We live in an age where we are increasingly challenged to prove that new methods of diagnosis and treatment save more lives, improve on established methods and save more money. In the short term they rarely do. A hospital wishing to implement new methods of diagnosing, treating and monitoring the treatment of cancer, may have increased costs for several years before the clinical benefit is evident (Purdy 2000a,b,c). Prototype equipment is expensive. Scientists with the appropriate skills are 'slowly grown' and costly to retain. A hospital challenged with simply balancing its budget is unlikely to make significant leaps in cancer care. Yet balance it must and the solution is a political one. More money alone does not buy good health care but it fosters retention of expertise and a climate of relevant translational research. Failed treatments are a waste of resources and are unethical. There must be a climate of confidence in technological progress in the period when proof is lacking. Those who can look back to times when governments 'looked around for ways to spend money' (e.g. the 1950s and 1960s) identify those periods with major steps forward which were born of focused effort, not 'proof before implementation' (e.g. the birth of X-ray CT, the beginnings of computer-based treatment planning).

The impact of new ideas

The impact on society may be assessed in three parts; on health services and health professionals, on the patient, and on the public. Recently, there have been efforts to quantify the outcome of conformal radiotherapy cost-wise (Tepper 1998). The health services must form close links with industry as the trend emerges towards industry-funded research and development in medical imaging and radiotherapy, not within its factories, but within hospitals. Expensive equipment should have a much higher duty cycle. It remains largely unused for 70% of the 24 hour day. To raise this level requires addressing the social and professional needs of radiographers, physicists and medical doctors. Unswollen staff numbers cannot ever raise the duty cycle. As hospitals establish internet sites, GPs (and patients) will become increasingly aware of competing options. This is an excellent way to raise their expectations but may fail to deliver if all they see are research efforts not bridging the gap to clinical reality. The education and training of medical doctors must itself have a greater involvement from physicists and engineers. Those with establishment anti-progress views must be gently cajoled into changing their minds. The education and training structure of medical physicists at present also leads to diverging career paths whereby most research scientists labour under increased pressure to win grants, publish and demonstrate novelty, usually only possible with 100% research activity, whilst many healthcare-funded physicists struggle to service existing techniques with little time for research. This is bad for both. Few see fit to resource translational science and this lack must be filled by the health services. This will in turn shorten the time gap from development to clinical impact. With the possible exception of X-ray CT, almost all clinical imaging and radiotherapy has developed over a timescale of decades rather than years, something to be regretted and changed. The educated, net-literate, patient will increasingly demand access to higher quality diagnosis and treatment. Hospitals may well need to employ contact staff who can knowledgeably and sensitively interface between the worried patient and technological solutions in flux. For the public at large the clinical implementation of new imaging and therapeutic techniques is incompatible with low taxation and the solution must be for those who clamour for better health care and are in paid employment to agree to be taxed more for it.

The breathless chase to keep abreast of ideas: rationale for this book and its place in a series

The physics of radiation therapy is now advancing at a pace which is hard to keep up with. Until a decade or so ago, the technical options for delivering radiotherapy were relatively few and progress relatively slow. All this has changed with the escalation of interest in, and activity developing, conformal radiotherapy and specifically intensity-modulated radiation therapy (IMRT). There is still a gulf between research activity and clinical practice but the latter will never come to fruition until those charged with its development have a thorough grasp of the techniques.

At the beginning of the 1990s there was no single textbook covering the full scope of these developments and the literature was growing apace. In 1993 I filled this void with 'Volume 1' of a book describing the physics of three-dimensional radiation therapy (Webb 1993). It took about two years to write this and most of what was covered is still fundamental theory and technique of validity today. To keep track of the deluge of ensuing literature I wrote a second 'Volume 2' which built on the first but also could be read independently of it (Webb 1997d). This writing occupied a further four years and the book appeared in 1997. 'Volume 2' had similar themes to the first but some particular in-depth developments. It homed in on conformal therapy. The two landmarks will hopefully stand the test of time. However, about this time it seemed to me that this was necessarily an ongoing task. The literature (at least to me) was getting out of hand and however fast I tried to assimilate it, it seemed to outdistance me. Others told me they felt the same. Particularly those newly entering the field felt the need for some lifelines of the kind these books provided. I was encouraged to 'keep going' and developed the strong feeling that the series would need continuation. Other books on a similar theme from different stables appeared during the mid 1990s and I welcome this.

My judgement (maybe wrong) is that the topics I covered in 1993 and 1997 cannot now, each and all, be followed and reviewed in depth by one person and certainly not within the confined space of one single textbook. The task is just too great and the end product would be too sketchy. My own interests have also polarized and specialized more and this 'Volume 3' homes in on a theme dear to my heart, the physics of optimizing CFRT and IMRT and delivering these forms of radiotherapy (figure P.1). I leave it to others hopefully to do a similar job for the other topics.

Steve Webb
June 2000

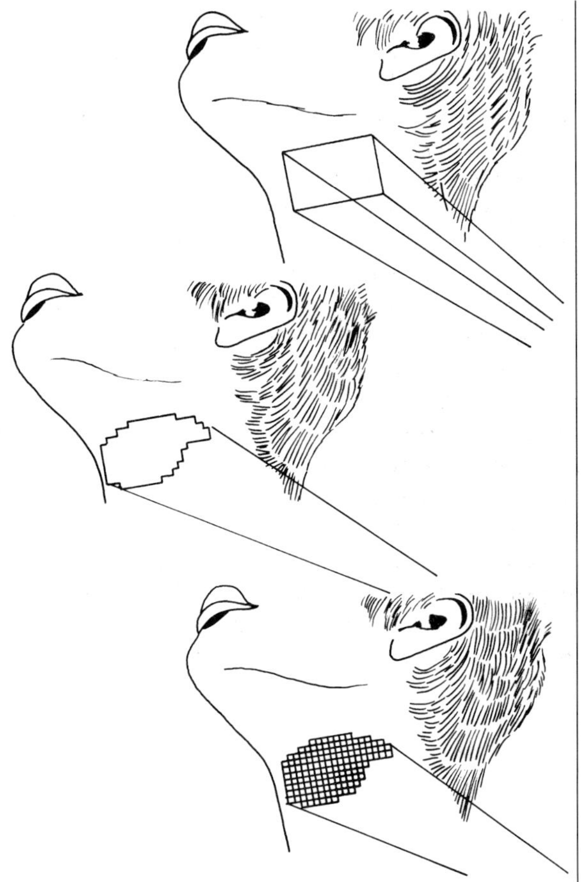

Figure P.1. *Illustrating the key differences between conventional radiotherapy (upper figure), conformal radiotherapy (CFRT) without intensity-modulation (middle figure) and CFRT with intensity modulation (IMRT) (lower figure). For almost a century radiotherapy could only be delivered using rectangularly-shaped fields with additional blocks and wedges (conventional radiotherapy). With the advent of the multileaf collimator (MLC) more convenient geometric field shaping could be engineered (CFRT). The most advanced form of CFRT is now IMRT whereby not only is the field geometrically shaped but the intensity is varied bixel-by-bixel within the shaped field. This is especially useful when the target volume has a concavity in its surface and/or closely juxtaposes OARs, e.g. as shown here in the head-and-neck, where tumours may be adjacent to spine, orbits, optic nerves and parotid glands.*

CHAPTER 1

IMRT: GENERAL CONSIDERATIONS

1.1. BACKGROUND TO THE DEVELOPMENT OF IMRT

1.1.1. Why conformal radiotherapy?

Conformal radiotherapy (CFRT) aims to exploit the potential biological improvements consequent on better spatial localization of the high-dose volume (Webb 1998c). CFRT may be broadly divided into two classes of technique: those which employ geometric fieldshaping alone and those which also modulate the intensity of fluence across the geometrically-shaped field (figure 1.1). The tenet is that by sparing more volume of organs-at-risk (OARs) the dose to the planning target volume (PTV) can be escalated. Despite studies which show that the application of the same treatment techniques to different patients actually gives different dose-volume histograms for the set of patients and thus in principle the possibility to customize the dose to the individual patient, this is rarely done in practice at present (Mc'Nee *et al* 1998, Nahum 1998). However, even protocol-driven trials (without individualization) are showing that an improved conformation correlates with an improved clinical outcome. For example, for prostate therapy, conformal blocking leads to less radiation induced proctitis and bleeding (Dearnaley 1995, 1998, Dearnaley *et al* 1999). The reduction of this late radiation damage was impressive although there was no reported improvement in early radiation acute reactions (Tait *et al* 1997, Tait and Nahum 1997, Carrie and Ginestet 1997, Teh *et al* 2000d). Whilst there is still controversy about predicting biological outcome (Peacock *et al* 1998, Roberts and Hendry 1998) the aim of conformal therapy is sound. Figures 1.2 and 1.3 show examples from the prostate and brain, respectively, of a concave surface when it requires the implementation of intensity-modulated radiotherapy. Figure 1.4 shows the need for conformal radiotherapy of the breast. Similarly 'conformal avoidance' is a term used to describe sculpting the dose to avoid OARs when this is more important than obtaining a good distribution of dose in the PTV; Aldridge *et al* (1998, 1999) have shown that this can sometimes require a large number of beams. The US National Cancer Institute QA programme for 3D CFRT is an example of the kind

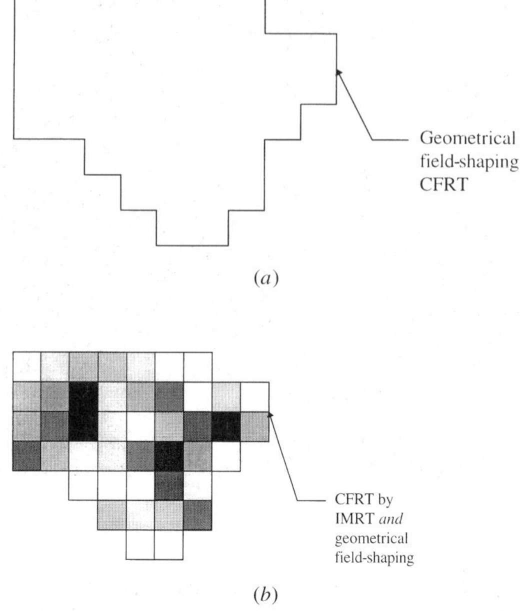

Figure 1.1. *Illustrates the two major classes of CFRT. The beam's-eye view of a hypothetical target is shown for just one field: (a) using simple geometric field-shaping alone; (b) using intensity modulation across this geometric field-shape. The grey levels reflect the intensity value with bright higher than dark. The latter is the kind of modulation which might result from inverse planning e.g. with the NOMOS CORVUS or the KONRAD treatment-planning system (see chapter 2).*

of database and analysis that is essential if the models and data for biological predictions of tumour control probability (TCP) and normal tissue complication probability (NTCP) are to be firmed up (Michalski *et al* 1998). There are also proposals for a similar programme to coordinate centres in the UK. However, some authors (e.g. Levitt 1999, Schulz 1999a) controversially dispute the claim of benefit for CFRT.

Schulz (1999a) has based his argument against conformal radiotherapy on the view that an improved cure could not be demonstrated to justify increased cost and complexity. His opinion has been challenged by Mohan *et al* (1999) who claims that there *is* evidence for increased cure, e.g. in the prostate against measurable (PSA change) criteria. They also argue that the development of new techniques would, at minimum, allow the generation of more data to support the dose-escalation hypothesis. They further argue for CFRT as a form of conformal avoidance whereby normal structures are saved even if there can be no demonstrable improvement in tumour control. They plead for an understanding

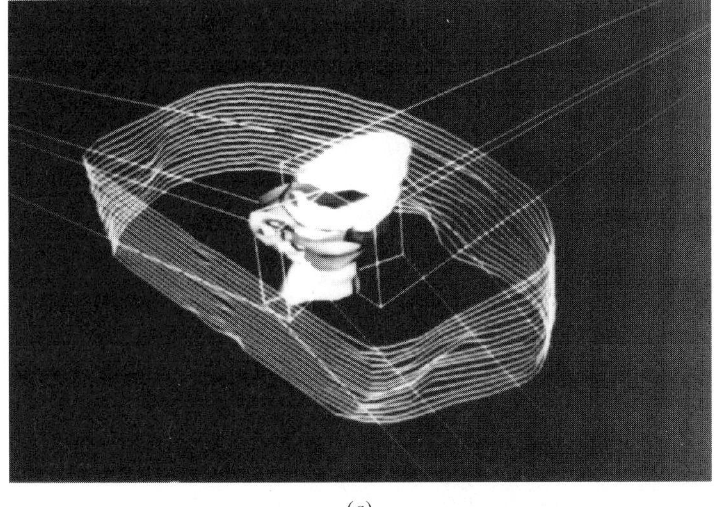

(*a*)

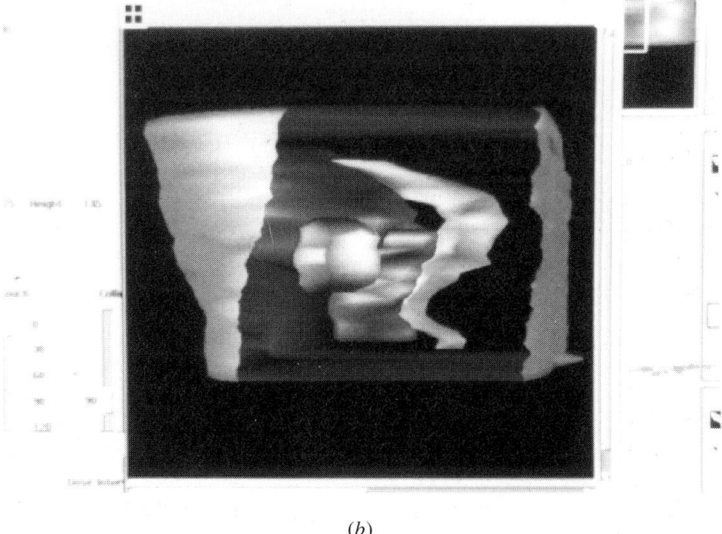

(*b*)

Figure 1.2. *(a) This shows a 3D rendering of the volume of the prostate, the adjacent bladder and rectum, which are OARs when treating the prostate. The rendering is formed in the planning system VOXELPLAN from contours in a series of transaxial slices created by CT scanning. It is clear that the concave nature of the surface of the prostate and the close proximity of the rectum in particular requires a concave dose distribution to be delivered to the prostate. This would not be obtained from the unmodulated field-shapes shown. Intensity-modulation is called for. (b) A similar clinical case viewed laterally extracted from the VOXELPLAN planning system.*

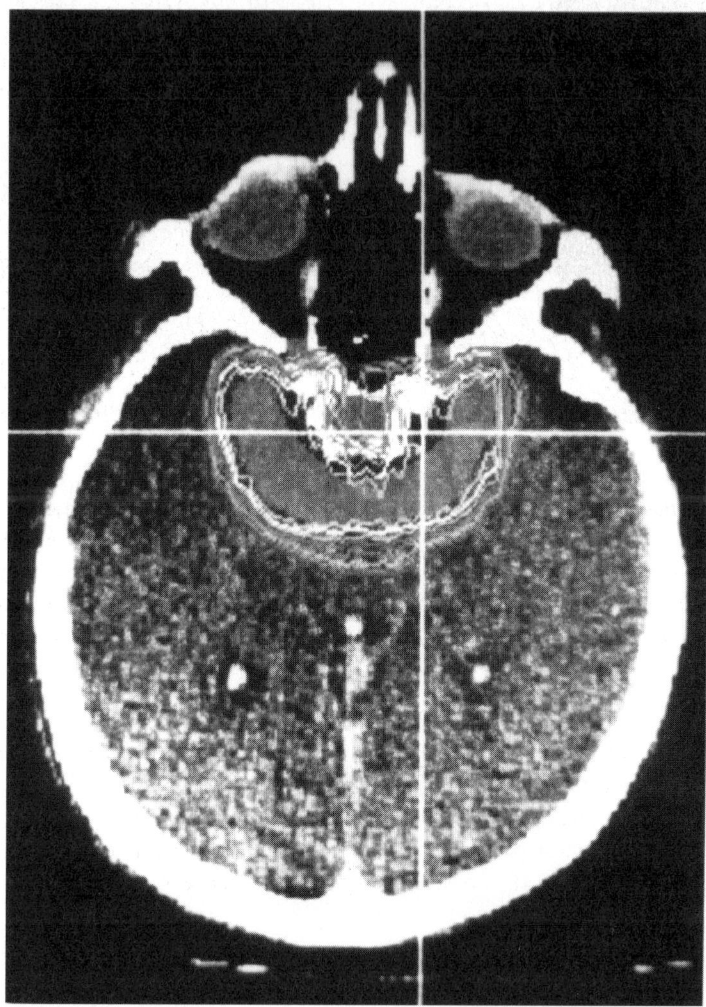

Figure 1.3. *Showing a transaxial slice through the brain indicating the requirement to treat a concave target region and hence the need for IMRT. (Courtesy of the NOMOS Corporation.)*

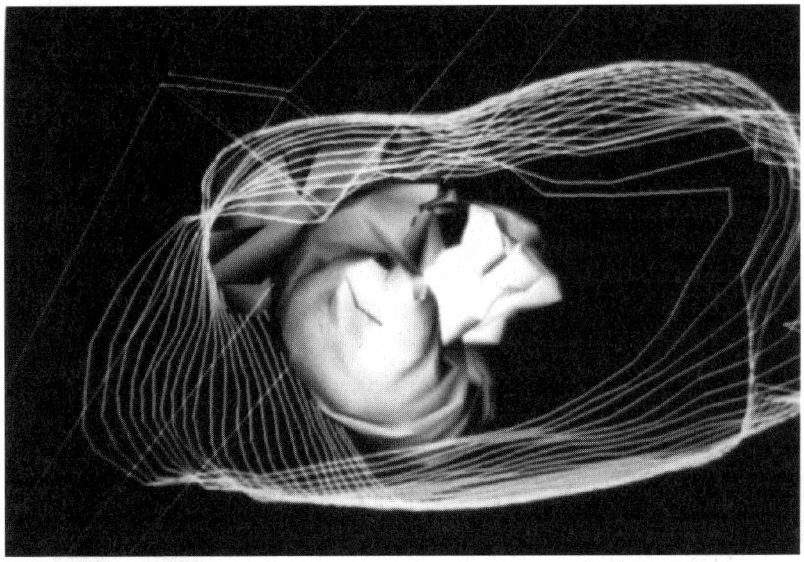

Figure 1.4. *This shows a 3D rendering of the volume of the breast together with the adjacent lung and heart which are OARs. The rendering is formed in the planning system VOXELPLAN from contours in a series of transaxial slices created by CT scanning. It is clear that the close proximity of the heart in particular requires a modulated dose distribution. This would not be obtained from the unmodulated field-shapes shown. Intensity-modulation is called for so the breast is uniformly irradiated with minimal dose to the OARs.*

that new technologies are inevitably inefficient and costly at first, but in time prove their worth. They cite many of the major developments in radiotherapy as having been developed more through vision than through trials. In a riposte, Schulz (1999b) remains unconvinced. From my view of the accruing data, and with the background outlined in my preface, I align with the critics of Schulz . This book is my contribution to the debate.

1.1.2. IMRT as a form of CFRT—old history

IMRT is a form of CFRT (Webb 1998d, Mohan 1998b). The period of history of IMRT has a length which depends on one's definition of IMRT. Table 1 shows the principal IMRT landmarks in a very broad class outline. If the period is meant to encompass only the serious attempts to produce automatic and arbitrary intensity maps by sophisticated electromedical equipment, then the history is relatively short, certainly not much more than ten years or so. However, intensity-modulated distributions were being produced many decades ago. A simple block produces a binary intensity distribution; the primary fluence is either present or (almost,

Table 1. *The principal IMRT landmarks.*

Date	IMRT landmark
Pre-1960	Primitive IMRT with blocks, wedges and compensators
1960	Proimos' gravity-oriented device
1982	Brahme *et al* proposed the solution for the wedged/blocked 1D IMB to give a uniform annular dose
1988	Brahme's paper on inverse planning for IMBs
1989	Webb proposed simulated-annealing inverse planning for IMRT
1991	Principle of segmented field therapy proposed (Webb/Boyer)
1992	Convery proposed the DMLC IMRT technique
1992	Announcement of the NOMOS MIMiC IMRT delivery system
1993	Concept of tomotherapy integrated spiral machine proposed by Mackie
1994	Bortfeld and Boyer made the first MSF MLC IMRT deliveries
1994	Principles of the DMLC IMRT delivery technique elaborated
1994→	Refinement of techniques and explosion of commercial interest in IMRT

excepting leakage) absent. A wedge produces a gradient of intensity across a space (figure 1.5). In the 1960s Basil Proimos and colleagues created high-dose volumes with concave boundaries by hanging attenuating blocks by gravity on an accelerator. As the gantry rotated, OARs were maintained in the shadow of the blocks. The idea is still being pursued today, and is now called the gravity-oriented device (GOD) (Proimos and Danciu 1997a, b). Chui (1998) is combining the use of a gravity-oriented shield (GOSh) with IMRT delivered by the dynamic multileaf collimator (DMLC) method (see section 3.1). However, in the early days, these techniques were cumbersome and did not lead to very uniform dose distributions in the PTV. Other methods to achieve conformal blocking of sensitive structure and conformal radiation of targets were pioneered by Takahashi in Japan, by Jennings and Green *et al* at the Royal Northern Hospital in London, and by Bjarngard and Kijewski *et al* in Boston (Purdy 1997).

The mainstay of intensity modulation throughout the decades has been the cast metal compensator. Apart from the tedium of casting this, its relative inflexibility to change between treatment fractions and the need to customize for each patient (thus keeping the maximum number of fields small), this device still achieves its objective (Harari *et al* 1998, Beavis *et al* 1999b, Meyer *et al* 1999a, b). Several groups have described the use of tissue compensators for constructing intensity-modulated beams (IMBs) for improving dose homogeneity of the breast. The work at the Royal Marsden NHS Trust is reviewed in section 4.1.13 (Evans *et al* 1995). Other studies have been reported by e.g. Carruthers *et al* (1999).

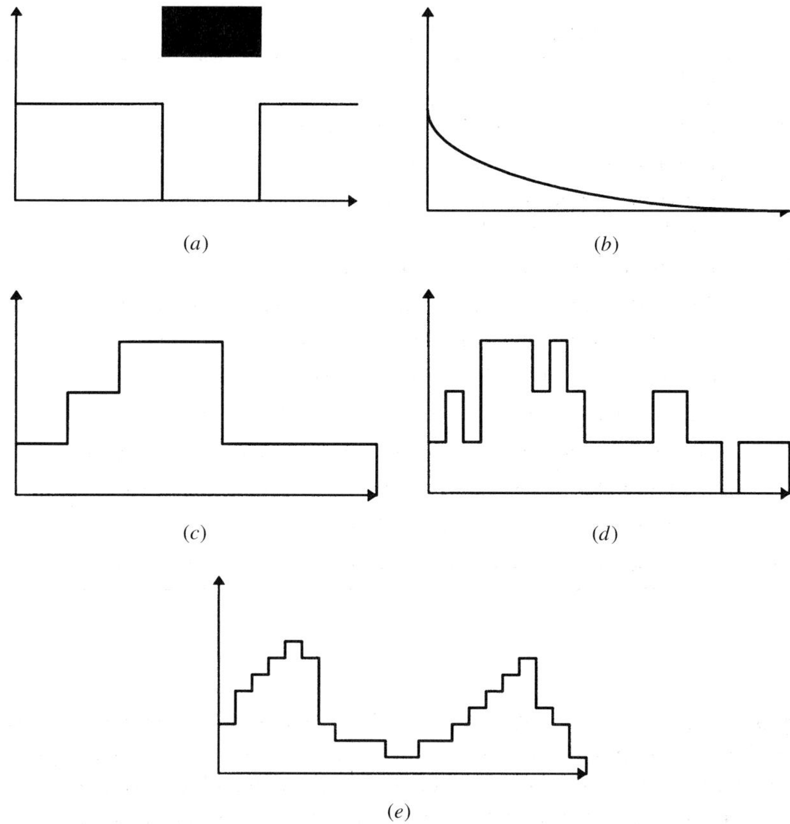

Figure 1.5. *IMRT of varying complexity: (a) 'binary on-off' (block); (b) wedge; (c) full modulation (coarse spatial and intensity scale); (d) full modulation (fine spatial, coarse intensity scale; (e) full modulation (fine spatial and intensity scale). The vertical axis in each case represents intensity or fluence and the horizontal axis represents distance. Here are presented just 1D IMBs corresponding to one line of a 2D modulation such as shown in figure 1.1b.*

All this delivery technology is 'old history' albeit with some modern applications. It is all also based on the concept of optimizing intensity patterns with *photons* (Oldham 1997). At present in the UK, but not in many other countries, the concept of intensity-modulated *proton* therapy is out of favour. Cole and Weatherburn (1997) summarized the plans for PROTOX based on a spot-scanning proton therapy gantry, requesting support for the UK proposal, but to date these are unfunded and the proposal is stalled. With the exception of the Facility at Clatterbridge, enthusiasm for high-energy protons is greater outside the UK. This book concentrates on photon IMRT.

1.1.3. IMRT as a form of CFRT—modern history and inverse planning

The 'modern history' of IMRT spans just a little more than a decade. It goes back to concepts introduced firstly by Brahme (1988) for inverse planning. Inverse planning creates, from the ideal dose prescription (or a statement of biological objectives), the 'best set' of IMBs for the problem (figure 1.5). In itself it does not usually determine a method of treatment *delivery* although constraints on the available delivery apparatus could be factored into these calculations (Cho *et al* 1998b, Gustafsson *et al* 1995, 1998, Webb *et al* 1998, Löf *et al* 1999). At least in the first ten years from 1988 inverse planning did not do this. The planning of IMRT has been posed as a separate task preceding considerations of delivery, and uncoupled from delivery constraints. Many planning techniques have been developed. These include analytical methods, stochastic (iterative) methods and hybrid methods. That there are so many (and there seems no end to the flow) reflects the fact that the problem is not uniquely solvable. The solutions depend on the goals and constraints which are set. This is discussed in sections 5.1 and 5.10.

Inverse planning under this name did not really start to be recognized until around the late 1980s. However, the concepts go back much further and can be recognized in the non-radiotherapy work of George Birkhoff (1940) and the optimization concepts of Newton (1970). Proimos and Danciu (1997a, b) have recently shown that the old idea of suspending in the beam, by gravity, lead-filled model shapes of the OARs can lead to significant protection of the organs and conformal shaping of the field. The idea goes back to the original work by Proimos in the 1960s (see Webb 1993, chapter 2, for a review).

Danciu and Proimos (1999) have made an experimental verification of the gravity-oriented absorber technique in conformal radiotherapy for cervix cancer. A cylindrical phantom of pressed wood was constructed containing four OAR areas. The cross section was photocopied and reduced to 65% of its natural size. Then the corresponding OAR areas were cut out by an electrical band saw and the holes so formed were filled with cerrobend material. This gravity-oriented device was coaxially attached to a rotating gantry such that, during the delivery of 360° rotation, the OARs were always in the shadow of their corresponding protector.

The most extreme form of intensity-modulated radiotherapy is that provided by the use of a gravity-oriented block. Frencl *et al* (1999) have developed a gravity-oriented block technique in Prague. The technique was implemented for an Orion 5 linac and isodose plans were prepared using the treatment-planning system TARGET-2. An arc-therapy technique for an irradiation of a tumour wrapping round a spinal cord was developed.

Cotrutz *et al* (1997) have demonstrated the method of planning for such modulated beam profiles and Cotrutz *et al* (1999a, b, 2000) have developed new ways of gravity-shielding leading to less dose inhomogeneity. Cotrutz *et al* (1999a) have given details of a new intensity-modulated arc therapy technique for delivering conformal radiotherapy. The technique relies on combining radiation from a number of annular sectors, each centred on nearby OARs with the OARs blocked by cerrobend. The role of each rotational beam is to give a constant dose level in the annular sector that fully covers each PTV. The feasibility of the technique has been demonstrated with several planning cases. So, in acknowledging this work, we see a link back to some of the earliest modulations created in the 1960s.

However, putting GODs to one side for now and returning to Brahme, it was he who first recognized that one could conceptually develop the process of intensity-modulated pencil beams from the gamma knife technique. The gamma knife would create a high dose at the convergence point (focus) of a large number of uniform-intensity pencil beams if the patient were actually in a circular water-bath in the device. If the patient were to be moved through the bath in small spatial steps then the same high dose could be deposited over a region in space. Now imagine taking away the water-bath. In this case, the pathlengths to the focus depend on the position of the patient in the machine and these vary as the position of the patient is changed. To now deposit a high uniform dose, the *intensity* of the radiation fields must be modulated from pencil to pencil and the familiar concept of IMRT emerges (Mackie 1997b).

Formally, a dose distribution $\boldsymbol{D}$ is related to a set of beamweights $\boldsymbol{w}$ via a matrix operator $\boldsymbol{M}$ through

$$\boldsymbol{D} = \boldsymbol{M} \cdot \boldsymbol{w} \tag{1.1}$$

where $\boldsymbol{M}$ is a matrix containing all the dosimetric links between each beam element and each dose voxel. The inverse operation

$$\boldsymbol{w} = \boldsymbol{M}^{-1} \cdot \boldsymbol{D} \tag{1.2}$$

carried out directly would not be very successful because the matrix $\boldsymbol{M}$ is sparse and ill-conditioned. Hence, iterative solutions are often employed, minimizing some cost function. Constraints on the problem lead to feasible solutions in beamweight-space. Some cost functions have local minima. One way to find out whether these exist is to repeat the search strategy from different initial estimates. If the same result ensues there are no local minima and a deterministic search strategy can be used. If, on the other hand, a set of different outcomes ensues then a non-deterministic search strategy, such as simulated annealing, must be used (Mackie 1997b, Webb 1995b, 1997c). The field of developing inverse-planning algorithms has become a 'cottage industry' and there seems to be no end to the flow of new techniques continuously being developed. The emphasis in new techniques is in:

(i) increased speed;
(ii) increased mimicking of clinical judgement (e.g. the use of cost functions based on dose-volume constraints or biological measures);
(iii) tailoring the inverse planning to the actual delivery technique; and
(iv) reducing the complexity of the solution.

Photon inverse planning with electron adjunct irradiation has even appeared (Lee *et al* 1997a). However, the bottom line is that radiotherapy still has to deliver energy to tumours by passing it through healthy tissue. So, in answering the question 'how good can IMRT ever get?' the answers tend to be 'how good do you want it?' and 'what are you prepared to pay (in effort, equipment complexity, delivery reliability and, inevitably, money)?' 3D planning for CFRT and IMRT is reviewed in detail in chapter 5.

1.1.4. Methods to deliver modern IMRT

To a great extent the planning methods have run ahead of delivery possibilities. However, this situation is changing and today there are at least six methods of delivering IMRT (Purdy 1996, Verhey 1999b, Webb 1999c, d, Boyer *et al* 2000a, b):

(1) the cast metal compensator;
(2) the use of multiple-static MLC-shaped fields (MSF/MLC) (figure 1.6);
(3) dynamic MLC techniques (DMLC) (figure 1.7) including intensity-modulated arc therapy (IMAT) (Yu *et al* (2000a, b);
(4) the use of the NOMOS MIMiC and the development of tomotherapy (figure 1.8);
(5) the use of a scanning attenuating bar (figure 1.9);
(6) the use of swept pencils of radiation.

These were discussed at some length by Webb (1997d) in 'volume 2'. Since that time, techniques 2–4 have become front runners and this volume concentrates on the extensive developments of these methods subsequent to that review. The captions to figures 1.6–1.9 summarize the key elements of each technique from a mechanical viewpoint, considering just primary fluence without leakage. The IMRT robot has been proposed and preliminary experiments have been carried out by Schweikard *et al* (1995, 1996). However, it is too early to say much about this method although we return to it in section 4.6.

1.1.5. IMRT in clinical practice

The historical development and modern practice of these IMRT delivery methods was reviewed in depth—as it had developed to the time of writing—by Webb (1993, 1997d). More brief and more general-scientist-oriented overviews are also available (Webb 1998c, 2000d). IMRT is now a technique considered mature enough to be an ASTRO course specialty (Mohan 1997). It is also a major component of the European School of Medical Physics, established in Archamps (near Geneva) in November 1998 and run annually in subsequent years. IMRT is now considered by its practitioners to be a mature radiotherapy technology and, as such, has become the subject of many teaching courses; in particular, the first European IMRT Winter School took place in Heidelberg from 9–11 December 1999. This School was sponsored by MRC Systems, one of the manufacturers of inverse treatment-planning computer systems and technology for delivering the dynamic multileaf collimator (DMLC) technique. However, the course was

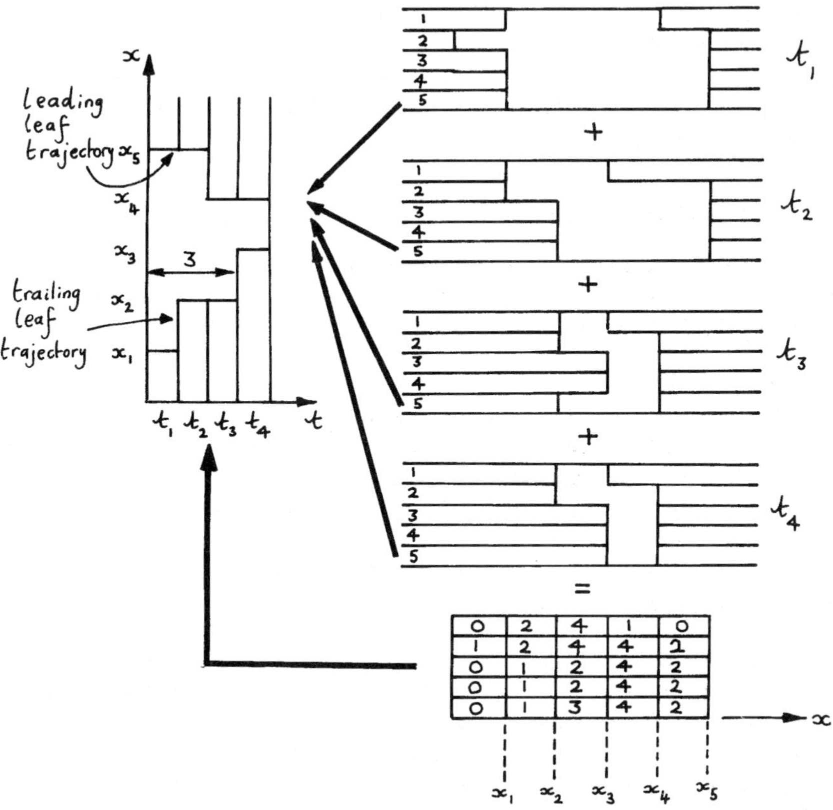

Figure 1.6. *Illustrating the multiple-static-field (MSF) technique for delivering IMRT. For illustrative purposes a five-leaf MLC is shown with leaves labelled 1, 2,…,5. Four field components (segments) at four times t_1, t_2, t_3, t_4 are shown. The radiation is off between segments. For illustrative purposes it is assumed that the irradiation fluence at each segment is the same and proportional to the time interval $t_n - t_{n-1} = 1$ unit, $n = 1,\ldots,4$. (At time t_0 the field is closed completely.) The leaves move positions between segments when the radiation is off. After the delivery of these four segments the 2D IMB shown at the lower figure has been delivered (assuming no leaf leakage and no scatter). To the left of the component diagrams is shown a type of trajectory diagram for the fifth leaf pair. The vertical axis represents distance along the direction of travel of a leaf pair; the horizontal axis represents time. The leaf trajectories are now a set of discrete positions changing at each time. From the trajectory diagram the intensity (primary fluence assuming zero leaf transmission and scatter) between any locations x_j and x_{j-1}, $j = 1,\ldots,5$ is the horizontal distance from origin to leaf, e.g. the intensity between x_2 and x_3 is shown as three units. The same data are entered into the fifth row of the 2D IMB. The other four leaf pairs would require different trajectory diagrams (not shown). In this example all leaf pairs execute what is known as the 'close-in' technique. However alternative patterns of leaf segments could be combined to give exactly the same 2D IMB and this is discussed in detail in chapter 3.*

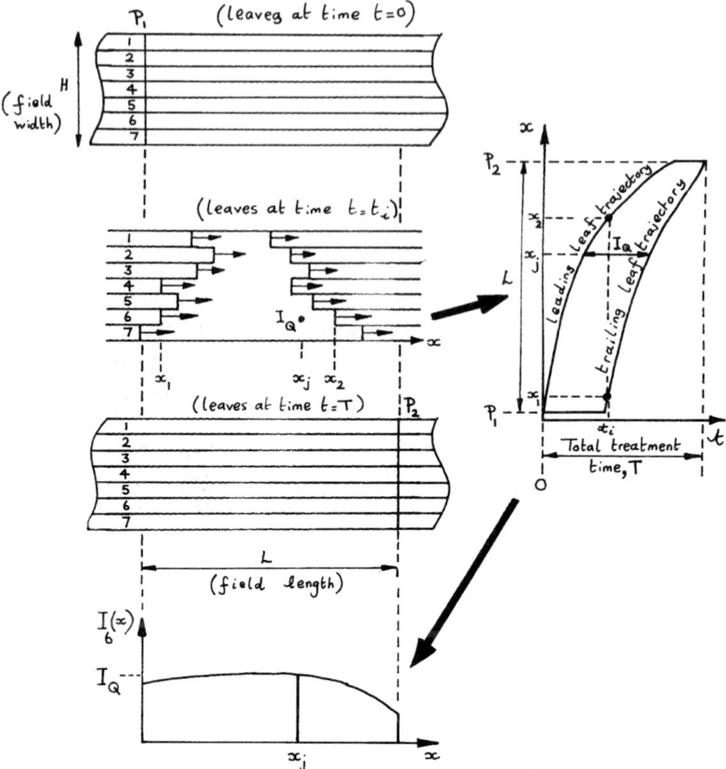

Figure 1.7. *Illustrating the concept of the dynamic MLC (DMLC) technique of delivering IMRT. For illustrative purposes a seven-leaf MLC is shown with leaves labelled 1, 2,. . . ,7. The leaves begin at a parked position with all leaves at the left of the field at P_1 at time $t = 0$ (upper figure) and subsequently move from left to right. The field width H is defined as the projected width of one leaf multiplied by seven. The leaves move as time progresses. The second figure from the top shows an arbitrary time $t = t_i$ at which the leaves have reached the positions shown. The area between the leading and trailing leaves (the open area) is irradiated (ignoring leaf penetration and scatter for now). The other areas are shielded. Finally (third diagram) both the leading and trailing leaves reach a final parked position P_2 at time $t = T$ (the total treatment time) after traversing a field length L. Each leaf pair executes its own trajectory and that for the sixth leaf pair is shown to the right. The vertical axis represents the distance x travelled and the horizontal axis represents time t. (Sometimes these trajectory diagrams are drawn with the axes reversed—it does not matter.) Both leading and trailing leaves move from P_1 to P_2 through distance L. A time t_i is shown at which the leading leaf is at $x = x_2$ and the trailing leaf is at $x = x_1$ as shown in the second diagram from the top. From the trajectory diagram the intensity I_Q (primary fluence assuming zero leaf transmission and scatter) at any location (e.g. $x = x_j$) is the horizontal distance between the two leaf trajectories. The trajectory diagram drawn actually corresponds to the gently varying IMB shown in the lowest diagram where the intensity at any position x is $I_6(x)$.*

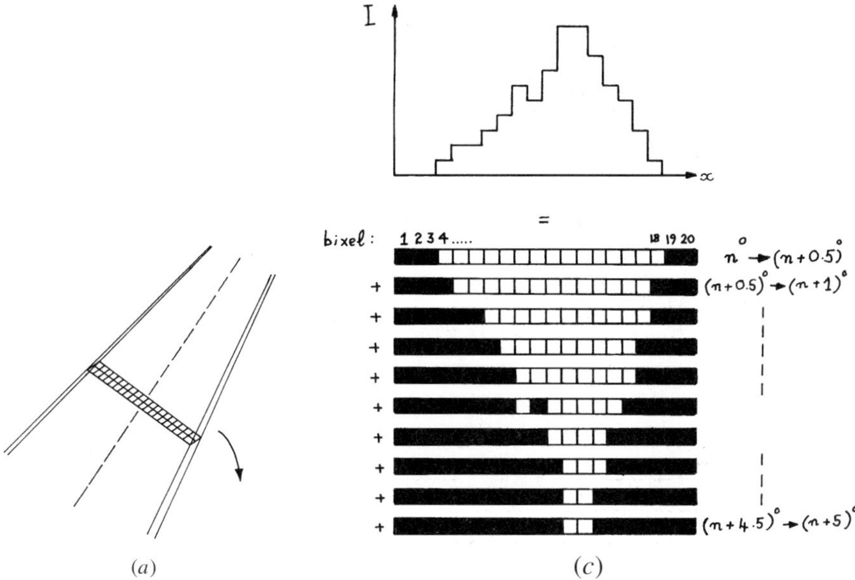

Figure 1.8. *Illustrating the concept of tomotherapy as performed by the NOMOS MIMiC machine to deliver IMRT. The radiation is collimated to a fan which is narrow in the longitudinal (superior–inferior) direction of the patient as shown in (a). The radiation passes through a collimator comprising two banks of 20 beam elements, each of which may have one of only two states—open or closed. The methodology to achieve this is discussed in chapter 2. The radiation fan and collimator rotate about the isocentric axis through an angle of 270°. Inverse planning has previously determined the 1D IMB for each bank at each of 5° intervals of azimuthal angle. One such 1D IMB is shown in (b). The MIMiC delivers this by changing the state of the beam elements (bixels) every 0.5° during continuous rotation. This (ignoring the negligible blurring) achieves the aim of delivering the IMB. The ten constituent states of the MIMiC bank are shown in (c) corresponding to the 1D IMB shown. Notice that superficially this is like a multiple-static-field MLC technique with an important exception. The MIMiC can generate islands of attenuation as shown in the sixth component for example. The success of the technique depends on rapid and reliable switching between open and closed states. The Wisconsin tomotherapy machine uses a similar binary collimator but with a different number of bixels, a single bank and continuous rotation as the patient is translated longitudinally creating a spiral trajectory.*

truly international with an international Board and Faculty. The School was also attended by many of the manufacturers of inverse treatment-planning systems and technology for delivering IMRT. A second Heidelberg IMRT School was held in

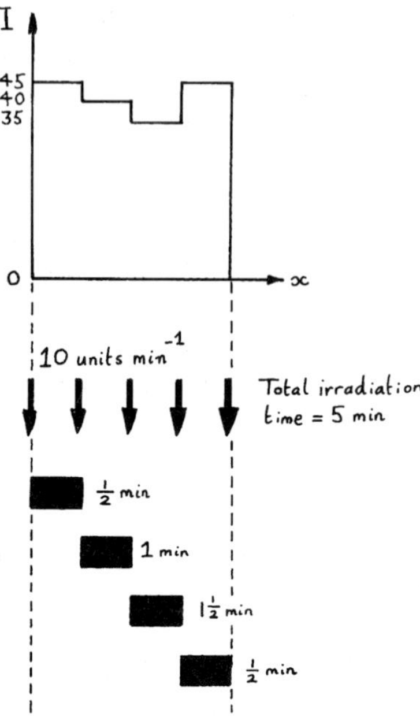

Figure 1.9. *Illustrating the scanning-attenuating-bar technique to deliver a 1D IMB. A radiation source is considered which delivers ten units of radiation per minute and the total irradiation time is 5 min. The 1D IMB in the upper part of the figure is required. This is achieved by scanning an attenuating bar whose width is equal to the bixel size in the IMB from left to right and allowing it to dwell for the times shown at each of four locations. Note that the total dwell time is less than the total irradiation time. This is a constraint and many 1D IMBs cannot be delivered because this constraint would be violated. Note this is a very conceptualized simple representation of the method which in practice can be considerably more complex.*

Heidelberg just prior to the 13th International Conference on the Use of Computers in Radiation Therapy (ICCR) in which approximately 40% of the papers were on IMRT.

The use of modern IMRT is presently confined to a relatively small number of research hospitals, perhaps of the order of 30–40 in the USA and three in the UK. In the UK, at present, there is relatively little clinical IMRT in practice, except by the first method referred to in section 1.1.4. However, there is a positive programme of research effort to bring IMRT into the clinic by techniques 2–4. In the USA, the situation is a little different with IMRT first established clinically by method 4 (Smith 1996) with methods 2 and 3 rapidly catching up (Boyer *et al* 2000a, b). The modern history of techniques 2 and 3 is intimately entwined with the development of the modern clinical MLC (Purdy 1996). The first commercial MLC in the UK was installed around 1992 and manufacturers are now considering its use for IMRT when designing the next generation of MLCs. The MLC is considered in detail in chapter 3. Purdy (1996) has remarked of IMRT that 'the era of IMRT has begun in earnest'. Coupled to this remark was the observation that, just as for the development of CFRT, so the clinical implementation of IMRT requires the establishment of a complete chain of processes, starting with inverse planning and going right through to verification. He also remarked that there have as yet (at least to the time of writing—1996) been no trials of the clinical efficacy of IMRT and therefore the development is predicated on an expected rather than a proven outcome, a theme amplified by Nutting (2000). The current world experience in IMRT is reviewed in chapter 4.

1.1.6. Reasons for IMRT at the present time

Why has IMRT assumed such a prominence today? It is an inescapable major theme of most radiotherapy physics conferences and it is here to stay. Indeed, May 1996 saw the very first conference in Durango, Colorado, entirely devoted to the discussion of IMRT, which resulted in the book edited by Sternick (1997b). Other 'schools' and 'courses' were appearing by 1998. The reasons are easy to identify.

(i) Clinicians require concave dose distributions in maybe 30% of clinical cases. These cannot be achieved without IMRT and IMRT offers a significant step-function leap in tumour control probability without compromising normal structures (figures 1.10 and 1.11).
(ii) Commercial manufacturers are making IMRT delivery technology available, even though it almost always requires further in-house development and customization to the local situation.
(iii) Computer control of radiation delivery is possible.
(iv) Inverse planning to determine IMRT distributions has reached maturity and can be performed in realistic times.
(v) 3D medical imaging by four modalities (CT, MRI, SPECT and PET) can more accurately determine the geometry of target and normal structures.
(vi) Techniques to verify and QA IMRT delivery are emerging.

The writing of history, and the determination of reasons for any development at a particular time, can be a subjective process but the above bare bones of fact are indisputable. There are many interpretations of precisely which influence affected which development. Thus, it is probably best to conclude that the last decade of the first century of radiation physics saw a synergistic development of fast computing, radiation delivery technology, computer control, inverse-planning theory and practice, 3D imaging and planning and a growing awareness of, and dissatisfaction with, the otherwise almost static nature of radiation therapy practice. For some, the progress in IMRT has not been fast enough whilst inevitably there

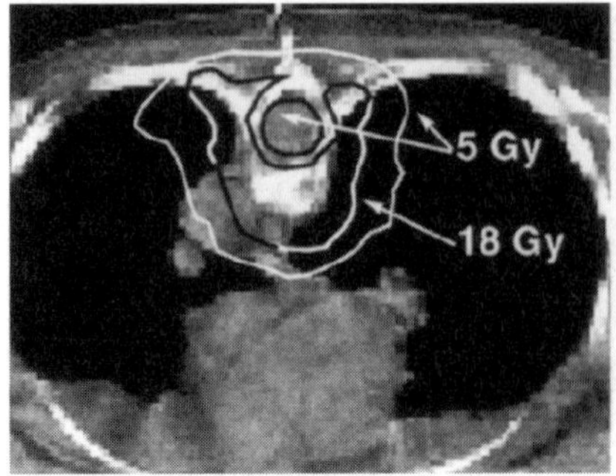

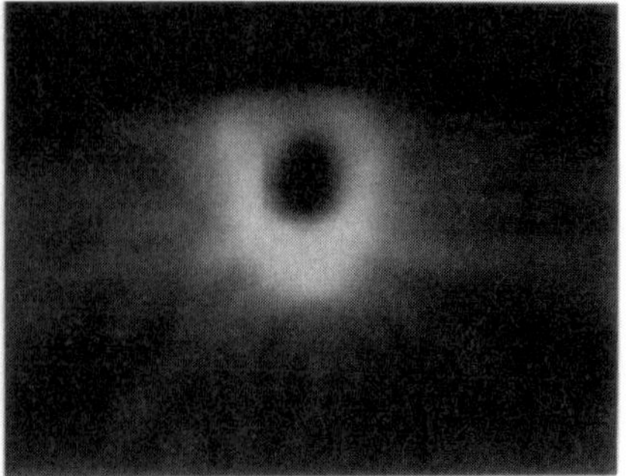

Figure 1.10. *A breast cancer, metastatic to T7, previously treated with a full course of spinal radiation. The patient continued with severe pain in the thoracic spine and was referred for palliative radiation. She was to receive 18 Gy in nine fractions to the target volume shown on the left. The delivered dose distribution is shown to the right. Note the achievement of a concave high-dose volume and protection of the spinal cord. (From Carol 1997a.)*

remain fierce sceptics of its need, practicability and safety (Schulz 1999a, b). The greatest challenge both in the UK and the USA is how to resolve doubts and develop safe IMRT practice against the background of the cost-cutting mentality of hospitals and some funding bodies. This topic was discussed by Webb (1997d) in his preface and is also the central theme of the review by Mackie (1997d). Mackie additionally postulated that, whilst trials to prove increased efficacy of IMRT may be hard to fund—and potential providers of new technology and therapy techniques usually require proof by trial—it may be easier to show that new techniques *do at least as well* which will get the IMRT show on the road at this perilous time of financial limitations in medicine.

1.1.7. Arguments for and against IMRT

The argument has sometimes been voiced that IMRT is potentially dangerous since it 'too tightly' conforms the high-dose distributions and that 'older' methods of conformal radiotherapy are safer. (Older is in quotation marks because in the 105 year history of radiotherapy, CFRT without intensity modulation is still quite new.) This is sometimes offered as a reason to not explore IMRT further. The same argument was offered 25 years ago against the introduction of CT into radiation therapy planning, that it would lead to too tight margins on high dose. In fact, the opposite has happened. CT has shown that the older techniques were actually more dangerous, missing the target sometimes, and CT has led to a widening of margins in some instances. Similarly, by conforming the high-dose volume to more complex shapes by IMRT it may be possible to *widen* the margin to allow for greater tolerance of location errors since IMRT inherently spares more of the

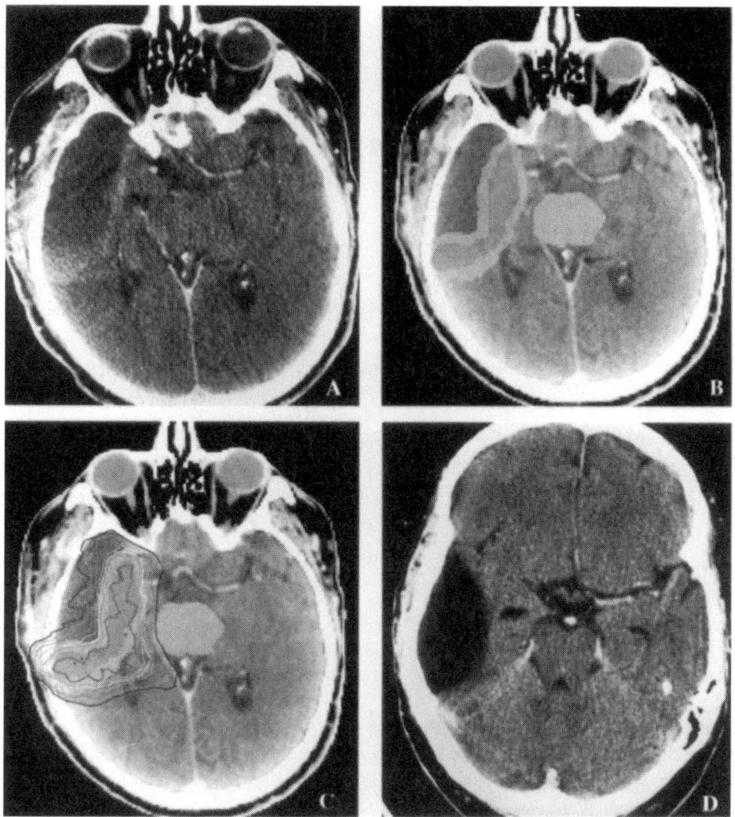

Figure 1.11. *IMRT for conformal boost in the primary treatment of high-grade glioma: (a) post-resection CT scan showing residual contrast-enhancing tumour along the cavity walls; (b) IMRT planning system axial image with the primary target volume (dark) and a surrounding expansion to CTV (less dark); (c) IMRT planning system axial image with isodose display for a prescribed dose to the primary target of 5 Gy × 7; (d) CT scan performed nine months after IMRT with no evidence of tumour recurrence. (From Wazer et al 1997.)*

adjacent normal tissue (Mackie 1997c). There is a danger attached to being a prophet of new ways of thinking. Sternick (1997a, b) has told of the young girl asked to write a brief account of the life of Socrates. She wrote: 'Socrates was a Greek philosopher who went about giving people good advice. They poisoned him.'

Of course, simple IMRT has been achieved for decades with wedges and cast metal compensators (Meyer *et al* 1999a, b). Planning for the use of such hardware can be done on a geometric basis alone. There are some who seriously argue that the new technology for developing IMRT is just not needed; neither is the complex planning (Sherouse 1997). Sherouse argues that the target must be small for there to be an advantage for conformal therapy, that beams should not be opposed, should be as widely spaced as possible and should have geometric symmetry to the degree possible. He has developed a vector technique to arrange this (see review and references in Webb 1997d, chapter 1). However, these methods do not always give the best dose distributions. Today, the development of sophisticated technology has an unstoppable momentum largely because the electrotechnical companies are making huge investments and we have a much deeper understanding of the possibilities of improving radiotherapy. At the same time there is a climate of cost-containment in hospitals. A danger is that the decision to implement or not to implement the benefits of conformal therapy may not be influenced by scientific arguments alone.

One of the difficulties of implementing IMRT clinically is that any particular IMRT delivery technique, which is perceived to be difficult or associated with complications, could damage the whole development of IMRT albeit by different delivery techniques. One might argue that the most successful delivery techniques are those which have been directly linked to specific planning computers, for example, the NOMOS developments linking CORVUS with the MIMiC (chapter 2) and the University of Ghent development linking UMPLAN with the ELEKTA DMLC (chapter 3). It should also be stressed that one must not confuse the ability to produce an IMB on a computer screen with the actuality of knowing that such a beam and its consequent dose distributions have been delivered clinically. It has been argued that perhaps we should have a menu of IMRT delivery techniques ranging from the most simple to the most complicated. It will be unlikely that a single technique, even with the use of a common piece of commercial equipment, will necessarily suit all clinical centres. We might argue that the development of IMRT is the very last thing which conventional accelerators would be able to provide for radiotherapy. So, it is very important to get it right and to prove that it works. More futuristic radiotherapy might rely on the use of robotic linear accelerators (chapters 4 and 5). Appendix 1A develops these themes from the ‘point and counterpoint’ approach.

1.1.8. Summary

In section 1.1 we have looked at the background to the development of IMRT as a specific form of CFRT and identified factors which have been important in establishing the method. The modern development has been placed in historical context and specifically the link to developments in inverse planning has been made. The main techniques for delivering IMRT have been identified and described in outline and these will be expanded upon in the rest of the book. The present clinical situation has been assessed and some factors, which potentially limit the future of IMRT, identified. A political dimension has been noted and, despite some adverse factors, the conclusion is drawn that IMRT has an unstoppable momentum.

1.1.9. A brief word on organization of material

In papers, the descriptions of IMRT developments usually home in on a specific geographical centre, and particular technical matters—especially delivery technique, specific manufacturer's equipment, specific patient data and tumour sites and links to specific inverse-planning, QA, measurement techniques, verification etc. For this reason, any extensive presentation such as this, requiring to organize and group material, has a range of options available. The contents pages provide the macroscopic choices I have made. However, some material could inevitably have appeared in alternative locations within the book, and so extensive consultation of the index is also advised.

1.2. HOW MANY IMBS ARE NEEDED AND WHERE SHOULD THEY BE PLACED?—THE USE OF COMPENSATORS

A general issue which, in the paradigm of disconnecting the planning of IMRT from its delivery is common to all delivery techniques, is that of how many IMBs are needed and where they should be placed. The optimum number of IMBs to deliver CFRT is infinite, spanning a full 4π volume in an unrestricted ideal (impractical) world. Some of the beams will be beneficial from all directions. Thus, the question of how many IMBs are needed for an optimum treatment must be rephrased to how many are needed to ensure that the dose distribution is little short of optimum. It might be expected that the situation is one of diminishing returns, with 'most conformality' achieved by determining the orientation of the first few beams, and then additional beams adding little more improvement (Boyer and Bortfeld 1997).

Stein *et al* (1997b) have studied the optimization of the *number* and *location* of IMBs for treatment of cancer of the prostate. They have created plans for equiangularly-spaced three, five, seven, nine and fifteen beam treatments for the separate cases when the dose to the PTV was 70, 76 and 81 Gy. For each plan, the IMBs were determined by the optimization technique of Bortfeld (1996) which uses a quadratic dose-based cost function and imposes penalty functions to control the dose in OARs. Scatter was included after optimization by using a convolution dose calculation. Plans were scored by dose-volume histogram (DVH) and also by calculating both TCP and NTCP. Biological scores were also computed using a statistically uncorrelated probability of uncomplicated tumour control (P_+) and also using a 'score function' (S) which penalizes large doses to OARs. The principal results were plots of TCP, NTCP, P_+ and S as a function of the number of beams for the three dose levels considered.

From these plots (figures 1.12–1.15) Stein *et al* (1997b) were able to deduce that:

(i) three-beam plans were acceptable at the 70 Gy dose level and no improvement was consequent on increasing the number of beams;
(ii) five-beam plans were required at the 76 Gy dose level;

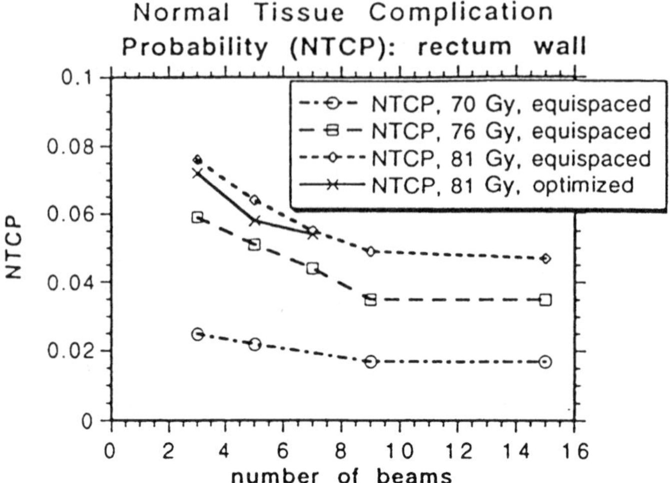

Figure 1.12. *Normal tissue complication probability (NTCP) for the rectum during treatment of the prostate is plotted as a function of the number of equispaced beams (3–15 beams) for 70, 76 and 81 Gy plans. NTCP versus the number of beams with optimized orientations is shown for 81 Gy plans and 3–7 beams as well. (From Stein et al 1997b.)*

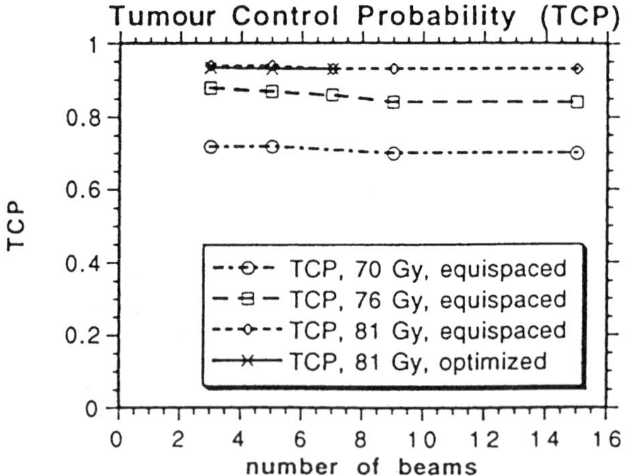

Figure 1.13. *Prostate tumour control probability (TCP) is plotted as a function of the number of equispaced beams (3–15 beams) for 70, 76 and 81 Gy plans. TCP versus the number of beams with optimized orientations is shown for 81 Gy plans and 3–7 beams as well. (From Stein et al 1997b.)*

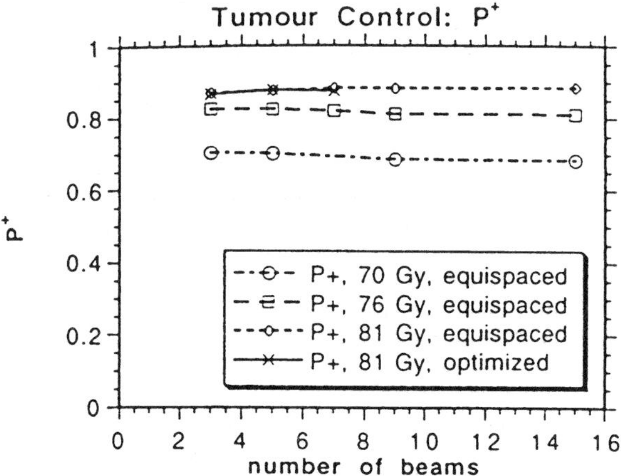

Figure 1.14. *The probability of uncomplicated prostate tumour control P_+ is plotted as a function of the number of equispaced beams (3–15 beams) for 70, 76 and 81 Gy plans. Here, P_+ versus the number of beams with optimized orientations is shown for 81 Gy plans and 3–7 beams as well. Rectum and bladder NTCPs are considered in the calculation of P_+. (From Stein et al 1997b.)*

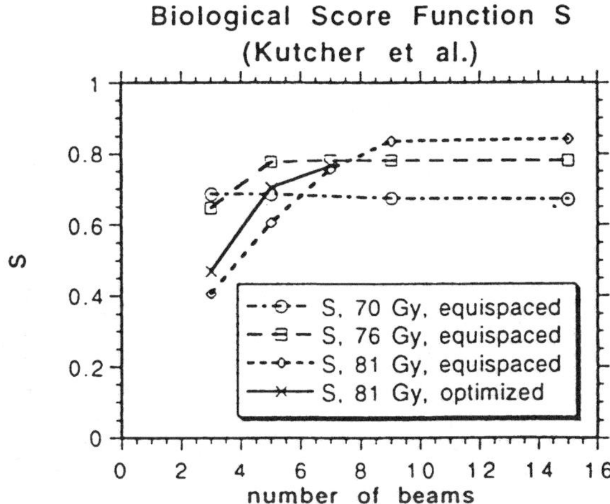

Figure 1.15. *The biological score function S is plotted as a function of the number of equispaced beams (3–15 beams) for 70, 76 and 81 Gy plans. Here, S versus the number of beams with optimized orientations is shown for 81 Gy plans and 3–7 beams as well. Rectum and bladder NTCPs are considered in the calculation of S. (From Stein et al 1997b.)*

(iii) nine-beam plans were required at the 81 Gy dose level;
(iv) the score function S was a more sensitive discriminant since P_+ did not control the NTCP values acceptably, giving equal weight to an $x\%$ change in TCP and NTCP;
(v) beyond the quoted numbers of beams the score functions levelled out.

The reason for the larger number of beams being needed as the dose is escalated is that the tolerance of a normal structure would be exceeded if fewer beams were used and it is necessary to spread out the dose in arrival space.

Stein *et al* (1997b) also optimized beam *direction* for plans with 81 Gy to the PTV for three, five and seven IMBs. They did this by either (a) an exhaustive global search or (b) fast simulated annealing (FSA). FSA used the zero-temperature tunnelling technique employing choice selections from a Cauchy distribution (Webb 1995b). The two methods gave the same results but the latter was 100 times faster and within acceptable clinical times (within an hour on a DEC ALPHA station). There was a small improvement in the treatment outcome when the location of three beams was optimized but no significant improvement in optimizing the locations of seven beams. Perhaps somewhat counter-intuitively they found that the preferred directions came from the direction of the rectum. A nice model example was given explaining that this was because, although the beams entered from this direction, the intensity in line with the rectum was low and the preferred direction enabled a better dose homogeneity to the PTV. The conclusions are in line with those of Söderström and Brahme (1993, 1995) although the latter made use only of P_+.

Stein *et al* (1997a) have compared the delivery of IMRT with a compensator and with multiple-static MLC-shaped fields (the MSF method—see section 3.3; see also figure 1.6). They conducted the inverse planning with the MRC Systems software called CONRAD (CONformal RADiotherapy [but also named after its originator Konrad Preiser—and sometimes also known as KONRAD] (Preiser *et al* 1997). CONRAD allows the variation of the bixel size and the number of intensity strata used in the optimization technique. It can thus be tailored to several methods of IMRT delivery (e.g. compensator or MSF) (Stein *et al* 1998). The IMRT was delivered with a Siemens Mevatron KD2. The clinical case was a clivus chordoma head-and-neck tumour. Optimum treatments were designed for either five or nine fields on appropriate fluence matrices of sizes which were different for each method of treatment delivery. Calculated and measured isodose distributions were compared favourably for the compensator method of delivery (figure 1.16). The MSF method, including corrections for leaf transmission and synchronization to remove tongue-and-groove artefacts (see section 3.2.2), required 30 subfields per orientation of the gantry. This resulted in overlong treatments (several hours) with the present status of equipment but a manufacturer upgrade was anticipated (Bortfeld *et al* 1997b).

It was concluded that five fields were the maximum which could possibly be delivered with the compensator method but that the conformality of the treament

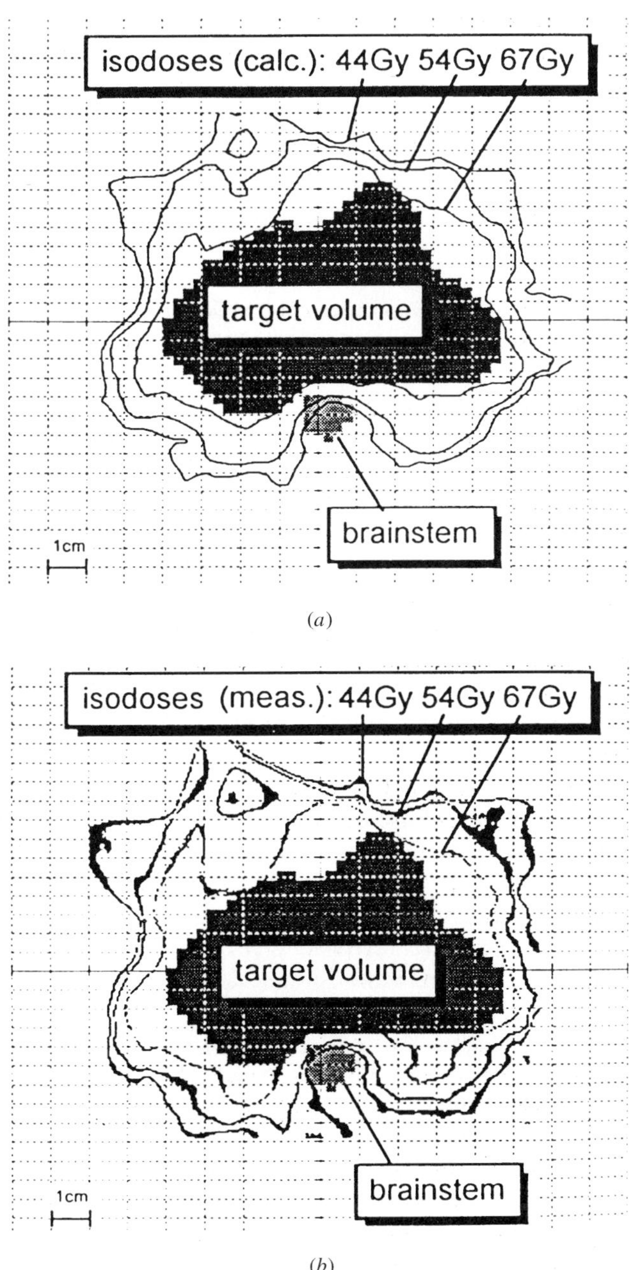

Figure 1.16. *(a) Calculated and (b) measured distribution of the five-beam compensator-delivered plan in a transverse plane showing the good agreement between the two. (From Stein et al 1997a.)*

could be improved if nine fields were delivered by the MSF method. Levegrün *et al* (1998) have verified the dose delivered through cast compensators designed using the CONRAD inverse-planning system finding the agreement was about 2%. Debus *et al* (1998) have reported the treatment of one patient at DKFZ with five fields, each with a cast lead compensator. The plan was created on CONRAD and the compensators, cast in layers, took four hours to make. Calculated and measured dose distributions agreed to 2% using a tissue-equivalent phantom and film and diodes. This patient had already received radical radiotherapy for a renal cell cancer to the vertebral body of T4 and so IMRT was designed to keep the radiation to the spine lower than 25% of the prescribed dose.

Bortfeld (1999c) has shown that the number of fields required for IMRT is effectively infinite if all fine details of the PTV are to be fitted. So the question of how many beams are required depends on the complexity of the target volume. Bortfeld also argues that it depends on the cost function used. When a biological cost function is used (see chapter 5) the number of fields can be quite small, beyond which no further improvement is seen. The question of how many beams are required cannot be separated from the problem of finding the optimum directions of incidence. Also, it is clear that for some tumours a smaller number of more appropriately positioned beams can give better results than a larger number of beams at suboptimal positions. This confirms work which has also been performed by Rowbottom *et al* (2000b).

Keller-Reichenbecher *et al* (1999b) have investigated the dependence of conformality on the number of fields and on the number of fluence strata for each field. They have studied two patients with head-and-neck tumours, and have used the inverse-planning system KONRAD and simple dose constraints without dose-volume constraints, given OARs with low volume effects. They first investigated three methods of fluence stratification:

(i) applied after ten iterations followed by stopping at the eleventh;
(ii) applied every iteration from the start; and
(iii) applied every second iteration.

The first technique yielded the lowest cost functions and was adopted for the rest of the study. Plans were prepared with a variety of combinations of numbers of beams (between three and 25) and numbers of fluence levels (between three, four, five and infinite). Dose distributions, DVHs and cost values were presented for all these options. Not surprisingly, the use of the highest number of beams and the infinite number of fluence levels gave the best conformality in terms of homogeneity of the PTV dose (but not necessarily organ sparing). However, they found that using quite a small number of beams and fluence levels (e.g. seven and five, respectively, or five and three) did not significantly degrade the outcome within the constraints set on this outcome. In summary, one may conclude that the issue is one of determining not the optimal conformality but the amount of conformality able to be sacrificed with minimal consequences.

The IMBs so generated were fed to an interpreter for the SIMTEC DMLC delivery scheme (see chapter 3) and it was found that on average the number of segments was 1.5–2 times the number of fluence levels. Given that the number of segments is the product of the number per beam and the number of beams, these become determined in practice by the available treatment time. At the time of this study (actually done in 1997) the SIMTEC inter-segment delay was some 20 s so it was very important to minimize the number of segments. This is becoming less of an issue with shorter inter-segment deadtimes.

We now group together, for convenience, a review of several studies which have resulted in the use of a compensator.

Clements *et al* (1999) have made use of the NUCLETRON/MRC system KONRAD IMRT calculation software to develop physical tissue compensators for implementation of IMRT for head-and-neck tumours. The compensator itself was made on an automatic milling machine. Frenzel *et al* (2000) have also used the KONRAD treatment-planning system to design compensators for intensity-modulated radiotherapy. A treatment plan with five fields was developed using KONRAD for a Rando-phantom. To verify the accuracy of the dose calculation the treatment plan was recalculated with VOXELPLAN. It was found that equivalent dose-volume histograms were obtained. The intensity maps with a grid size of 1×1 cm at the isocentre were then exported to two milling machines to create compensators. The first was an ACD5 from Par Scientific and the second was called CNC-Flachbettanlage GFM4433 from Isel-Automation. The milling machines milled a styrodur block which was filled with a steel granulate and compressed. The compensator was combined with a conventional shielding block made of MCP96. The fluence maps were measured with a BIS710 fluoroscopic megavoltage imaging system manufactured by Wellhöfer Dosimetrie. The measured fluence maps were then compared with those predicted by KONRAD. Subtraction images showed that individual pixel differences of up to 80% could occur but that most pixels did not deviate from calculated values by more than 5%. The delivered dose distributions for each individual field were measured using a prototype megavoltage imaging system ADAS manufactured by Wellhöfer Dosimetrie and also using film. Finally, the internal dosimetry in the Rando-phantom was measured by using 198 TLD dosimeters in one disk of the phantom to get a narrow grid of dose measurements. A close agreement was observed between calculation and measurement but it was important to position the phantom accurately, particularly for comparing regions where there is a high dose gradient.

Popple and Rosen (2000) have presented a quite novel method of delivering multiple-field IMRT using ‘a’ compensator. Essentially, all five compensators for IMBs from five gantry rotations are built into one physical device (figure 1.17a) which is rotated and shifted so that, for each different field, a different part of the compensator comes into view, thus delivering a different IMB (figure 1.17b). This retains some of the inherent security attached to the use of compensators but at the same time employing a measure of automated computer control .

Meyer *et al* (2000) have presented an analysis of the practicalities of

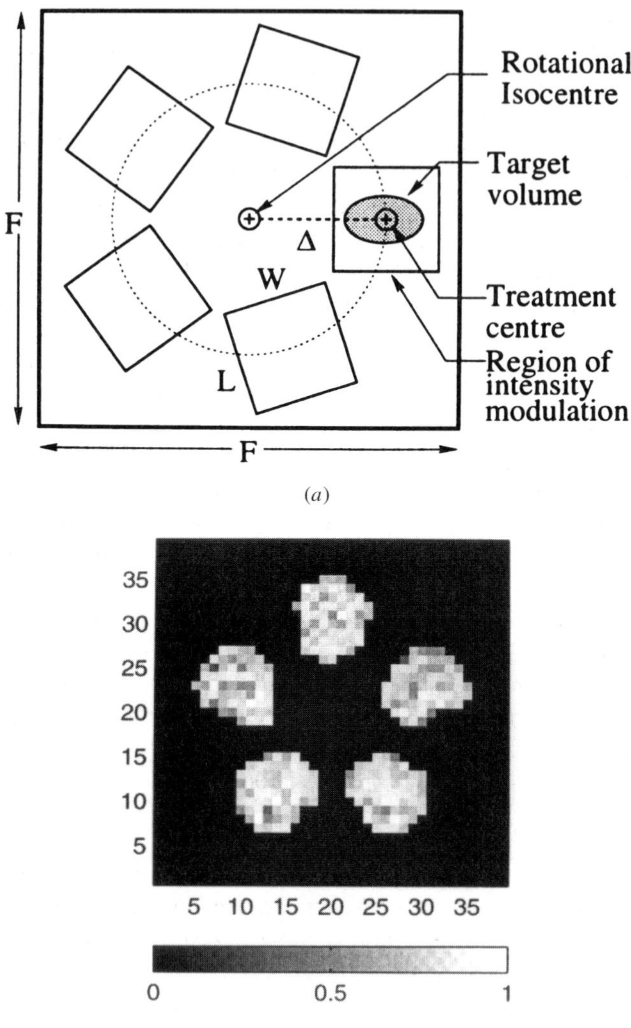

Figure 1.17. *(a) Treatment of five intensity-modulated fields using a single attenuator. The area outside of the intensity-modulated region is blocked using a combination of the secondary jaws and the MLC. (b) Example transmission map for a five-field prostate case. (From Popple and Rosen 2000.)*

delivering IMRT using a compensator. They specifically focused on treatment-delivery inaccuracies and showed that the fluence inaccuracy is a precise function of the X-ray linear attenuation coefficient, the radius of the assumed spherical milling drill, the 'step over distance' (that is, the lateral offset between longitudinal passes of the drill) and the angle that the intensity profile makes to the horizontal. They specifically developed expressions for the inaccuracy arising for generating an oblique slope and a valley-type feature and they study these both in the lateral and in the longitudinal direction. A very useful table gives the limiting values for the angles of the IMB profile as a function of the physical constraints.

Ma *et al* (1998d) actually made nine (Ellis-type) compensators out of extruded brass strip for a clinical nine-field prostate IMRT treatment and concluded that, whilst time-consuming, this form of IMRT was clinically feasible. Thompson *et al* (1998) and Basran *et al* (1998) have also manually constructed compensators for IMRT.

Now, returning to the question of how many fields, Schlegel (1999) has shown that of the order of five IMBs is adequate for IMRT of the prostate, with each beam sub-divided into five levels of fluence. IMRT is being implemented clinically at DKFZ Heidelberg, and inverse planning and the DMLC IMRT technique are now considered to be simpler than originally thought, but still a challenge.

Shepard *et al* (1997) have come to somewhat different conclusions. They have modelled the familiar problem of a 'C'-shaped target containing an OAR in its concavity. The study was performed in MATLAB using pre-computed pencil-beam photon sets. They varied the number of beam angles, the width of the collimator bixel and also studied the effect of varying the target position within a circular contour. They concluded that the degree of conformality improved as the collimator bixel size decreased (2 mm being best), that it was necessary to have at least eleven orientations but that there was almost no dependence on target location within the field. Some of this work is reminiscent of much earlier studies by Webb (1991a, b).

Shepard *et al* (1999) have republished this material. The heart of their work is that the pencil-beam dose distributions are pre-computed using convolution/superposition methods and are incorporated in an optimization environment within MATLAB which is able to take advantage of a number of different optimization techniques. The dose kernels are also available on their website. In this study, however, they simply used the importance-weighted quadratic dose cost function, independent of the size of the PTV and OARs. They repeated that uniform beam irradiation and simple segmented IMRT were inferior to full IMRT. The improvement in PTV dose uniformity and OAR dose sparing as the collimator size decreases and the number of beams increases is somewhat obvious and was known nearly a decade ago (Webb 1991a, b). Some new results indicate that dose optimization for inhomogeneous phantoms could start with the beamweights computed for the corresponding homogeneous phantom.

Richter *et al* (1997) compared five- and nine-field irradiation with IMBs for two cases; the first a convex PTV pituitary gland and the second a concave

PTV for a typical head-and-neck treatment. Planning was done by the Bortfeld technique implemented within KONRAD. It was shown that, for the former case, five-field treatment was actually a little superior in sparing the eyes but the latter technique did benefit, but even then only marginally, by the use of nine fields. The Würzburg rotation therapy method was almost as good. Richter *et al* (2000) concluded fixed-field IMRT was somewhat superior to rotational IMRT.

De Meerleer *et al* (2000) have shown reduced rectal and bladder toxicity with three-field prostate IMRT compared with five-field irradiation.

Verhey *et al* (1997a, b) have studied the IMRT problem for complex lesions of the head and neck. They used the NOMOS CORVUS treatment planning system and, based on dose distributions achieved, found that, for a fixed number of intensity levels, more beam directions were better than fewer, selected beam directions were better than random beam directions, and IMRT was better than the best plan produced with simple modulation. Indeed Pickett *et al* (1997) described a reduction in rectal dose from implementing fixed-field IMRT. An interesting concept has been offered by Xia *et al* (1998), from the same group, that IMRT might be *combined* with more conventional conformal radiotherapy (non-IMRT) to improve dose distributions without compromising treatment delivery *time*.

Boyer and Bortfeld (1997) have compared the dose distributions generated by nine intensity-modulated fields at fixed 40° gantry angle separations with the corresponding distributions generated by CORVUS for the NOMOS MIMiC continuous-arc delivery. They allowed the former IMBs to have up to 30 quantization levels of intensity. In the study of 25 clinical cases, three of which were presented, they concluded that the use of nine fixed fields was not inferior to the arc technique. A second study reported in the same paper showed that there was little improvement using more than nine static fields but that reducing the number to three was detrimental. An earlier study by Webb (1994a) reached a different conclusion but, as pointed out by Boyer and Bortfeld (1997), this was because the number of intensity quanta in the fixed-field method was kept low. In fact, Webb (1994b) in a fuller study found results in agreement with those of Boyer and Bortfeld (1997).

Haas *et al* (1998) have studied the optimum orientation of beams for a coplanar 2D IMRT problem (among other considerations). They developed cost terms reflecting: (i) conformality to the PTV; (ii) minimum exposure to the OARs; and (iii) minimizing the high dose to the other healthy tissues. With the number of beams fixed to a chosen value they used a genetic algorithm to evolve the set of beams with the optimum properties and confirmed that for a concave PTV the conclusion of Stein *et al* (1997b) held good that a (partly blocked) beam was allowable through the rectum (figure 1.18). There was also only a marginal improvement when increasing the number of beams from four to five. With plans comprising five beams there was little point in optimizing further the beam orientations. Haas *et al* (1999) have built the algorithm into a workstation for inverse planning (see also Haas 1999). Pugachev *et al* (1999, 2000) have shown that this may be true for the prostate but may not necessarily be true for other

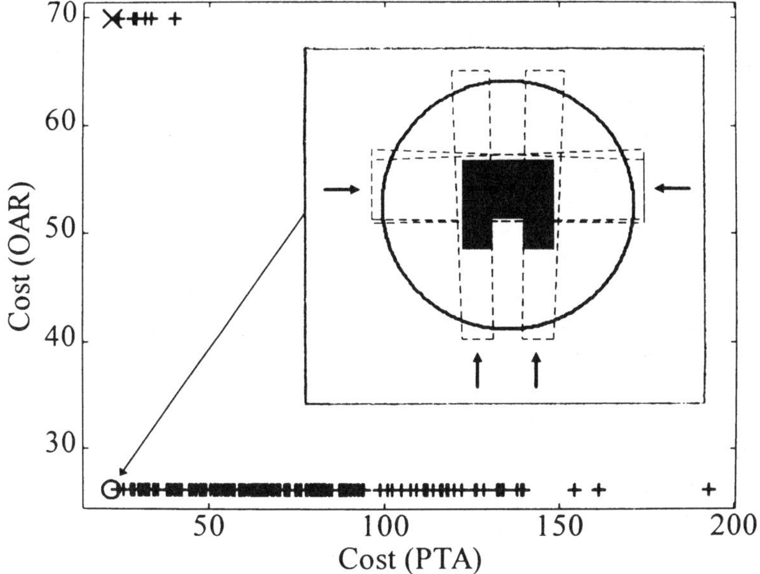

Figure 1.18. *Illustrating the solution space for a three-field plan involving a 'C-shaped' PTV and an OAR within the concave region; the inset shows the best compromise solution allowing split beams. (From Haas et al 1998.)*

tumour sites (see section 5.7). Löf *et al* (2000) show the improvement in IMRT by optimizing beam directions for few beams.

Sauer and Bratengaier (1999) have shown that tumour control for a head-and-neck and a lung case could be achieved using no more than seven IMB ports. The optimization technique developed used a dose-based cost function and it was considered that there was no need for the use of radiobiological parameters. More details were given by Sauer *et al* (1999).

Johnson *et al* (1999a) have come to the conclusion that acceptable IMRT can be delivered by between five and nine coplanar fields with no more than ten intensity levels per field and they have introduced the DMLC technique in Chicago using a Varian system.

Yu *et al* (2000b) have shown that clinical IMAT with just a few (e.g. three) rotations often generates acceptable dose distributions. This corresponds to just the associated few intensity levels. IMAT has taken many years to come to the clinic after its initial invention.

In concluding this review of specific studies which address the issue of how many IMBs are needed one might state some generalities. Firstly, we may note that stating generalities is itself fraught with difficulties. The studies usually concentrate on a particular tumour type and conclusions from one tumour type may not apply for all types. It has been known since the early days of IMRT that the conformality improves as the number of beams increases (Webb 1989) when the PTV is of a *highly concave shape*. It is because the PTVs that arise in practice often have less demanding shapes that the benefit flattens out with diminishing returns as the number of beams is increased. Also, most studies have maintained the coplanarity of the central axis of beams and there have not been many in-depth studies of whether one could use even fewer beams if they came from non-coplanar orientations. Again, the outcome would be dependent on the geometry of the PTV.

No-one has ever made the ultimate inverse-planning investigation to determine the truly optimum solution because this is, I maintain, an impossible job. The reason is that all the variables in an optimization problem are coupled—beam energy, orientation, bixellation, beamweights, beamshapes etc. A *truly optimal* plan would be that which resulted from allowing a completely free variation in all these parameters, something never done in practice. Hence, what usually happens is that workers fix some parameters and search for the best values of others left free. A consequence is that there is an incomplete message delivered about how many beams are needed. It depends on what else you choose to fix and what problem is under consideration. We shall return to this theme in chapter 5.

We now turn our attention to inspecting the several ways in which IMRT can be delivered in practice.

1A. APPENDIX: IMRT—POINT AND COUNTERPOINT

Table 1A. *IMRT is a subject which, as well as being technically complex requiring detailed reference to the literature for a good understanding, also gives rise to a number of debatable 'conversation points'. Some are listed here in the form of 'point and counterpoint'. Longer debates on some of these issues appear in the body of this and my previous two texts. The reader is invited to take up a position...*

Point	Counterpoint
The development of IMRT should be left to industry alone.	Collaboration between industry, universities and hospitals has been shown to lead to more focused, problem driven, solutions.
IMRT is 'research fun'; it is not needed clinically.	Many clinical problems do not need IMRT. Possibly 30% do and IMRT is the only way to increase local control or provide conformal avoidance.
The development of IMRT requires hospitals to expand their facilities and employ more staff at a time when governments are urging cost cutting.	This may be true in the initial stages. But the effort is worth it. Hospitals rarely cost in that a failed treatment is a wasted high cost. IMRT can improve the local control rate. More expensive successful therapy is better than cheaper unsuccessful therapy. We should invest more not less in healthcare.
IMRT generates too tightly conforming dose distributions. We do not know the tumour extent that well and the patient may move. Conventional therapy is safer.	3D multimodality medical imaging is being linked to CFRT and IMRT. This will become standard. Clinicians will learn a new PTV-defining skill. Movement will be measured, quantified and controlled. Margins might actually be increased as volumes take more complex shapes.
IMRT is too hard to learn and implement.	There is clearly a learning process. For this reason there should be a coordinated approach to educate, train, and retain staff who have chosen to work on IMRT. These skills are not widespread and do not come cheaply.
Conformal radiotherapy, the MLC and IMRT are very new.	This common misconception arises because the practical implementation is indeed new. Clinical implementation is even newer. However the MLC was described in the 1950s, conformal radiotherapy in the 1960s and IMRT concepts go back at least to 1980.

Table 1A. *(Continued.)*

Point	Counterpoint
If IMRT is so old why have I only recently become aware of it?	I cannot remember when I first heard or read the specific mnemonic 'IMRT' but it was probably around 1996. I think (but do not know for sure) that the term has stuck by usage rather than by edict. Many of us were writing the words in full long before that. Prior to this many competing terms vied for the accepted title. Some of these were: intensity-modulated beams (IMBs), beam intensity modulation (BIM), intensity-modulated radiotherapy (IMRT), fluence modulation, beam modulation, dynamic modulation, dynamic therapy, dynamic collimation, moving field (or beam) therapy, tracking therapy. IMRT seems to me a good descriptor. I have even heard it used as a verb as in 'I am going to IMRT my patient'.
Inverse planning is quicker than forward planning.	Possibly false. Inverse planning concentrates attention on specifying constraints and goals rather than trial and error. There are still choices to be made by humans. Batch processing reduces the time a human is actually present.
Treatment plan optimization is achievable.	The optimum solution can never be attained because the potential search space (orientations, beamweights, energy, particle type) is too large. Constrained customization is feasible.
Since we do not entirely trust TCP and NTCP models we cannot do IMRT aiming to improve biological outcome.	Dose distributions arising from IMRT can be assessed in surrogate dose terms just as non-CFRT plans have been evaluated for a century. This is true whether or not biological or dose-based cost is the optimized quantity. The calculation of biological outcome probabilities can be viewed as additional information.
We should not do IMRT because it has not been proved to be better than conventional therapy.	This is a familiar 'catch-22'. There has to be some act of faith to establish enough centres and clinical trials to make the assessment. A wise sage might say that pretty well nothing in the technical development of radiotherapy has ever been established by trial to be better (e.g. planning itself, use of high energy beams, image based planning...). Most has come by faith and subsequent experience.

Table 1A. *(Continued.)*

Point	Counterpoint
There are too many 'paper planning' IMRT studies.	These can establish that IMRT will at least do no worse than conventional therapy. Those centres which have then implemented IMRT have found fewer toxicities despite dose escalation.
IMRT requires a large number of beams with fine-scale intensity modulation and fine-scale spatial sampling.	This may be true but it may not. This depends on the problem. Early investigators (including me!) made the case for complex IMRT, partly to demonstrate expected benefits and partly because the theory of IMRT planning ran ahead of the practical development of delivery. Times have changed. We are all now asking 'how good do you want your IMRT to be?' and then seeing what 'limited' or 'partial' or 'simplified' IMRT can achieve.
The quest for the ultimate IMRT technique should be stopped.	The 'ultimate IMRT' may be practically unachievable. However, the knowledge is worth having for a specific planning problem in order to know how close 'simpler IMRT' comes. As with much in life, you get what you pay for.
IMRT is hard to verify.	Hard but not impossible. There are five 'classes' of possibility: (i) mechanical monitoring; (ii) QA with phantoms; (iii) *in vivo* dosimetry; (iv) portal fluence measurement and input fluence estimation; (v) dose reconstruction from recorded measurements of detected and/or measured input fluence.
The choice of IMRT technique is too wide. How does a centre know what to invest in?	Innovation always leads to choice. Choice also depends on more factors than mere technical performance. For example, in the USA the MIMiC IMRT was initially much more popular than in Europe. This may have taken in fiscal issues. In Europe, including the UK, the MSF and DMLC technique received much attention and lead the way. There is a case made for providing some independent guidance to purchasers. Presently, most IMRT courses are linked to specific products or sponsored by specific industry even when not billed as such.

Table 1A. *(Continued.)*

Point	Counterpoint
IMRT is only useful for rare cancers such as head-and-neck cancers and brain tumours.	IMRT of the breast can lead to better dose homogeneity and spare heart and lung. IMRT of the prostate could become routine.
The dose models for IMRT are not accurate.	Once a simple dose model has been used to create intensity-modulated fields, a full Monte-Carlo dose calculation can be done just as for unmodulated radiotherapy.
The IMRT plan created by inverse planning is modified by the IMRT delivery mechanism.	This can be true but it is now possible to calculate the effects of most of the delivery mechanisms. It is also possible to factor these into the inverse planning itself.

CHAPTER 2

ROTATION IMRT: TOMOTHERAPY

As already indicated (section 1.1.4) the delivery of IMRT is possible by several techniques and, during the past decade, three have been front runners. These form the subject of the next three chapters although, as will become clear, some overlap and possible confusion of terminology leads to a precise classification being unreachable (see e.g. section 2.2). In this chapter, we shall consider the two methods by which IMRT is delivered via an arcing gantry equipped with a multivane collimator residing in a narrow fan beam of radiation. These could be classified as 'rotation IMRT' and share similar features albeit with important distinctions. The first method is IMRT via the use of the NOMOS MIMiC equipment and the second is the development of the University of Wisconsin machine for 'tomotherapy'. Given that the word 'tomotherapy' means 'slice therapy' it could also be used generically to embrace the first method.

2.1. NOMOS MIMiC DELIVERY

PEACOCK is the name of the IMRT system marketed by the NOMOS Corporation whose founding scientist and former Managing Director was Dr Mark Carol. The original name of the Company was MEDCO, founded by Mark Carol and John Friede (the current Managing Director). In 1993, the name changed to NOMOS, rooted in the Greek for 'to learn, understand and solve'. The name PEACOCK incorporates both the planning system (formerly known as PEACOCKPLAN, now known as CORVUS) together with the treatment delivery system, the MIMiC. The name PEACOCK was chosen because some dose distributions computed by inverse planning have the appearance of the 'eye' feature in the tail of the peacock (figure 2.1). The Corporation's logo (which is a trademark) makes use of this similarity (figure 2.2). The feathers also open a little like the vanes in the special collimator (called the MIMiC) which NOMOS have developed to deliver IMRT (figure 2.3). It is stated (Mueller 1996, Boston 1997) that Carol first had the idea in 1975 at a time when computer control was just not up to the job. But following the publication of the simulated-annealing method of IMRT computation (Webb 1989)

Figure 2.1. *The tail feathers of a peacock showing the familiar 'eye' pattern. (From NOMOS Corporation brochure.)*

Figure 2.2. *A dose distribution delivered by the NOMOS MIMiC used as the corporation's logo since it is similar to the 'eye' pattern in the tail feathers of a peacock. This is a trademark of the NOMOS Corporation. (From NOMOS Corporation brochure.)*

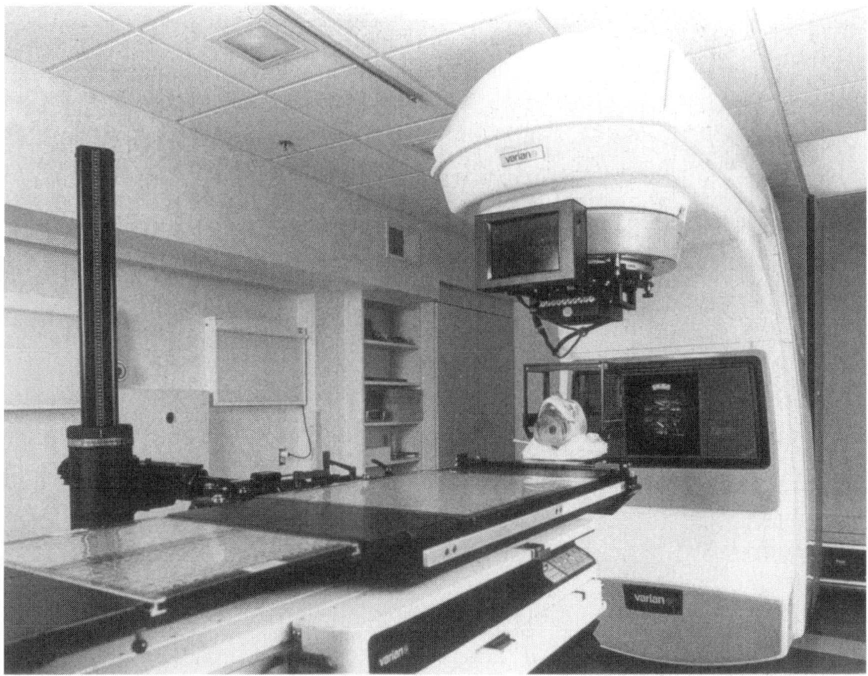

Figure 2.3. *A view of the NOMOS MIMiC attached to a Varian accelerator also showing the CRANE. (From NOMOS Corporation brochure.)*

he revisited this idea, starting in 1992. This is flattering to the work done at the Institute of Cancer Research but the commercial product was entirely the work of the NOMOS Corporation [1]. The final clearance for the hardware was given by the Food and Drug Administration (FDA) in March 1995, and for the planning software in April 1996. A detailed review of IMRT, mainly concentrating on the NOMOS technology, has been written by Sternick *et al* (1998). The development was also reviewed by Webb (1997d).

[1] The announcement of the development of the MIMiC was dramatic and somewhat unexpected by the radiotherapy (physics) community. This was possibly our fault since scientists are usually unprepared for developments which come from a hitherto unknown group, and at the time MEDCO was little known amongst radiotherapists and physicists. I have often told this somewhat personal story. In the autumn of 1992, I was in Geneva for a conference on 3D radiotherapy held at the World Health Organisation. I was scheduled to lecture on inverse planning (the date was Tuesday, 20 October 1992). Having given my lecture and retaken my seat, Mark Carol stood up and said "you have heard how we can plan for and would want to deliver IMRT; now I can show you apparatus to do it based on this concept". His lecture had been scheduled to immediately follow mine. He gave what I believe was the first announcement of the MIMiC and showed photographs. It was one of those moments one sensed would have long-term impact. If I remember correctly, Carol was encouraged to over-run well into a scheduled morning coffee break and also given a second unscheduled session in the afternoon, so much interest was there in this new apparatus.

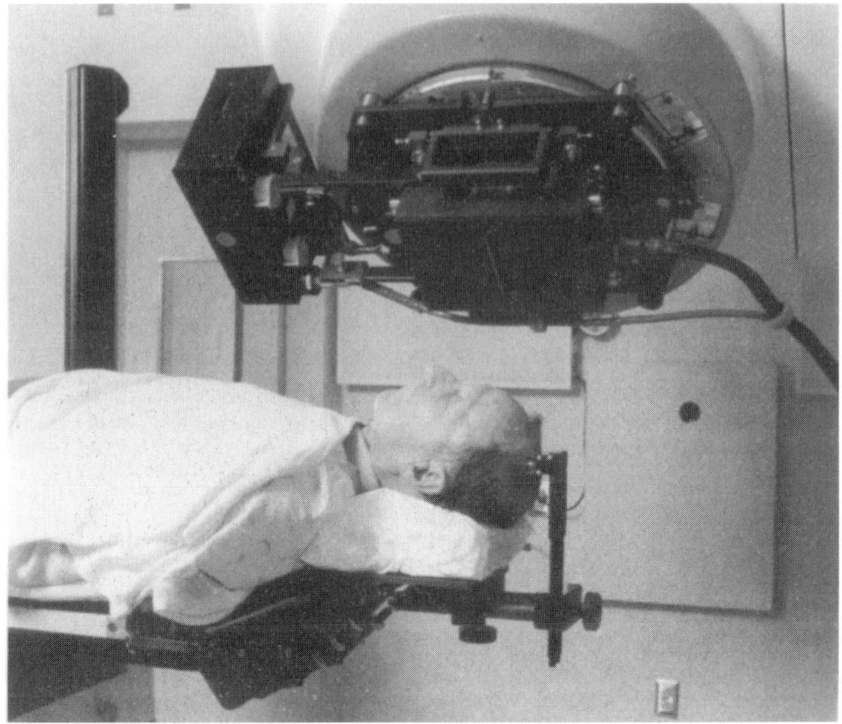

Figure 2.4. *The NOMOS MIMiC and the controller with the NOMOGrip, TALON and CRANE. (From Curran 1997).*

The multileaf intensity-modulating collimator (MIMiC) is a special-purpose collimator, retrofittable to any linear accelerator, and capable of delivering IMRT by tomotherapy, i.e. by treating one slice of the patient at a time. The apparatus is as follows. For each gantry orientation relative to a transaxial plane through the patient, the radiation is collimated to a slit, narrow in the superior–inferior direction and wide enough to span the slice of the target in the plane orthogonal to this. This radiation slit delivers radiation to a transaxial slice of the patient (figure 2.4). Into this slit may move, at right angles to its long axis, a series of short stubby vanes from rest locations outside the slit. By varying the dwell-time of the vanes within the slit, a one-dimensional IMB is created (figures 2.5 and 2.6). The whole tumour is treated by repeating the process for abutting slices.

The MIMiC comprises two banks of such stubby collimators which can enter each of two separate adjacent slits of radiation. The NOMOS MIMiC vanes are sometimes referred to as 'leaves' or 'binary shutters'. The banks are independently controlled by a pressurized air supply driven by an ultra-quiet air compressor. Air is fed to the MIMiC via an umbilical which rotates with the gantry. A second

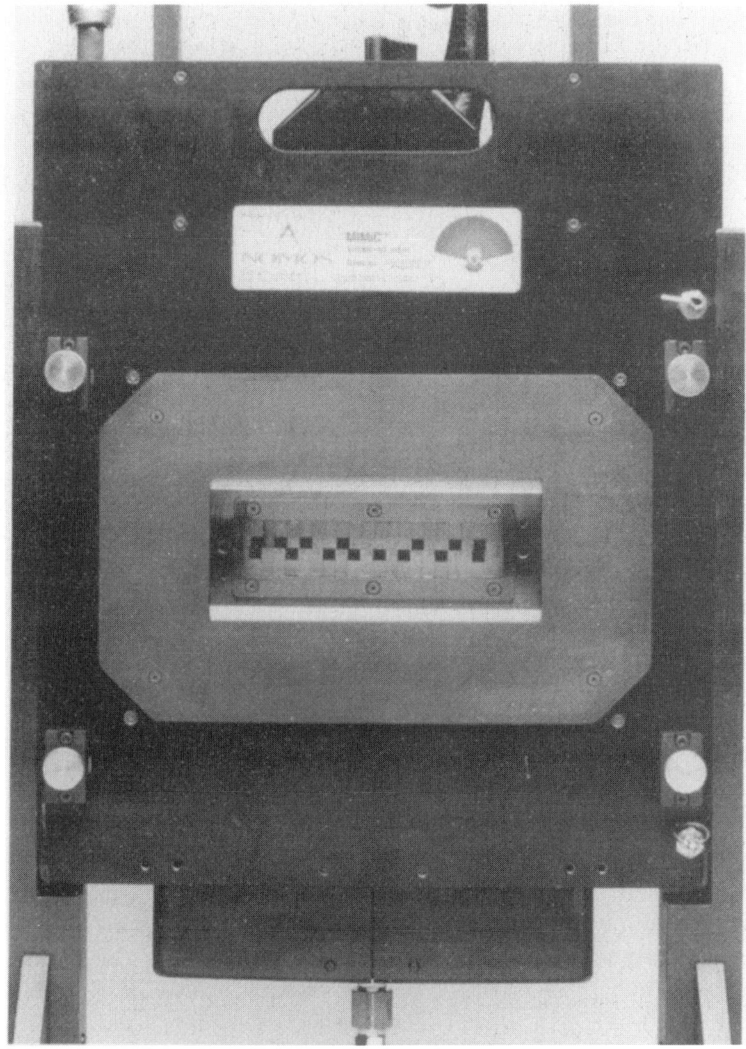

Figure 2.5. *A close-up of the MIMiC showing some vanes open and others closed. The two banks are clearly visible. The NOMOS Corporation peacock-eye logo is visible on the face-plate. (From Curran 1997.)*

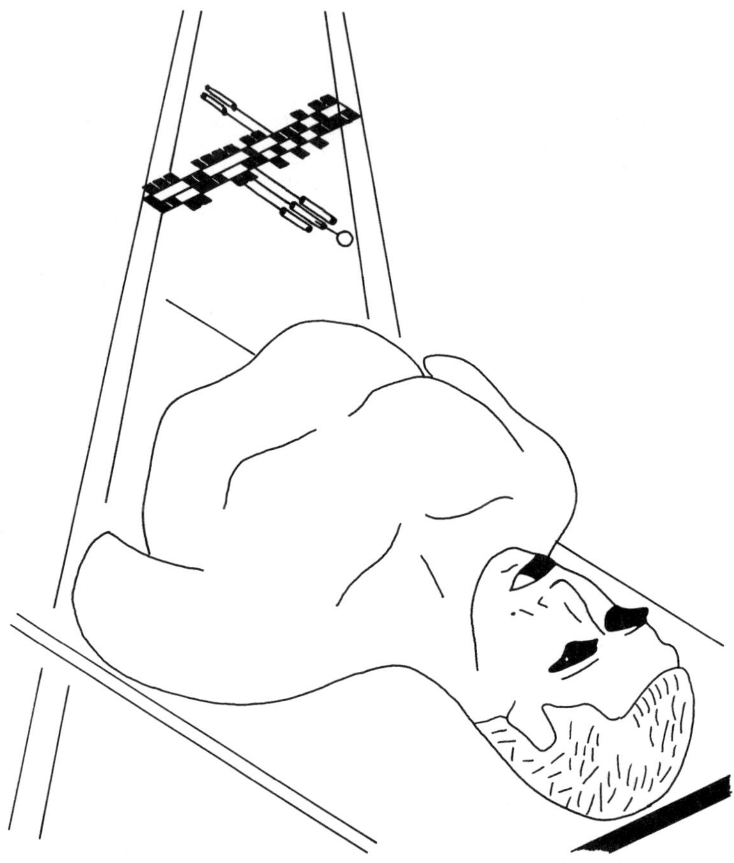

Figure 2.6. *A three-dimensional schematic diagram showing the overall arrangement of the beam and its beam elements (bixels) in relation to the patient. Some of the bixels are shown open (white patches in the beam) and some of the bixels are shown closed (black patches in the beam). The bixels are created by vanes of finite thickness but, to avoid confusion, the vanes themselves are not shown. Instead, schematically, the bixels are shown connected to an actuating mechanism which can open and close them. The figure is not to scale.*

umbilical carries the 5 V and 12 V power supplies. In practice, the treatment delivery can take place with the gantry moving but planning assumes that, for each configuration of the collimator, the gantry is stationary at angular intervals of 5°, then moved with the radiation off to the next orientation for delivery. The vanes can switch very fast but it has been shown that the switching rate does affect the dose output (Tsai and Engler 1999).

The MIMiC is attached externally to the head of the linear accelerator and movement of the leaves is controlled by an on-board computer (figure 2.4). MIMiC deliveries are computed as a set of 55 static fields at 5° intervals over a 270° arc (avoiding irradiation through the couch support). The MIMiC leaf positions are updated every 0.5° allowing pencil beams at a particular (nominally stationary at 5° intervals) gantry position to vary in intensity in 10% steps (see also figure 1.8). The gantry position is known independently of the linac's measurement by a separate computer, known as the MOCK, which senses measurements from two inclinometers to better than 0.1°. The MIMiC does not measure the radiation output since this is assumed to be properly measured and governed by the accelerator's own systems (Curran 1997). The sequential transaxial slices are ensured to be accurately abutting via the use of the CRANE (see later) and recently even non-coplanar arcs have been created with sequential slices (Salter and Hevezi 1999).

PEACOCKPLAN is a specialist 3D planning system developed by NOMOS for creating the prescription files to drive the MIMiC. The planning involves several stages including:

(i) acquiring and entering CT image data;
(ii) registering the data to a coordinate system;
(iii) drawing the tumour and normal-tissue anatomy on the images;
(iv) prescribing the doses to each volume and the relative importance of achieving these for each volume;
(v) optimizing the dose calculation by iterative simulated annealing;
(vi) displaying, analysing and approving the results.

All these stages are described in detail by Curran (1997). The details of how the planning is done—specifically the cost functions optimized—have evolved since the system was first launched in 1992. PEACOCKPLAN is now incorporated into a more generalized planning system known as CORVUS (the collective noun for a group of ravens) which also has tools to create delivery prescriptions for other methods of delivering IMRT, such as the DMLC technique (see chapter 3 and figure 1.7).

The MIMiC was conceived at NOMOS by Mark Carol, who has often been at pains to present the essential concept of the device in simple analogy terms for all to understand. At the Durango (Colorado) Conference on IMRT in May 1996, he first described the irradiation to be like determining the best set of lighting conditions which will show up the actors (tissue sites) in a stage play ('intensity theatre') in different degrees of illumination (dose). Various parts of the stage are highlighted differently. Needless to say, the players must be in their expected positions for

Figure 2.7. *The actor is on stage, his head illuminated by a series of beams which also illuminate other areas (one is shown). The beams from a light source are like the beams from an X-ray source passing through the MIMiC. By varying the intensities and directions of the beams, the overlap intensity can be adjusted as required. Of course, the actor must stand in the overlap area during the performance. (With acknowledgements to a story from Mark Carol.)*

the method to succeed so their positions in the real performance (treatment) must match those at rehearsal (planning/simulation) (Carol 1997a) (figure 2.7). Low (1999, 2000) has amplified this verification theme of IMRT using the NOMOS MIMiC. In the same review, Carol (1997a) gave a second analogy. Now the aim is to use bombs to destroy a military target (tumour). But the target is close to the homes of ordinary citizens (normal tissue). Old fashioned blanket bombing (non-conformal radiotherapy) is inappropriate. New weapons are developed (IMRT) in which damage is inflicted all along the delivery path not just at the end-of-range. By dropping bombs from varying angles (tomotherapy) and of varying powers (intensity) the damage might be concentrated on the target and avoided elsewhere. The limiting factor might be that the target may not be in the same place at bombing (therapy) as at surveillance (diagnostic planning) and the normal structures may move about (lack of immobilization). Yet a third analogy is that IMRT is like applying heat to a cup of coffee in such a way as to heat the cup but not the coffee. The expected performance of IMRT in future years is opined by Carol (1997b) in a tract subtitled 'IMRT in 2003'. As 2003 arrives and passes, it will be interesting to observe whether his foretellings are borne out, just as those of us who studied George Orwell's book '1984' passed though that year noting some predictions to have materialized and others to have been short of the mark.

2.1.1. MIMiC dosimetry

A study by Webb and Oldham (1996, 1997a, b) modelled the radiation delivery (see also Webb 1997a, b). The treatment-planning stage comprises determining the set of IMBs which would generate a dose distribution best matched to a dose prescription. In its simplest form this is usually done by restricting the computation to the set of one-dimensional IMBs in a two-dimensional transaxial plane, of mathematically zero width in the orthogonal longitudinal direction, which generate a two-dimensional dose distribution in the corresponding transaxial plane of mathematically zero width. That is, some particular CT slice is chosen in which to define a prescribed dose distribution but in inverse planning this slice is regarded as mathematically of zero width and aligned with the midline of a bank of bixels so no account is taken of what happens away from this plane. In determining the IMBs there is required to be some model of the dose deposited in this two-dimensional transaxial plane per unit fluence in each bixel. In many previously reported techniques for calculating IMBs, this model has been very simple. A more realistic treatment-planning system makes use of the elemental dose distribution from a single open bixel which includes a factor describing the profile within the transaxial slice of the form shown in figure 2.8.

Note that this profile describes the behaviour, in this plane, away from the central axis of the radiation passing through each separate bixel. When planning is carried out in the above-defined transaxial plane the corresponding profile in the superior–inferior direction is not used. Thus, the planning process computes the intensities of each bixel assuming that the radiation from each bixel could be delivered independently and sequentially in time with all the other bixels closed. The planning process has no option but to make this assumption since to assume any other beam profile would require unknown knowledge of the final set of bixel intensities. The result is a set of bixel intensities describing IMBs which are precisely locked to the corresponding two-dimensional dose distribution but with the assumption that each element of radiation is (time-wise) independently delivered. This problem would arise with any inverse-planning technique which has to have at its heart some model relating elemental dose distribution in a transaxial plane to fluence for a single open bixel. In the past such considerations have received relatively little comment. Thus, the planned dose distribution is that which would be delivered by time-independent delivery of each elemental intensity. As is clear from figure 2.8, the term 'independent' refers only to the independence of delivering radiation from each bixel time-wise. Radiation from any one bixel is not confined spatially to the line of sight of that bixel but, via the profile of the elemental dose distribution, deposits dose beyond the line-of-sight region.

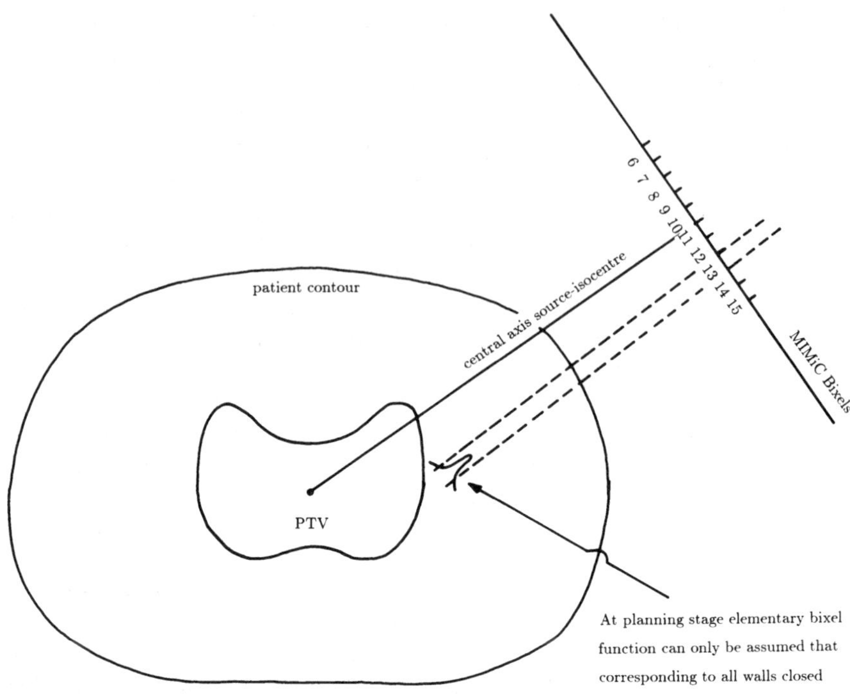

Figure 2.8. *A transaxial view of a typical treatment-planning problem involving creating a conformal dose distribution to a planning target volume (PTV) with concavities in its outline. To the upper right is shown one (of many) orientations for the intensity-modulated slit beam, consisting of numbered bixels. A schematic of the elemental profile factor, in the direction along the bank of bixels, is shown assuming that all bixels, except one delivering radiation, are closed. The profile factor is the same in the orthogonal direction also. This determines the elemental dose distribution which is used at the treatment-planning stage.*

There is an unavoidable fundamental difference between the planning model and the delivery technique. We are now in a position to view the problem which arises. Each of the classes of delivering IMBs does so in different ways. In each class there is some simultaneity in the delivery of the intensity through each bixel. For example, a compensator delivers fluence through all bixels at once. The multiple-static field technique (see section 3.3) superposes static fields of different combinations of open bixels but with groups of open bixels irradiated simultaneously. The dynamic multileaf collimator (DMLC) technique (see section 3.1) also departs from the concept of independent bixel delivery. We shall study each of these situations in turn as we meet the techniques in detail. Of course, although independent bixel delivery would perfectly match the planning situation, it would be highly inefficient and is totally ruled out in practice.

A collimator such as the MIMiC delivers radiation to all bixels simultaneously, but with bixels open or closed for varying fractions of the irradiation time as required to create the desired intensities. Consider figure 2.9 where just three bixels are shown of a single 1D IMB. To deliver the fluence to bixel 13 whose intensity is I_2 (greater than intensity I_1 in bixel 14, but lower than the intensity in bixel 12) two components are added together. For a period of time to give intensity I_1, the open bixel 13 finds itself adjacent to two open bixels, 12 and 14. It thus has an approximately flat profile in the direction along the bank, i.e. *not* the profile assumed at the planning stage. Then, for a further time, sufficient to give intensity $I_2 - I_1$, bixel 14 is closed and so the radiation profile for bixel 13 for this component is as drawn (also *not* the profile assumed at the planning stage). The result of delivering these two components will be a contribution to the dose distribution which is different from that which would be obtained if the dose had been delivered from bixel 13 alone, with bixels 12 and 14 in the closed state.

We have thus illustrated the fundamental problem that the dose distribution actually delivered will depend on the relative states of the neighbouring bixels and will be different from that determined at the planning stage. Only if the elemental dose distribution from a single open bixel had the property that wide fields were precisely the superposition of such distributions would the treatment distribution exactly equal the planned distribution. In general, without some manipulation, this is not the case. Although this simple illustration with figures 2.8 and 2.9 has been two dimensional, the analysis of Webb and Oldham (1996, 1997a, b) was a full three-dimensional analysis taking into account the states of both banks of bixels. They quantified and commented on the magnitude and trends in the difference between the distributions planned and those delivered at treatment time. To do this they have developed a 'component-delivery' (CD) model.

For each orientation of the gantry and for each of the forty bixels in the two banks, the radiation is delivered in CD mode by considering the relationship between the intensity of that bixel and that of its three nearest neighbours. The four intensities are ranked. The task is then to decide: (i) the number of component irradiations; (ii) the values of the component intensities; and (iii) the appropriate wall conditions determining the profile factors which will be applied to each component delivery. These decisions depend solely on the order in which the four intensities are ranked. It was shown that when, as in the NOMOS PEACOCK planning system, the elementary bixel functions are 'stretched', the difference between the planned and the delivered distributions is minimized but never goes away altogether, due to the intrinsic nature of the delivery. A study by Langer (1988) also addressed the coupling of adjacent bixel states.

Balog *et al* (1999b) have performed a detailed study of the output of the NOMOS MIMiC collimator which is being used on the University of Wisconsin 'tomotherapy workbench' (see section 2.3). For this experimental model of what

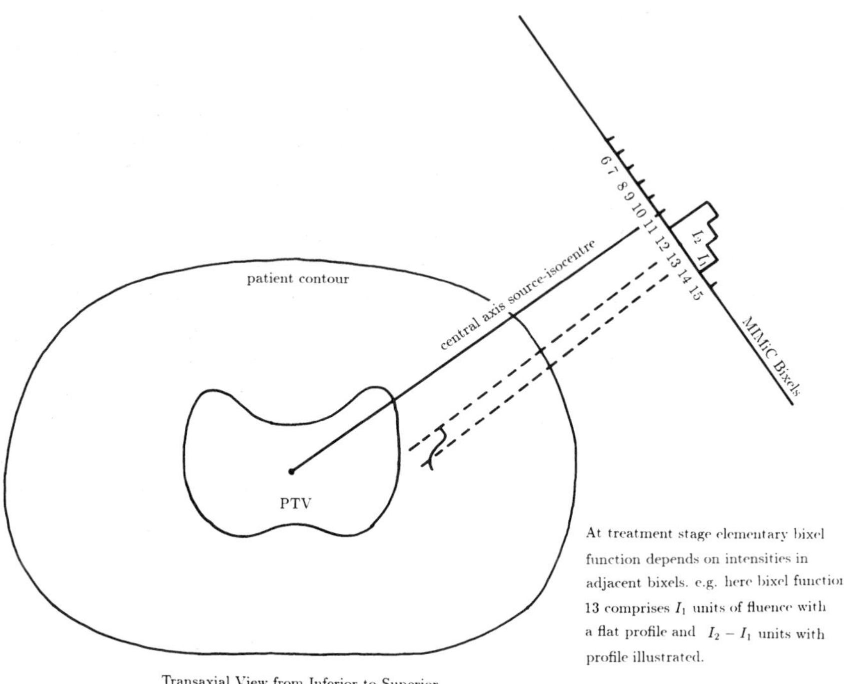

Figure 2.9. *A transaxial view representing the situation corresponding to figure 2.8 at the time of treatment delivery. To the upper right is shown one (of many) orientations for the intensity-modulated slit beam, divided into numbered bixels within one bank, and three arbitrary representative bixel intensities. At the treatment stage the elemental profile factor for any particular bixel depends on the intensities in adjacent bixels. This is termed 'component delivery—CD-mode'. For example, in delivering the intensity I_2 from bixel 13, first I_1 units of radiation will be delivered with a flat uniform profile in the direction along the bank of bixels, since the adjacent bixels, 12 and 14, will also be open. This is followed by delivering $I_2 - I_1$ units of radiation with the profile illustrated, because bixel 14 is closed for this second delivery component. The orthogonal profile will depend on the status of the corresponding bixel in an adjacent bank. Thus, the dose distribution will be different from that which would correspond to delivering each bixel component individually (which is termed 'independent vane—IV-mode' and corresponds to the planning situation in figure 2.8).*

will become the Wisconsin spiral tomotherapy system, only one bank of the MIMiC vanes are in use—all that is necessary. This group have a convolution–superposition (C/S) dose model in their in-house developed inverse-planning system and also have a separately constructed MIMiC controller. Hence, this implementation is different from the PEACOCK system but the basic MIMiC dosimetry is common to both applications. The Wisconsin planning method does cater for tissue inhomogeneities. Energy fluence is used to predict TERMA (inhomogeneity corrections included) and dose is computed via C/S kernels (also incorporating inhomogeneity corrections).

By comparing measured and predicted percentage depth dose curves, Balog *et al* (1999b) determined that a 4MV Mohan spectrum accurately models the spectrum of the GE Orion 4MV accelerator which is the basis for the workbench. From this it was determined and measured that the dose profile along the fan-beam direction increased by about 5% at the outer bixels (figure 2.10). Calculated and measured dose profiles along the slit also agreed well for a configuration with every other set of two MIMiC vanes closed. Measurements of simultaneous versus sequential delivery of two bixels showed that the latter exhibited a 32% tongue-and-groove (TG) underdose with a FWHM of 2.1 mm at a depth of 1 cm. The integrated dose decrement due to this TG effect was measured as 4.9% of the total dose resultant from two leaves without such an effect.

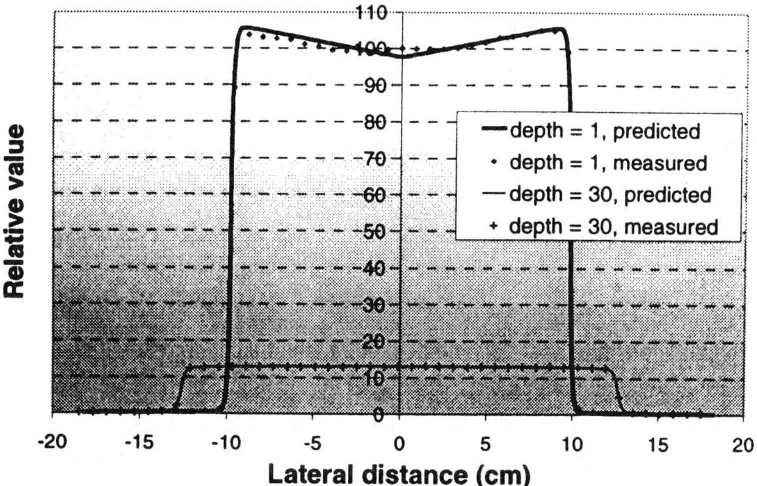

Figure 2.10. *The measured versus the predicted dose profile for all MIMiC leaves open at a depth of 1 cm and a depth of 30 cm. The measured profiles were obtained with an ion chamber in a water tank. The profile distances are in the fan-beam direction. (From Balog et al 1999a.)*

Balog *et al* (1999b) concluded that the energy fluence modelling from a 1D MIMiC is an 'involved but entirely manageable problem', the above two factors being the only two which significantly change the energy fluence per bixel, and that both can be calibrated. The fact that the status of adjacent bixels directly alters the dosimetry for any particular bixel is the same observation as made and then modelled by Webb and Oldham (1996, 1997a, b).

Seco *et al* (2000) have further studied the effect of what we might call the 'MIMiC interpreter', i.e. the difference between computed and delivered dose due to the physical behaviour (constraints) of the MIMiC. They built a model inverse-planning code, based on the method of Bortfeld, in which the IMBs and corresponding dose distribution were built iteratively, but every so many iterations the full dose calculation was performed taking into account the physical behaviour of the collimator. It was determined for a model problem that doing this every 2000 iterations was consistent with creating a conformal dose distribution and led to a 'well-behaved' progression of the cost function (no huge changes between 'before' and 'after' cost functions). Seco *et al* (2000) also applied a median window filter every so many iterations and showed that this increased the smoothness of the IMBs beneficially. This is similar to the work of Webb *et al* (1998) for the DMLC technique (see chapter 4). It was found to be advantageous to synchronize the two intervention processes, or at least to weakly synchronize them.

Holmes *et al* (1998) and Curran *et al* (1998) have shown that the CORVUS treatment-planning system uses a pencil-beam kernel fitted to a small number of measurements. Low and Mutic (1998) have also reported on the dose calculation algorithm. The profiles are contained in an instrument data file. CORVUS can, as well as performing the inverse planning and driving the MIMiC, also give the velocity profiles for the technique of IMRT delivery known as the DMLC method (see section 3.1). This is a significant change in the policy of the NOMOS Corporation which initially was to concentrate on a 'one-product' (the MIMiC) method of delivering IMRT, but in later years moved to supporting all three main IMRT delivery methods. For IMRT delivered by the Varian and by the Siemens DMLC systems the measured dose was reported to be between 2–4% at variance with the computed dose distributions. The configuration was not reported but the reason for the lack of agreement may be because of the above problem. The error shrank to only 0.5% for non-IMRT radiotherapy.

An issue of some concern in IMRT with the NOMOS MIMiC is the question of the size of potential dose inhomogeneities at the matchline between successively indexed fields (Xia *et al* 1999b, Dogan *et al* 1999, Linthout and Verellen 1999). Multiple couch indexes are needed because the MIMiC only irradiates a slice of a width of 1.68 cm from each gantry rotation. Hence, for tumours of larger axial extent, multiple rotations are performed with the couch indexed, creating abutted dose distributions. The indexing of the couch is provided by a special piece of apparatus known as the CRANE (figures 2.3 and 2.11) capable of ensuring the indexing is accurate to 0.1–0.2 mm, a level of stability not found on ordinary treatment couches (Curran 1997). Carol *et al* (1996) have shown, by calculation not

Figure 2.11. *A view of the NOMOS MIMiC attached to a Varian accelerator. (From NOMOS Corporation literature.)*

measurement, the importance of accurately indexing the fields with high precision; large dose errors being the result of otherwise small indexing errors. Balog *et al* (1998) have shown that very small errors on the couch longitudinal index can lead to large dose errors, and have compared the mode of operation of the NOMOS MIMiC with helical tomotherapy performed by the Wisconsin machine (see section 2.3).

Lin *et al* (1998) have conducted a study for four patients of the effect of systematic misalignment of the gantry and/or the couch position when IMRT is delivered using the NOMOS MIMiC planned by CORVUS. They found that the effect of systematic variations was more pronounced than the effect of random variations. It was also apparent that the gantry angle misalignment did not lead to errors which were substantial (3° change in angle leading to 3% change in dose parameter) but the couch misplacement was very important (25% change for 3 mm movement).

However, even if the fields are perfectly indexed slice-by-slice, Low and Mutic (1997) have shown that dose inhomogeneities will arise in the abutment regions if each gantry arc is not over a full 360° rotation. This is often the requirement, e.g. irradiation of the head-and-neck may well wish to avoid direct beam entry into the spinal cord. Low and Mutic (1997) arranged to irradiate a

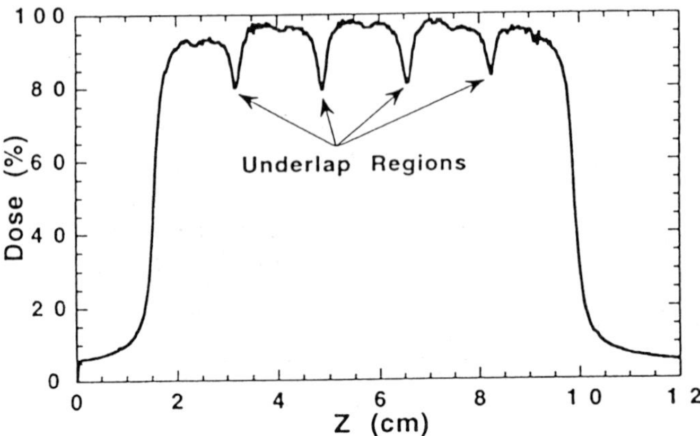

Figure 2.12. *Example of the longitudinal (z) dose profile measurement corresponding to a distance of 7.1 cm above the isocentre and in the vertical plane. There were five rotations with four abutments in which cold regions of approximately 15% were observed. This is due to an incomplete (not 360°) rotational arc. (From Low and Mutic 1997.)*

phantom with a 290° arc symmetrical about the vertical room axis with the beam avoiding the angles from below the couch. The phantom was a water-equivalent cube containing cylindrical target volumes placed either centrally or equidistant above, below, to the right of and to the left of isocentre. Films were placed within the cube horizontally at strategic positions to measure the dose profiles longitudinally and transaxially.

It was found that there was virtually no change of dose longitudinally for all measurements in the horizontal central plane. Divergent fields from left and right compensate each other. However, there *was* a dose inhomogeneity longitudinally in the *vertical plane* varying from 6% hot spots below the horizontal to 15% cold spots above the horizontal (figures 2.12 and 2.13). This effect was ascribed to the lack of symmetry of irradiation about the horizontal plane. In summary, the divergent fields from above are not compensated by corresponding divergent fields from below. However, the effect is not properly built into the PEACOCKPLAN planning system.

Low *et al* (1999b) extended this study and concluded that hot and cold spots could best be avoided by using large arcs of rotation and small leaf bank width settings for the MIMiC. They also measured that the on-axis abutment error was 25% mm^{-1} compared with the 10% mm^{-1} value quoted by Carol *et al* (1996), and based on calculations not measurements. Low *et al* (1999b) concluded that proper use of the CRANE could lead to acceptably small dose errors. They also presented a novel way of halving errors in which, on odd numbered days, the

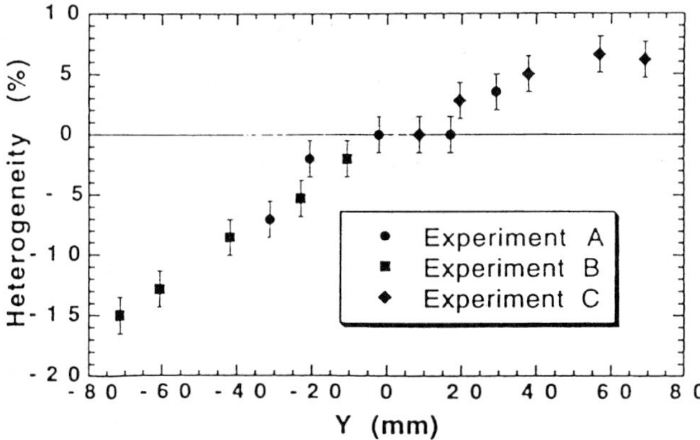

Figure 2.13. *The intrinsic dose inhomogeneity as a function of position in the vertical plane. y increases below the isocentre ($y = 0$) and decreases (negative values) above the isocentre. This is due to an incomplete (not 360°) rotational arc. (From Low and Mutic 1997.)*

treatment is delivered as planned, but, on even numbered days, the MIMiC settings are advanced by one leaf bank width. This would distribute the abutment errors throughout the volume rather than that they be always at the borders of each indexed slice pair. Random interfraction variations in patient positioning can also lead to a smoothing out of the dose inhomogeneity.

Verellen *et al* (1997a, b) have immobilized the head-and-neck region using a thermoplastic cast with individualized earplugs and the use of fiducial markers for localization. They then treated head-and-neck tumours using the NOMOS MIMiC IMB-delivery system. Using phantoms, they measured doses using TLD, alanine and film and found the measured doses using alanine were within 1% of the predicted dose values for different treatment schedules (figure 2.14). TLD measurements were slightly less accurate (as much as 4% difference between calculation and measurement). These were point-dose measurements recorded for simulated treatments that involved multiple couch longitudinal translations. Comparisons of specific transaxial planes have shown that calculated and digitized film distributions gave isodoses which were spatially correlated to within 2 mm. Surface dose and patient film dosemeters were shown to have low values. The immobilization procedure detected translational errors of 1 mm and rotational errors of 1° in most cases. Where the error was larger, corrections were made by comparing sagittal radiographs at treatment with those at planning simulation stage. At the time of this report, eleven patients had been treated, nine lesions of the head-and-neck and two intracranial lesions (see also chapter 4 for an update on this clinical work).

Verellen *et al* (1998) noted that in their earlier study the use of the

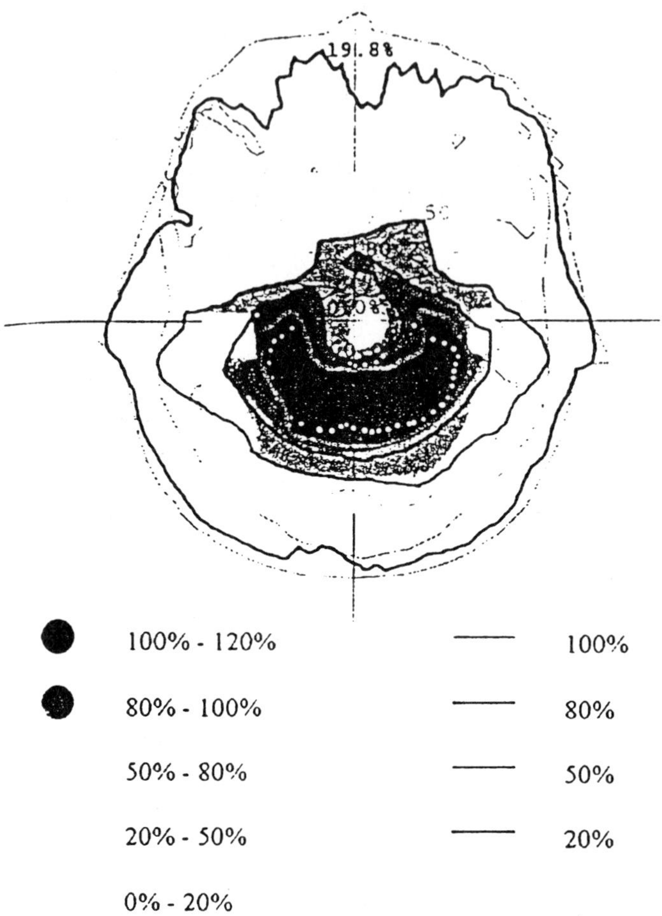

Figure 2.14. *Superposition of an axial plot of isodose lines generated by the IMRT planning system (the 20%, 50%, 80% and 100% lines are shown) and the corresponding film irradiated in the anthropomorphic phantom. The film has been calibrated against ionization measurements and only the regions (0%, 20%), (20%, 50%), (50%, 80%), (80%, 100%) and (100%, 120%) have been made visual. The 100% is equal to 0.5 Gy. The superposition of both images was made by matching the bony structures. The positioning accuracy shows from the perfect match of the target contour that was generated by the PEACOCKPLAN (black line) and the target that was marked on the film prior to irradiation (white dots). (From Verellen et al 1997a; reprinted with permission from Elsevier Science.)*

thermoplastic mask to immobilize the patient led to positioning errors of the order of 3 mm and 2° rotational errors. In a new study, anatomically relevant fiducial landmarks were introduced to ensure that the target localization becomes independent of the immobilization technique. Individualized silicon-based earmoulds and a customized bite block were designed for each patient. Lead beads placed in each of these provided a reference for room-based lasers. It was established that the precision of this positioning method was 0.8 mm. Mean systematic and random errors were thus reduced to better than 1 mm. As shown by Carol, this leads to an acceptable dosimetric error.

Verellen *et al* (1998) have also studied the effect of excluding or including leakage dose in the IMRT calculation. Each of the MIMiC leaves leaks by about 0.5% in accord with the manufacturer's specification. They have shown that the ratio of measured (analine and TLD) doses to calculated doses could be in error by up to 100% in very low-dose regions in OARs for a complex PTV with a concave shape if leakage were ignored but the error was substantially reduced when a leakage contribution was included. According to the report, the manufacturer was about to release a new version of the software taking this effect into account.

Another interesting concept of rotational beam-intensity modulation might be grouped here. This is the idea of delivering IMRT using a Cobalt machine executing many revolutions. During each revolution 36 lead blocks, supported by a ring, move into and out of the field selectively blocking the radiation according to plan. The radial intensity of the beam is modulated in 10% increments when ten rotations are made (Leybovich and Zakharchenko (1997). This is because a block is either present or absent for each rotation. Longitudinal modulation would be an extension of this concept. The equipment is the subject of a Russian patent (Leybovich and Zakharchenko (1993). In this device (figure 2.15) an assembly is mounted at each one of regular angular intervals around a ring (20 in total). Each assembly comprises a series of wedged slots which can accept wedged absorbers focused to the source. If the absorber is present then the radiation is attenuated and *vice versa*. The presence or absence of absorbers can be changed for each rotation of the source. By making N rotations and N selected sets of absorbers, the intensity can be modulated in steps of $1/N$. In this respect, the device functions (conceptually) somewhat like the NOMOS MIMiC in practice. Zakharchenko (1997) has described a similar technique with a static ring of modulating absorbers.

Brahme and Lind (1999) have proposed a ring system of cobalt-60 sources which is a more efficient way of delivering multiple pencil beams. Using P_+ as an indicator of response it was shown (for a specified problem) that P_+ plateaued at about ten beams.

Schreiner *et al* (1998) have also investigated the possibility of constructing a tomotherapy machine using a ^{60}Co source. To do this they used a conventional Cobalt machine to conduct experiments with a rotating table-top CT jig, coming to the conclusion that a machine based on this source would be feasible.

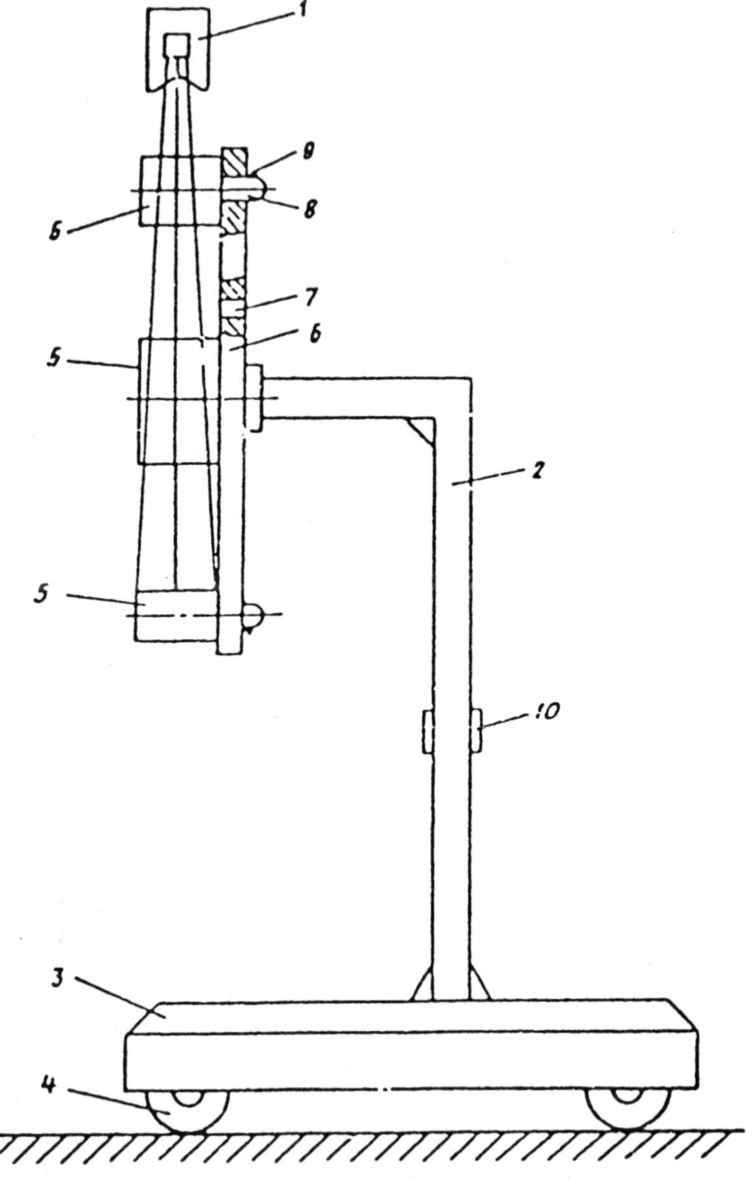

Figure 2.15. *A way to produce tomotherapy embodied in a Russian patent. '1' is an X-ray source which rotates around a circle. '5' are the positions of absorbers creating intensity modulation by the method described in the text. (From Leybovich and Zakharchenko 1993.)*

2.2. DEVELOPMENTS IN IMRT USING THE NOMOS EQUIPMENT REPORTED AT THE 12TH ICCR AND RELATED STUDIES

The NOMOS Corporation presented a large number of papers at the 12th ICCR, Salt Lake City, in May 1997, bringing the radiotherapy community up-to-date with the developments at this Corporation, which in many ways is functioning as a self-standing radiation therapy physics and technology development centre. The main initiatives are reported here. Whilst it is unusual to focus a review section on a particular conference, the cohesion of this approach was much commented upon at the time and so is reflected here. Of course, there has been a 13th ICCR more recently (in fact two weeks before this book was delivered to the publishers) and references to that are scattered within these reviews.

Some of these papers indicated that NOMOS was diversifying its product range and scientific activities beyond providing planning and delivery using the MIMiC. The planning system is also now coupled to the alternative method of delivery known as the dynamic multileaf collimator (DMLC) method. This forms the topic of sections 3.1 and 3.2. For a quick introduction see also figure 1.7. When writing about IMRT techniques it is becoming less easy than it was to view them entirely independently and so inevitably there is some classification difficulty. Ahead of detailed discussion in sections 3.1 and 3.2 it is sufficient to say that the DMLC technique involves creating sets of 1D IMBs (comprising a 2D IMB) by moving the leaves of a conventional MLC.

Hill *et al* (1997) described how the 3D planning system CORVUS supports the development of DMLC therapy via the sliding window technique (see section 3.1). Firstly, IMBs are computed in the same way as for driving the MIMiC. Each 1D IMB is mapped to a pair of MLC moving leaves. The IMB is then converted into a series of unidirectional static leaf sequences via the standard Bortfeld–Boyer method (Bortfeld *et al* 1994a). This is a form of 'interpreter'. Then, there are two possibilities.

(i) The static sequences are translated into dynamic sequences in which each segment end point is derived from the static leaf positions. The intensity delivered for each pencil beam is thus smeared (figure 2.16). Smearing occurs when the slope of the trajectory of the trailing leaf is different from that of the leading leaf. The method also increases the treatment time slightly.

(ii) MLC leaves are either stationary or moving at constant velocity. This ensures zero blurring since all segments have the same slope (figure 2.17) but occurs at the expense of an even larger increase in treatment time.

Several problems with the DMLC technique have been identified by Hill *et al* (1997):

(i) the usable MLC field size is twice the leaf overtravel;
(ii) leakage must be taken into account;
(iii) the interpreter must ensure that adjacent leaves from opposite banks must not overlap (except for the Varian MLC);

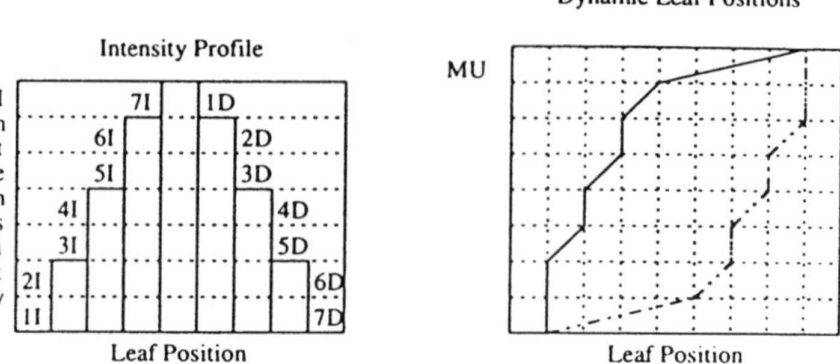

Figure 2.16. *Example of the DMLC algorithm with blurring. Each segment end point is derived from the static leaf positions. The intensity delivered for each pencil beam is smeared depending on the shape of the intensity profile. Smearing over the intensity distribution will occur when the slope of the trajectory of one leaf is different from the slope of the trajectory of the opposing leaf. This dynamic algorithm also pays a slight penalty in MU over the static algorithm. (From Hill et al 1997.)*

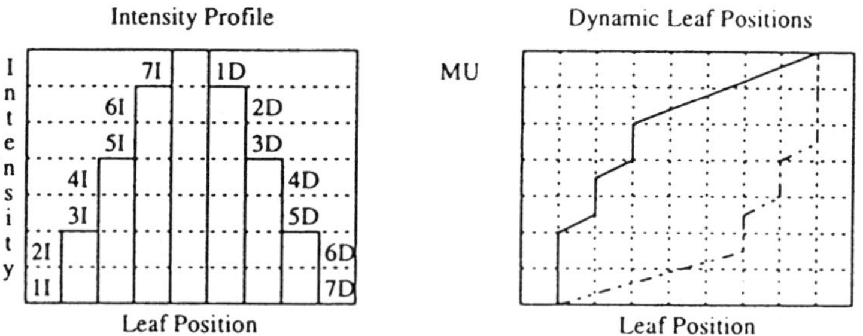

Figure 2.17. *Example of the DMLC algorithm without blurring. MLC leaves are either in motion at maximum speed or stationary. The advantage of this algorithm is that the intensity profile is not blurred. The disadvantage is that the number of monitor units in overhead is increased by the product of the distance leaves travel with the ratio of dose rate to leaf speed. (From Hill et al 1997.)*

(iv) if the MLC is mounted on a carriage that can and must move to deliver large field sizes, then the IMB distribution is broken down into two separate deliveries;
(v) the amount of time a leaf-pair junction is in a field is minimized by moving the junction under a jaw once that IMB has been completed. This minimizes leaf-edge transmission arising with some manufacturers' MLC which have rounded leaf sides;
(vi) the algorithm searches for the minimum number of segments to reduce treatment time;
(vii) the tongue-and-groove problem is solved by synchronization (see section 3.2.2) provided the leaf sides are not rounded.

CORVUS presently supports the Varian and Siemens DMLC and is being upgraded to include an interpreter for the Elekta DMLC. Xing *et al* (1999c) reported comparisons between CORVUS-generated dose distributions and experimental measurements (see section 3.3 for details). Hill *et al* (1997) have shown nice examples of 2D IMRT distributions delivered with the Siemens MLC (figure 2.18).

Holmes *et al* (1997) have addressed the problem of leakage through the MLC leaves in the DMLC technique. CORVUS creates each 1D IMB profile by inverse planning, in which the elemental dose distribution from one single open bixel stands on a zero-leakage pedestal. This arrangement couples the desired dose delivered per 1D IMB to an expected dose distribution. Hence, when the IMB is actually delivered by a DMLC via its decomposition into a series of static MLC shapes (see section 3.3), there is a leakage contribution of $x\%$ ($x \simeq 2$) through each closed leaf portion. As a result, the delivered dose distribution does not exactly match the experimentally measured one. Holmes *et al* (1997) have shown that one could *re-compute* the expected dose distribution by creating a 'mirror set' of each MLC static contribution, in which open and closed spaces replaced each other and the 'open' part was weighted by $x\%$. Thus, the number of computed components is double minus 1. With this modification, predicted dose distributions match the measured ones (figures 2.19 and 2.20). However, an alternative is to create pseudo-IMBs from the IMBs via the methods of Spirou and Chui (1994) and use *these* to create the static MLC fields. Then, the delivered distribution will match that planned without this extra step being needed.

The heart of an inverse-planning system is its dose computation and (for the MIMiC) Kania *et al* (1997) and Bleier *et al* (1997) have discussed how to obtain the elemental dose distributions for the bixel from broad-field dose measurements. The method was validated by then creating large field distributions from the finite size pencil beams (FSPBs) and showing that these match experimentally determined distributions. Nash *et al* (1997) have shown how an octree decomposition of voxel sizes can be used to partition the dose space so that the optimization is accelerated. This particularly facilitates the use of DVH-based objective functions in CORVUS (Carol *et al* 1997a, b, Carol 1997b, see also chapter 5). This objective function

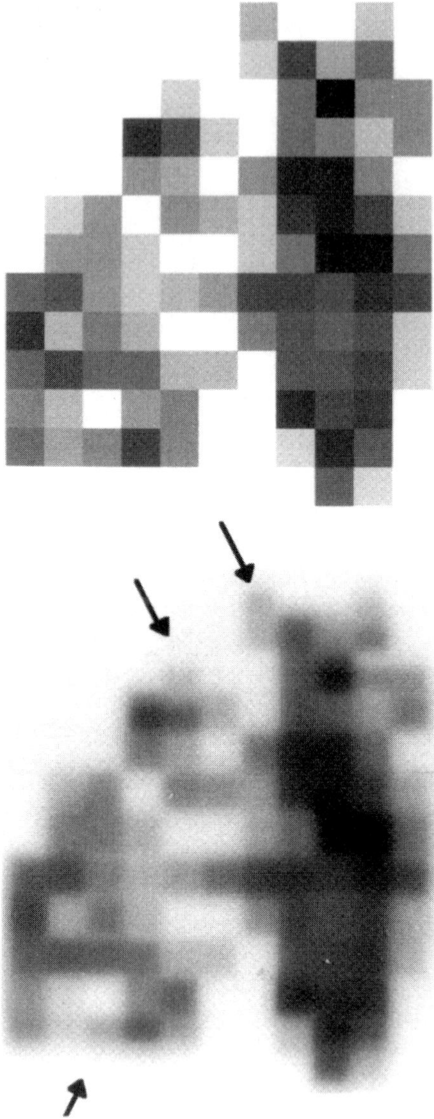

Figure 2.18. *Delivery of intensity modulation with the Siemens MLC. The top figure is the 2D intensity distribution created by CORVUS. The modulation was converted into instructions for a DMLC delivery technique. The bottom figure is the dose distribution measured from a film positioned near* d_{max}. *The areas highlighted exhibit a cropping of the pencil beams in the in-plane direction, believed to be caused by the tongue-and-groove effect. (From Hill et al 1997.)*

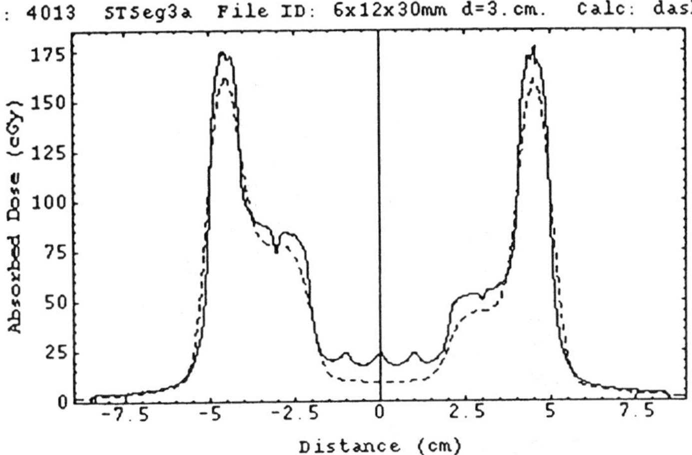

Figure 2.19. *A comparison of a calculated and a measured profile through a 2D IMB computed by CORVUS and delivered using a Varian MLC. The effect of neglecting leakage in dose calculation is shown for this highly modulated beam. The dashed curve is the calculation and the full curve is the measurement. The dose discrepancy in the valley is approximately −5% relative to the maximum dose but 20% in relative terms. (From Holmes et al 1997.)*

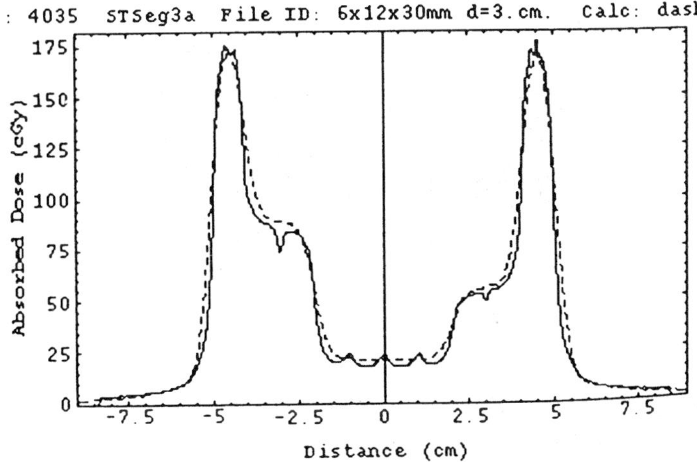

Figure 2.20. *The recomputation of dose including a 2% leakage showing a much better agreement between calculation and measurement compared with figure 2.19. (From Holmes et al 1997.)*

is specified as a clinician would specify the constraints on forward planning and, as such, removes the necessity for a learning and subjective phase of determining importance factors in the cost function for inverse planning. Carol *et al* (1997a) have made the point that this replaces one sort of learnt experience with another. The goal is to automate the inverse planning. The present versions of CORVUS support these concepts. Sternick *et al* (1997) have shown that the energy of the photon beam is virtually irrelevant for IMRT and so it is perfectly acceptable, and at lower cost, to have just a single 6MV linac available for IMRT work. Genetic algorithms have been studied by Rasmussen *et al* (1997) to try to reduce planning times.

The verification of the accuracy of selecting the PTV and ensuring that this is in the correct position at the time of treatment is an important issue. Riker *et al* (1997) have studied issues of image registration for MR and CT data and have shown that the metric of 'goodness of fit' must not be used interactively to adjust the registration parameters. Campbell *et al* (1997) have developed a prototype ultrasound device for localizing the prostate in the treatment room (figure 2.21). The NOMOS name for this equipment is the BAT (Carol 1997b) (BAT stands for Beam-mode Acquisition and Tracking). It comprises a scanner (known as the BAT belfry) attached to a robotic articulated arm (known as the BAT wing) and it returns an image of the prostate onto which the computed dose distribution in the corresponding plane may be superposed (Harnisch *et al* 1998). If there is a mismatch, the system determines the rotations and translations which must be effected to bring the two into registration (figure 2.22). Following registration, the process is repeated as a quality assurance check. Harnisch *et al* (1998) have shown that for 16 consecutive treatments of the prostate, couch shifts were generally within 1 cm. Only 6% of the shifts exceeded 1 cm for the five patients studied. Markman *et al* (1999) have tested the BAT using an ultrasound phantom with an embedded hypoechoic region and radio-opaque spherical target model. The phantom was deliberately misaligned and then the solid-body transform computed to bring the target back to the required position. It was claimed that for the

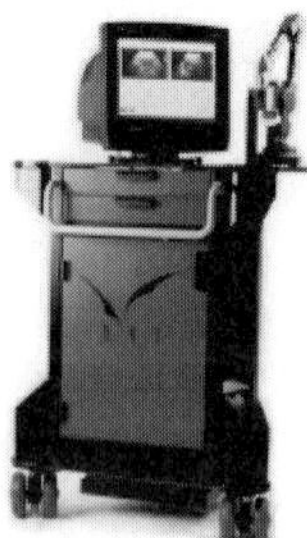

Figure 2.21. *The NOMOS BAT ultrasound positioning device. (From the NOMOS website.)*

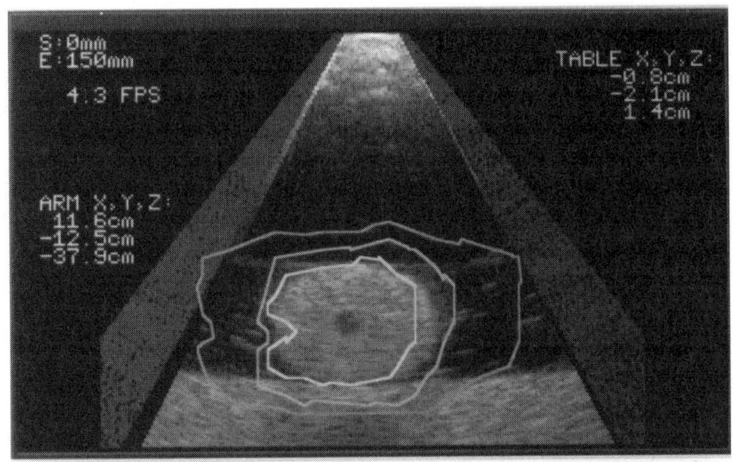

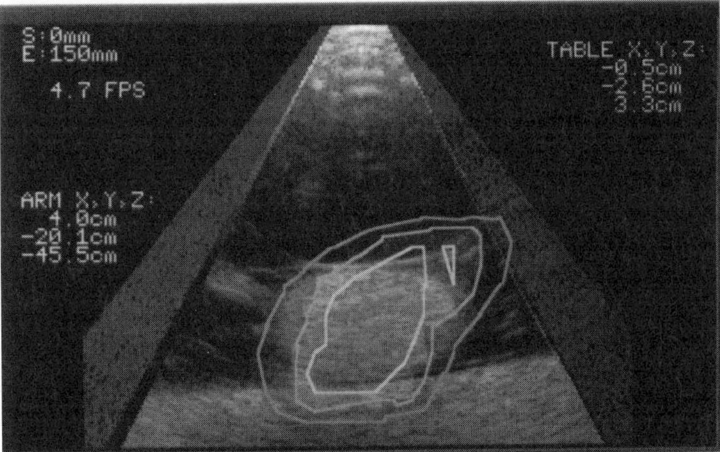

Figure 2.22. *Ultrasound images of a prostate phantom with overlayed dose, axial (top) and sagittal (bottom) views. The dose overlay washes have been outlined for clarity since they are not visible without colour. They represent 50%, 80% and 90% of the hot-spot dose. The coordinates for couch offset and arm position of the BAT are shown. (From Campbell et al 1997.)*

phantom the achieved accuracy was 0.5 mm in all directions. Lalonde *et al* (1999) determined that submillimetre accuracy was achievable. Pollack (1999) reviewed a study which has shown a good correlation between the localization of the prostate using the BAT and the corresponding data on prostate position obtained through repeated CT scans. The use of the BAT required about ten minutes extra time at the time of treatment and of course did not require the patient to go to another machine nor to receive radiation dose. Whilst agreement was generally good, some discrepancies were observed and put down to this transfer. Typically, the r.m.s. movement increments for the prostate, fraction-by-fraction, are quite small but this belies that individual movement increments can be large enough to require a margin of 1.5 cm if BAT-driven relocation is not performed. Margins of this size have been shown to be correlated with higher rates of rectal bleeding. Thus, Pollack (1999) has concluded that the use of the BAT is: (i) less invasive than other localization methods (seeds, rectal balloon etc); (ii) avoiding patient transfer to repeated CT scanning; (iii) likely to lead to less rectal complications. McNeeley *et al* (1999) also used ultrasound prostate relocalization. Lattanzi *et al* (1999a, b, 2000) have shown that the prostate localization with the BAT correlates well with CT localization and can replace it as a fraction-by-fraction relocator. Willoughby *et al* (2000) have shown the BAT can generate a good measure of prostate position when compared with CT measurements. Beyer *et al* (2000) have shown that whilst mean deviations in prostate location may be small, individual daily deviations can be large. McGary and Grant (1998) on the other hand correct positional errors using a fiducial system and lasers without the use of the BAT. Chen (1997) has also discussed non-megavoltage methods of imaging the patient at the time of treatment, including ultrasound and also video techniques.

Mohan *et al* (1999) have performed hypofractionated IMRT to the prostate using 28 fractions to deliver 70 Gy. They used just five intensity-modulated fields and at each fraction adjusted the position of the minimally-immobilized patients by using the BAT. Twenty-seven patients were treated this way and no difference in toxicity was observed between this regimen and a more conventional irradiation technique. Mohan *et al* (2000) have reported that 51 patients were treated with the Varian DMLC technique. This particular form of IMRT used a shorter treatment course with increased dose per fraction.

Finally, the 12th ICCR Conference saw the first written report of the '2D MIMiC'. The concept was described at the earlier Durango (Colorado) Conference, in May 1996, but the written report did not appear until late 1997 (Carol 1997b). The prototype device delivered beamlets of 8 mm square size at the isocentre and the distensible membranes (balloons) could be filled and emptied in 200 ms dictating 10° gantry angle steps for modulation. A 2D IMB could be created by the addition of two 2D IMB components, one with the 2D MIMiC rotated 90° about its central axis relative to the other (figure 2.23). In addition to the generation of 2D IMBs, the device can create 'islands' of zero dose (Dawson *et al* 1997). It was postulated that this device could be regarded as the ultimate 'complete radiation tailoring system' since not only could it provide 2D IMRT with the radiation varying in intensity over a grid of bixels in any generally uncorrelated way, but also it could act as: (i) a wedge; (ii) a block; (iii) a compensator; and (iv) a beam-off device. The spatial resolution of the prototype was coarse and development seems to have ceased, believed due to difficulties with the lifetime of the membranes.

It is always hard to pin down precise statistics concerning the number of patients treated worldwide by any new technique but, at the Durango conference, Carol (1997a) stated that 124 patients had been treated in seven institutions with the MIMiC by 1 April 1996. These comprised 45 malignant intracranial lesions, 33 benign intracranial lesions, 31 metastatic brain lesions, 14 malignant extracranial lesions and 3 benign extracranial lesions. The most common pathologies were malignant astrocytoma grade 3/4 (37 cases), meningioma (18) and pituitary adenoma (8). Woo *et al* (1997) have reported on the clinical experience at Baylor. Sternick and Carol (1997) have reported that the NOMOS MIMiC had, by the summer of 1997, treated several hundred patients for tumours of the prostate, head-and-neck, CNS and thorax. Reported clinical outcomes have indicated that

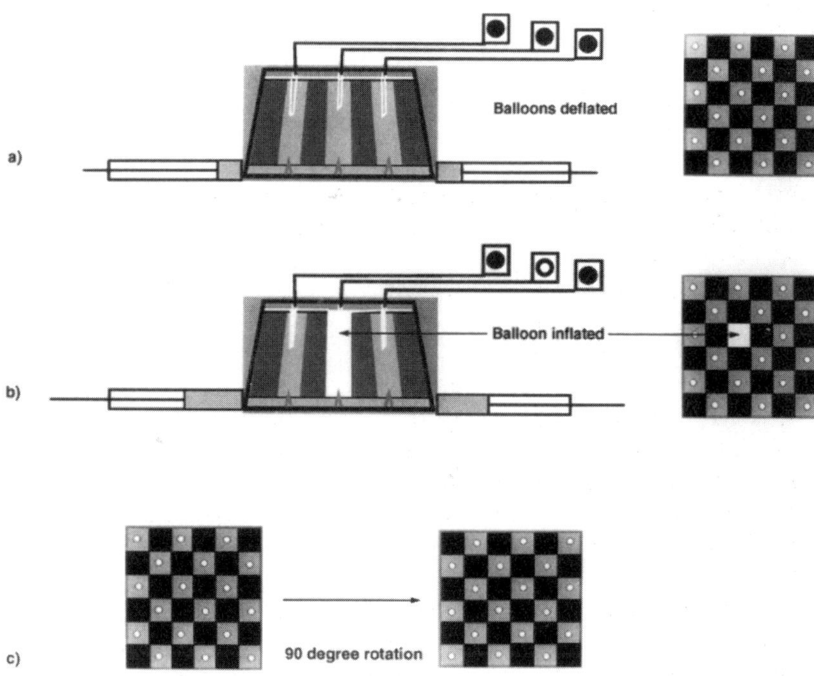

Figure 2.23. *The principle of the 2D MIMiC, sometimes known as the 'volume box'. Inflation of a balloon in a bixel creates an open bixel. A 2D IMB is created by dual irradiation with the 2D MIMiC rotated by 90° between irradiations (c). (From Carol 1997b.)*

IMRT improved normal tissue structure sparing whilst simultaneously allowing target dose escalation (e.g. Smith *et al* 1998). By August 1998, the number of completed treatments had risen to 1753 worldwide (Bleier 1998) and by January 2000 there were more than 65 clinical centres using the MIMiC and more than 6000 treated patients (Bleier 2000). No doubt later and larger statistics will be reported but it is still worth recalling these earlier figures since they give an indication of the rate of growth of the technique. It is also worth recording that IMRT with the NOMOS MIMiC preceded the first IMRT treatment using the DMLC method and, at the turn of the millennium, this method had still treated more patients than any other competing IMRT technique. The detailed clinical picture with respect to the use of the MIMiC is presented in chapter 4.

2.3. THE TOMOTHERAPY MACHINE AT THE UNIVERSITY OF WISCONSIN

Tomotherapy, first reported in the summer of 1992, has been under development for many years at the University of Wisconsin (UW). Mackie *et al* (1999b) have explained that the idea of tomotherapy was formulated in Wisconsin in the late 1980s but was shelved due to the limitations of the slice-by-slice method (the analogy with early CT). The patents for the multivane collimator were pursued and the idea of the multivane collimator was (and still is) used under licence by the NOMOS Corporation. When helical CT became established, the Wisconsin group regenerated interest in tomotherapy. A prototype dedicated unit is being assembled based around a GE Advantage Hi Speed gantry specially designed to have an 85 cm bore (Mackie 1997a, b, c, d, e, Fang *et al* 1997, Reckwerdt *et al* 1998, Mackie *et al* 1999a, b, 2000). This rotates at between 1 and 10 r.p.m. and all power and communications are provided by slip rings. Helical tomotherapy will be delivered via a single slit aperture with 64 vanes each projecting to 6.2 mm at a 85 cm isocentre. Each vane is 10 cm high and sides have a tongue-and-groove design. The gantry will house a compact 6MV Siemens S-band linac with a gridded gun. These leaves will modulate fan beams from 0.5 to 5.0 cm width (patient longitudinal axis) with a 40 cm length in the transaxial plane at the isocentre. The length of the accelerator is 30 cm. The magnetron, radiofrequency load and linac are cooled by blowing chilled air on the stationary side to water-to-air radiators on the rotating side. The bremsstrahlung target is a button of tungsten free to rotate in a stream of water. The field is not flattened so head scattering is minimized but primary fluence non-uniformity requires calibration.

Leaf verification is particularly easy since the leaves can be in only one of four states, either open, closed, opening or closing. Unlike the NOMOS MIMiC in which leaves have to come to a closed state every 5°, the leaves may stay open if they are open in sequential segments. The clinical system is expected to be completed in the year 2000 (Mackie 1999b, Mackie *et al* 2000).

The ring gantry will be equipped with a 738 element xenon CT detection system to intercept the exiting beam. This could be used to reconstruct CT

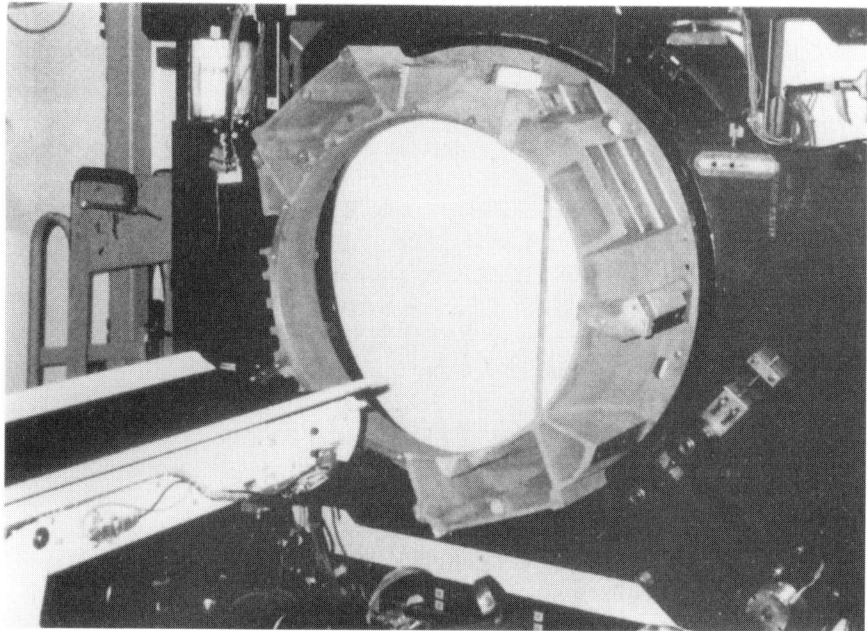

Figure 2.24. *The clinical helical tomotherapy prototype being assembled at the University of Wisconsin. (From Mackie et al 1999b.)*

sections for tomographic verification and also for transit dosimetry reconstructing the delivered dose distribution (Mackie *et al* 1997). Figure 2.24 shows the clinical helical tomotherapy prototype being assembled at the University of Wisconsin. At the time of taking this photograph the main components had not been mounted on the rotating slip ring gantry. A diagnostic CT scanner, also to be on the gantry, records projection images and, without reconstructing an image of the patient, it is possible to align the projection datasets at diagnosis/planning and at tomotherapy in order to correct motion artefacts. This method makes use of a cross correlation which, using data with synthetic offsets, has been shown to be able to give the determination of the offset to better than a pixel (Fitchard *et al* 1997). The method has also been in use on the UW tomotherapy workbench. Fitchard *et al* (1998a, b, 1999) have shown that the offsets introduced artificially into experimental data can be removed by cross correlating the projection sinogram with that of a reference sinogram. The UW workbench is a dedicated experimental facility with a MIMiC collimator with home-built control, a 4MV (effective energy 1.36 MeV) GE Orion linac and a rotating spiralling phantom table. The source-to-isocentre distance is 93 cm and source-to-detector distance is 128.5 cm. The detector (General Electric Medical Systems CT detector, about 1% efficient) has 738 channels although only the central 500 are used (Kapatoes *et al* 1999). The detector has been modelled using Monte-Carlo EGS4/BEAM (Keller *et al* 2000).

Balog *et al* (1997) and Reckwerdt *et al* (1997) have developed a pencil-beam model to describe the elemental fluence of the single bixels in this machine. Whilst the planning code makes no use of the specifics of radiation delivery, post optimization dose calculation does. Balog *et al* (1997) have shown the correctness of the beam model by comparison with measured dose profiles for various combinations of open bixels. For example, two adjacent bixels delivered together or separately would lead to a quite different distribution (figure 2.25). Similar observations have been made by Webb and Oldham (1996, 1997a, b, 1998) in the context of the NOMOS MIMiC (see section 2.1.1).

Balog *et al* (1999a) have studied the tongue-and-groove effect for the UW prototype tomotherapy machine in detail. They concluded that, because the prototype collimator has more leaves than the NOMOS MIMiC, it will indeed contribute a potential underdose but that this can be avoided in practice by ensuring a leaf delivery sequence which avoids the sequential opening and shutting of adjacent leaves. Interestingly, the possibility of defocusing the whole collimator by about 1/4° was investigated.

Tomotherapy should present less of a matchline problem since the point response function is triangular, and triangular functions are very forgiving in terms of movement errors in the longitudinal direction (Mackie 1997c).

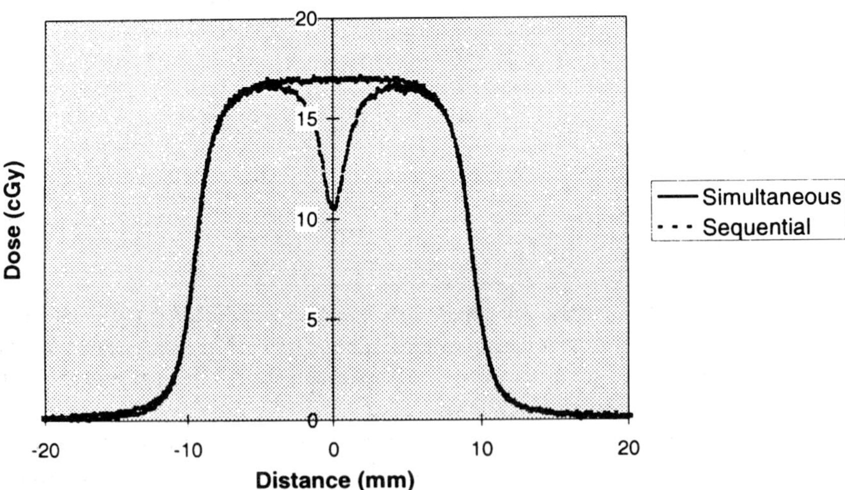

Figure 2.25. *The resultant dose profile when 20 MU are delivered to two leaves of a 1D MIMiC simultaneously versus the resultant dose profile when 20 MU are delivered sequentially. (From Balog et al 1997.)*

Holmes and Yu (1999) have presented a general formalism for optimizing tomotherapy which includes the physical limitations of scatter and leakage through closed vanes in the iterative inverse-planning process. Hence, the developed aperture opening functions correspond to those which, when used for actual delivery, will deliver the planned dose distribution. This was achieved by recognizing that the physical limitations could be built into the matrix operation which links beam space to dose space. At each iteration the aperture functions are created and the dose distribution computed including scatter and leakage. This then forms the basis for the next iteration towards the optimal aperture functions.

Mackie (1999a) has compared developments in radiotherapy practice with evolutionary developments. He observed that in evolution, species tend to converge to common successful types with similar morphology. However, there is also the phenomenon of 'radiance', the tendency to accentuate differences in order to gain a competitive advantage. Mackie feels that after 50 years of convergence, there is now some tendency for radiance in the field of radiotherapy. Special-purpose IMRT devices are an example of radiance in radiotherapy. His own tomotherapy system, under development by the Tomotherapy Corporation, is one such example (Mackie *et al* 2000). Another is the Cyberknife in which a compact x-band linac is wielded by an industrial robot (see chapter 4). These developments move away from conventional radiotherapy, in which 90% of the time co-planar beams are used, with electrons less than 10% of the time, and exploit the observation that when IMRT is delivered using an arcing technique the beam energy is not particularly important.

Since, at the time of writing, the prototype machine has yet to be finished, the clinical cases presented for the tomotherapy delivery have so far been simulations. Reckwerdt *et al* (2000) have shown that really excellent conformality could be obtained for a variety of tumour sites with more than acceptable sparing of the appropriate OARs. These tumour sites were breast, prostate, mesothelioma and nasopharynx.

2.4. VERIFICATION OF IMRT BY THE NOMOS MIMIC AND THE WISCONSIN MACHINE

The issue of verification of IMRT delivered via the NOMOS MIMiC has been addressed by several authors (e.g. Low 1999). Dynamic IMRT dose measurements require that the entire delivery sequence be completed for each measurement. Low *et al* (1997a, b, c, 1998d) have made measurements of 3D dose distributions using the special 'box phantom' provided by NOMOS (Curran 1997) (figure 2.26). This contains a series of 20 parallel rectangular films closely packed between slices of water-equivalent material. When computed and measured dose distributions were compared; the agreement was found to be within 1.4% in the high-dose low-gradient region and within 2% in the lower-dose region. It was necessary to adjust the dose calculation to take account of the leakage radiation. Spatial localization was very good, with all experiments agreeing within 2 mm of calculation.

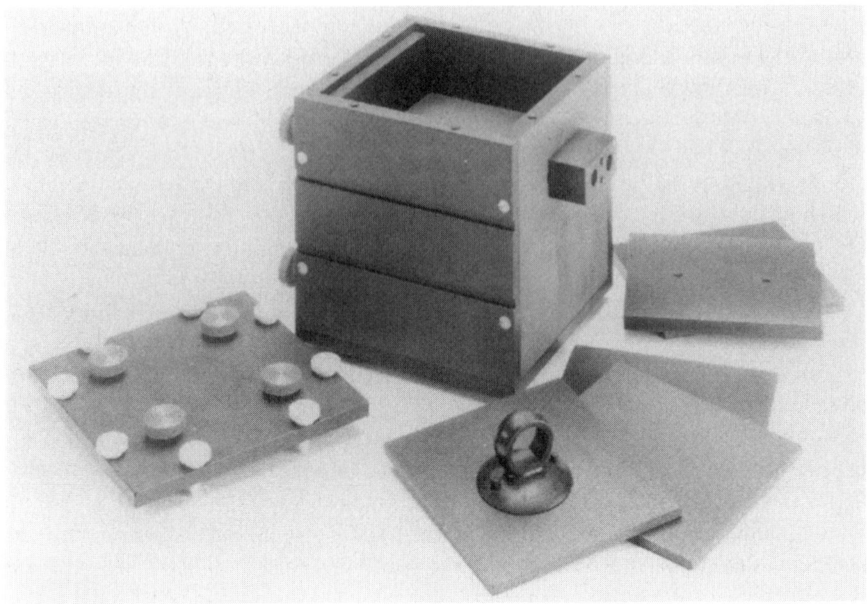

Figure 2.26. *The NOMOS 'box phantom' used to verify IMRT treatment. (From Sternick and Sussman 1997.)*

Low *et al* (1997a) have commented that, for IMRT to become more widely available, better QA tools must be provided by the manufacturers since the data handling is quite cumbersome. Low *et al* (1998c) have developed a phantom of the same external shape as the NOMOS phantom but which is also able to take TLD chips and a cylindrical ion chamber. Conversely, Tsai *et al* (1998) used film sandwiched between the layers of the Rando anthropomorphic phantom.

Low *et al* (1998a) have quantitatively compared several calculated and experimentally measured dose distributions for a series of test irradiations with the NOMOS MIMiC. These made use of the same box phantom (NOMOS call this the Target Box) described above. They found that it was necessary to carefully apply PEACOCK calibration factors and also factors to account for the variability of delivery of MUs with gantry angle. However, provided this was done, the agreement was quantitatively very good, both in terms of absolute dose in Gys and the spatial agreement of the isodose lines in complex plans with concave targets. The main limitation on agreement of absolute dose was the lack of ability of the calculation to cater for scattered radiation and leakage. Spatial localization was accurate to 2.0% and absolute dose to 3.5%. Agreement did not depend on the degree of conformality of dose but became worse with the increasing number of couch indexes corresponding to phantoms of long longitudinal extent.

Low *et al* (1999c, d) have presented the results of a detailed comparative study of planned and delivered IMRT dose distributions using the NOMOS MIMiC. Much of the paper describes the immense care that must be taken with experimental procedures when quality assurance of IMRT is under investigation. Their method made use of the box phantom which can be CT scanned, can accept a set of carefully spaced films and can also take an ionization chamber and TLD chips. Several phantoms were planned, some with complex targets and some with multiple couch indexes. The measured doses were corrected for the varying output of the accelerator and then, for specific locations within the phantoms, the ratio of the measured and the calculated dose was found. This ratio depended on the target geometry, on the number of couch indexes (more so) and also on the degree of conformality forced by the inverse treatment-planning. No single factor was found to be acceptable but Low *et al* (1999c) settled on a value of 1.075 for this 'PEACOCK calibration factor' for subsequent clinical use. The factor is greater than one (measured doses are larger than calculated doses because the calculations do not include the leakage and scatter). The delivered dose distributions were then divided by the appropriate factor, thus determined, before measured and calculated relative dose distributions were compared. Once this was done it was established that the difference between the two was always lower than 3% at a point of 3 mm (fixed isodose), considered acceptable by standard criteria. Hossain *et al* (1999) have shown why ion chamber dosimetry cannot be effectively used for verification of IMRT delivered with the NOMOS MIMiC.

McNutt *et al* (1997) have presented a different method of dose verification of spiral tomotherapy. The Wisconsin Machine computes a CT image of the patient at the time of treatment. A measurement is also made of the exit dose at the plane of a portal imager. These data are sufficient to determine the actual dose delivered (D_{act}). The algorithm begins by assuming the exit primary energy fluence (Ψ_0) is the same as the measured exit dose. This is then backprojected into the CT volume to determine the terma everywhere. Convolution/superposition is then carried out to determine a first best estimate of the exit dose (D_1). The next best estimate of exit primary fluence is then formed by multiplying the first estimate (Ψ_0) by the ratio of the measured to first-estimated total exit dose (D_{act}/D_1). This process cycles, iteratively yielding the best estimate of the actual dose delivered throughout the patient. McNutt *et al* (1997) applied the method to 72 IMBs at 5° intervals on the tomotherapy workbench at Wisconsin.

Olivera *et al* (1998) have described how the maximum likelihood (ML) technique can be used for three quite distinct purposes with respect to the Wisconsin tomotherapy machine. It has been used to reconstruct megavoltage CT images of a neck phantom with improvement demonstrated over filtered backprojection, and many features that are visible on a kilovoltage CT image also visible on the MVCT image. It was also used to reconstruct a map of dose from measured transmission data. Computer experiments were made with variable added noise and the method was shown to be robust. Finally, the ML technique was also used for the initial dose optimization to determine the fluence profiles. Two model distributions were studied: the familiar 'C' shape with an OAR in the concavity and a torus entirely surrounding an OAR. Good dose distributions were obtained, although it was emphasized that the method still required extension to take account of dose-volume constraints and penalties. Ruchala *et al* (1999c) have shown that MVCT images can be made on the UW MVCT system with about 10 cGy of dose, which demonstrate a spatial resolution of 3 mm and a contrast resolution of about 2%. Ruchala *et al* (2000) have shown that MVCT reconstructed contrast was linear with electron density. They have also shown that the MVCT contrast reconstructed at low dose (7 cGy) was linear with MVCT contrast reconstructed at 175 cGy, thus enabling the concept of low-dose imaging pre-therapy.

Kapatoes *et al* (1998) have used the 738-element xenon chamber to compare measurement with the prescribed fluence sinogram for the Wisconsin tomotherapy machine. Ruchala *et al* (1998) have alternatively proposed using MVCT with images reconstructed by an ML algorithm which they claim is better than filtered backprojection. Ruchala *et al* (1999a) have described how MVCT images can be reconstructed from incomplete data collected on the Wisconsin tomotherapy machine. The data are necessarily incomplete because the irradiation is of the PTV only rather than of the whole body which would be required for acquisition of a complete dataset. Ruchala *et al* (1999b) have shown how the leakage radiation through the closed leaves of the NOMOS MIMiC can be used to reconstruct MVCT images leading to the concept of no-extra-dose position verification. Ruchala *et al* (2000) have shown comparative images of MVCT reconstructions of a German shepherd dog head made under three conditions: (i) before or after the therapy using helical scanned open fields (7 cGy); (ii) during the therapy using a 'flash' of the fully-open MIMiC (2.5 cGy); and (iii) during therapy using just the leakage radiation. The images decreased in quality from (i) to (iii) due to the decreased number of photons. As the Wisconsin machine comes to fruition we may expect to see more definitive statements on quality control and verification.

Kapatoes *et al* (1999) have given details of the delivery verification technique for the UW prototype IMRT delivery machine. The method is also applicable to the NOMOS MIMiC slice-by-slice delivery and indeed it is postulated, though as yet not developed in detail, that a similar method would be workable for the DMLC IMRT technique (see chapter 3). The technique is based on using a one-dimensional portal detector to determine the delivered fluence for each 'projection' (irradiation from a particular gantry angle) during tomotherapy. This is called the verification sinogram and is computed from the treatment detector sinogram. It is then compared with the planned fluence sinogram. It could also be used to compute a tomographic image of dose by techniques similar to those developed at the Royal Marsden NHS Trust by Hansen *et al* (1996). The verification method is of the fourth 'class' as summarized by Webb (1997d). Provided the patient does not move during the treatment the delivery is verified on a pulse-by-pulse basis. If a CT image dataset has also been taken then dose reconstruction can be performed. The general schema is shown in figure 2.27.

The technique assumes to first order that the measurement $s_i\ (t)$ at the ith detector element is a weighted linear sum of the input energy fluence $\psi_j\ (t)$ emitted

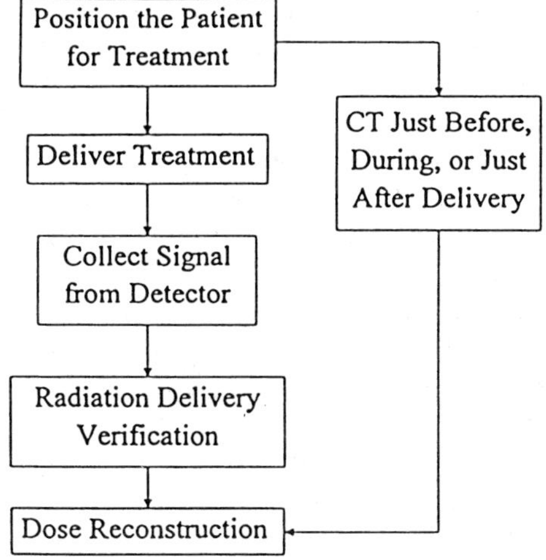

Figure 2.27. *General block diagram of a treatment that includes radiation delivery verification and dose reconstruction. The signal collected in the exit detector is used to perform radiation delivery verification. The output of this process, the incident energy fluence distribution, and a CT of the patient in the treatment position are used to accomplish dose reconstruction. (From Kapatoes et al 1999.)*

from the jth temporal collimator leaf through

$$s_i(t) = \sum_j D_{i,j}(t)\,\psi_j(t) \tag{2.1}$$

where $D_{i,j}(t)$ is the contribution to the ith detector element from the jth MLC leaf for unit fluence. In vector form

$$s(t) = D(t)\,\psi(t)\,. \tag{2.2}$$

Provided the system can be considered linear, the verification sinogram is created from inverting this equation, i.e.

$$\psi(t) = D^{-1}(t)\,s(t)\,. \tag{2.3}$$

The verification sinogram $\psi(t)$ is then compared with the planned fluence sinogram.

The method requires the measurement of all the components of the matrix D. Figure 2.28 shows that this requires the evaluation of the response of each of the i detectors when just one of the MLC vanes is opened and the process is repeated for all the vanes. This needs to be done for the gantry in each of the locations that it will take up during IMRT delivery. The measurements need to be done for

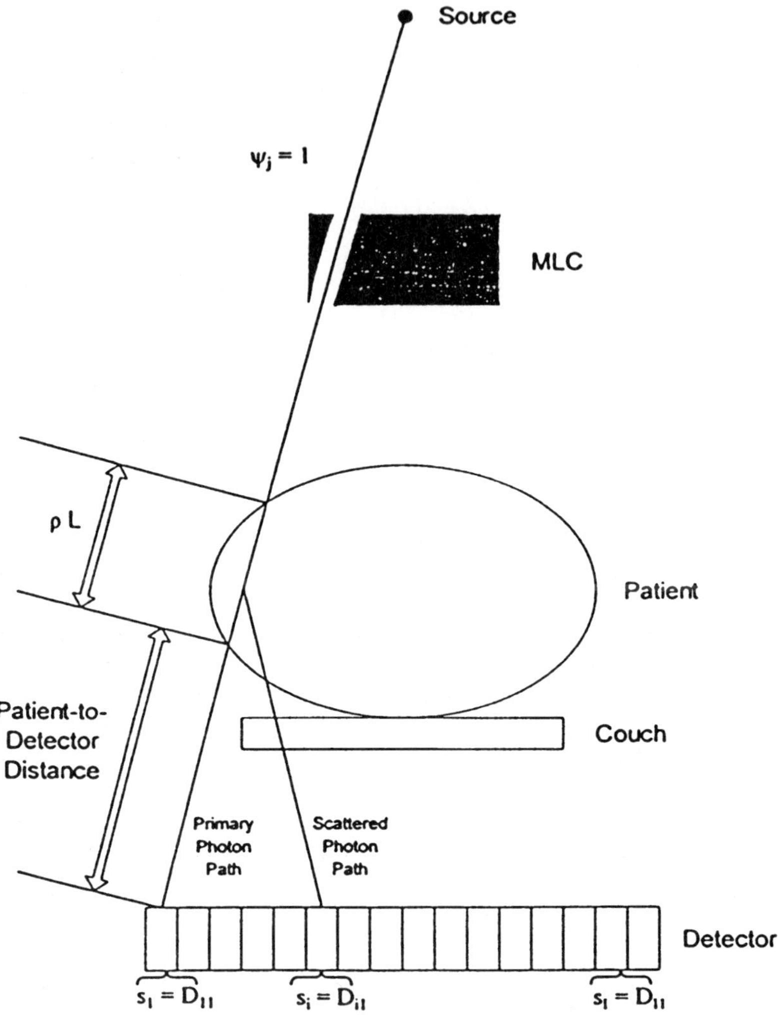

Figure 2.28. *Diagram of components displaying the meaning of one column of the D matrix. A source is delivering radiation through the opened first leaf of an MLC implying that the energy fluence for that leaf is unity. The signal detected will primarily depend on the radiological pathlength (ρL) and patient-to-detector distance. The signal at a detector will be due to both primary and scattered photons. In tomotherapy, the source, MLC, couch and detector are integrated as one unit. The MLC and detector allow for a unique opportunity for radiation delivery verification. (From Kapatoes et al 1999.)*

each patient fraction and (of course) with the patient in position. Kapatoes *et al* (1999) have however explained that it may be possible to make use of a stored reference dataset of values and also that, even if the measurements are made at each fraction with the patient in position, the dose delivered is quite small and can be subtracted from the tomotherapy-delivered dose. It is clear that the total number of components of D measured is N_D =(number of detector channels) $\times$ (number of MLC leaves) $\times$ (total number of projections) $\times$ (number of slices) which can be (say) $500 \times 20 \times 72 \times 50 = 3.6 \times 10^7$ elements, needing 144 Mbytes of memory.

Kapatoes *et al* (1999) have shown that the linearity assumption is not strictly true. It is necessary to make two major corrections. The first is to subtract the detector dark current and the contribution due to collimator leakage. (If this is not done then inversion equation (2.3) generates negative fluences.) A more important and much more complex correction is to take account of the fact that the delivery of radiation with several elements of the detector simultaneously open is not the same as that with the elements delivered sequentially, as studied in detail by Webb and Oldham (1996) (see section 2.1.1). There is a larger interleaf signal when all are delivered together than if leaf-by-leaf delivery were performed. This can, however, be accounted for by factoring the known ratio of tongue-and-groove-penumbra effect into the detected values. However, a catch-22, this requires knowledge of the leaf patterns (which is the outcome of the inversion). So, the inversion has to be done *twice*. Firstly, it is done without knowing and including this effect. This gives a qualitative fluence map which indicates the status of the vane opening. Knowing this, the correction for the effect can be made to the measured data and a *second* inversion performed. This, of course, gives the same qualitative pattern for the input fluence but now also gives the correct quantitative value. This is actually only a first-order correction assuming that only nearest-neighbour correction needs to be done. These two correction procedures essentially linearize the process.

The UW tomotherapy workbench has been used to illustrate the success of the method. Figure 2.29 shows a block diagram of the approach in summary. It has been found that the method can reconstruct the planned fluences to within 2%.

Kapatoes *et al* (2000) have shown examples of the method in action to demonstrate its ability to pick-up a deliberate mispositioning of an irradiated dog cadaver.

Olivera *et al* (2000) have commented that the ‘usual’ way to correct for observed mismatches between the true and desired patient position (as evidenced by the methods described above) is to move the patient to the correct position. However, a new proposal is to leave the patient in the incorrect position and to modify the intensity modulation. Olivera *et al* (2000) have shown that linear shifts can be removed by appropriate resampling of the delivery sinogram. Examples have shown that the success of this approach was independent of the magnitude and direction of the offsets. It was postulated that, though in its infancy, this method might become able to correct for movement actually during a radiation fraction.

In summary, verification and quality control of tomotherapy has been

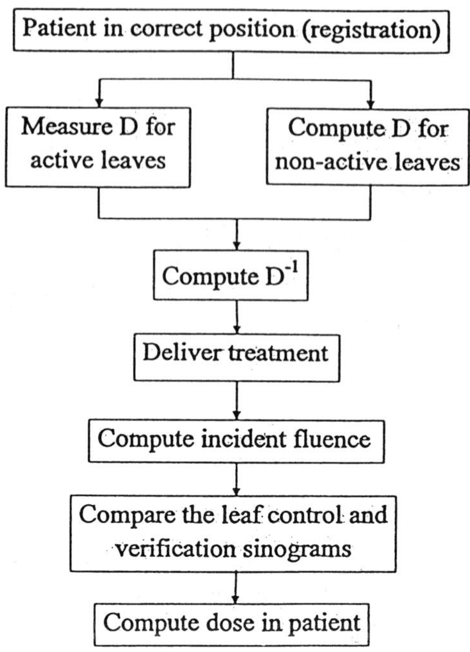

Figure 2.29. *Block diagram of a radiotherapy treatment that includes radiation delivery verification and dose reconstruction. (From Kapatoes et al 1999.)*

addressed largely by comparing predicted to measured volumetric dose distributions. Statements are made on the % dose accuracy at a point and on the geometric closeness of calculated and measured isodose contours. Alternative information can be gathered on the fidelity of movement of the vane systems and the resulting projection data, intermediate to dose calculation. The subject is revisited again in section 4.4.

CHAPTER 3

IMRT USING A MULTILEAF COLLIMATOR

3.1. IMB DELIVERY USING A MULTILEAF COLLIMATOR

In this chapter we turn our attention to the delivery of IMRT using a multileaf collimator (MLC). IMBs may be constructed using a sequence of static MLC-shaped fields in which the shape changes between the delivery of quanta of fluence, the so-called multiple-static-field (MSF) technique. Alternatively, the leaves may define changing shapes with the radiation on, the so-called dynamic multileaf collimator (DMLC) technique. However, classification is messy because the latter can operate in a step-and-shoot mode in which essentially the 'movement' is between MSF locations. This chapter is organized in order to present the DMLC technique first, including this possibility, and then we shall return at the end (section 3.3) to the *pure* MLC-MSF technique, which includes specifically the delivery of a *few* MSF configurations. This chapter is the longest because these two techniques have been subject to the greatest expansion in recent years.

3.1.1. Multileaf collimation

Apart from the specialist equipment designed to deliver IMRT by rotational-arc methods (chapter 2), the mainstay of IMRT delivery is the MLC (Galvin 1999b). Multileaf collimators were first commercially introduced in Europe, then in the USA (Klein 1998). There have been several reviews of the history of their introduction and their use (Webb 1993, 1997d). The first use of the MLC was to act as a replacement for cast blocks to perform geometric field shaping. Their use for intensity modulation has grown out of this.

It is often remarked that the MLC is entirely able to replace cast blocks. This statement usually requires some qualification. For example, Adams *et al* (1999b) have made a comparison between the ability of the MLC and a shaped block to spare normal brain tissue when irradiating either spherical tumours in a model phantom or specific patient cases. They found that the MLC treats on average

14% more normal brain to greater than 50% dose and 17% more normal brain to greater than 80% dose. The range was large and uncorrelated with tumour volume. It had more to do with the precise shape of the PTV and whether an integer or non-integer number of leaves were required for each beam's-eye view. Bedford *et al* (2000c) have also shown that the MLC stepped edges can cause over-irradiation of the prostate boost phase with a three-field technique.

There is no complete audit of the number of MLCs in use worldwide nor of the number in use for IMRT. However, a recent North American survey was carried out to establish the pattern of MLC use and this included questions about MLC use for IMRT (Klein *et al* 1999a). Approximately 250 centres had been equipped with an MLC between 1992 and 1997, thus having five years' experience. An Assessment of Technology Subcommittee of the AAPM polled these 250 centres. Replies were received from 108 centres; some centres with more than one MLC. The survey was restricted to three major manufacturers and the following were represented in this return: Elekta (13), Siemens (16), Varian 52-leaf (60), Varian 80-leaf (45), total (134). The breakdown of returns was at least 40% for all manufacturers. So, a fairly obvious first finding is the predominance of Varian equipment in the USA. Many centres purchased an MLC for IMRT use. However, only 10% were using it at the time of the survey, though 78% anticipated doing so. Only 26% of the centres which justified the purchase of the MLC on the basis of IMRT were actually carrying out IMRT. However, 97% of the centres which justified an MLC purchase for IMRT anticipate using it for this purpose in the near future. The rest of the survey covered use for conventional (non-IMRT) radiotherapy. The clinical sites were listed, the methods of file transfer, actual use compared with anticipated, purchase justifications and user satisfaction. Klein *et al* (1999a) concluded that, not surprisingly, the MLC is still being used mainly as a field-shaping block-replacing tool. However, its use for IMRT is slowly escalating.

3.1.2. Quality control of MLCs for static use

Hounsell and Jordan (1997) have presented the results of their quality assurance of the Elekta Oncology Systems (EOS)—formerly Philips Medical Systems–Radiotherapy—MLC. A new quality assurance programme has been instituted at the Christie NHS Trust, Manchester but this is still within acceptable limits of staff burden. The Elekta MLC was the first MLC installed in the UK (Leadbetter *et al* 1997). It has considerably shortened the time taken to create newly shaped fields between treatment fractions.

Mubata *et al* (1997b) have described a programme of quality assurance which the use of a Varian MLC makes mandatory. They described checks which should be made at different time intervals, including calibration of leaf position, leaf carriage skew, checks on the light field, the radiation centre, the shape digitizer and carriage sag.

Eilertsen (1997) has made a detailed study of the performance of the Varian MLC in conjunction with a Varian PortalVision electronic portal imaging device

(EPID). The main conclusion of a complex paper was a method to determine whether the individual leaves were in the correct position using the image data from the EPID. Clear distinctions must be made between the light field, the (diode) measured radiation field and the fluence map, converted to dose, obtained from the liquid-filled EPID.

3.1.3. MLC equipment developments including microMLCs—use for the DMLC technique

There are now several commercially available MLCs for implementing the dynamic MLC (DMLC) technique. These will be reviewed in section 3.2.3. Despite these, several University Hospitals are building their own equipment and these are now summarized. Figure 3.1 shows the main geometric and radiation features of a (any) MLC.

Schlegel *et al* (1997a, b) have developed a microMLC for creating small geometrically-shaped fields for brain stereotactic radiotherapy. The collimator has 40 leaf pairs made of tungsten, each with a width at the isocentre of 1.6 mm, depth 9 cm, maximum overtravel 2.4 cm, maximum leaf speed 1.5 cm s^{-1} creating a maximum field size of 7.3×6.4 cm^2. Each leaf is equipped with its own motor and the motors are arranged in a fan-shell pattern (figure 3.2). The leakage reduction mechanism for the Heidelberg motorized microMLC is the following. On each side, the four leaves in the middle of the MLC have tongues and grooves with a thickness of 0.3 mm. This reduces the leakage in the critical area in the middle of the microMLC to below 2%. All the other leaves need no tongues and grooves, as they are not focused to the source. The microMLC is designed to deliver IMBs for irradiating concave target volumes. The control software of the MLC-host computer was written in Visual basic and is running under WINDOWS-95 on a Pentium PC. Absolute positioning measurements showed the accuracy was 0.2 mm and the worst interleaf leakage was 2.5%. The system is being duplicated for clinical test phases in Cologne and Würzburg. As well as using the specially constructed control unit, this group are investigating whether the electronics of a commercially available integrated large-field MLC could also be used for operating the microMLC (Schlegel *et al* 1997c, Föller *et al* 1998). The prototype Heidelberg motorized microMLC is marketed by MRC Systems GMBH via the Stryker–Leibinger Company.

Küster *et al* (1997) have used the Monte-Carlo programme GEANT from CERN to model the design of an 'intermediate size' MLC, the so-called 'computerized large overtravel collimator (cLOC)'. This will have 70 tungsten leaves of height 5 cm with a large overtravel of 7.5 cm in the isocentric plane. GEANT is an alternative to EGS4 which operates faster for this application. Several possible forms for the leaf face were evaluated, concluding that a curved face was best giving a transmission penumbra in agreement with the geometrical penumbra. The former was computed with GEANT and the latter by considering the amount of attenuating material at each leaf position and some mean X-ray

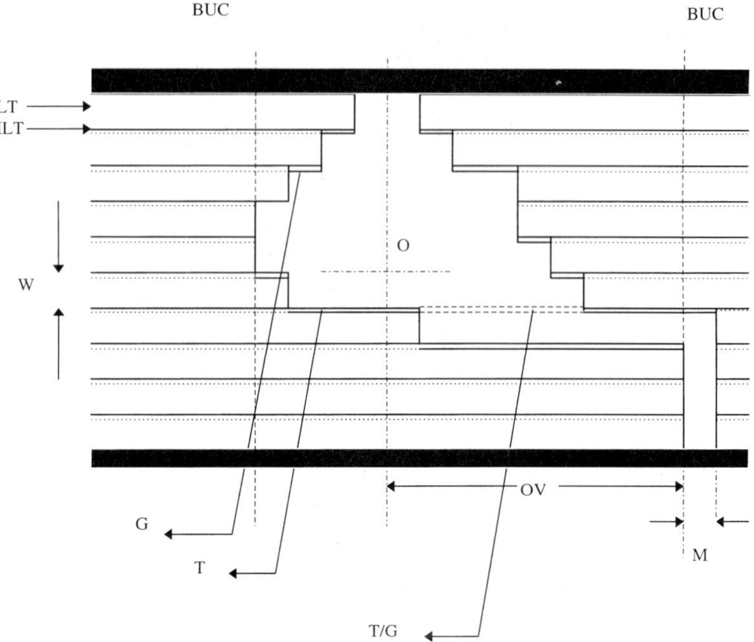

Figure 3.1. *A schematic diagram showing the basic parameters defining an MLC. The figure shows just ten leaf pairs of what would in practice be a larger set. The number N of leaf pairs is a defining parameter. The leaves project to a width W at the isocentre (many MLCs have all leaves the same width but others have been produced with wider leaves at the extremities). O is the radiation field centre and the principal field axes are shown dotted-dashed. The distance OV represents the maximum overtravel that a leaf may make. The distance M represents the minimum distance that adjacent leaves may approach each other. T labels one of several leaf tongues shown and G labels one of several leaf grooves shown. T/G labels one of the regions defined by the overlap of the tongues and the grooves. LT labels the position at which the leaf transmission would be determined and ILT labels the position at which the interleaf transmission would be determined. The MLC is usually backed up by conventional jaws (lines labelled BUC [back-up collimator] and black boxes) in line with (BUC) and orthogonal to (black boxes) the direction of leaf movement. For static-MLC collimation these usually envelope the field (as shown here) and for the DMLC IMRT technique one pair (that labelled BUC) can track the most-leading and most-trailing leaves whilst the other pair (black) envelopes the field in the direction orthogonal to leaf movement. This allows the reduction of the leakage through shielded regions even more than by leaves alone. For some DMLC techniques the BUC jaws are used, alternatively, in conjunction with the leaves to produce the modulation and overcome limitations imposed by the minimum leaf gap M.*

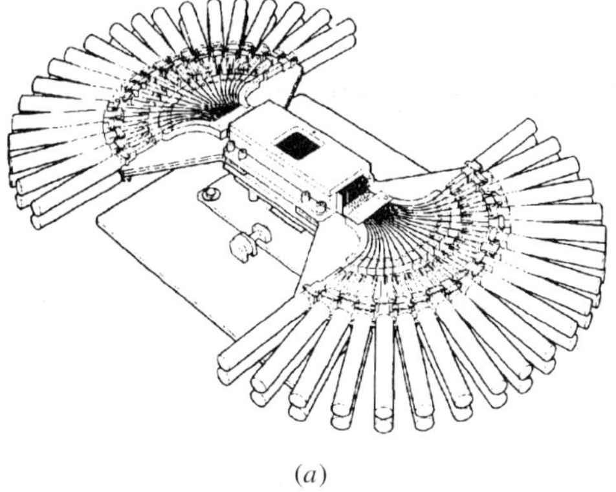

(*a*)

(*b*)

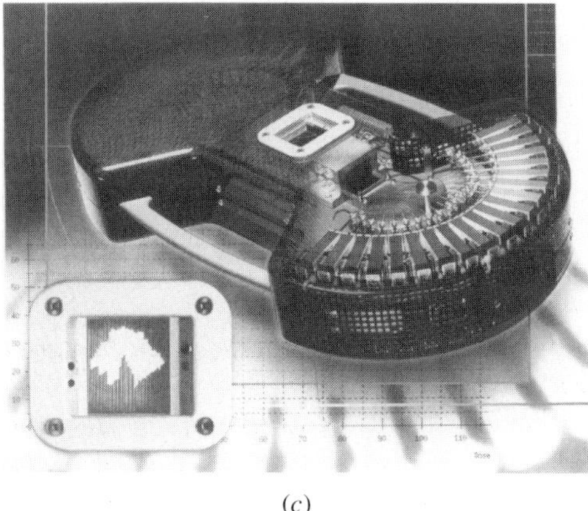

(*c*)

Figure 3.2. *(a) Design drawing for the motor driven MLC built in Heidelberg. (From Schlegel et al 1997a.) (b) Photograph of microMLC (covers off). (Courtesy of Professor W Schlegel.) (c) Photograph of assembled microMLC. (From MRC Systems literature.)*

attenuation coefficient. The peak interleaf leakage for leaves focused to the source was an unacceptable 20% but when the two middle leaves were defocused by 1° with respect to this and then the other leaves aligned parallel to these, the leakage reduced to a peak of 8% with an average of 2.6% (figure 3.3).

Hädinger *et al* (1997, 1998) have described the control of an MLC for DMLC therapy. The MLC was constructed by DKFZ Heidelberg and implemented in Würzburg. It has 35 leaf pairs with a leaf width at an isocentre of 5 mm, a leaf

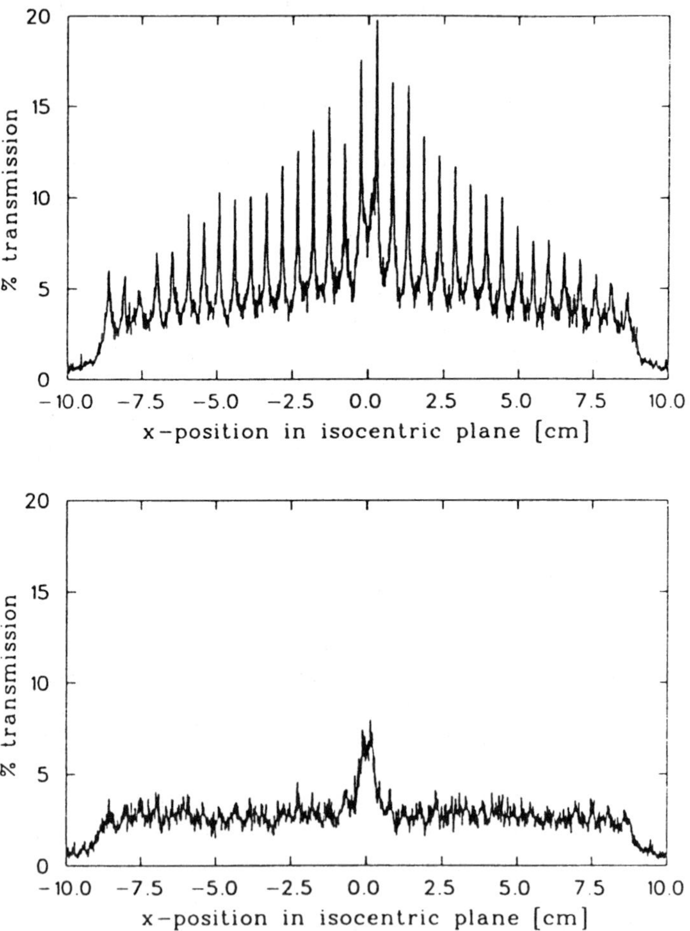

Figure 3.3. *Leakage radiation through an intermediate size collimator (the cLOC MLC) with focused leaves (top) and leaves defocused by an angle of 1° (bottom), expressed as a percentage of the maximum absorbed dose at the reference axis. (From Küster et al 1997.)*

height of 5 cm, a maximum overtravel of 7.5 cm and a maximum leaf speed of 2.0 cm s^{-1}. It is single-focused and operated by programmable microcontrollers. The linac jaws follow the outermost leaves to minimize radiation transmission through the collimator.

A computer controlled microMLC is also manufactured by the BrainLab Corporation. This fits into the gantry tray slot and is compatible with Varian Model-C accelerators. The direction of leaf movement of the microMLC is orthogonal to that of the standard leaf sized MLC. The central leaves project to 3 mm at the isocentre and can be used to geometrically shape small fixed-gantry-angle fields for stereotactic radiotherapy (figures 3.4 and 3.5). The microMLC has been used to treat patients clinically at the Charité Hospital, Berlin (figure 3.6) (Cosgrove *et al* 1998a, b).

The physical characteristics of the BrainLab MLC have been measured by Xia *et al* (1999a). The 26 leaf pair microMLC easily attaches to the accelerator by means of a transfer trolley. The leaf widths at 100 cm SAD isocentre were 3 mm for the central 14 pairs, 4.5 mm for the next three leaf pairs and 5.5 mm for the outer three leaf pairs on each side giving a maximum field size of 10 × 10.2 cm. This design, whilst not being the ideal proposed by Nedzi *et al* (1993), nevertheless is more flexible than an MLC with fixed-size large leaves. Each leaf overtravelled by 4.69 cm. Rotated to its neutral position the leaves move in the same direction as the upper Varian accelerator jaws and at right angles to the direction of the conventional MLC, allowing a combination of two MLCs to provide complex collimation (figure 3.5). Figure 3.7 shows some typical circular fields generated.

Xia *et al* (1999a) have shown that the percentage depth doses varied smoothly and predictably with field size and were no more than 2% different from the 4 MV percentage depth doses for the same field sizes formed by conventional jaws. Output factors were measured for square and circular fields and varied smoothly and predictably with leaf position, varying by only 2% until the fields reached 7 cm equivalent square. After this, for smaller fields, the output factor rapidly decreased due to the leaves shielding the flattening filter which generates scattered X-rays. The curves for both square 1 × 1 cm field and circular 1 cm diameter field reach a value of approximately 0.87. It is arguable whether the output factor should be defined for fixed or field-tracking jaw positions, and the output factor depended somewhat on this choice.

Penumbra was measured to be comparable to photon blocks and smaller than observed for the conventional MLC. Interleaf leakage was 2% and intraleaf leakage was 1.3% based on film measurements through closed leaves. By studying abutting fields, leaf positioning precision was found to be better than 1 mm. From these observations Xia *et al* (1999a) concluded that the method of conventional field dosimetry could be applied for the dose calculation in the irregular fields formed by the microMLC.

Cosgrove *et al* (1999) have also performed a detailed dosimetric study of the BrainLab m3 microMLC attached to a Varian Clinac 2100C accelerator. They reported the feature that the ends of the leaves are milled to three angled straight

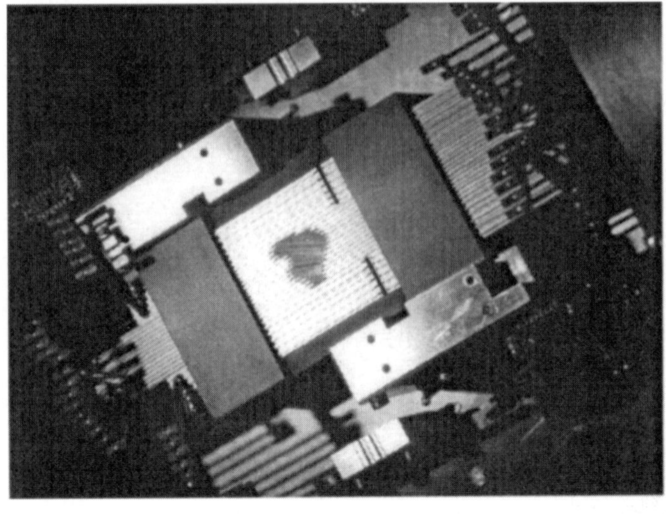

(*a*)

(*b*)

Figure 3.4. *(a) Picture of the (covers-off) BrainLab microMLC. (From BrainLab website.) (b) End-on view. (Taken on a trade stand in Berlin.)*

Figure 3.5. *The BrainLab MicroMLC leaves are orthogonal to the 'conventional' Varian MLC leaves. This figure shows a shape collimated using both the MicroMLC and the conventional MLC. (Courtesy of Dr V Cosgrove.)*

edges each covering a third of the leaf edge. The central straight edge enables opposing leaves to touch when closed. The other two straight edges correspond to the field divergence at full extension and full retraction. As a result of this feature it was reported that the penumbra did not vary with field size, the 80–20% being 2.4 mm at d_{max} for a 2×10 cm^2 rectangular field. The effective penumbra increases as the leaf widths increase and depends on the angle which the leaves make to the Y-jaws. The mean leakage between leaves was found to be 2.8% and that through the leaves was 1.9%. The leakage through the leaf ends depended on the position within the field. When the abutment was central it was on average a very large 15% but by moving the abutment 4.5 cm off axis this dropped to 4.5%. Since this can always be done for a given collimation this was deemed to be acceptable. It was determined that the presence of the m3 MLC did not change the dependence of the isocentre on gantry angle, what little sag there was being the same, with and without the MLC. Output factors and depth dose curves were the same as for the corresponding circular stereotactic collimators.

Benedict *et al* (1999) have shown that IMRT delivered by the DMLC technique using a BrainLab microMLC delivered dose distributions which were superior to those achievable with multiple arcs for brain tumours with complex shapes. The dose in the PTV was more homogeneous and there was better normal-tissue sparing. The same finding was reported by Boccuzzi *et al* (1999).

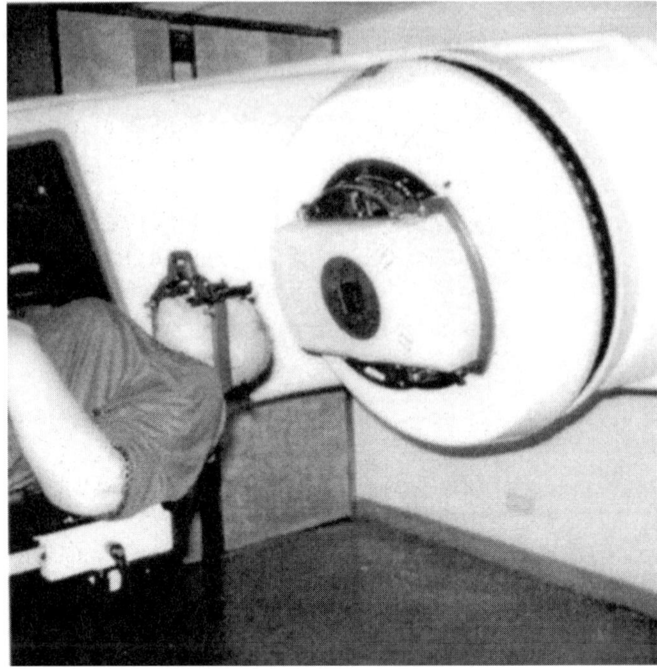

Figure 3.6. *The BrainLab MicroMLC attached to a Varian accelerator being used to treat a patient at the Charité Hospital, Berlin. The patient is immobilized in a stereotactic frame. (Courtesy of Dr V Cosgrove.)*

Watson *et al* (2000) have combined the use of the BrainLab microMLC with the Varian enhanced dynamic wedge to produce superior (compared to absence of wedge) dose distributions for fixed-field irradiation of certain brain targets.

As well as the microMLCs constructed by the University of Heidelberg (marketed by MRC Systems GMBH) and by the BrainLab Corporation, a third has been constructed by the University of Texas (marketed by Radionics). This has 15 pairs of leaves, each projecting to a width of 4 mm at isocentre, with independent motors lying in an arc with the leaf sides having a step edge to create a 'tongue-and-groove' arrangement. The leaves travel in a plane. The accuracy of leaf positioning is to within 0.5 mm (Shiu *et al* 1997a, 1998, Bues *et al* 1999a). Shiu *et al* (1997b) have made a detailed comparison between the use of the University of Texas microMLC and the use of fixed-radius circular collimators for performing stereotactic radiotherapy, the former outperforming the latter with respect to improved dose-volume histograms. Dong *et al* (1997) have presented the model for computing the dose from this MLC. Bues *et al* (1999b) have shown that the use of the microMLC for stereotactic radiotherapy of intracranial tumours with nine non-coplanar static IMBs created by the KONRAD treatment-planning

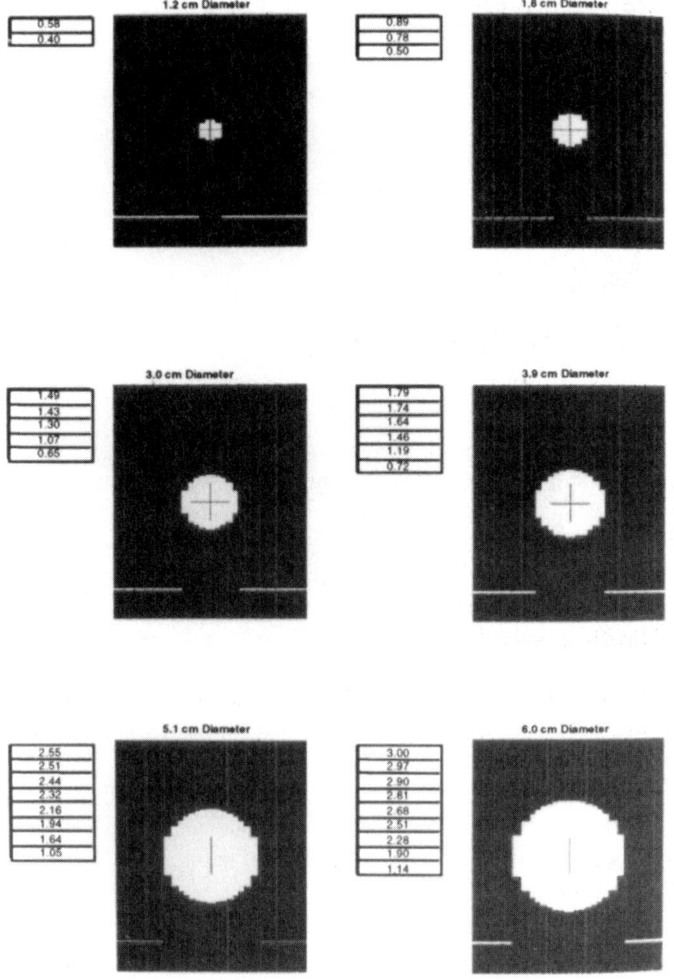

Figure 3.7. *An illustration of the placement of microMLC leaves for circular fields showing the variable leaf widths. (From Xia et al 1999a.)*

system, consistently outperformed rival treatment techniques using a conventional-size MLC or jaw-collimated conformal arc plans. MicroMLCs have been reviewed by Bortfeld *et al* (1999).

Bues *et al* (2000) have presented a new method to model the fluence produced by a Radionics miniature MLC when delivering an intensity-modulated field. They have shown that it is not adequate to convolve the fluence profile with a single convolution kernel due to nonlinearities introduced by the segmentation process.

The map models the finite size of the radiation source, collimator scatter, penumbra from the leaves, both parallel and perpendicular to the direction of leaf motion, and finite transmission through the leaves. Whilst a discrepancy of some 20% occurs in the former technique, using this new technique they were able to compute the delivered fluence within an accuracy greater than 3%.

Meeks *et al* (1999a) have measured the dosimetric properties of a miniature MLC from Wellhöfer Dosimetrie, Schwarzenbruch, Germany, attached to a Philips SL75/5 accelerator at 6MV. This is an intermediate size MLC which has 20 double-focused leaf pairs each projecting to 4.5 mm at isocentre. The leaves can overtravel by 2 cm and have a maximum leaf speed of 1 cm s^{-1} and a height of 8 cm. The maximum transmission through the leaves was 0.4% and through the leaf sides was 0.5%. The maximum transmission between leaf ends when touching was 1.2%. Consequently, it was determined that the backup collimators could remain at the envelope of the largest set field (9 $\times$ 9 cm). This considerably simplified dosimetry because it was found that even for the most eccentric rectangular fields, and also for circular fields, the equivalent square output factor applied to within an accuracy of about 1.5%. Percentage depth doses also did not vary by more than 1% from the corresponding values set by jaws or blocks. The penumbrae in the leaf direction, and orthogonal to this, were identical due to the double-focusing and were considerably sharper than for an alloy block or linac jaws (e.g. 80–20% penumbra was 4.0 mm for the microMLC, 4.5 mm for the alloy block and 5.7 mm for the linac jaws for a field size of 8.1 $\times$ 8.1 cm).

Pignoli *et al* (1997) have described the development of a 'one-off' DMLC unit. This is based on a collaboration between the Istituto Tumori of Milano and the 3D-Line Company. The MLC has 16 pairs of leaves, double-focused, projecting to a width of 6.25 mm at the isocentre, a maximum leaf speed of 5 mm s^{-1} and an over-axis travel of 10 mm at the isocentre. A later version (3D-Line Company Literature) (figure 3.8) has 48 double-focused leaf pairs, a maximum leaf speed of 1 cm s^{-1} and overtravel of 25 mm.

Stereotactic brain lesions larger than about 3 cm in extent are usually treated not by arc therapy with a single circularly collimated beam but with multiple MLC-shaped fields in which the shape is adjusted to the BEV of the PTV from each gantry orientation. Generally this is a static process. A group in Milan have proposed that the accelerator be used in arc-therapy mode with the shape made by the MLC, previously described, dynamically adjusted as a function of gantry inclination (Loi *et al* 1998). The adjustment is made by an MLC which is motorized and which can change its shape automatically via a prescription downloaded from a planning computer. They call this a dynamic MLC, a rather confusing description since the term is generally used to imply leaf movement across a field to create an intensity modulation (section 3.1.4). In the case of the Milan machine there is no intensity modulation but the leaves do move dynamically to different field geometries as the gantry rotates. The gantry is equipped with an inclinometer to indicate the gantry orientation and correlate this with the required shapes (figure 3.9). In this sense the machine has some similarities with the NOMOS MIMiC.

Figure 3.8. *The 3D-Line miniMLC DMLCv.IV. (From company literature.)*

The MLC in question can collimate a maximum field size of $10 \times 10\ \text{cm}^2$ and the leaves move on an arc (double-focusing) so that the penumbra does not depend on leaf position. The measured penumbra is 2.7 mm and 3.1 mm in the direction of leaf motion and perpendicular to this (determined from film measurements). It was built by 3D-Line Srl (Milan) and fitted to a Varian Clinac 2100C linac. The control unit stores up to 10 arcs each with 40 different shapes. The angular resolution of the inclinometer is 0.5°. Leaf transmission was less than 1%; leakage between leaves was less than 2.5% and the maximal leakage through abutting leaves was 5.5%. The accuracy of leaf positioning was about 0.2 mm. The advantages claimed were high spatial resolution, small uniform penumbra, capability to track BEV of PTV and leaf position fidelity.

A similar study has been reported by Cardinale *et al* (1998a) in which the shape of the field defined by a microMLC was changed as a function of

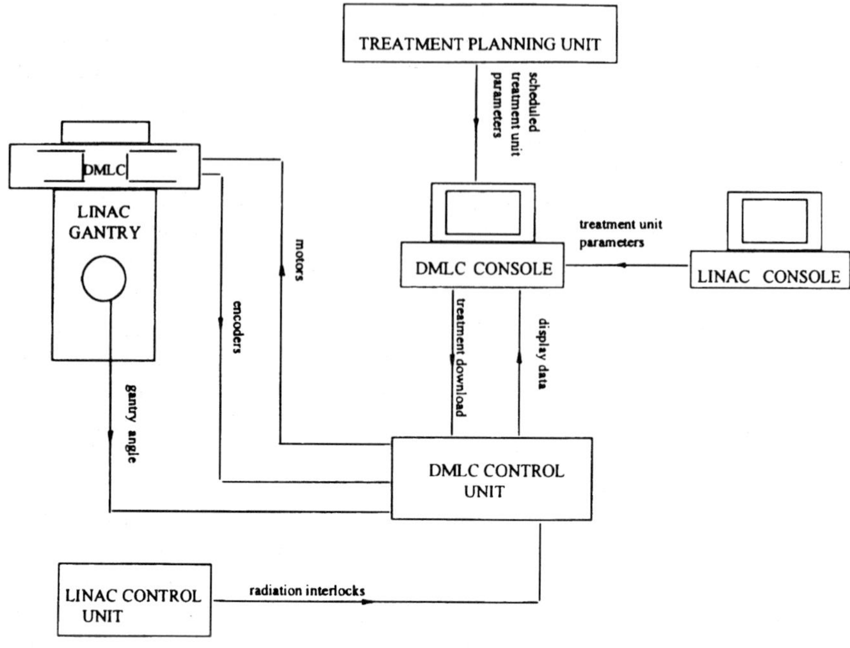

Figure 3.9. *Shows the flow chart by which the leaf positions of an MLC are changed dynamically as the gantry rotates. An inclinometer measures the gantry location and the shapes as a function of this orientation are downloaded to the MLC from a planning unit. (From Loi et al 1998.)*

gantry position within the arc with dosimetric advantages. Boyer *et al* (1998a, b) implemented the same method experimentally using the BrainLab microMLC attached to a Varian Clinac 2100C linac. The field shapes were adjusted every 10° of gantry arc to the beam's-eye view of the target. The method has also been implemented by Kokubo *et al* (1998) for 117 patients. In their implementation the leaves of the MLC were adjusted positionally every 2° of gantry arc.

Yi *et al* (1999b) have concluded that IMRT delivered using a microMLC can significantly improve the radiation therapy of non-spherical brain tumours when compared with more conventional irradiation methods. Hoban *et al* (2000) have also concluded that a judicious choice of beam directions combined with micro-collimation can improve IMRT.

A miniMLC has recently been put on the market by the DIREX Company called the DIREX AccuLeaf (Bortfeld and Oelfke 1999). This has 48 leaf-pairs defining a maximum field size of 11×10 cm^2 with a maximum leaf speed of 2 cm s^{-1}. It has four banks of leaves in two orthogonal moving sets, one above the other. It is battery operated. Bortfeld and Oelfke (1999) have shown that conformality of IMRT can improve if the leaf width is decreased for the MLC. The inverse planning programme KONRAD was used for seven evenly spaced coplanar IMBs and it was shown that the conformality improved as the leaf width changed from 1 cm to 5 mm and ultimately to 1.6 mm, the latter yielding a clearly higher target dose homogeneity. They argue that so-called microMLCs and miniMLCs should have sufficient overtravel to be in the range of half the maximum field size and that the leaf transmission should be less than 2%. To decouple neighbouring leaf pairs, the MLC geometry should be at least single-focused and, because positioning errors can cause matchline problems, the positioning errors should be smaller than 0.5 mm. Bortfeld and Oelfke (1999) give a useful table showing the properties of five microMLCs. These are: the Stryker Leibinger (MRC Systems), the BrainLab m3, the Radionics, the 3D-Line (Wellhöfer), and the Direx AccuLeaf. These collimators differ in their number of leaf pairs, their field size, their overcentre travel, their leaf width, their leaf transmission, their maximum speed, their clearance to isocentre, their total weight and their geometric design. Scherer *et al* (1999) also plan IMRT with a microMLC.

Yi *et al* (1999a) have studied the effect of varying the step sizes in DMLC IMRT for an MLC with 1 cm wide leaves, coming to the not surprising conclusion that the conformality improves with smaller step sizes. Interestingly, and in line with other observations, there was no energy dependence of IMRT conformality. Thinner MLC leaves were also shown to be advantageous. Yi *et al* (2000) have compared the use of 10 $\times$ 10 mm, 5 $\times$ 5 mm and 2 $\times$ 5 mm bixel sizes for IMRT of a brain tumour, prostate cancer and a lymphoma. They have used the CORVUS inverse-planning system and have shown that the best conformality always followed a reduction in the bixel size. The improvement in conformality was greatest when there were few beams rather than many. However, they cautioned that 'the smaller the better' is not always valid. Reduction to 2 $\times$ 5 mm led to worse results because of the increased MUs required and thus increased leakage.

Braunstein *et al* (1999) have also concluded that cranial targets are better conformed with MLCs with small leaf widths and also have decided that 8–12 beams were adequate for most planning problems.

Cheung *et al* (1999) have studied the efficacy and limitations of using standard and microMLCs in the treatment of nasopharyngeal carcinoma, coming to the conclusion that standard-size MLCs are suitable for most cases unless the PTV is very close to the spinal cord, in which case microMLCs can provide collimation equivalent to the use of a block. They show that it is important to choose the appropriate MLC orientation with respect to the tumour and they give the necessary penumbra data to assist the choice.

Varian introduced in 1998 an MLC with 60 pairs of leaves. The inner 40 pairs project to 0.5 cm at isocentre to create the central 20 cm of field and the outer 20 pairs project to 1 cm at isocentre. The collimator is known as the 'Millennium MLC-120'. Figure 3.10 shows that the hybrid MLC can create a field with better collimation to a curved edge. The maximum leaf overtravel is

Figure 3.10. *Shows how the Varian Millennium MLC-120 can produce a finer-scale field shaping than large-leaf MLCs. (From Varian product literature.)*

20 cm. The maximum leaf extension from a bank is 15 cm. Leaf and interleaf transmission is advertised as 2–3%. The MLC-120 can be used in IMRT DMLC mode either to deliver step-and-shoot or with continuous modulation.

3.1.3.1. Pseudo-microMLCs

Sun and Meng (1997) have presented an interesting concept to create a 'virtual' MLC with a projected leaf size of 5 mm at isocentre. The virtual MLC uses two field shapes, each collimated by a conventional MLC whose projected leaf width is 1 cm. For the first, the collimator centre is aligned conventionally with the field centre but it is offset by 5 mm for the second field in the direction perpendicular to the direction of leaf motion, thus creating by superposition a virtual MLC of a projected leaf width of 5 mm. It was claimed that this significantly reduced field edge penumbra especially when collimating 45° edges. Steinberg (1999) has presented the same concept, the basis for high-definition intensity (HDI) marketed by Siemens. Welch (1999a) has presented the same concept as implemented for an Elekta MLC. Welch (1999b) has shown how a technique called 'time-share' may be used to create fields which have edge structures smaller than the width of any one leaf using a conventional MLC. The idea is simply to move the whole collimator 0.5 cm perpendicular to the direction of leaf movement and deliver a pair of fields that add up to a field shaped as if it had been shaped by an MLC with leaves of a 0.5 cm width. The logical extension to other simulated field widths is obvious. The phased-field technique has been evaluated for pituitary treatments on a Rando phantom by Sharif *et al* (1999). Svatos *et al* (1999) have proven that the concept can be reinvented indefinitely. This concept of combining several fields using a conventional MLC has its roots in the work of Galvin *et al* (1996). They were the first to show experimentally that one could deliver a field as a series of n subfields each with $1/n$ of the total dose and with each subfield indexed by $1/n$ of the leaf width relative to the previous subfield. In doing this the leaves would be adjusted to the appropriate new envelope of the PTV. Galvin *et al* (1996) made experimental measurements with $n = 1, 2, 3, 4$ and showed that indeed this reduced the 'waviness' of the isodose lines at d_{max} and at deeper depths, that it reduced the 90–20% and 80–20% penumbras relative to those for the full leaf width single field, but that, even so, the penumbras were significantly greater than they would have been with a cerrobend block. Galvin *et al* (1996) also discussed methods of portal verification of such an arrangement by combining multiple EPID images of subfields. Bortfeld *et al* (2000) have shown, using an analysis in frequency space and sampling theory, that the sampling interval at 6 MV is 1.5–2 mm. However, they also showed why the use of an MLC with double this size leaf width *but the same sampling interval* would only slightly degrade the fluence distribution.

Siochi (1999b) has argued that this would be very difficult to implement in the context of intensity modulation and has proposed an alternative. In this, two 2D modulations are combined. One is created with the direction of leaf motion orthogonal to the other. Each leaf movement is quantized in steps of 5 mm. Hence, four 5 × 5 mm pencil beams fit into each 1 × 1 cm square. Film dosimetry has shown the feasibility of the concept but no apparatus has been built, neither can any inverse-planning system currently cater for this. Siochi (2000) has provided the optimization algorithm for the leaf segment setting for the virtual microMLC.

Evans and Partridge (2000) have developed a technique to generate a 2D IMB, of arbitrarily small pixel size, using a conventional MLC with a projected leaf width much larger than the required small pixel size. This is achieved by sweeping the leaves (of normal width) in two orthogonal directions such that the resultant 2D IMB is the sum of the 2D IMBs created from the one-dimensional leaf sweeps. The quantization of the leaf sweep in the direction of the sweep can of course be as small as desired and the outcome is only limited by the finite penumbra and the accuracy of placing the leaves in the required locations. The algorithm for computing the component sweeps is in two stages. Firstly, imagine the 2D IMB is broken up into squares whose side size is equal to the projected leaf width. These squares then contain $n \times n$ smaller bixels. A filter first equalizes the gradient in the direction of each leaf sweep. i.e. firstly for (n) columns the gradient between adjacent columns is set to the average of this gradient over n

rows. Then for (n) rows the gradient between adjacent rows is set to the average of this gradient over n columns. At any stage, if negative values occur the whole row or column containing those negatives has an offset added to raise the minimum to zero. This filtering process, which of course changes the resultant 2D IMB, is cycled for many iterations until the gradients are no longer changing by more than a small preset amount. The outcome of this filtering stage is then decomposed into two orthogonal sweeps. i.e. for any beam element $a_{i,j}$ the appropriate sweeps are x_j and y_i such that $a_{i,j} = x_j + y_i$. There is an infinity of solutions to this and so the solution is chosen which delivers the smallest fluence in equal x and y parts. This is applied to all the blocky subregions. The algorithm is summarized in figure 3.11. Evans and Partridge (2000) then applied the algorithm to four model 2D IMBs, an edge, a circle (representing CFRT) and a modified sinc function and a hemisphere (representing IMRT). They computed the r.m.s. differences between the ideal distribution (that on the finest grid spacing), the practical 1D sweep with the large projected leaf width and then the 2D sweep resulting from the algorithm. They also delivered the distributions to film. It was found that the r.m.s. errors decreased as the pixel size decreased. The component deliveries were quite spiky and therefore efficiency of delivery dropped by about a factor of two.

Williams (1999) has developed a new concept for generating MLC-shaped fields with fine spatial resolution without the need for a purpose-built microMLC. The device makes use of a parallel-grid collimator, 10 mm in pitch with 2.5 mm wide slits. There is a plurality of slits and the grid is placed in front of the conventional MLC and then successively indexed by one, two or three slit-widths with respect to its original position (figure 3.12). After four separate sequential exposures, each providing a slit field 2.5 mm wide separated by 7.5 mm of almost zero exposure, the full shaped field will have been delivered but will be bordered by step-like edges with a 2.5 mm instead of a 10 mm quantitization (figure 3.13). A logical extension of the concept provides for varying the pitch and slit-width to provide variable resolution and variable efficiency. Advantages of the device are that the maximum field size remains 40×40 cm, that the device can be located on the shadow tray of the linear accelerator and that the device can be relatively light. The so-called pseudo-microMLC has been patented and its properties are being studied.

3.1.4. The DMLC technique

Since its introduction for replacing configured blocks as a field shaper, the MLC has assumed a new role as the potential creator of IMBs by the DMLC technique. This is the method in which the leaves of an MLC are instructed to move to a sequence of locations during the irradiation. Each leaf pair defines a 1D modulation (creates a 1D IMB) and these can differ between leaf pairs. However, the concept is that all required leaf pairs move together, at the end of which a 2D intensity modulation has been created (Boyer *et al* 2000a, b). There are numerous different implementations of the technique and many of the basic concepts of the method were reviewed in

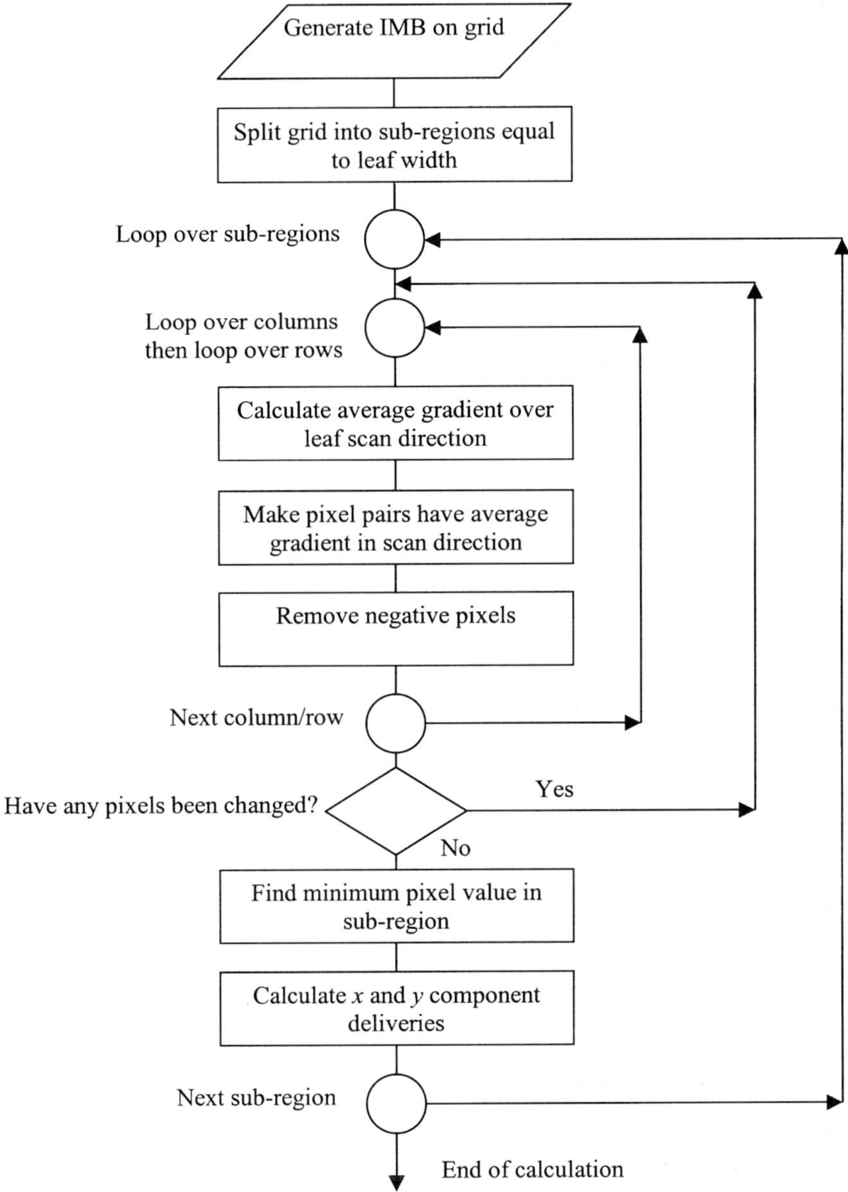

Figure 3.11. *Flow diagram showing the algorithm for decomposing an IMB into two orthogonal components. The process goes from top to bottom except where indicated by an arrow. (From Evans and Partridge 2000.)*

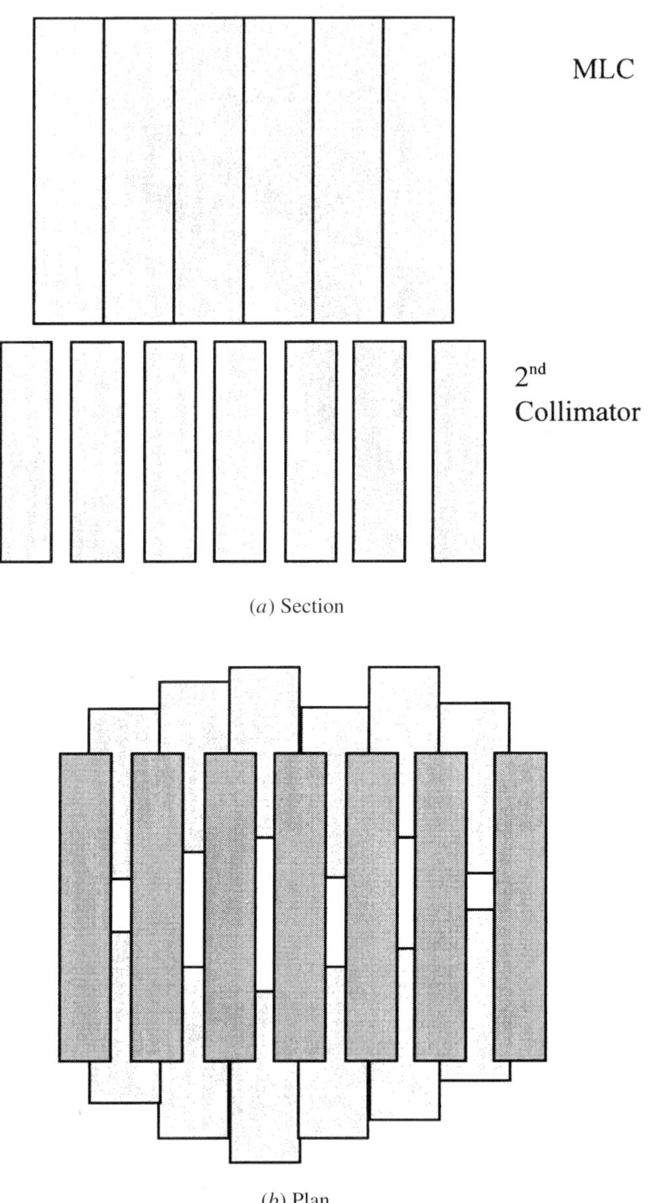

Figure 3.12. *Illustrates the principle of the Williams variable resolution MLC—see text for details. (Courtesy of Dr P Williams.)*

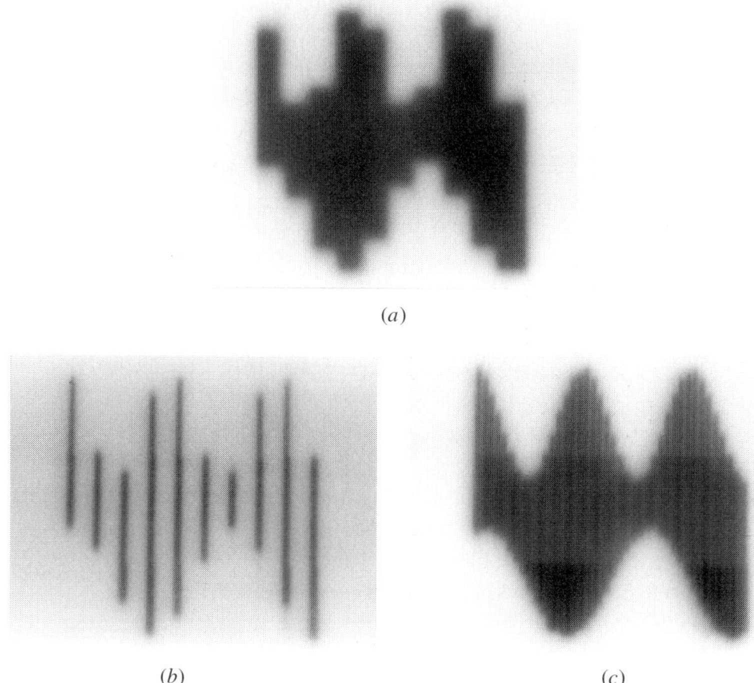

Figure 3.13. *Illustrating the difference in spatial resolution between collimating a sinewave field using a conventional MLC and the same field created using a combination of the conventional MLC and the Williams collimator. (a) Sinewave collimated by conventional MLC. (b) Mask to find, by measurement, the best slit width to produce a uniform dose with four step increments. The mask had three widths; 2 mm at the end defining the most positive part of the sine wave, 2.5 mm in the middle and 3 mm at the end defining the most negative part of the sine wave. (c) The result of four superposed irradiations with a shift of 2.5 mm between each irradiation. As expected the mask with 2.5 mm aperture was about right. (Courtesy of Dr P Williams.)*

depth by Webb (1997d, chapter 2). One key concept is that of an 'interpreter', a piece of computer code which turns the desired intensity modulation into a set of instructions to drive the leaves of the MLC. Another key concept is that of 'control points' being the set of intermediate positions taken up by the leaves at each fractional part of the total irradiation time. We shall discuss these shortly.

Estimates of the importance of the MLC in this dynamic role are beginning to appear, mainly from North America (Klein *et al* 1998). The technique is also now being evaluated in comparison with its competitors. For example, Stein *et al* (1997a) have compared the use of compensators and multileaf modulation

(DMLC) using the Varian system. Compensators were shown to be adequate for less than five beams and to take a long time to manufacture, but the treatment time was short. On the other hand, the DMLC technique preparation time is short, delivery time is very large (at present, depending on the manufacturer and the software version release number) but the method easily copes with a larger number of fields. It was shown that the use of nine static-gantry DMLC fields gave a nicely conformal result for a clivus chordoma case protecting the brainstem (see also section 1.2). Park *et al* (1999) have compared the DMLC technique with the use of a compensator for making a 1D modulation to compensate for the outline variation in the head-and-neck region.

3.2. ISSUES IN DELIVERING IMBS VIA THE DMLC TECHNIQUE

IMRT is often said to 'waste photons' (Phillips *et al* 1998) since photons are being delivered which are being deliberately attenuated for part of the treatment time. Hence, leakage is an issue and treatment times must be minimized. This is a criterion inherent in 'interpreters'. It is also an issue when designing the shielding for a room in which IMRT will take place (Mutic *et al* 1999). Boyer *et al* (2000a) have pointed out that MLCs were never designed for DMLC therapy and so are generally too thin. Another key issue is that accelerator manufacturers differ in whether they will allow 'interdigitation' (figure 3.14), the phenomenon whereby one trailing leaf might advance beyond the position of either adjacent leading leaf. The Elekta MLC requires the maintenance of a minimum gap between leading and trailing leaves and their adjacent counterparts. The Siemens MLC allows the leaves to close and to align with their adjacent opposite counterpart, but not pass. The Varian MLC allows full interdigitation, i.e. leaves may pass their opposite adjacent counterpart. A further phenomenon is the so-called 'tongue-and-groove' effect whereby the fluence delivered in the beam's-eye view of the tongue-and-groove interlocking region of the leaf sides may become smaller than that in the two adjacent 'channels' defined by the adjacent leaf pairs. These concerns recur in the following discussion of the DMLC technique.

3.2.1. *Leaf setting in DMLC therapy*

Boyer (1997) and Boyer and Strait (1997) have presented a particularly intuitive way of understanding the algorithm for determining the leaf trajectories from a 1D IMB profile. This analysis is applied to just one 1D IMB (so ignores for now adjacent leaf constraints; it gives the ideal leaf velocity profile). It considers the generation of primary fluence so ignores scatter and leakage. Consider figure 3.15*a* showing a 1D IMB, $I(x)$, with two maxima and a local minimum. The figure can be divided into the four regions shown depending on whether the IMB is increasing or decreasing. The curve can be interpreted as the time-position graph $T(x)$ $(=I(x))$ for the trailing leaf, with the horizontal axis the time-position graph for the leading leaf since the vertical difference between the two at any position

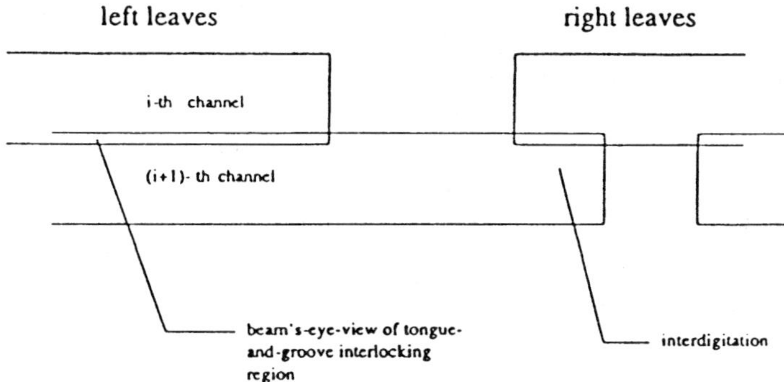

Figure 3.14. *Two leaf pairs of a MLC are shown, the ith and the $(i+1)$th channels. The leaves are shown interdigitating, an arrangement forbidden, for example, on the Elekta MLC but allowed on the Varian MLC. The beam's-eye view of the tongue-and-groove interlocking region is also illustrated.*

x is the intensity. The problem with this interpretation is that it requires the trailing leaf to move backwards in time and the leading leaf to move with infinite velocity! However, both problems can be removed by applying transformations. These transformations are as follows. In the region where the IMB is decreasing a reflection operator is applied so that the curves become, in region (i)

$$T'(x) = T(x) \tag{3.1}$$

in region (ii)

$$T'(x) = T_1 - [T(x) - T_1] \tag{3.2}$$

in region (iii)

$$T'(x) = T(x) \tag{3.3}$$

in region (iv)

$$T'(x) = T_2 - [T(x) - T_2] \tag{3.4}$$

where T_1 and T_2 are the half-maximum values to the left of the two negative-gradient regions. This transforms the graph to the form shown in figure 3.15*b*. Next, the third and fourth parts of the graph are lifted by a shift operator ΔT so that, in region (i)

$$T''(x) = T'(x) \tag{3.5}$$

in region (ii)

$$T''(x) = T'(x) \tag{3.6}$$

in region (iii)

$$T''(x) = T'(x) + \Delta T \tag{3.7}$$

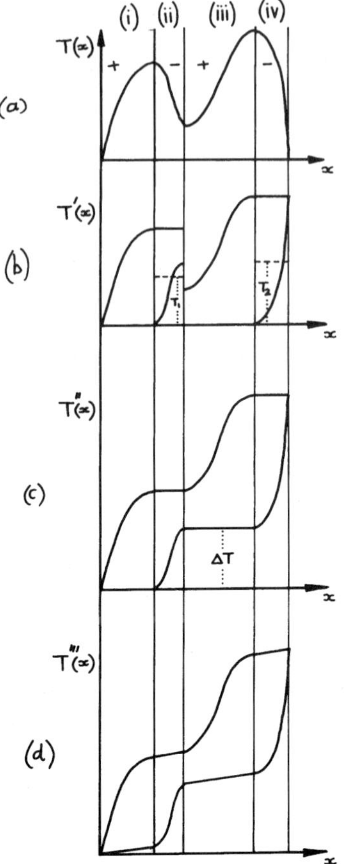

Figure 3.15. *Figures which show how a 1D IMB can be transformed graphically into the required idealized leaf motions for the DMLC technique—see text for details. (The diagrams have been adapted from those in Boyer and Strait 1997 and were personal communication from Professor Art Boyer.)*

in region (iv)

$$T''(x) = T'(x) + \Delta T. \tag{3.8}$$

This removes the time reversal and creates figure 3.15*c*. To remove the requirement for leaves to move at infinite velocity apply a shear operator in all four regions

$$T'''(x) = T''(x) + \frac{x}{\hat{v}} \tag{3.9}$$

with $\hat{v}$ the maximum leaf speed. This creates figure 3.15*d* which is a familiar pattern of leaf movement. Now, since $T(x) = I(x)$ the intensity profile, we have by differentiating $T'''(x)$ in regions (i) and (iii) where the fluence profile is increasing

$$\frac{\mathrm{d}T'''(x)}{\mathrm{d}x} = \frac{1}{v(x)} = \frac{\mathrm{d}I}{\mathrm{d}x} + \frac{1}{\hat{v}} \tag{3.10}$$

in regions (ii) and (iv) where the fluence profile is decreasing

$$\frac{\mathrm{d}T'''(x)}{\mathrm{d}x} = \frac{1}{v(x)} = -\frac{\mathrm{d}I}{\mathrm{d}x} + \frac{1}{\hat{v}}. \tag{3.11}$$

Rearranging, we have the familiar equation that where the profile is increasing

$$v(x) = \frac{\hat{v}}{1 + \hat{v}(\mathrm{d}I/\mathrm{d}x)} \tag{3.12}$$

and where the profile is decreasing

$$v(x) = \frac{\hat{v}}{1 - \hat{v}(\mathrm{d}I/\mathrm{d}x)}. \tag{3.13}$$

These equations give the velocity of the trailing leaf in the region where the profile is increasing (the leading leaf is moving at maximum velocity $\hat{v}$) and gives the velocity of the leading leaf in the region where the profile is decreasing (the trailing leaf is moving at maximum velocity $\hat{v}$). The equations are those derived by Stein *et al* (1994), Stein (1997), Svensson *et al* (1994) and Spirou and Chui (1994) by a little more complicated mathematics (see a review in Webb 1997d, chapter 2). The derivation by Boyer and Strait (1997) is particularly graphic. They have also shown how similar transformations could be made to give the required leaf positions for the multiple-static-field method of delivering a 1D IMB. The effect of errors in leaf position placement was also determined in terms of the actually-delivered IMB. Boyer (1997) proposed to measure the movement of leaves in real time using a CCD camera viewing a Gadox screen through a water cylinder (see section 3.2.8).

These are the ideal equations yielding the minimum treatment time for just one pair of beams. Modifications are necessary when more than one pair of beams are moving, due to machine constraints (see section 3.2.4). These additional considerations have led to the development of so-called 'interpreters' (see section 3.2.4).

3.2.2. *The 'tongue-and-groove effect'*

The sides of the MLC leaves interlock in a tongue-and-groove (figure 3.14). Hence, when a field is comprised of matched components in which either the tongue or the groove projects into the open part of the field, the resulting junctioned field may have an underdose in the region of interlock.

Sykes and Williams (1998) have measured the magnitude of the tongue-and-groove effect for two Philips accelerators, each operating at two different energies. To do this they double-exposed films to two matched fields in which one was the complement of the other (leaves open in the second where closed in the first and *vice versa*). They have also conducted separate experiments in which the tongue-and-groove regions were comprised of either adjacent leaves or adjacent plus opposite leaves. They found small differences in the configurations (machine, energy, depth) examined but these were not as significant as the overall observation that the dose reduction in the tongue-and-groove region was some 15% to 28% spread over a width of 3.8 to 4.2 mm (figure 3.16). This is in stark contrast to the theoretical prediction that the effect is of the order of 80% over 1 mm and is explained by the effects of the finite size of the focal spot and the presence of electron transport. Also, they have commented that, in practice, patient movement between fractions and patient scatter will also reduce the effect. Nevertheless, in DMLC therapy it is an effect which can be entirely removed, as we shall now see.

A potential problem with the DMLC delivery technique is underdosage of the beam's-eye view of the interlocking region of the leaf sides. In general, any two adjacent pairs of leaves track across the field differently, out of step, and so from time-to-time during their traverse, non-full-depth leaf sides are exposed to the radiation. The result of this is that this interlocking region can be underdosed (figure 3.17). In dynamic DMLC therapy, the definition of the 'tongue-and-groove' effect is that the fluence below the beam's-eye view of the interlocking region

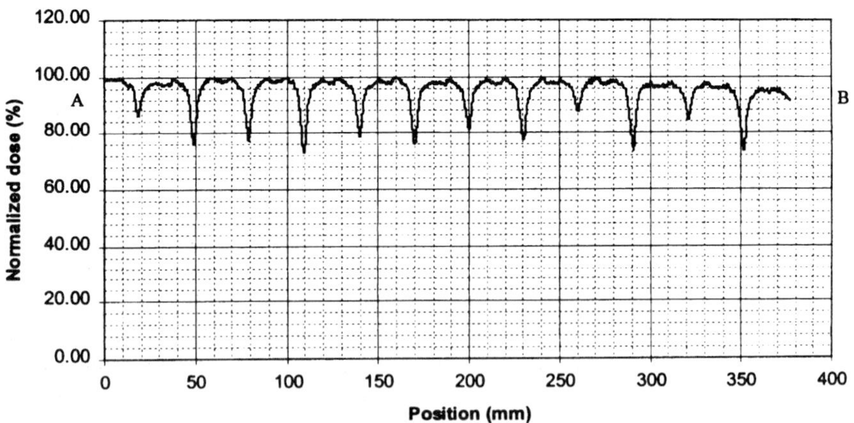

Figure 3.16. *A profile across a tongue-and-groove film when adjacent leaves create the tongue-and-groove effect. The measurement was made on film at 6 MV at d_{max} and scanned with a nominal resolution of 85 microns. It shows that the magnitude of the effect is smaller than the theoretical prediction but spread over a wider area. (From Sykes and Williams 1998.)*

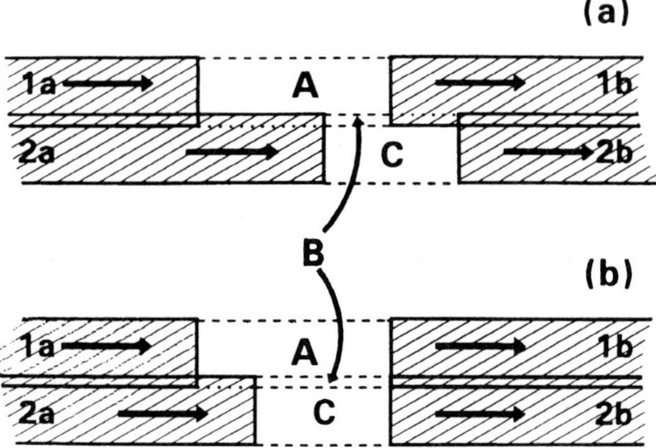

Figure 3.17. *Dynamic multileaf collimation. Two beam's-eye views of the MLC; only two leaf pairs are depicted. All leaves move with the same constant velocity from left to right (see arrows). (a) Due to the tongue-and-groove arrangement between the two leaf pairs, the delivered intensity in points in the overlap region B is substantially lower than in points in regions A and C. (b) Due to a proper synchronization of the leaf pairs the tongue-and-groove underdoses are avoided. (From van Santvoort and Heijmen 1996.)*

becomes less than that either side in the adjacent channels at any one position in the direction of leaf motion across the field.

However, a solution to the tongue-and-groove problem has been provided by Van Santvoort and Heijmen (1996). This solution makes use of the fact that the *absolute* time difference between the passage of the leading and the trailing leaves in a pair is irrelevant, only the *difference* between the two times determining the intensity. So, the leaf trajectories can be iteratively adjusted so that at each point along the trajectory the smaller absolute time gap is always inside the larger absolute time gap (so-called leaf synchronization) (figures 3.17 and 3.18). In the original solution provided by van Santvoort and Heijmen (1996), the partial transmissions through the full-depth leaves and through the partial-depth leaf sides were set to zero. Hence, the intensity in the beam's-eye view of the interlocking region was *equal* to the smaller of the two adjacent intensities. Webb *et al* (1997a,b), Webb (1998d,e) and Stein (1997) have provided the full solution when these non-zero finite transmissions were also included. In fact, this leaves the synchronization solution intact but leads to the intensity in the beam's-eye view of the overlap region being anywhere between the two adjacent track intensities (depending on the transmission values and the leaf trajectories). This methodology also leads to a solution known as partial synchronization (figures 3.19, 3.20, 3.21 and 3.22). Because the technique alters the total treatment time upwards it is necessary to iteratively loop the calculations. Stein (1997) has proved

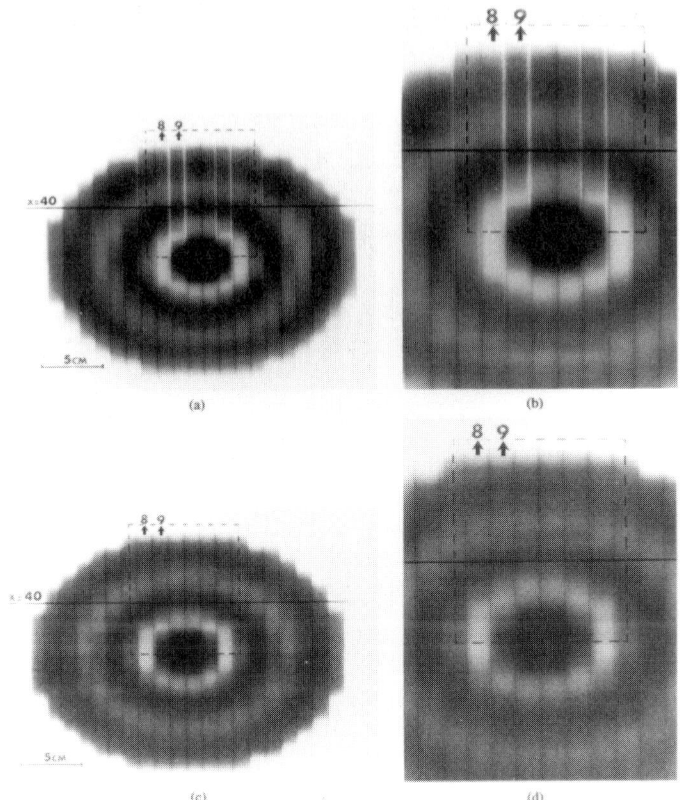

Figure 3.18. *Irradiated films for a Mexican-hat intensity-modulated field: (a) realization with non-synchronized leaf trajectories; (b) the marked area in (a) enlarged; (c) dynamic multileaf collimation with synchronized leaf trajectories; (d) enlargement of the marked area in (c). Note that the underdose has been removed by the process of synchronization. (From van Santvoort and Heijmen 1996.)*

that elimination of the tongue-and-groove underdose by full synchronization is also a sufficient condition to eliminate interdigitation (but not *vice versa*) (see also Keller-Reichenbecher *et al* 1997). Appendix 3A gives the detailed mathematics of the synchronization and partial synchronization solution. It is important to recognize that, provided the (small) increased treatment time can be tolerated, the 'tongue-and-groove problem' in DMLC IMRT is totally solved.

Dirkx *et al* (1997b, c, 1998, 2000), Heijmen and Dirkx (1998) and Heijmen *et al* (1997) have presented a different iterative method which also takes account of collimator scatter and finite leaf transmission. In their method the leaf trajectories

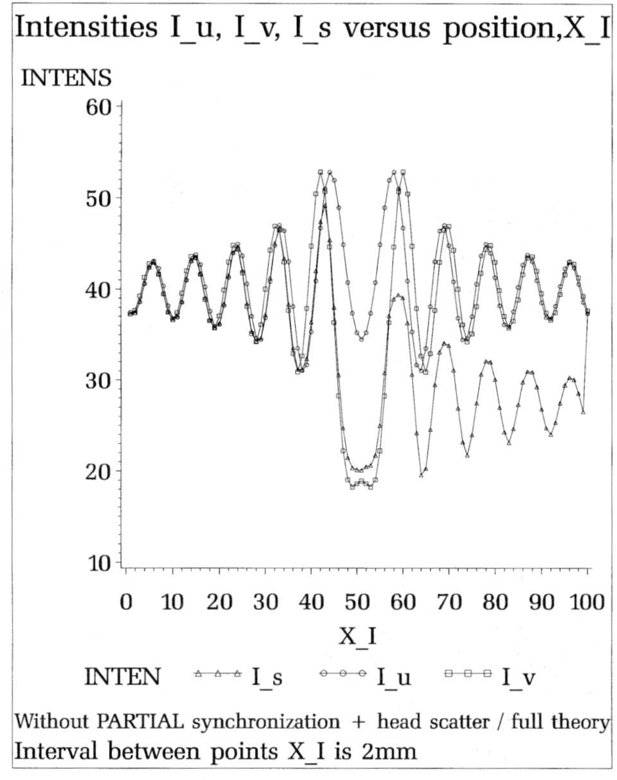

Figure 3.19. *The intensity profiles in the u-channel (I_u) and v-channel (I_v) (u and v defined as two adjacent channels in a DMLC sweep) together with that (I_s) in the beam's-eye view of the interlocking tongue-and-groove channel when synchronization is not applied and the beams define part of a Mexican hat 2D IMB. Note the severe tongue-and-groove underdose.*

were synchronized without accounting for these effects and the delivered fluence was computed with the effects. This was compared with the required prescription fluence. The number of monitor units at a point x was then modified iteratively until the delivered fluence, with the effects included, matched the prescription. The process had always converged to acceptable discrepancy by ten iterations and often by less. Dirkx *et al* (1997b, c, 1998) showed a nice agreement between calculations and measurements of dose in a plane (computed by convolving the intensity profiles with a pencil beam). It was found that some tuning of the agreement was possible by adjusting the value of the transmission of the full-depth leaves. A value of 2% was used for the double-focused MLC attached to the Scanditronix MM50 racetrack microtron.

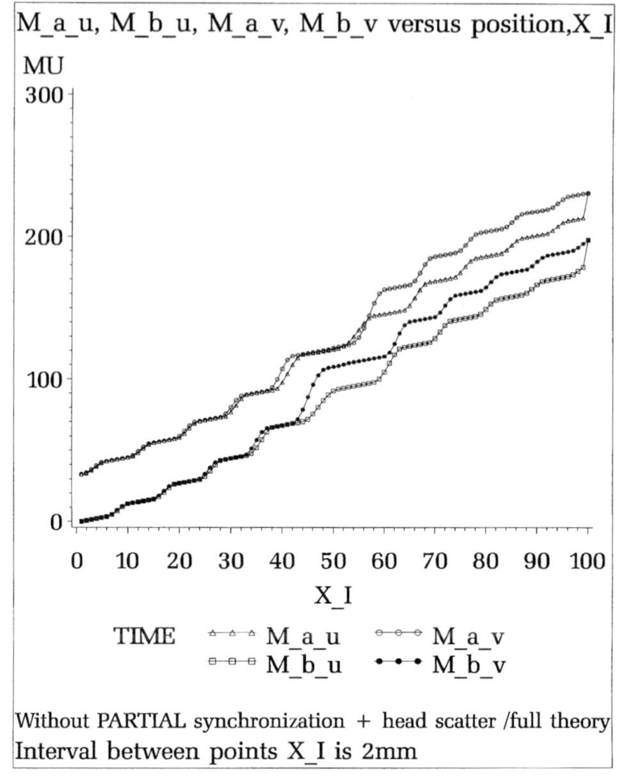

Figure 3.20. *The leaf trajectories for the two channels in figure 3.19 when synchronization is not applied.* $M_{b,u}$, $M_{a,u}$, $M_{b,v}$ *and* $M_{a,v}$ *describe the cumulative beam-on time (in MU) for the arrival of, respectively, the b and a leaves, of the leaf pairs respectively labelled by u and v.*

The studies by Webb *et al* (1997a, b) and Webb (1998d, e) allowed the profile of the leaf sides to vary in depth creating a situation in which the tongues and grooves were not of equal attenuation. Even so the tongue-and-groove effect can be removed during DMLC therapy as described above. Yu (1998a) has further commented that the equal-depth arrangement actually yields the worst tongue-and-groove underdose when matching two *static* fields. He has shown a better arrangement in which the junction between two leaf sides is a slanting plane crossing the vertical plane through the centre of the interlocking region at exactly half depth. This yields a lower integral tongue-and-groove underdose, being the intersection of two exponential curves. Specifically, the half-depth leaf leads to a transmission $\sqrt{(T)}$ where T is the transmission of the full-depth leaf (figure 3.23). Alternatively, he has proposed an 'integral sign' intersection profile which, again,

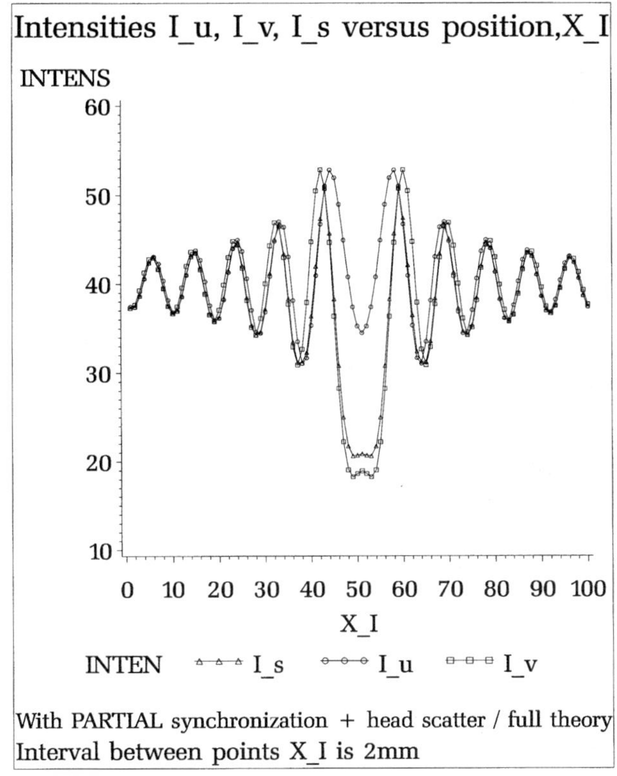

Figure 3.21. *The intensity profiles in the u-channel (I_u) and v-channel (I_v) (u and v defined as two adjacent channels in a DMLC sweep) together with that (I_s) in the beam's-eye view of the interlocking tongue-and-groove channel when either partial or full synchronization is applied and the beams define part of a Mexican hat 2D IMB. The underdose has vanished.*

is at half depth at exactly half interlock region width. This instead leads to a 'V-shaped' underdose profile. Yu (1998a,b) and Yu and Sarfarez (1998a, b) have made measurements of the tongue-and-groove effect for the Elekta MLC and also Monte-Carlo calculations of the underdose profile for other geometries. It was confirmed that electron transport reduces the underdose but spreads it over a wider area, in line with the observations of Sykes and Williams (1998).

Symonds-Tayler and Webb (1998) have also presented a design for an improved leaf side arrangement which can remove the tongue-and-groove effect from maximally-efficient MSF deliveries (see section 3.3). In this proposal, the height of the stepped edge of the leaves is reduced to that necessary to produce 51% transmission (assuming full-depth leaf leakage is 2%) as shown in figure 3.24*d*

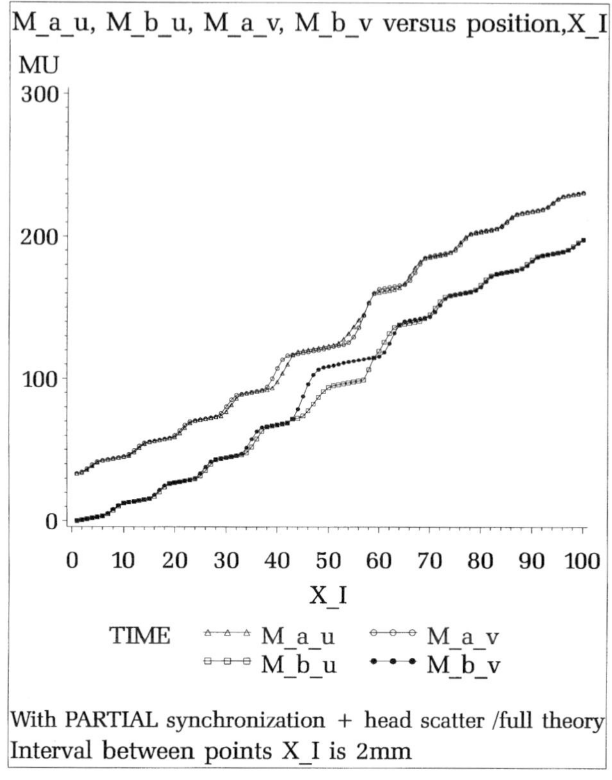

Figure 3.22. *The leaf trajectories for the two channels in figure 3.21 when partial or full synchronization is applied. $M_{b,u}$, $M_{a,u}$, $M_{b,v}$ and $M_{a,v}$ describe the cumulative beam-on time (in MU) for the arrival of, respectively, the b and a leaves, of the leaf pairs respectively labelled by u and v.*

which is halfway between the leakage and open transmission. Then there would be neither an underdose nor overdose due to the tongue-and-groove effect, as shown in figure 3.24f. However, if nothing were done about the large air gap between closed leaves, then this would give rise to an unacceptable leakage of about $(51\%)^2 = 26\%$. Consequently, a set of filler blades must be introduced in the gaps between closed leaves. The front of each filler blade must project just as far as the front of the more retracted adjacent leaf. Various schemes have been proposed.

3.2.3. Industrial development

Three electromedical technical companies are actively developing tools for DMLC IMRT therapy: Varian, Siemens and Elekta. Interestingly, the open availability of

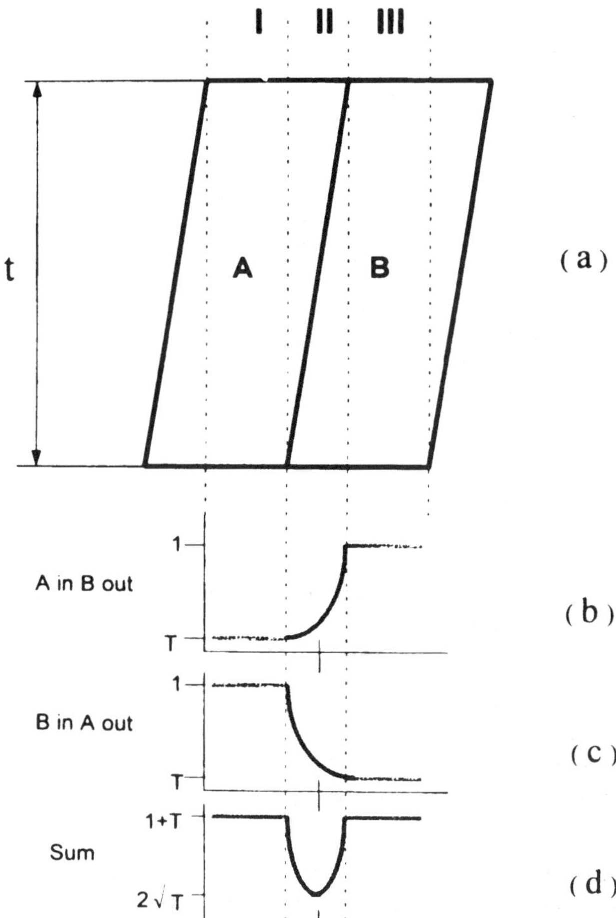

Figure 3.23. *(a) Schematic illustration of a new MLC leaf design, where the overlapping area is divided diagonally into two triangles. The primary photon intensity when the side of a leaf defines a field boundary (b,c) and when the two fields of equal exposure are abutted at the junction of two leaves (d). (From Yu 1998a.)*

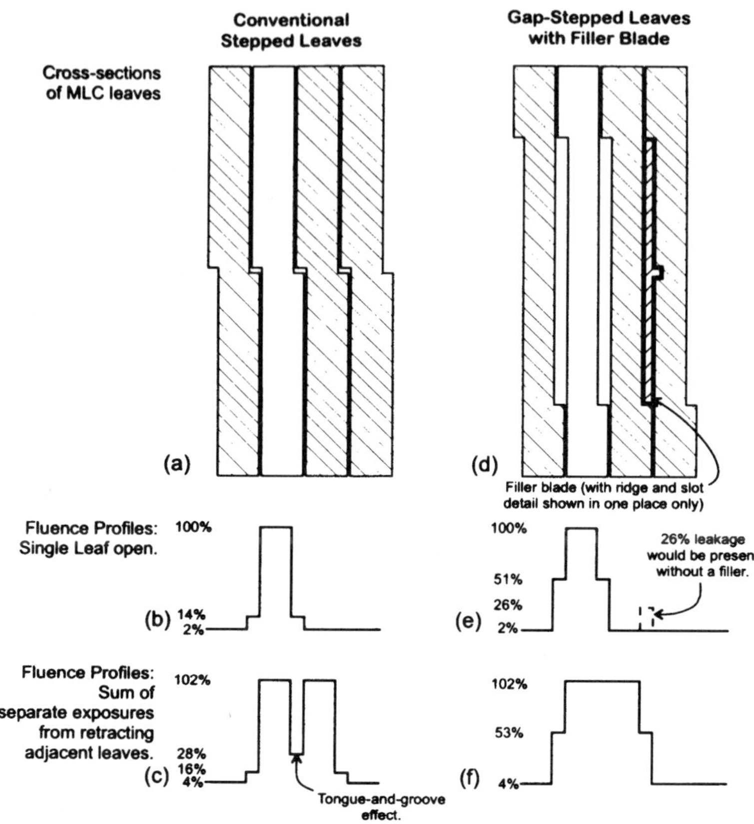

Figure 1. See text for explanation.

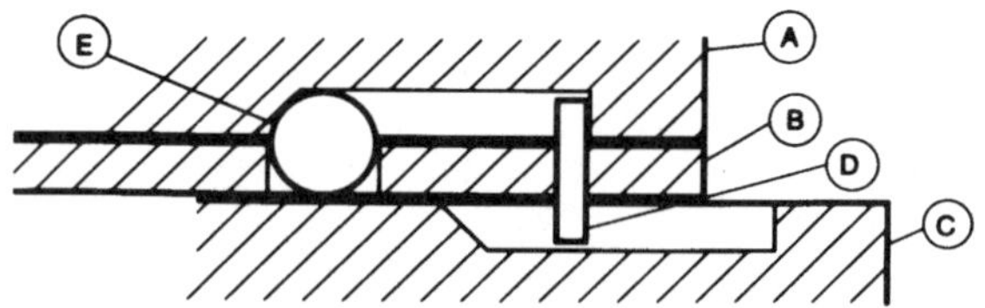

Figure 3.24. *A proposed arrangement to reduce the tongue-and-groove effect—see text for details. (From Symonds-Tayler and Webb 1998.)*

the research literature does not appear to have been a factor in limiting commercial enterprise. Clearly those developments which have been patented by the individual companies have been taken through this process between the submission and publication stages of the related papers[1]. This is in accord with the philosophies expounded in the 'point and counterpoint' by Antonuk *et al* (1999).

3.2.3.1. Varian DMLC technique

Varian market equipment which is FDA approved, and patients have been treated with the equipment at the Memorial Sloan Kettering Cancer Center, New York. The DMLC IMRT technique has been used clinically at the Memorial Sloan Kettering for over 300 patients (Ling *et al* 1996, Burman *et al* 1999). Ling *et al* (1998) argue that this is an example of art combining with science, a theme developed further by Mijnheer (1998). Fuks (1998) has reported that by March 1998 the number had risen to 1050 in the dose-escalation study of treating prostate cancer. A dose of 64.8 Gy was given to 9% of the patients, 70.2 Gy to 25%, 75.6 Gy to 42%, 81 Gy to 22% and 86.4 Gy to 2% of the patients. The method uses the segmented treatment table to set the leaf positions. Patients with prostate cancer have now been treated with IMRT for the full treatment dose. This contrasts with the treatment during the experimental phase when only a boost dose was delivered (Ling *et al* 1996).

The Varian method is in the grey area for nomenclature description in that it is truly a DMLC technique, requiring no intervention between the delivery of the subfields, but also, since the radiation is off between segments, it is strictly a multiple-static-field (MSF) technique with a large number of segments (the so-called 'step-and-shoot' technique). It allows interdigitation (Varian 1997). The Varian DMLC IMRT technique is being developed, as well as at the Memorial Sloan Kettering Hospital, New York, at the University of Berne and at the Charité Hospital , Berlin.

André *et al* (1997) have also used the Varian DMLC to deliver test irradiation patterns to films using the sliding window technique. Specifically, these modulations have involved moving the carriage which supports the MLC. This is because the leaves have certain limiting tolerances. They must not be more than 14.5 cm out of the carriage position; they must not move more than 16 cm over the centre position; the backup jaws must not move more than 2 cm over the centre; the carriage must be out of the radiation field. The position of the leaves was ordered by 200 control points with linear interpolation between them. At any one step, one of each leaf pair moved at the maximum speed of 1.5 cm s^{-1} in order to minimize the treatment time so interleaf leakage was minimized. The modulations delivered were more complex than those which would be met in clinical practice.

[1] Information concerning the development of the Elekta product was available to me prior to public release as a member of the Elekta International Consortium. Only material now in the public domain through later peer-reviewed papers, conference reports and the publicly-available Elekta house magazine *Wavelength* is included here. References to Consortium meetings are included simply to date-stamp certain developments.

Spirou *et al* (1999) have shown that for a Varian MLC it is necessary to deliver large-field IMRT using split segments. It is possible to split the fields either in space or in time or both, and they show that by combining 'split at segment' with 'split at time' this technique can create any field, albeit with reduced efficiency.

3.2.3.2. Siemens DMLC technique

Siemens have also developed the 'step-and-shoot' method and implemented it at University of California, San Francisco (UCSF). Experimental studies have also been made at Heidelberg (Bortfeld *et al* 1997b) (see also section 1.2). Siochi (1997) has patented a method (the leaf sequencer is known as IMFAST and the delivery module is known as 'SIMTEC' **SI**emens **I**ntensity **M**odulation **TEC**hnology) to determine the minimum number of fields shaped by an MLC to give a specified 2D IMB. Once again the 'classification' is as a DMLC technique but is delivered as a series of 'step-and-shoot' segments. This patent has been assigned to Siemens. The method works by first subtracting any baseline intensity as the first field component and then creating a set of subfields in which a unit of intensity is either on or off at each bixel, the unit not necessarily being 1MU. Then an iterative algorithm, such as simulated annealing, moves around the non-zero bits, at each stage inspecting whether two patterns have become identical and if so combining them. If ten sequential moves do not alter the number of field components the algorithm terminates. The algorithm arrives at a minimum-number set of fields. This is *not* the set with the minimum number of MUs (i.e. maximum MU efficiency). The set are then ordered to minimize the leaf movements between each static-field component. Siemens claim that IMFAST and SIMTEC involve no special procedures to calibrate and check the MLC and no special beam-modes. Each segment is fully recorded as it is concluded, so that interrupted treatments could be continued (Siemens 1998). Siemens advertise this SIMTEC method of delivering elements of fluence using a coloured figure showing an assembly of Duplo™ blocks, the well-known Danish children's toy. There are other interpreters developed for the 'step-and-shoot' DMLC method with the Varian MLC and these are reviewed in section 3.3. Siochi (1998a) has compared the above leaf-sequencing algorithm with others e.g. those based on a 'power-of-two' sort (see section 3.3), and sweeping window. It was found that, when fed with 100 random test cases and then 49 modulations created as part of clinical practice, the Siemens algorithm always gave the most efficient delivery time. The method incorporated elimination of tongue-and-groove underdoses.

Siochi (1999c) has given a detailed specification of the algorithm. The goal is to minimize the total treatment time, τ. This comprises the sum of the total beam-on time, the record-and-verify overhead for each component (at time of his writing 18 s) and the total time for leaf travel. Thus,

$$\tau = \sum_{i=1}^{n} \frac{M_i}{d_r} + \sum_{i=2}^{n} \mathrm{Max}\left(V_t, \frac{\mathrm{Max}\left(\mid x_i^j - x_{i-1}^j \mid \right)}{v} \right) \tag{3.14}$$

is minimized, where d_r is the doserate, M_i is the number of monitor units for the ith segment, V_t is the record-and-verify overhead. v is the leaf speed, x_i^j is the position of the jth leaf at the ith segment, and the argument for the second Max function extends from $j = 1, \ldots, 58$. This equation is simply the sum of the beam-on times and the sum of the overheads where the overhead may be provided either by the record-and-verify time or by the time taken to move leaves, whichever is the greater. The basis of Siochi's algorithm is a combination of two processes. The first is 'extraction' of a set of shape matrices (matrices with only 1s [radiation on] and 0s [radiation off]) which can be delivered efficiently. The second part is (what he describes as) 'rod pushing' and is the creation of further shape matrices in which all leaf pairs move between matrices unidirectionally. An example of 'rod pushing' is shown in figure 3.25 corresponding to the delivery of the simple 1D modulation (12321) = (11100) + (01110) + (00111). This particular case is a restatement of the Bortfeld *et al* (1994a) 'leaf sweep' decomposition.

Siochi's algorithm acknowledges that the extraction of fundamental shapes and subsequent rod pushing may not be at all obvious. Thus, an optimization routine tries several sequences and evaluates each one according to equation (3.14) finding that with the minimum delivery time. The optimization requires that three very important constraints are satisfied. The first is that, for the Siemens' MLC, interdigitation is forbidden. This provides constraints on the patterns which may lie in adjacent rows. Secondly, in any one row alone, the pattern must only comprise

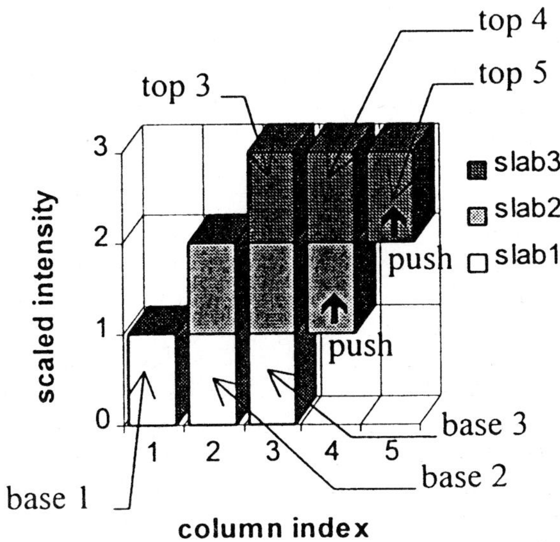

Figure 3.25. *The 'rod pushing' concept, part of the basic segmentation process. (From Siochi 1999c; reprinted with permission from Elsevier Science.)*

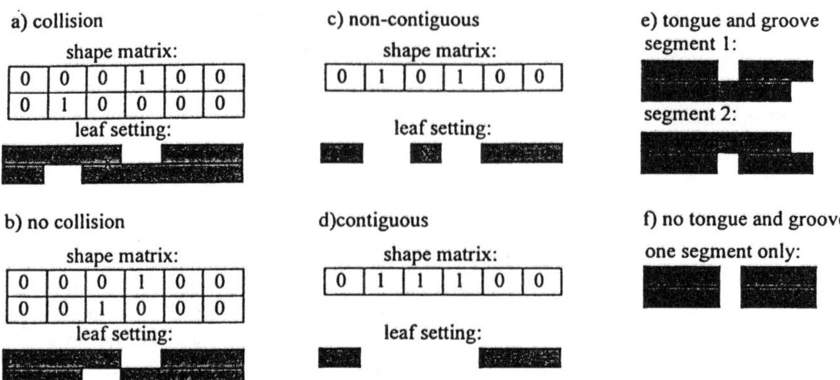

Figure 3.26. *Leaf setting and shape matrix constraints applied to the optimization algorithm: (a) Collisions between opposing and adjacent leaves; (b) no collision; (c) island blocks between leaves cannot be created, hence the configuration in (d) must be used. The leaf settings in (e) create tongue-and-groove underdose effects whereas (f) shows that a different combination can eliminate them. (From Siochi 1999c; reprinted with permission from Elsevier Science.)*

contiguous ones with surrounding zeroes since the MLC leaf pair clearly cannot form 'islands' of zeroes. Thirdly, the patterns must not generate tongue-and-groove underdose. Hence, the algorithm caters for these constraints during its selection of the optimum set of multiple-static-field components (figure 3.26).

In examples investigated it was found that the largest time savings resulted from those patterns which were most complex. Because of the record-and-verify overhead and the leaf movement time, minimizing the number of segments does not necessarily minimize the treatment time. The time savings obviously decreased when all the constraints were applied. For example, missing out the tongue-and-groove constraint and considering only collision avoidance led to faster deliveries. The fully optimized solutions were always more efficient than rod-pulling alone. The algorithm fully considers the system-related quantities. It is important to note that the algorithm was specifically designed to take account of the features of the Siemens MLC operating in step-and-shoot mode. Siochi (2000) has presented an optimization scheme to segment the virtual MLC optimizing the number of segments.

Dominiak *et al* (1998) have interfaced the NOMOS planning system CORVUS to SIMTEC running on a Siemens PRIMUS accelerator for IMRT. Comparison between PLanUNC generated dose distributions and Siemens MD2 delivered dose distributions have also been made. Comparisons between these calculated and delivered dose distributions have shown these to be close in general but differing when the DMLC technique creates small fields (Chang *et al* 1998b,

2000). This is known to be a difficult dosimetric problem to overcome (see section 3.2.6). Ting *et al* (1999) have also presented a leaf-sequencing algorithm for DMLC and, whilst not yet optimized to deliver the minimum number of leaf sequences, it gives a comparable number to other algorithms which have been so optimized.

3.2.3.3. Elekta DMLC technique

Elekta Oncology Systems (EOS)—formerly Philips Medical Systems–Radiotherapy—are developing their DMLC technique via an International Consortium (figure 3.27) which includes academic staff from the William Beaumont Hospital, Royal Oak, the Institute of Cancer Research and Royal Marsden NHS Trust, Sutton, The Antoni van Leeuwenhoek Ziekenhuis, Netherlands Cancer Institute, Amsterdam, the University of Ghent, Belgium, the Christie NHS Trust, Manchester and the University of Washington at Seattle (Elekta 1998a, de Neve *et al* 1998a, de Neve 1999, Rad 1999). The Consortium grew from a smaller base of sites which first met in 1994 with the formal establishment of the Consortium in 1996. Subsequent to the centres shown in figure 3.27*a*, two other centres have joined: the Thomas Jefferson University and the University of Maryland (figure 3.27*b*). At the time of writing, the University of Tubingen is being considered as the ninth member. The development of the Elekta DMLC technique has been described (Elekta 1998b) and the Elekta DMLC IMRT specification is given in Elekta (1999a).

There are several unresolved issues concerning the DMLC technique on which the Elekta Consortium is working. These include the following.

(i) Since there are many ways in which a 2D IMB can be decomposed, is there a way which is least sensitive to leaf-positioning errors?

(ii) Should the machine constraints on the leaf positions (interdigitation, overtravel, leakage, tongue-and-groove effect) be (a) built into the inverse planning or (b) ignored in inverse planning and put into an interpreter?

(iii) Are smoother IMBs (Webb *et al* 1998) less sensitive to leaf position errors? How would the implementation of smooth-beam constraints in inverse planning for 2D IMBs affect the tongue-and-groove effect? Can such constraints be implemented within the inverse-planning systems CORVUS and KONRAD?

(iv) Is there a dosimetric mismatch when, for a 2D IMB, and in any given leaf pair channel, the left leaf for one component is at the same position as the right leaf for another component? (the so-called 'along-the-leaf-direction problem').

(v) How many fluence levels should be utilized per beam; and is it preferable to predetermine fluence quantization during inverse planning or should this be applied *a posteriori* to whatever IMBs the inverse planning throws up (Bortfeld 1998)?

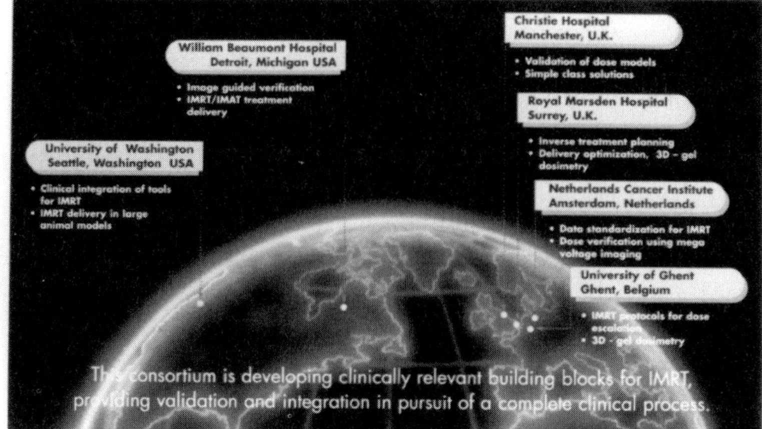

(*a*)

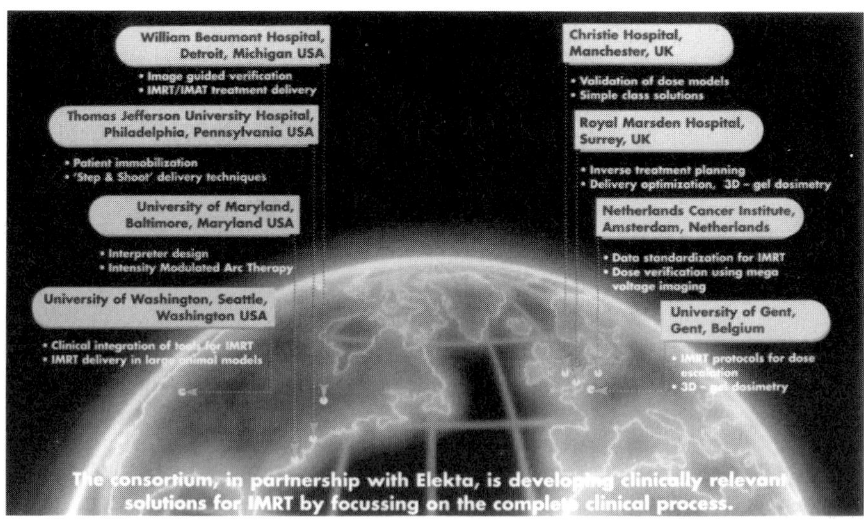

(*b*)

Figure 3.27. *(a) A promotional poster made by Elekta Oncology Systems to advertise their Consortium as it was in 1998. (b) Subsequently, two other centres have joined the Consortium by early 2000.*

(vi) Are IMBs with few fluence quanta significantly worse that IMBs with a full spectrum of fluences?
(vii) Can one develop 'smart IMRT' ('partial IMRT?') in which a few gantry angles, a few fluence levels and a coarse bixel interval apply? Is conformality compromised by simplicity?
(viii) Can forward planning ever generate IMBs which are as good as those generated by inverse planning?
(ix) Is any one of the many developed interpreters (see section 3.2.4—Yu, Convery, Sharpe, Budgell, Galvin, Kuterdem...) superior to the others and if so by what criteria? Is this superiority dependent on the clinical task? Has the problem been thoroughly evaluated including all the above issues?
(x) What should be the relationship between 'home-grown' interpreters and the commercial treatment-planning systems?
(xi) Is step-and-shoot delivery preferable to dynamic delivery or *vice versa*?

Some of the studies made by the Consortium members are reported in other sections of this chapter. The policy of Elekta is to develop IMRT by the DMLC method by working closely with universities and hospitals where much of the primary research is carried out, some funded by Elekta. The Elekta DMLC technique is clearly based on the design features of the Elekta MLC and this has been reviewed (admittedly from the Elekta perspective) in comparison with the MLCs available from the other manufacturers (Elekta 1997).

3.2.4. The interpreter

All DMLC techniques require an 'interpreter' to convert each 1D IMB profile into an appropriate set of leaf positions as a function of time. There is an infinity of possible solutions to this problem depending on what is being optimized. Several of the earliest methods optimized the delivery time (see a review in Webb (1997d, chapter 2) and the graphic illustration of the algorithm by Boyer and Strait (1997) (see section 3.2.1)). However, several teams have constructed their own interpreters partly to satisfy themselves of the DMLC principles and partly to achieve different aims. These generally include taking account of certain physical limitations of the MLC itself which might disallow the solution with minimum time. We have already met one interpreter in section 3.2.3 when discussing the Siemens product. We now review many more of these interpreters.

3.2.4.1. Christie NHS Trust, Manchester interpreter

The Christie NHS Trust, Manchester interpreter (Budgell *et al* 1997, 1998b, Wilkinson *et al* 1998) is modular, written in Fortran and determines both leaf and jaw positions for dynamic delivery. There is no compromise of the required modulation within the irradiated field. To achieve this there *is* some deposition of dose outside the required field. The maximum field size is 25×25 cm^2, limited by the 12.5 cm overtravel. The resolution along the leaf direction of travel is

1 mm. The X-jaws (the jaws defining the field perpendicular to the direction of leaf motion) stay at a constant position and the Y-jaws track the most-lagging trailing leaf and the most-ahead leading leaf. There is no check for zeroes in the field. The interpreter maps the primary fluence only and leaves the treatment of scatter to the dose model. The interpreter checks that no two leaves in any pair come closer than 10 mm and that the leading leaf of one pair does not collide with the trailing leaf of one of the adjacent pairs. (This is known as avoiding 'interdigitation'). The interpreter works in absolute monitor units (MUs). The leaves all finish together (sometimes called synchronization, although, as we have seen, this word has another important meaning for DMLC therapy in the solution of the problems introduced by the tongue-and-groove effect (see section 3.2.2 and Webb *et al* 1997a, b, Webb 1998d, e). The Christie NHS Trust, Manchester interpreter has all the left-most and right-most leaves starting in line. This interpreter deliberately puts a small non-zero dose everywhere in the field in order to overcome the leaf collision restriction which specifies that no leaf may come within a specified distance of its opposing leaf, nor the two opposing leaves in the two adjacent channels.

The constraint imposed by the Elekta MLC of not permitting leaf interdigitation (figure 3.28) provides a challenge to develop an interpreter to avoid the eventuality. Details of the Christie NHS Trust, Manchester interpreter have been described at some length by Budgell *et al* (1998b). The interpreter was developed to generate a system of control points spaced at equal MU intervals through the exposure. The positions of the leaves and backup collimators are monitored 12.5 times each second to ensure they remain within tolerance. Given the maximum overtravel of 12.5 cm across the midline, modulations which include both positive and negative gradients may only be achieved within a maximum field width of 25 cm.

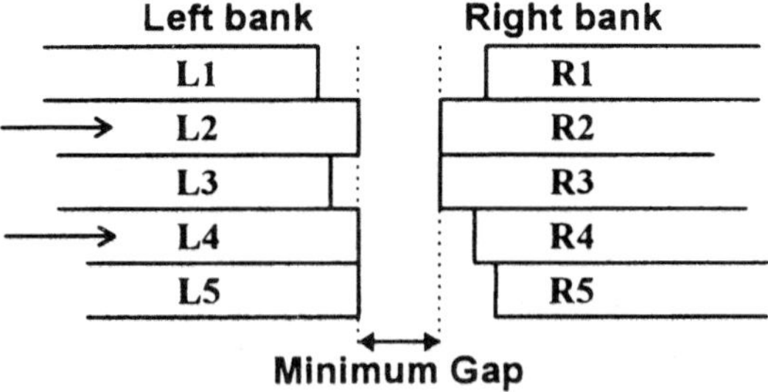

Figure 3.28. *The minimum gap constraint in the Elekta interpreter. Leaves in the left bank must not encroach within a preset distance of either the directly opposing leaf or the opposing adjacent leaves. This has limitations for delivering a 2D IMB to an irregularly shaped field (see e.g. figure 3.29). (From Budgell et al 1998b.)*

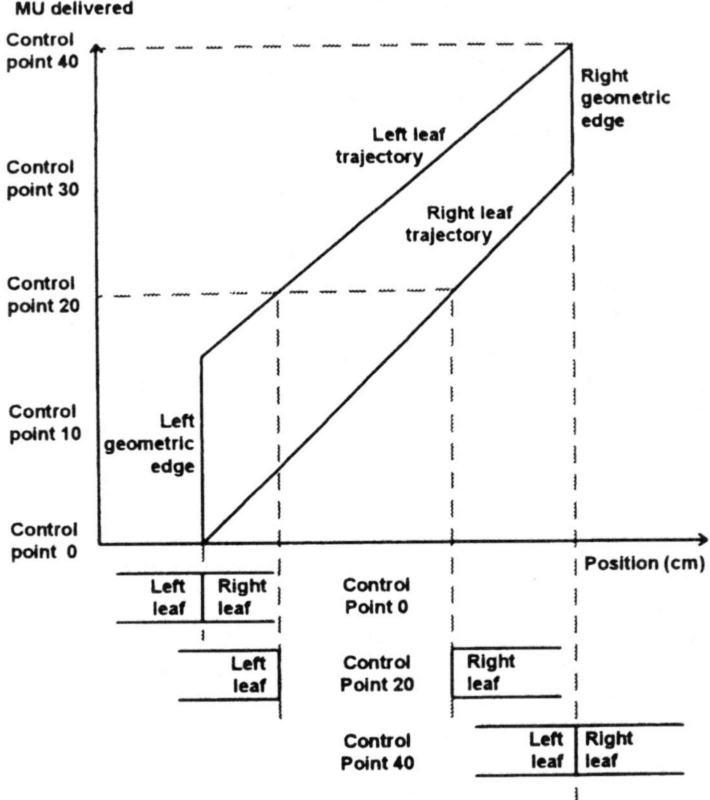

Figure 3.29. *A pair of leaf trajectories that would generate a wedged intensity profile. This scheme is prohibited by the Elekta interdigitation constraint. (From Budgell et al 1998b.)*

The problem posed by the interdigitation constraint is graphically obvious from figure 3.29. This shows the leaf trajectories required to generate a wedged intensity distribution across the field and shows that this requires both left and right leaves of a pair to start and stop at the same location, a situation prohibited by the Elekta MLC interdigitation constraint. This happens whenever an IMB starts with a rising edge e.g. as shown in figure 3.30. Two solutions to this problem are shown in figure 3.30. In figure 3.30*a* is shown the prohibited arrangement. One option is therefore to start the left leaf 10 mm behind the field edge (figure 3.30*b*). This avoids violating the interdigitation constraint but delivers some unwanted dose outside the field edge. In figure 3.30*c* an alternative arrangement is shown in which the right leaf starts 10 mm inside the field edge leading to an underdose in the 10 mm margin. This might seem a problem which is not too important for any

a)

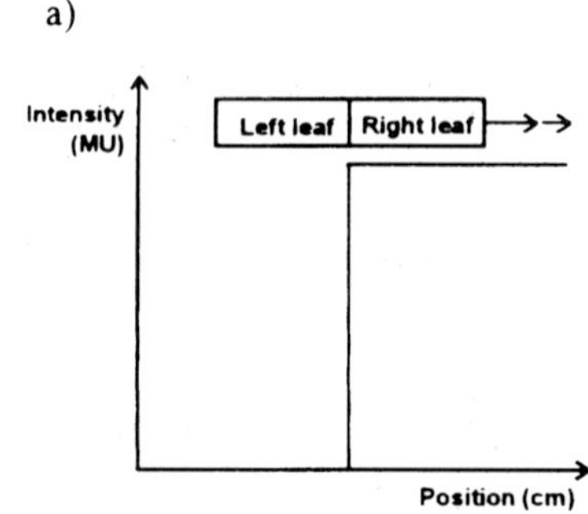

b) c)

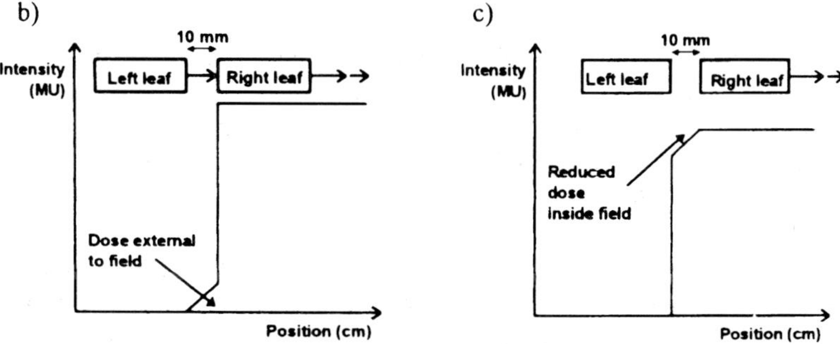

Figure 3.30. *Strategies for avoiding collision with the directly opposing leaf at the geometric edge of the beam. The leaf starting positions in (a) are prohibited as opposing leaves are in contact at the field edge. Those in (b) do not restrict the modulation inside the geometric field edge but deliver unwanted additional dose beyond. Those in (c) result in no additional dose beyond the geometric edge but the modulation is compromised inside the edge. (From Budgell et al 1998b.)*

1D IMB but as soon as the adjacent leaf interdigitation constraint is also observed it becomes obvious that, when two adjacent IMBs are not the same, the region of overdose or underdose can increase in magnitude and lateral extent. The reader is referred to Budgell *et al* (1998b) for a very detailed graphical discussion. It soon becomes apparent that it is not the width of the minimum leaf gap which is the main restriction but the relative geometric edge positions in adjacent leaf pairs.

Budgell *et al* (1998b) have solved the problem by developing an interpreter based on 'rectangular edge synchronization' (RES). In this technique all the right leaves start at the left edge of a rectangle encompassing the 2D IMB; all the left leaves start 10 mm back from this i.e. 10 mm out of the field. They all start their movements when the beam is switched on. The left leaves then move as quickly as permitted to their ideal starting positions. This has the effect (see figure 3.31) of

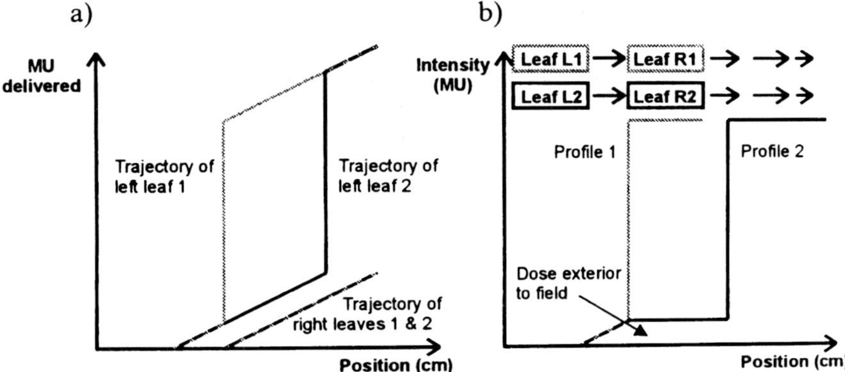

Figure 3.31. *Leaf trajectories (a) and the resulting intensity profiles (b) illustrating how the dose outside the left geometric edge may be reduced to a fixed level by RES without compromising the modulation. This is the technique applied when the geometric edge of the field is a complicated shape. (From Budgell et al 1998b.)*

increasing the intensity everywhere within the rectangle but outside the geometrical field edge to the same low fluence whilst having no effect within the high-fluence region. The maximum fluence outside the geometric edge is always less than would occur without RES. The value of this minimal intensity is given by

$$\text{minimal intensity (MU)} = \frac{\text{minimum leaf separation (mm)} \times \text{MU rate (MU s}^{-1})}{\text{maximum leaf velocity (mm s}^{-1})}. \tag{3.15}$$

For a monitor unit rate of 400 MU min^{-1}, a maximum leaf speed of 2 cm s^{-1} and a minimum leaf gap of 10 mm, the excess dose is that delivered by 3.33 MUs. In practice, Budgell *et al* (1998b) allocate 5 MUs to reduce the burden on maximum leaf velocity. This also represents the minimum that can be delivered to central islands by this method. A second feature of the RES method is that some profiles must be delivered with lower than optimum efficiency in order that all the left pairs arrive together at the right field edge. Two very useful side effects of these operations are: (i) the optimum use of backup collimators so reducing radiation leakage and (ii) the effective elimination of the tongue-and-groove effect.

Budgell *et al* (1998b) applied the method to 40 test cases and also performed measurements. They found that the average intensity outside the field edge and the maximum intensity outside the field edge were consistently better by about a factor of two when the RES algorithm was used in comparison with another unspecified interpreter. The integral intensity is larger, since the minimum is spread over a large area, but this does not matter since the regions close to the PTV are shielded from high values and the regions far away probably do not matter much. The interpreter was determined to be robust.

The Elekta DMLC is operated in 'step-and-shoot' mode via the specification of 'control points', being the positions of the leaf pairs and the number of MUs to be delivered at each position before movement to the next. The number of control points is a variable. The fewer used the more inaccurate will be the realization of the required IMB. Budgell (1997, 1998) has studied the variation of the discrepancy between desired and delivered IMB as a function of the number of control points and has concluded that about 50 are required, the error not decreasing as this number increases.

The effect of misplacement of the MLC leaves during a DMLC delivery has also been studied by Budgell *et al* (1998a, 2000b). They described two types of error: (i) the leaves are linearly offset from their required position; (ii) the leaf setting calibration has a gain error. Then, they delivered a 2D IMB in which a leaf gap created a flat profile by scanning and showed that the predicted dose errors matched both simulated dose measurements and actual experimental measurements for a range of leaf trajectory gradients (MU min^{-1}), the errors increasing linearly in proportion to this gradient. It was recommended that sweeping a small gap should become part of a regular QA programme. In practice the effects of leaf position inaccuracy are greatest for highly modulated, low-dose intensity profiles.

Budgell (1999) has given details of both the theoretical calculation, the modelling prediction and the experimental determination of the r.m.s. error in delivering 2D IMBs by the MSF technique and by the DMLC technique as a function of the number of control points specified. It is possible to deduce that the theoretical maximum r.m.s. error when delivering a single 1D IMB by the MSF technique with control points equally spaced in fluence interval is given by

$$\text{r.m.s. error (\% of maximum intensity)} = \frac{1}{2\sqrt{(3)} \times (\text{number of fields})}. \quad (3.16)$$

There is no simple relationship for the DMLC technique but it was shown that the r.m.s. error lies between zero and

$$\text{r.m.s. error (\% of maximum intensity)} = \frac{1}{\sqrt{(3)} \times (\text{number of control points} - 1)}. \quad (3.17)$$

From this it is easy to see that for a large number of control points the two predictions will converge; however they may diverge widely for a low number of control points. Given that the maximum number of MUs and the peak number are related through an efficiency (or inversely a modulation scale factor) it is apparent that the above two expressions remain functionally the same when expressed as a percentage of the peak intensity. The curves simply scale. Budgell (1999) also argued that since the number of control points simply specifies a sampling density, the relationship between the r.m.s. error and the number of control points can also be elucidated by studying the frequency spectra of the modulations.

These theoretical analyses are limited in that they only relate to the single 1D leaf trajectories. However, the 1D modulations are constructed from two coupled

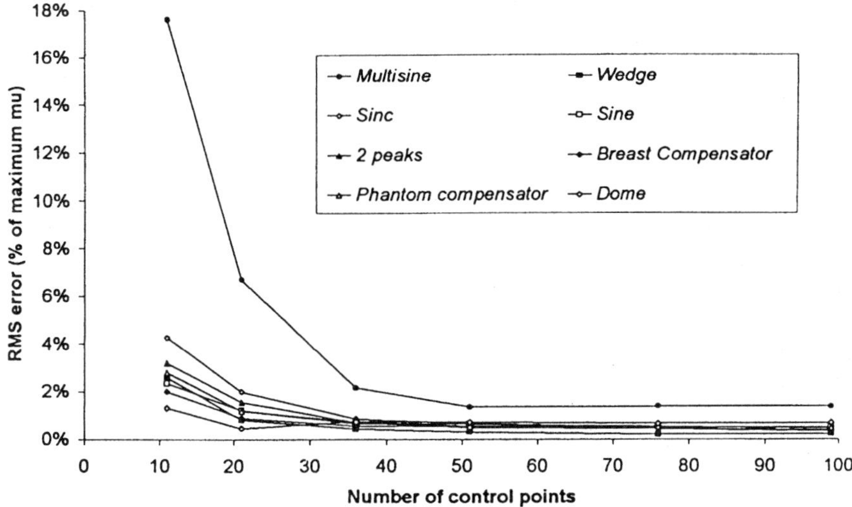

Figure 3.32. *The r.m.s. error versus number of control points for a set of test prescriptions. (From Budgell 1999.)*

leaf trajectories (right and left leaf pair). Also, MLC constraints and the specific interpreter utilized will affect the determination of r.m.s. error. Hence, Budgell ran a series of test cases, different modulation patterns sampled with varying number of control points, to determine a modelled prediction of the r.m.s. error taking these effects also into consideration. Finally, he made experimental measurements which also factored in the effects of radiation transport.

The test modulations, when modelled, led to variations with the number of control points falling on curves which lay between the extremes of zero and that predicted by equation (3.17) exactly as expected (figure 3.32). Moreover, it could be determined that provided an r.m.s. error of 2% was acceptable only the most complex modulations required 50 control points (figure 3.33). Simpler modulations such as a wedge or a breast compensator could be achieved with this level of accuracy with about 20 control points. A similar analysis showed that 25 were adequate for MSF deliveries.

Budgell (1999) pointed out that there were other factors which needed to be balanced. For example, the more control points are used in MSF deliveries, the smaller the number of monitor units per control point and the more radiation intersegment 'deadtime' occurs. Also, combinations of modulation, required for example to perform therapy with combined IMBs, involve a combination of the errors from individual profiles.

Mott *et al* (1998a, b, c) have shown that the DMLC technique may be applied to deliver IMBs which have been arrived at by a simple technique which

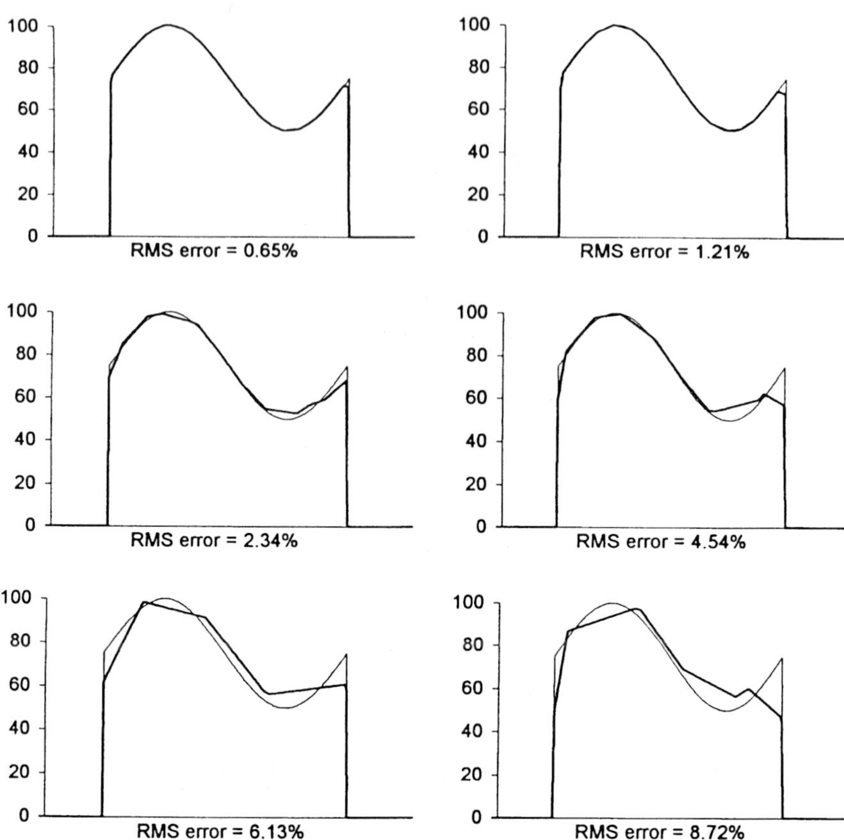

Figure 3.33. *Computed monitor unit fluences overlaid on input intensity profiles for a sine modulation function at six different values of r.m.s. error. (From Budgell 1999.)*

compensates for the change in body surface as radiation is transmitted to an isocentric plane. Mott *et al* (1999) have explained that the technique is required to be iterative to take account of head scatter, leakage and phantom scatter. The methodology is shown in figure 3.34. Firstly, the depths to all points in an isocentric plane are computed, together with the primary and scatter doses. These are then inverted to create the required modulation. The modulations are passed to the interpreter to create the leaf sequences and then the doses are recalculated to include head and phantom scatter. The process cycles to adjust the trajectories until the required outcome is achieved.

The method was tested using a very challenging anthropomorphic phantom (figure 3.35) designed to obtain a uniform dose to the baseplane. That this can be achieved is evident from figure 3.36 in which it can be seen that 10%

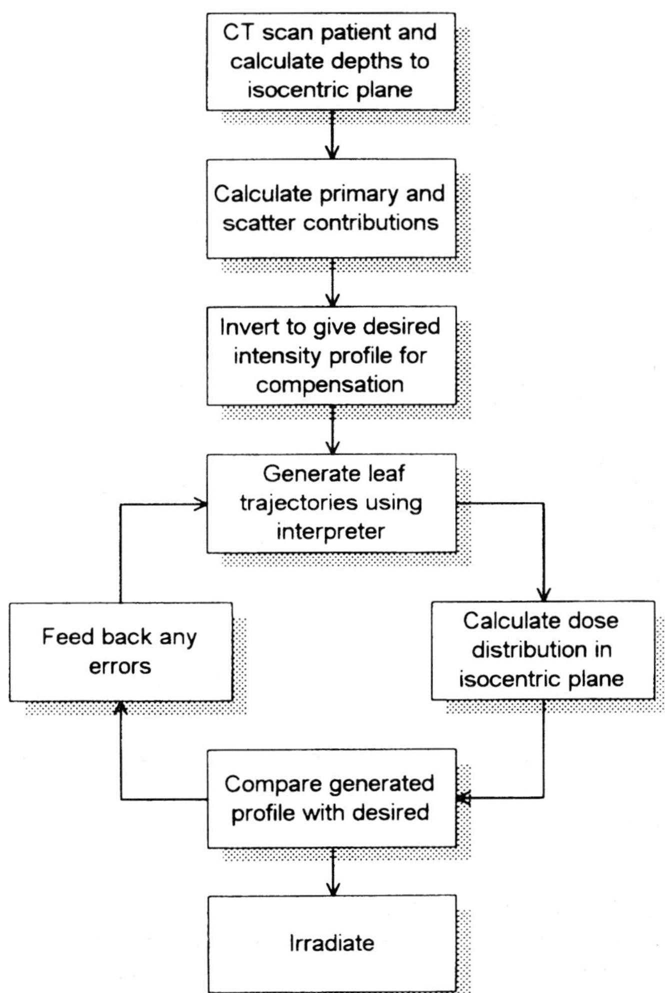

Figure 3.34. *The steps required for the production of a compensated field to be delivered by dynamic multileaf modulation using the Manchester technique. (From Mott et al 1999; reprinted with permission from Elsevier Science.)*

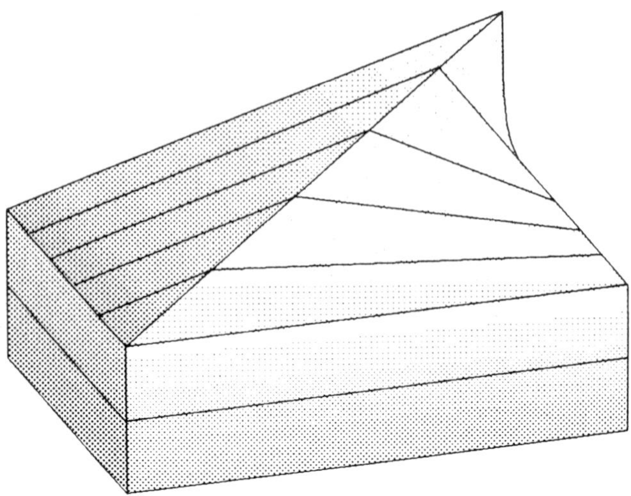

Figure 3.35. *A polymethylmethacrylate phantom designed for testing the Manchester compensation technique. The base of the phantom is* 20×20 *cm*^{2} *and the peak is 10 cm high. The shaped portion is then placed on top of two 5 cm thick sheets of the same material and the goal was to produce a uniform dose between these two sheets (see figure 3.36). (From Mott et al 1999; reprinted with permission from Elsevier Science.)*

inhomogeneities without compensation can be removed with compensation. Mott *et al* (1999) went on to show that four-field compensation to a Rando phantom could generate a more uniform high-dose volume to a central region, compensating for contour variations.

The Manchester group have used the DMLC technique in a novel way (Wilkinson *et al* 1998, Hounsell 1998a, Elekta 1999b)). They first compute the compensation required to turn the patient into a box-shape. This then allows the application of 'class solutions' in which the shape of the tumour is used to determine further intensity-modulation. The combined modulation is the product of the two separate modulations. This approach means that a set of standard modulations can be used corresponding to tumour shapes in a box and the only individualization is in 'turning each patient into such a box'. This avoids altogether the need for patient-tailored inverse planning (figure 3.37).

The technique of the Christie NHS Trust has been further described by Mott (1999). It has been used as a form of patient compensation to improve the dose homogeneity in the treatment of bladder cancer. The treatment technique has ethical committee approval and is regarded as a good way to get IMRT into the clinic. Only four out of the twenty fractions are delivered by IMRT in 'local-service mode' at 100 PRF. Computed dose distributions showed the 90% contour

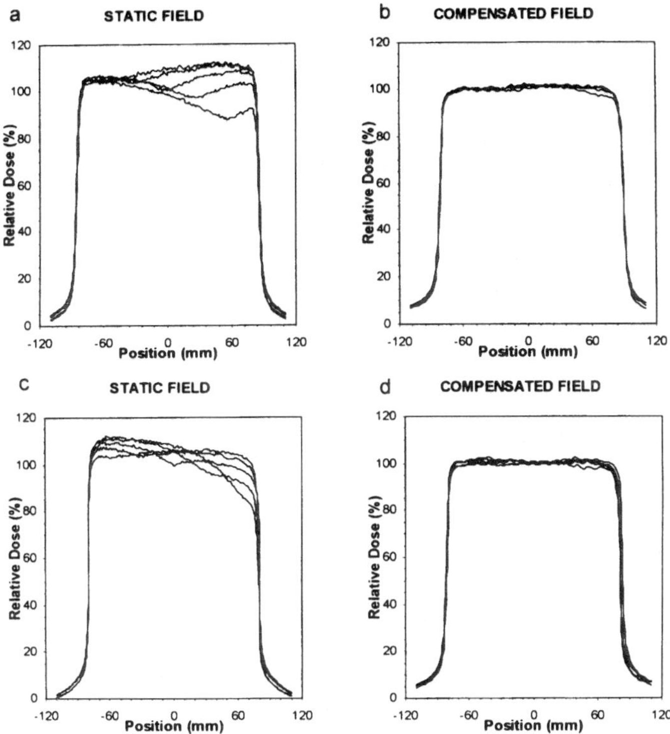

Figure 3.36. *(a) Dose profiles along the direction of leaf motion in the phantom irradiated with the uncompensated beam. (b) Dose profiles along the direction of leaf motion with the compensation. (c), (d) ditto but across the direction of leaf motion. (From Mott et al 1999; reprinted with permission from Elsevier Science.)*

nicely enclosing the PTV. Doses were verified using an absolute-dose phantom. The protocol involved a dummy run ahead of treatment. Absolute doses were confirmed to 0.6%. The efficiency of the treatment depended on the patient but generally lay between the reciprocals of 1.5 and 1.7. Up to two interrupts occurred per beam and 20–25 min were needed for the entire treatment. It was noted that the time to deliver could be reduced if the pulse repetition frequency could be increased to 400, if the leaf speed could increase, and if the megavoltage imaging could be streamlined.

An interesting, not too dissimilar, idea was put forward by Hristov *et al* (1999) who have shown that, whilst inverse planning for three-field prostate therapy could be tuned, a set of fixed inverse-planned beams have led to significant improvements.

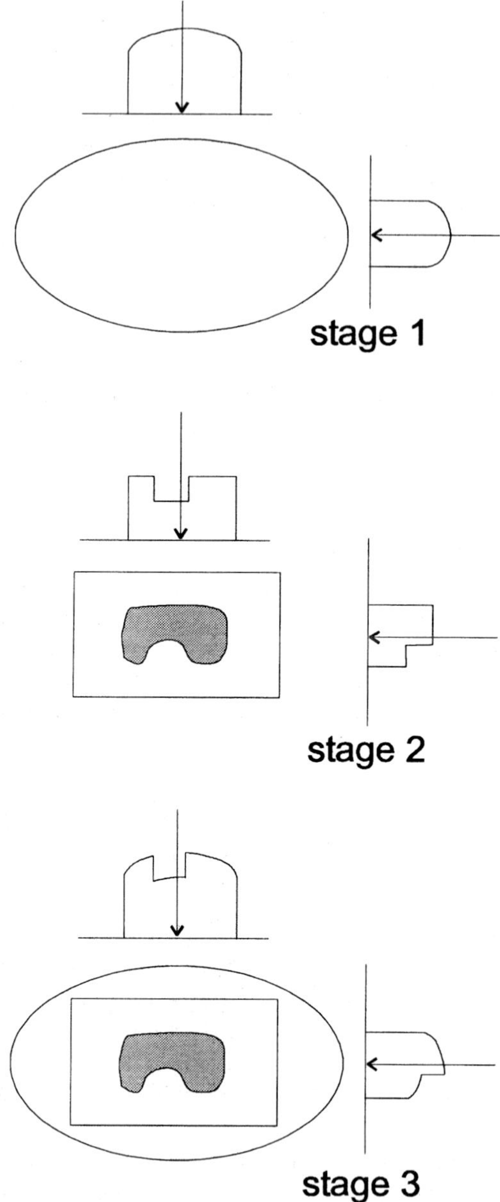

Figure 3.37. *The combining of customized compensation and a class solution for IMRT. At stage 1 the compensation is made for the patient shape to turn the patient into a box shape. In stage 2 the modulation is performed for the target shape and in stage 3 the two modulations are combined. (From Wilkinson et al 1998.)*

Budgell *et al* (2000a) and Mott *et al* (2000a) have described and presented the results of implementing the complete chain of IMRT delivery using the DMLC IMRT technique at the Christie NHS Trust, Manchester. By November 1999, seven patients had been treated using the dynamic MLC for four out of their 20 fractions and the improvement in homogeneity of dose to the bladder was observed using the IMRT compensation procedure. The delivery efficiency varies between the reciprocals of 1.37 to 1.71 comparing favourably with that of a 20° wedge factor of 1.51. It was concluded that the methodology paved the way for the introduction of more complex clinically beneficial IMRT techniques.

3.2.4.2. Royal Marsden NHS Trust, Sutton, interpreter

We now turn to the approach taken at the Institute of Cancer Research and Royal Marsden NHS Trust where an alternative interpreter method has been developed (Convery 1997a,b, 1998, Convery and Webb 1998 and Convery *et al* 1998). This interpreter creates each 1D IMB using a combination of both moving leaves and moving Y-jaws, again seeking to generate a match to just primary fluence, ignoring scatter and leakage. It was designed as a genuine dynamic therapy interpreter. It checks that no interdigitation occurs and yet can create islands of zero primary fluence by placing a left leaf and a right Y-jaw at the same location along the direction of travel (or right leaf and left Y-jaw) (figure 3.38). The tongue-and-groove underdose is completely avoided by appropriate synchronization (which tends to require little extra time because avoiding interdigitation already tends to

Figure 3.38. *Minimum leaf separation requirements on the Elekta MLC. In (a) the most leading leaf must not only maintain the minimum separation with the opposite right leading leaf but also with the right leaves adjacent to this one. (b) shielding a region during dynamic collimation with the backup diaphragm whilst maintaining the necessary minimum leaf separation. (Regions in black are shielded by both an MLC leaf and a backup diaphragm (jaw), regions in dark grey by an MLC leaf only and in light grey by a backup diaphragm only.) (From Convery and Webb 1998.)*

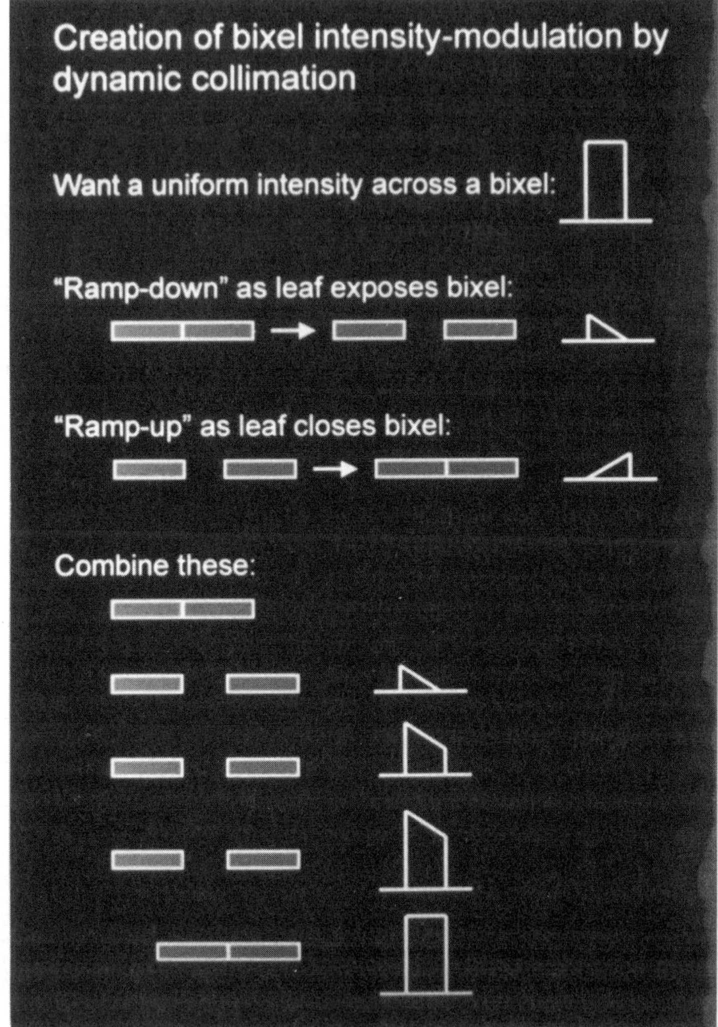

Figure 3.39. *This diagram shows how the combination of a leading leaf moving at maximum speed and a trailing leaf covering the same space at maximum speed but at a later time can give rise to a 'flat' bixel intensity. (Courtesy of Dr D Convery.)*

create almost synchronized solutions). The interpreter outputs the leaf positions at a series of control points and can be fed to the Elekta MLC. Leaves, when in motion with the radiation on, move at maximum speed and so create bixels with uniform fluence throughout each bixel corresponding to the assumption made within the NOMOS CORVUS treatment-planning system for which the interpreter was initially tailored (figure 3.39).

Convery has applied this interpreter to 2D beams resulting from the CORVUS inverse-planning system and delivered the 2D IMB to a film at d_{max}. There were large intensity variations from one bixel to another, something which the MIMiC does not mind (and so CORVUS tends to produce) and the DMLC technique can cope with, albeit at reduced efficiency, there being a very large number of field segments. There are critics of the 'spiky IMBs' generated by this system (e.g. Galvin (1999a) who claim that they are not intuitively understandable. However, (i) the dose consequences can be observed via isodose plots and DVHs; (ii) the satisfaction of dose-volume constraints can be checked; (iii) apart from any 'lack of comfort' concerns, the spiky form of the 2D IMBs is dosimetrically irrelevant so long as the distributions can be accurately delivered, as Convery has shown.

Figure 3.40 shows an example 2D IMB computed by the NOMOS CORVUS treatment-planning system. Figure 3.41 then shows the component fields computed by the algorithm developed by Convery and Webb (1998). In turn, figure 3.42 shows the resulting 2D intensity modulation delivered by the Elekta MLC. One particular feature of the Convery interpreter is that leaves and jaws are both used to control the modulation, indeed can even deliver zero primary

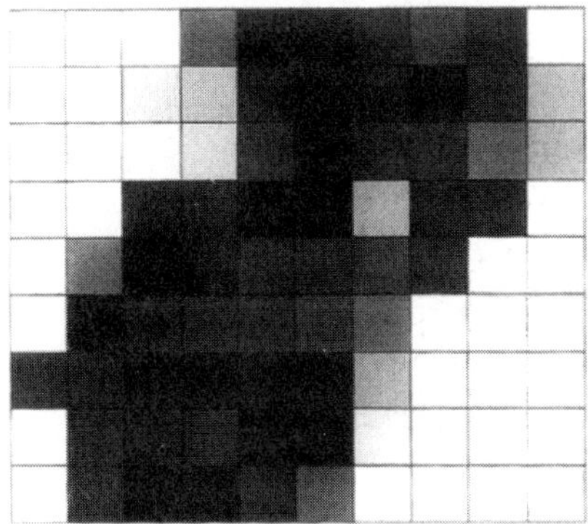

0	0	0	6	10	10	9	7	8	1
0	0	2	3	9	10	9	10	8	3
0	0	0	3	7	9	8	8	5	3
0	0	9	9	10	10	3	8	7	0
0	5	10	9	7	7	6	7	0	0
0	10	8	8	7	6	5	0	0	0
8	8	10	10	9	9	3	0	0	0
0	7	8	8	10	10	1	0	0	0
0	7	10	10	8	5	0	0	0	0

Figure 3.40. *An example of a discrete IMB shown as (a) grey scale and (b) the matrix of relative intensities. The intensity distribution is for a posterior oblique field of an eight-field prostate plan with 10% intensity stratification and* 1×1 *cm beam elements at the isocentre plane, planned with the CORVUS planning system. (From Convery and Webb 1998.)*

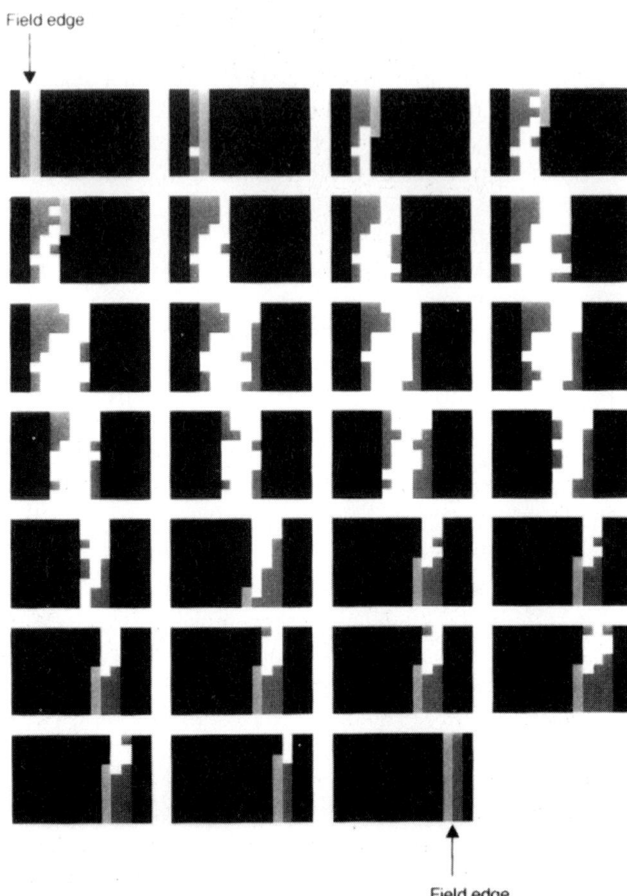

Figure 3.41. *Control point sequence derived to deliver the relative intensities shown in figure 3.40. Regions in black are shielded by both an MLC leaf and backup diaphragm, regions in dark grey by an MLC leaf only and in light grey by a backup diaphragm only. The left and right field edges are marked. During dynamic delivery, leaf and diaphragm positions vary between control points. (From Convery and Webb 1998.)*

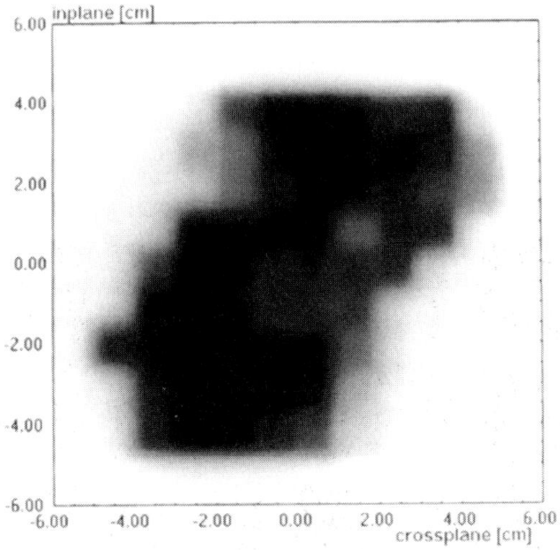

Figure 3.42. *An example of a 2D intensity modulation, presented as a grey scale image, delivered by the Elekta MLC and measured at peak depth at the isocentre for the dynamic prescription shown in figure 3.41. (From Convery and Webb 1998.)*

fluence islands, and overcome the limitation that the Elekta MLC does not allow interdigitation. The interpreter also ensures that there is no leaf collision. Kuterdem *et al* (1999) have described the solution of the identical problem in Seattle for the Elekta DMLC technique. This also makes use of both leaves and jaws.

Figure 3.43 shows a three-level intensity pattern with 1×1 pixel size designed at the UCSF as part of a multi-segment treatment of a head-and-neck tumour. The figure also shows the port film of the patient with the delivered intensity pattern superimposed for pre-treatment verification when the delivered intensity pattern has been determined and delivered using the multiple-static-field technique. Verhey has argued that the multiple-static-field technique with a large number of field segments can arguably be described as a DMLC technique. Convery's interpreter can also generate the appropriate MSF components for 'step-and-shoot' DMLC IMRT. An issue, however, with step-and-shoot may be the somewhat long wait time between the startup of each segment, due to the need to stabilize the radiation source (at least with the earliest clinical versions of the Elekta software).

An issue to be resolved is whether the zeroth control point should instead be the largest open field so that portal image verification can precede the delivery of IMRT with the first step then being a 'move-only' step to the left-most position. A second issue is that if the modulation is of small amplitude standing on a high-dose pedestal then it makes sense for the first step to be a large open-field irradiation with

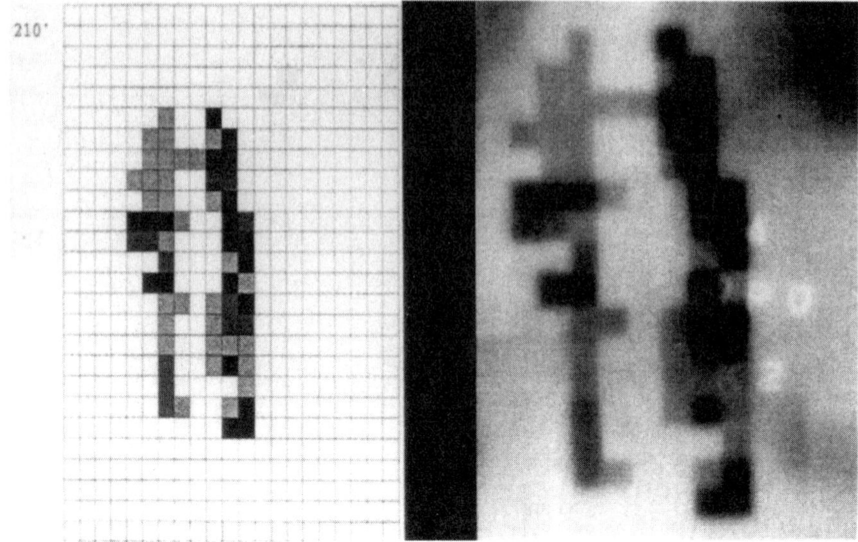

Figure 3.43. *On the left is a three level intensity pattern with* 1×1 *cm pixel size designed for one of four field directions used in an inverse-planned IMRT treatment of a tumour of the head-and-neck. On the right is a port field of the patient with the intensity pattern superimposed for pre-treatment verification. (From Verhey 1999a.)*

the second step being a 'move-only' step to the left-most position prior to starting the DMLC part to achieve the modulation. This would be the case for example for breast radiotherapy where only a small 'tuning' modulation is required (Chang *et al* 1998a, b, 1999b, Lo *et al* 1998, 2000, Evans *et al* 1997a, 1998, Dubal *et al* 1998) (figure 3.44). It is now well-known that the homogeneity of dose delivered to the breast improves as account is taken of the full geometry, through the use of either planning using a contiguous CT dataset (Beckham *et al* 1997) or the design of tissue compensation based on the use of electronic portal imaging (Evans *et al* 1997b, 1998, Dubal *et al* 1998). At the Royal Marsden NHS Trust the number of such IMB increments is usually small (of the order of four) and the intensity fluence increments are equal (see section 4.1.13). However, Beavis *et al* (1999c) have shown dosimetric advantages of using unconstrained and therefore non-uniform fluence increments.

3.2.4.3. William Beaumont Hospital, Royal Oak, interpreter

Yu *et al* have developed another interpreter at the William Beaumont Hospital, Royal Oak, Detroit. This was the first interpreter developed (Yu *et al* 1995). It has a graphical user interface. A second new interpreter at the William Beaumont

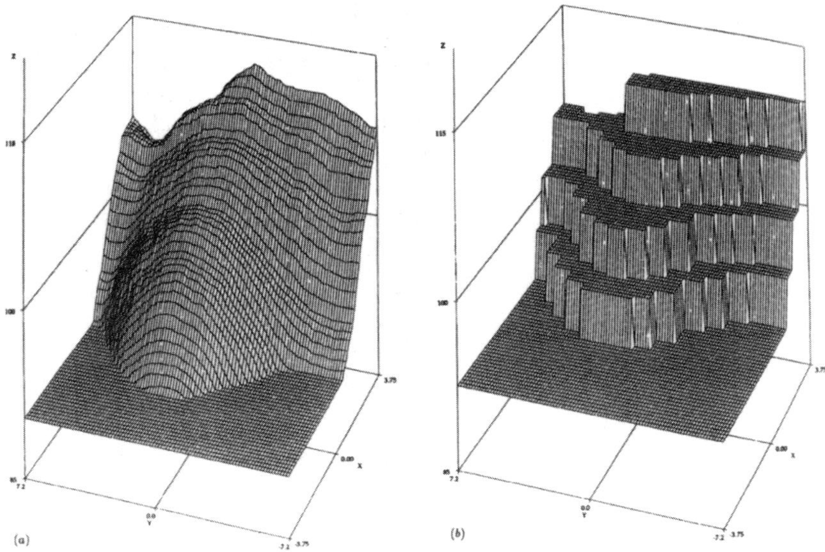

Figure 3.44. *(a) An example of an ideal intensity-modulated beam profile for treating the breast tangentially (x and y axes are length and width respectively in the radiation field, z is the desired intensity). (b) the ideal intensity-modulated profile discretized to be delivered using shaped MLC fields. The base level is just a continuation of the lowest required intensity level to the region outside the patient, hence the base level is half of one increment higher for the discretized intensity distribution than for the ideal intensity distribution. 85% of the total dose can be delivered with an open field and then the smaller fields can be delivered using the MLC to shape the fields. (From Hansen et al 1998a.)*

Hospital aims to minimize the number of field segments (Wu *et al* 1999a). The planning system KONRAD is used to generate unconstrained intensities. These are then binned to a finite number of intensity levels *a posteriori*. (For example, Stein (1999) and also Keller-Richenbacher *et al* (1999a, b) have shown that between five and seven beams and between three and five intensity levels, resulting in seven to twelve segments for each beam, can suffice to achieve a treatment plan quality that is almost equal to the one expected for arbitrary IMBs. Clearly this statement depends on the planning problem being addressed and is the subject of much current research.) Within each fluence stratum an intelligent sort is made to assign the original generated fluences to a best set of a small number of fluence increments. Finally, the leaf sequencing is performed *a posteriori* to satisfy the Elekta MLC machine constraints on leaf placement. Then the dose distribution corresponding to these IMBs is replanned inside the ADAC PINNACLE system which uses a slightly different beam model from KONRAD (Wong *et al* 1998b, Sharpe *et al* 1998). Laub *et al* (2000a) have described the use of the Hyperion treatment-planning system for planning colorectal cancer, with similar post-planning quality assurance. An overview of the approach from the Elekta perspective has been written (Elekta 1998c).

Brabbins and Yan (1999) have given further details of the William Beaumont Hospital IMRT delivery technique for treatment of the prostate as outlined in the above two abstracts and the publication from Elekta. The PTV is constructed making use of multiple CT scans taken at intervals during the first week of treatment to determine a 'confidence-limited PTV'. Five-field IMRT has been developed and, by May 1999, 35 patients had been treated. Planning studies have shown that the use of the confidence-limited PTV reduced the underdose in the clinical target volume from a massive 40% seen if a simple half centimetre margin had been used to a value of around 5%. Treatment planning has been made using dose-volume constraints using KONRAD. The dose-volume contraints were as follows. Less than 5% of the rectal wall volume should receive a dose greater than 75.6 Gy. Less than 30% of the rectal wall volume should a receive a dose greater than 72 Gy. Less than 40% of the rectal wall volume should receive a dose greater than 65 Gy. Less than 50% of the bladder should receive a dose greater than 75.6 Gy. Additionally, the aim was to achieve less than 10% dose inhomogeneity using 18 MV radiation and five beams spaced at 72° intervals. Typically, after the re-segmentation described above, all two-dimensional IMBs have small segments and typically there are between 35 and 45 beam segments per treatment. The whole treatment takes between 20–25 min with the accelerator operated in 'clinical treatment mode'. This centre is also developing IMRT of the breast (Elekta 1999d).

3.2.4.4. The University of Washington, Seattle, interpreter

The interpreter at the University of Washington, Seattle, has been described by Kuterdem *et al* (2000). It is capable of sequencing leaf patterns both dynamically and as multiple-static-fields to cope with varying IMB types. For the MSF technique the first collimation is to the widest field component to allow for portal imaging. The remaining control points are then set to minimize the total leaf travel. The number of fluence levels is a variable and spatial subsampling can be used if desired. The dynamic interpreter sets the leaf velocities according to the familiar equations (section 3.2.1), synchronizes movements to eliminate tongue-and-groove underdose and extensively uses the backup diaphragms to overcome the limitations imposed by the gap constraints. The method can supersample bixels to yield finer spatial resolution. Kuterdem *et al* (2000) compared the two methods with varying choices of parameters for two IMB sets. They tabulated the r.m.s error, TG underdose (impossible to eliminate completely with the MSF method—see section 3.3.2), the efficiency and the treatment times. They concluded that dynamic treatment would be significantly more time-efficient but that this was largely due to the large intersegment wait time (about 6 s at the time of the report) for the MSF method.

3.2.4.5. *Ma's interpreter*

Yet another algorithm for setting the leaves of a DMLC technique has been published by Ma *et al* (1998a). This considered simply the realization of the primary fluence in a single 1D IMB and at this stage ignored interdigitation constraints and tongue-and-groove effects. The algorithm assumed that the IMB could be specified as a series of stepwise changes of intensity as a function of position as is usually given by an inverse-planning algorithm, e.g. that in the NOMOS CORVUS treatment-planning system. $(N + 1)$ positions x_0 to x_N are defined, together with the intensities I_0 to I_N between these positions. I_j is the intensity between x_{j-1} and x_j. The leaves sweep from x_N to x_0 with $x_i > x_{i+1}$ by sign convention (figure 3.45). The 'integral fluence constant' is simply the area under this histogram-like IMB. It may be written as

$$\phi = \sum_{j=1}^{N} I_j \left(x_{j-1} - x_j\right) \tag{3.18}$$

or alternatively, through some algebra, as

$$\phi = \sum_{j=0}^{N} a_j x_j \tag{3.19}$$

where $a_j = I_{j+1} - I_j$ is the *change* of intensity between I_j and I_{j+1} and $I_0 = I_{N+1} = 0$. One can also express

$$I_j = \sum_{i=0}^{j-1} a_i. \tag{3.20}$$

This simply says the value of the IMB at location j is the sum of the changes to intensity along all positions up to that point as is physically obvious and necessary.

If one then expresses the positive values of a_i as $a_k^{(+)}$ and the negative values of a_i as $-a_l^{(-)}$ then one may write

$$I_j = I_j^{(+)} - I_j^{(-)} \tag{3.21}$$

where

$$I_j^{(+)} = \sum_{k=1}^{j^+} a_k^{(+)} \tag{3.22}$$

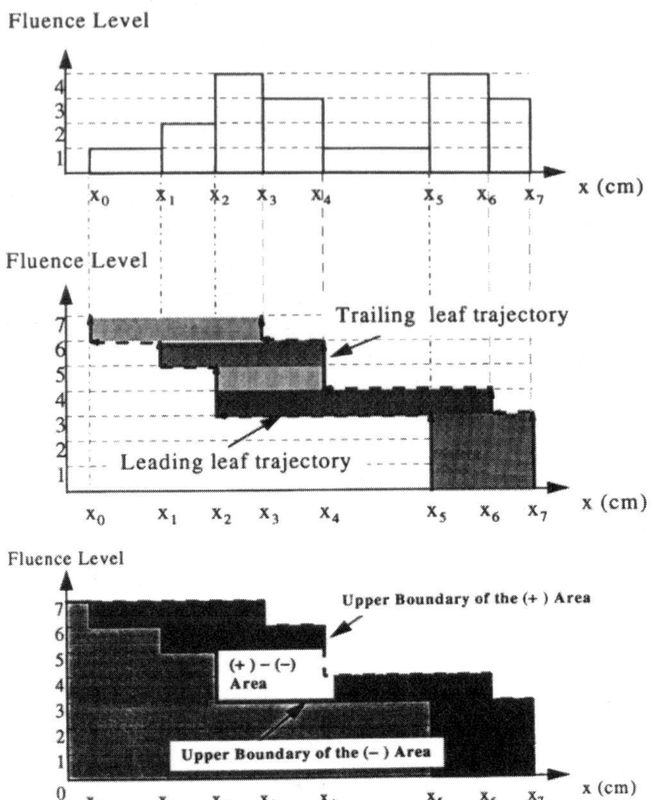

Figure 3.45. *An example of the intensity profile and leaf trajectories for the beam delivery using the algorithm of Ma et al (1998a). The upper diagram shows the desired intensity modulation and the lower diagram shows the trajectories of the leading and trailing leaves. The coordinate system has been reversed between the specification of the equations and the construction of the trajectory diagram (Ma 1998f) (From Ma et al 1998a.)*

is the trajectory of the trailing leaf and

$$I_j^{(-)} = \sum_{l=1}^{j^-} a_l^{(-)} \tag{3.23}$$

is the trajectory of the leading leaf where j^+ and j^- are the total number of rising and falling a_k or a_l values respectively up to point j such that $j = j^+ + j^-$. These equations completely specify the trajectories of both leaves. If $x_k^{(+)}$ defines the locations where the fluence is rising and $x_k^{(-)}$ defines the locations where the fluence is falling, then the coordinates $(x_k^{(+)}, I_k^{(+)})$ define the trailing leaf trajectory and $(x_l^{(-)}, I_l^{(-)})$ define the leading leaf trajectory. Figure 3.46 shows a numerical example.

Ma *et al* (1998a) have shown that this algorithm: (i) avoids any leaf collision between opposing pairs; (ii) has the optimum MU-efficiency; (iii) is equivalent to the optimized velocity-modulation method; and (iv) can be generalized from 'step-and-shoot' to the DMLC technique. They show a nice numerical example which shows that, despite the moderately complex mathematics, the visualization of the outcome is straightforward (figures 3.45 and 3.46). (It should be noted on reading this paper that the coordinate system has been reversed between the specification of the equations and the construction of figure 3.45 (Boyer 1998b)). In fact, the leaf trajectories in 'step-and-shoot mode' look exactly the same as the multiple-static-fields delivered via the Bortfeld–Boyer technique. This is not surprising because the continuous case is really no more than an extension of the discrete case with small intervals. For reference, note that the authors have agreed (Ma 1998f) that there are a lot of algebraic errors in this paper which confuse the understanding, albeit that the algorithm is correctly stated.

Ma *et al* (1999a) have developed their algorithm further to generate the leaf movement patterns which lead to a minimum beam delivery time. Like Budgell *et al* (1998b) they call this algorithm 'synchronization' since all leaves start and finish together when a 2D IMB is delivered. It is important to note that this is a quite different use of the same word from the mathematical technique which generates the patterns which remove the tongue-and-groove underdose. In fact, Ma *et al* (1999a) state at the end of their paper that the tongue-and-groove underdose is specifically *not* removed by their method. This is not surprising because we know that the synchronization solution has to lead to an increased delivery time and this is specifically being disallowed in the algorithm by Ma *et al* (1999a). The goal of the algorithm was to minimize the subfield variations of a leaf sequence while keeping the total monitor units and beam delivery time minimum for all leaf-setting sequences. The total monitor units for delivering the two-dimensional leaf-setting sequence is minimized once each individual MLC leaf pair is adjusted to the largest value of its individual monitor units. The derivation begins as follows (from Ma *et al* 1999a).

The total intensity 'volume' V is given by

$$V = \sum_{i=1}^{L} \omega \int_{x_0}^{x_N} I_i\,(x)\,\mathrm{d}x \qquad (3.24)$$

where $I_i\,(x)$ is the intensity value at the position x_j along the midwidth of the ith MLC leaf pair, ω is the leaf width, L is the total MLC leaf pair number, x_0 and x_N are the starting and ending positions for all leaf pairs (let x_0 be *high* and x_N

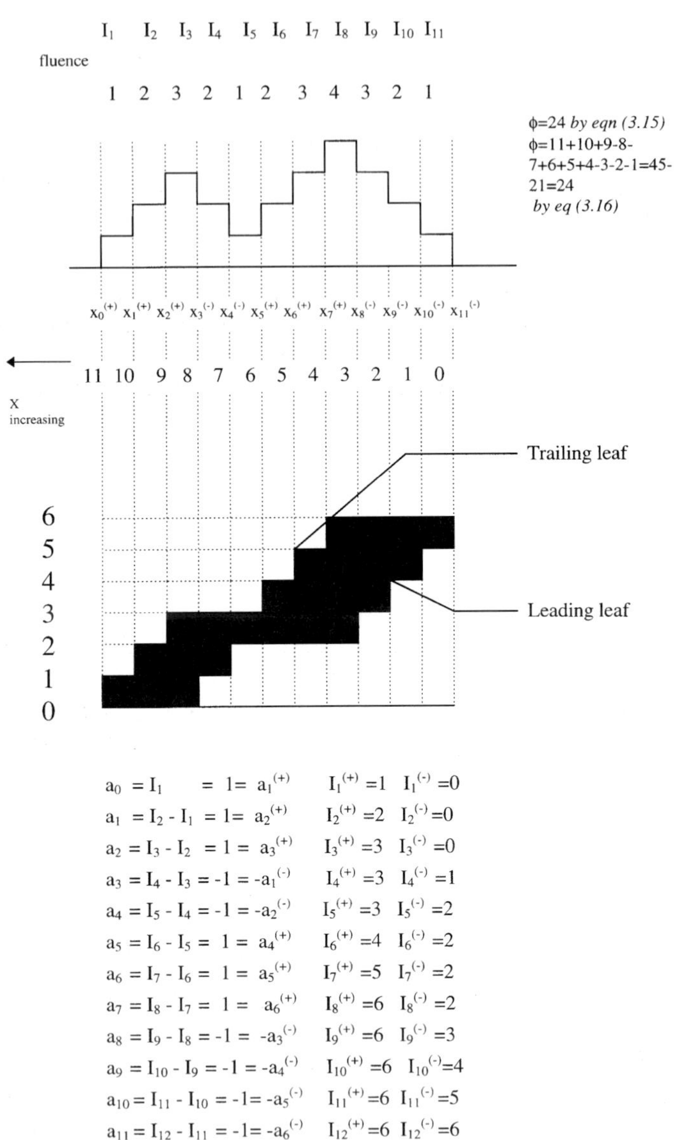

Figure 3.46. *This figure also illustrates the technique of Ma et al (1998a). It may be readily appreciated that the data shown are derived from equations (3.18) to (3.23). The leaf trajectory is shown corresponding to the fluence pattern desired. In this figure the leaves move from left to right, the opposite convention to stated in the paper but consistent with the equations.*

be low). If x is discretized into N intervals then we have

$$V = \sum_{i=1}^{L} \sum_{j=1}^{N} \omega I_{ij} \times \left(x_{j-1} - x_j\right) \qquad (3.25)$$

where I_{ij} is the discrete intensity matrix. Now, Ma *et al* (1999a) put this in matrix notation. They introduce matrix operator W of N rows and $(N+1)$ columns where the elements are

$$W_{ij} = \delta_{ij} - \delta_{i+1,j} = \delta_{ij} - \delta_{i,j-1} \qquad (3.26)$$

and δ is the Dirac symbol. W is a matrix with 1s down the main diagonal and -1s down the diagonal one above this and 0s everywhere else. Its inverse W^{-1} has 1s in the upper right triangular space down to and including the main diagonal and 0s everywhere else. With vector x being $(x_0, x_1, \ldots, x_N)^{\mathrm{T}}$ and vector y being $(\omega, \omega, \ldots, \omega)^{\mathrm{T}}$ (T representing transpose) one can write

$$V = y^{\mathrm{T}} I \left(Wx\right) = y^{\mathrm{T}} \left(IW\right) x = y^{\mathrm{T}} Ax \qquad (3.27)$$

where the coefficient matrix A is defined as

$$A = IW \qquad (3.28)$$

and contains the *changes* in intensity along each profile. The elements of I are always positive whereas those of A can be positive and negative. The leaf-setting sequence is thus defined from the matrix A rather than I (reminiscent of the 1D analysis above). Matrix A is decomposed into two other matrices via

$$A = A^{+} - A^{-} \qquad (3.29)$$

where the elements are specified by $A^{+}_{ij} = A_{ij}$ if A_{ij} is positive, and zero otherwise. Similarly, $A^{-}_{ij} = -A_{ij}$ if A_{ij} is negative, and zero otherwise. Thus, we may see that

$$V = y^{\mathrm{T}} Ax = y^{\mathrm{T}} \left(A^{+} - A^{-}\right) x = y^{\mathrm{T}} A^{+} x - y^{\mathrm{T}} A^{-} x. \qquad (3.30)$$

Ma *et al* (1999a) now observe that the accumulated intensity volume is the difference between a positive integer calculated from A^{+} in the first term and a positive integer calculated from A^{-} in the second term. They state that intuitively the first term could represent the volume under a trailing leaf set and the second term a volume under a leading leaf set. Mathematically, the leaf-setting algorithm determines I_{T} and I_{L} at position x where both I_{T} and I_{L} are positive and monotonically increasing, i.e.

$$I = I_{\mathrm{T}} - I_{\mathrm{L}}. \qquad (3.31)$$

Now by inverting equation (3.28) we have

$$I = AW^{-1} = \left(A^{+} - A^{-}\right) W^{-1} = A^{+} W^{-1} - A^{-} W^{-1} \qquad (3.32)$$

so that the trajectories are

$$I_{\mathrm{T}} = A^{+}W^{-1} \tag{3.33}$$

and

$$I_{\mathrm{L}} = A^{-}W^{-1}. \tag{3.34}$$

Matrix operator W^{-1} simply sums the elements of each row of A^{+} or A^{-} so the row elements of I_{T} and I_{L} are monotonically increasing and are the monitor units for each subfield of the leaf-setting sequence. Ma *et al* (1999a) show that this leaf-setting sequence *generates the minimum beam delivery time.* Finally, they specify how to obtain the leaf shapes. The shape of a subfield at the instant when the accumulated monitor units is m is specified from the expression

$$O\,(m) = (I_{\mathrm{T}} - m\Lambda)^{(+)} - (I_{\mathrm{L}} - m\Lambda)^{(+)} \tag{3.35}$$

where $O\,(m)$ is the shape of the open subfield that is mapped by all the positive +1 elements in the matrix. Λ is a unit matrix whose elements are all 1. The symbol (+) denotes the act of replacing all positive elements of the bracket matrix with 1 and all other elements with 0. The trailing leaf positions are specified by the junction between 0s and 1s in the first term in equation (3.35) while the leading leaf positions are specified by the junction between 0s and 1s in the second term of equation (3.35). This completes the specification of the *unsynchronized* 2D leaf-setting algorithm.

The next task is to create the leaf synchronization. This is more complicated and here we give just the essence. An arbitrary matrix S is introduced so that

$$I = I_{\mathrm{T}} - I_{\mathrm{L}} = (I_{\mathrm{T}} + S) - (I_{\mathrm{L}} + S) = I'_{\mathrm{T}} - I'_{\mathrm{L}}. \tag{3.36}$$

Clearly the primed alternative leaf-setting trajectories generate the same 2D IMB. The purpose of leaf synchronization is to find a matrix S so that the monitor units for the primed leaf trajectories are the same for all leaf pairs. This becomes an optimization problem and Ma *et al* (1999a) show how to minimize the change in the areas of each subfield for each segment of the trajectory. The algorithm has been implemented for a Varian accelerator and MLC. However, it does *not* provide a solution for the Elekta MLC nor for the Siemens MLC because these have additional constraints on the leaf movements (no interdigitation) and these have as yet *not* been accounted for in the algorithm. Neither does the algorithm take account of dosimetry—it does not remove the tongue-and-groove underdose. Ma *et al* (1999a) demonstrated, using a 2D IMB generated from the CORVUS inverse-planning system that both the unsynchronized and the synchronized algorithm replicated the required 2D IMB fluence.

The formalism (sign convention) in the paper of Ma *et al* (1999a) is incorrect (although the concepts *are* correct of course) and has been written correctly here. Now we give an example using the convention here.

Suppose there are just $L = 2$ leaf pairs and $N = 2$. Then, equation (3.25) gives

$$V = \sum_{i=1}^{2} \sum_{j=1}^{2} \omega I_{ij} \times \left(x_{j-1} - x_j\right). \tag{3.37}$$

So

$$V = \omega \times \left[I_{11}\left(x_0 - x_1\right) + I_{12}\left(x_1 - x_2\right) + I_{21}\left(x_0 - x_1\right) + I_{22}\left(x_1 - x_2\right)\right]. \tag{3.38}$$

So in matrix form

$$V = (\omega \quad \omega) \times \begin{pmatrix} I_{11} & I_{12} \\ I_{21} & I_{22} \end{pmatrix} \times \begin{pmatrix} x_0 - x_1 \\ x_1 - x_2 \end{pmatrix}. \tag{3.39}$$

Now instead, using equation (3.27) we have

$$V = y^{\mathrm{T}} I W x = (\omega \quad \omega) \times \begin{pmatrix} I_{11} & I_{12} \\ I_{21} & I_{22} \end{pmatrix} \times \begin{pmatrix} 1 & -1 & 0 \\ 0 & 1 & -1 \end{pmatrix} \begin{pmatrix} x_0 \\ x_1 \\ x_2 \end{pmatrix} = (\omega \quad \omega) \times \begin{pmatrix} I_{11} & I_{12} \\ I_{21} & I_{22} \end{pmatrix} \times \begin{pmatrix} x_0 - x_1 \\ x_1 - x_2 \end{pmatrix} \tag{3.40}$$

which is exactly the same as equation (3.39). Then, using equation (3.28)

$$A = IW = \begin{pmatrix} I_{11} & I_{12} \\ I_{21} & I_{22} \end{pmatrix} \times \begin{pmatrix} 1 & -1 \\ 0 & 1 \end{pmatrix} = \begin{pmatrix} I_{11} & I_{12} - I_{11} \\ I_{21} & I_{22} - I_{21} \end{pmatrix}. \tag{3.41}$$

It can easily be seen that

$$AW^{-1} = \begin{pmatrix} I_{11} & I_{12} - I_{11} \\ I_{21} & I_{22} - I_{21} \end{pmatrix} \times \begin{pmatrix} 1 & 1 \\ 0 & 1 \end{pmatrix} = \begin{pmatrix} I_{11} & I_{12} \\ I_{21} & I_{22} \end{pmatrix} = I \tag{3.42}$$

but only when W^{-1} is defined so

$$W^{-1} = 1 \qquad \text{if} \qquad j \geq i \qquad \text{otherwise} \qquad 0 \tag{3.43}$$

(the opposite convention from that given in Ma *et al* 1999a).

Finally we compute field components. Suppose a particular numerical example

$$I = \begin{pmatrix} I_{11} & I_{12} \\ I_{21} & I_{22} \end{pmatrix} = \begin{pmatrix} 4 & 6 \\ 3 & 1 \end{pmatrix} \tag{3.44}$$

so from equation (3.41)

$$A = IW = \begin{pmatrix} 4 & 6 \\ 3 & 1 \end{pmatrix} \begin{pmatrix} 1 & -1 \\ 0 & 1 \end{pmatrix} = \begin{pmatrix} 4 & 2 \\ 3 & -2 \end{pmatrix}. \tag{3.45}$$

So from equation (3.29)

$$A^{+} = \begin{pmatrix} 4 & 2 \\ 3 & 0 \end{pmatrix} \tag{3.46}$$

and

$$A^{-} = \begin{pmatrix} 0 & 0 \\ 0 & 2 \end{pmatrix} \tag{3.47}$$

from which using equations (3.33) and (3.34)

$$I_{\mathrm{T}} = A^{+}W^{-1} = \begin{pmatrix} 4 & 2 \\ 3 & 0 \end{pmatrix} \times \begin{pmatrix} 1 & 1 \\ 0 & 1 \end{pmatrix} = \begin{pmatrix} 4 & 6 \\ 3 & 3 \end{pmatrix} \tag{3.48}$$

and

$$I_{\mathrm{L}} = A^{-}W^{-1} = \begin{pmatrix} 0 & 0 \\ 0 & 2 \end{pmatrix} \times \begin{pmatrix} 1 & 1 \\ 0 & 1 \end{pmatrix} = \begin{pmatrix} 0 & 0 \\ 0 & 2 \end{pmatrix}. \tag{3.49}$$

Note $I = I_{\mathrm{T}} - I_{\mathrm{L}}$ and that the row elements are monotonically increasing and give the monitor units for each subfield calculated below. From this we find from equation (3.35)

$$O\,(m = 1) = O\,(m = 2) = \begin{pmatrix} 1 & 1 \\ 1 & 0 \end{pmatrix} \tag{3.50}$$

and

$$O\,(m = 3) = \begin{pmatrix} 1 & 1 \\ 1 & 1 \end{pmatrix} \tag{3.51}$$

and

$$O\,(m = 4) = \begin{pmatrix} 1 & 1 \\ 0 & 0 \end{pmatrix} \tag{3.52}$$

and

$$O\,(m = 5) = O\,(m = 6) = \begin{pmatrix} 0 & 1 \\ 0 & 0 \end{pmatrix}. \tag{3.53}$$

Summing these six matrices recovers the matrix I, i.e.

$$I = 2 \times \begin{pmatrix} 1 & 1 \\ 1 & 0 \end{pmatrix} + \begin{pmatrix} 1 & 1 \\ 1 & 1 \end{pmatrix} + \begin{pmatrix} 1 & 1 \\ 0 & 0 \end{pmatrix} + 2 \times \begin{pmatrix} 0 & 1 \\ 0 & 0 \end{pmatrix} = \begin{pmatrix} 4 & 6 \\ 3 & 1 \end{pmatrix} \tag{3.54}$$

as required. This completes the analysis of the example.

3.2.4.6. *Other interpreters*

Agazaryan *et al* (1999) have described yet another interpreter for the DMLC technique. This one (in Los Angeles) takes its IMB input from a homemade inverse-planning system and then solves the usual problems to predict leaf positions. It will support several MLC manufacturer and linac manufacturer combinations. An iterative method corrects for leaf transmission. Synchronization is built-in. It will support fully dynamic operation and also step-and-shoot.

Agazaryan *et al* (2000a, b) have developed a leaf sequencer for a NOVALIS linear accelerator manufactured by BrainLab, equipped with an m3 microMLC. The algorithm corrects the transmission through MLC leaves by an iterative method and leakage between opposing and neighbouring leaves is also minimized.

Individual leaves are synchronized to reduce the tongue-and-groove effect. The interpreter will support a variety of IMRT delivery modes. In general, delivery is dynamic but can switch to step-and-shoot in instances where the multi-leaves cannot meet tolerance specifications. IMRT verification is performed using film dosimetry and an amorphous silicon 2D array of detectors. A special cylindrical phantom was constructed from dry water into which ISORAD-P diodes could be embedded. Inversely calculated profiles are applied to the phantom by doing forward calculations and comparing with measurements. This is analogous to the technique implemented in DKFZ Heidelberg .

Beavis *et al* (1999a) have observed that the step-and-shoot and dynamic interpreters provide different approximations to the required fluence profile. For example, a dynamic delivery can never deliver an entirely flat intensity profile whereas a series of step-and-shoot fields can. They have developed therefore a hybrid interpreter in which parts of the field are delivered using the step-and-shoot technique and other parts are delivered dynamically. They have shown that this leads to a better agreement with the required fluence profile for a number of clinical cases.

Beavis *et al* (2000) have given further details of what they call the 'slide-and-shoot' technique. They firstly explain that they have modified the Bortfeld *et al* (1994a) multiple-static-field technique to allow non-uniform spatial and fluence intervals. This alone improves the accuracy of the sequencing but as expected slightly increases the treatment time. They then combined this modified multiple-static-field technique with a dynamic delivery, provided the intensity beam does not have plateaux or points of inflection. They show that the combination of these two techniques can yield a significant increase in the agreement between a continuous fluence distribution and that delivered.

3.2.4.7. Comparison of interpreters

We have reviewed six different interpreters. They were under development at much the same time. These interpreters were initially built as standalone code but efforts are underway to build some of them into well-known commercial treatment-planning systems (TPSs). At present, these include the NOMOS CORVUS system, HELAX system, ADAC system and CMS system. This is necessary so the output of the TPS can directly communicate with the accelerator control. At the present time it is probably fair to say there is no agreement whether one interpreter is superior to the others. This is because each has been tailored to slightly different MLC environments and also because, given the implementation on different MLCs, there have been no comparative experiments. Elekta Consortium members have been making intercomparisons but as yet have not produced public results.

Geis *et al* (1999) have compared different interpreters as implemented by Siemens for the DMLC technique studying efficiency and precision.

3.2.4.8. Interpreting within the planning process

Cho *et al* (1998b) have solved the problems of the physical limitations of the DMLC technique (interdigitation, maximum leaf overtravel, minimum leaf gap, leaf maximum velocity) by including the known constraints for the Elekta DMLC *into the inverse-planning algorithm itself* rather than applying these constraints *a posteriori* to already-developed IMBs through an interpreter. One might compare this methodology with that of Gustafsson *et al* (1998) who did a similar thing for the MSF technique (see section 3.3). The philosophy behind the work of Cho *et al* is that if the constraints of interdigitation, maximum leaf overtravel and leakage are not included in the inverse planning, then IMBs can be generated which are impossible to deliver with the Elekta DMLC with maximum efficiency. If these constraints are added *a posteriori* the efficiency of the technique can suffer.

Cho and Marks (2000) gave the detailed algorithm which is based on the method of projection onto convex sets. The iterative algorithm for establishing the IMBs and the leaf trajectories alternates between dose-space and position–time-space. The so-called ‘MLC-sensitive’ optimization was engineered for the Elekta MLC. The minimum leaf gap between opposing leaves within the same channel and adjacent channels was set to 1 cm, the maximum leaf speed to 2 cm s^{-1} and the dose rate to 400 MU min^{-1}. These limitations were applied as ‘hard’ (non-violable) constraints. The minimization of tongue-and-groove underdose was conversely applied as a ‘soft constraint’. A specific multislice planning problem was artificially constructed knowing it would generate violations of the interdigitation prohibition and minimum gap constraint in a version of the code in which the MLC-constraints were switched off. Example trajectories were shown demonstrating these features. An attempt was then made to correct these imperfect IMB profiles for collision and maintain the required gap by post-optimization slowing down of the trailing leaves or speeding up of the leading leaves within the allowed limits. It was then shown that this leads to a worse conformality than if the constraints were included in the algorithm iteratively during convergence to the leaf trajectories.

It may be commented that this work assumed the collimation was by leaves alone and did not take account of leaf leakage, head scatter and penumbra. It was noted that the alternative interpreter from Convery and Webb (1998) could of course avoid the problem *a posteriori* by also using the jaws. Some comment on the possible increased radiation transmission through jaws alone was also made but Convery *et al* (2000a, b) have shown that this can be corrected also in an iterative feedback loop (see section 3.2.6.1).

Fraass *et al* (2000) have developed a technique called automated dose-based conformal (ADBC) field shaping in which the optimization included the specification of the MLC leaf positions performing the delivery. The technique was built into UMPLAN and led to fast computation and more conformal dose distributions.

Alber *et al* (2000) have also constructed an inverse-planning code called

Hyperion in which the constraints of the treatment delivery were incorporated, e.g. for the DMLC technique the maximum leaf speed constraint was included. For the compensator the maximum thickness constraint was included. These were incorporated as hard physical constraints in a modular inverse-planning system. Other constraints, such as the desirability to have smooth IMBs (see also Laub *et al* 2000a and also section 4.2.1) were incorporated as soft constraints. A feature of this modular approach was to recompute dose distributions *a posteriori* using Monte-Carlo code. Hyperion is the name of the most famous poem of the poet Friedrich Hölderlin who died in Tubingen on 7 June, 1843 and who espoused many early Greek ideals. This trend for naming planning systems follows from NOMOS CORVUS (a group of ravens) and DKFZ CONRAD (CONformal RADiotherapy and after the DKFZ physicist Konrad Preiser (hence KONRAD in Germany) who worked on the coding.

Kolmonen *et al* (1999, 2000) have developed a new approach for IMRT using step-and-shoot MLC methods. Leaf positions and irradiation times of different subfields are optimized directly without computing intensity-modulated distributions over treatment fields. This enables the implementation of MLC constraints in the optimization algorithm. The optimization algorithm was able to find a feasible treatment plan, although there was no guarantee that the globally best plan had been found.

The work of Keller-Reichenbecher *et al* (1999a) to incorporate the sequencer into the inverse-planning system KONRAD is discussed in section 4.1.6 (see also section 3.3.4 for related considerations).

3.2.5. The emulator or virtual DMLC

It is useful to be able to 'see' the expected movement of the leaves ahead of treatment. This would enable, for example, the planner to note whether unrealistically small segments arose. It could check the anti-collision algorithms in the code and also the location of the backup jaws. It could facilitate a calculation of the efficiency of the DMLC technique which is defined as the ratio of the summed open areas on a position–time graph to the total area. Both Yu *et al* (1995) and Convery (1997a, b) have created such an 'emulator' or 'virtual DMLC'. Convery's code accepts data as output by the Yu *et al* (1995) interpreter and also in the Elekta Oncology Systems' binary format. Both emulators can operate step-by-step 'freezing' the DMLC motion, or in continuous looped motion. The emulators also produce useful hard-copy (figure 3.47). The Siemens DMLC system also displays the field segments.

The emulators also allow a viewer to see the differences between the output of different interpreters on the same input IMB dataset. These differences may arise due to the different optimization strategies which have been incorporated.

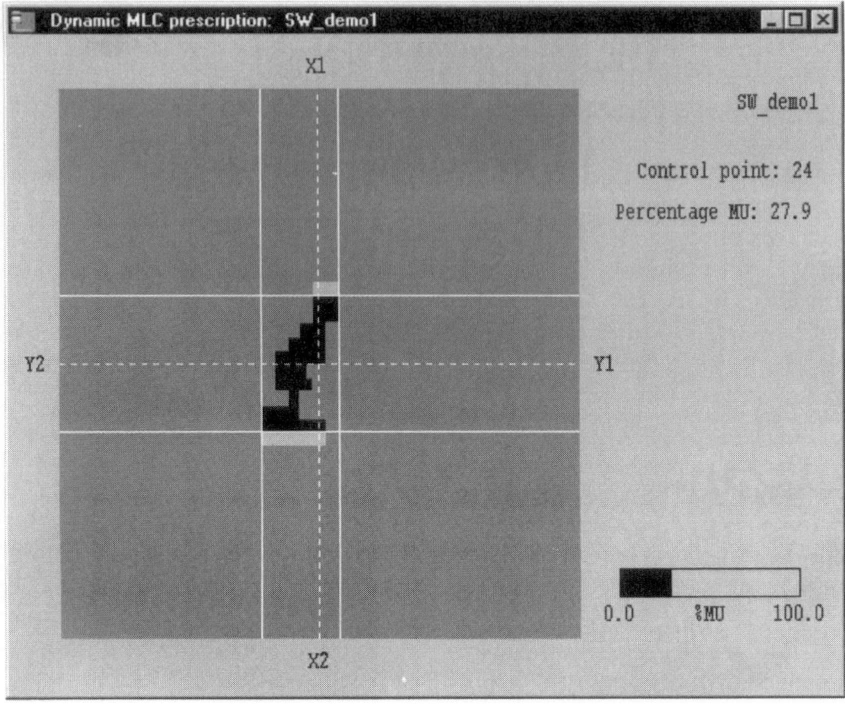

Figure 3.47. *An example of the virtual DMLC built by Convery. This shows the leaf positions at just one control point from the delivery of a 2D IMB. Dark grey indicates the presence of MLC leaves; light grey indicates jaws and black indicates the open field. The full curves indicate the position of the jaws and the dotted lines are the field central axes. (Courtesy Dr D Convery.)*

3.2.6. Modelling the dose delivered by the DMLC technique

3.2.6.1. Modelling head scatter and leakage

Most interpreters choose to determine the leaf positions corresponding to a primary fluence distribution as generated by some inverse-planning system. To model the actual dose delivered, it is necessary to consider both primary, scattered and leakage radiation (Palta 1998). Convery and Webb (1997) have modelled the head scatter as a double Gaussian and then determined the scatter at each dose-point by calculating the fraction of the head-scatter function 'seen' at that point for each leaf position and summing (figure 3.48). The model was validated by showing firstly that it generated the correct output factors for static MLC-shaped fields. Convery (1997a) highlighted the necessity to model all three contributions, primary, head-scatter and phantom-scatter components, into the calculation of dose distributions

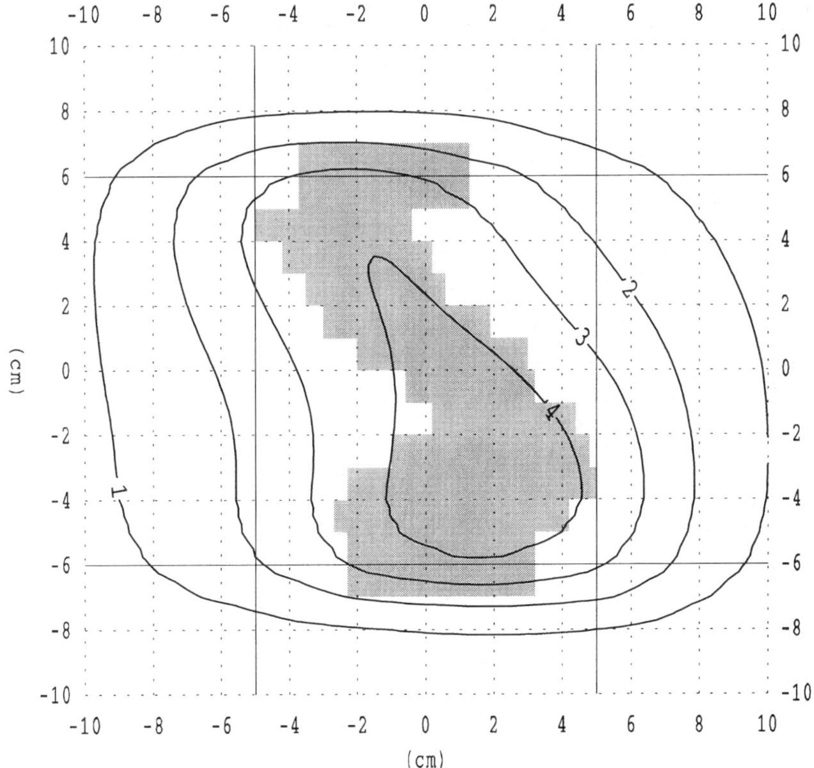

Figure 3.48. *The head-scatter intensity distribution as a percentage of the pure primary in a plane parallel to the MLC and through the isocentre. The shaded region shows the MLC aperture projected on to the isocentre plane and the thin solid lines show the positions of the backup diaphragms and main collimator jaws. The contour lines are at 1%, 2%, 3% and 4% of the primary fluence. (From Convery and Webb 1997.)*

delivered by DMLC therapy. Comparisons were presented between computed and measured dose distributions.

Convery *et al* (2000a, b) have shown that if the 'raw IMBs' from the CORVUS inverse-treatment-planning system are delivered using the Elekta DMLC technique and leaf-sequencing files from this interpreter, that the delivered dose from any particular IMB, measured using film dosimetry, does not exactly match the calculation. This is because the CORVUS system cannot know about the transmission effects and the head scatter. Convery *et al* (2000a, b) have developed a technique whereby these physical effects are built into an iterative loop to convert the raw fluence maps into modified fluence maps (and hence corresponding modified DMLC leaf sequences) such that when these are delivered, and the effects

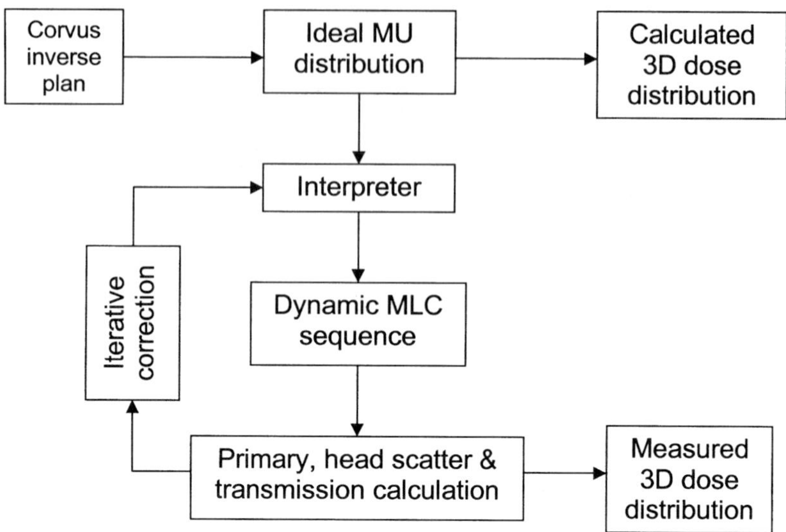

Figure 3.49. *Overview of the iterative correction procedure applied so that delivered DMLC fluences match those computed by CORVUS. (From Convery et al 2000a, b.)*

of head scatter and leaf transmission are included, the total fluence matches that computed by CORVUS. Correspondingly, the delivered 3D dose distributions then also match the predicted 3D dose distributions. The algorithm is summarized in figure 3.49, and figure 3.50 shows a comparison of the calculated and measured 3D dose distributions.

Hounsell and Wilkinson (1997a, b) and Hounsell *et al* (1999) have a similar technique (figure 3.51). Details can be found in Hounsell (1999). An exponential scatter model was used with two parameters adjusted to get the correct dose for static square fields. Then the dose from various IMRT fields was computed and shown to agree with measurements to better than 1%. These fields included the moon-shaped fields which are common in DMLC therapy. Convery and Webb (1997) have shown that the total scatter in typical DMLC therapy is no more than about 5%. So it could be argued that quite a high inaccuracy on estimating the scatter calculation is tolerable, given the small proportion of the total radiation which is due to scatter.

Saxner *et al* (1997) have much the same model for head scatter. The contribution of scattered radiation to the treatment field is computed from the area of the scattering source visible by each point in that field. Ahnesjö *et al* (1998) have shown that accurate dose and MU calculations require a 3D computation of the effect of the DMLC collimation. Siochi (1998c, 1999d) also analysed this effect. Siochi (1998b) then postulated that the scatter could be factored into the calculation by adjusting the MUs per segment to avoid the iterative looping that

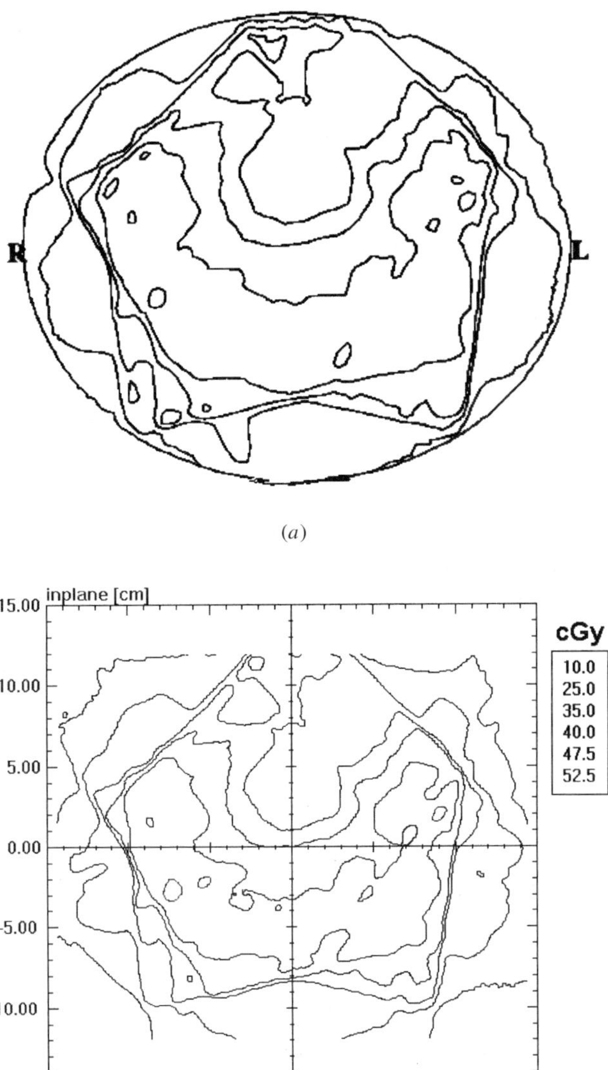

(*a*)

(*b*)

Figure 3.50. *(a) Calculated and (b) film-measured dose distributions showing the effect of including all the delivery physics in the computation. (From Convery et al 2000a, b.)*

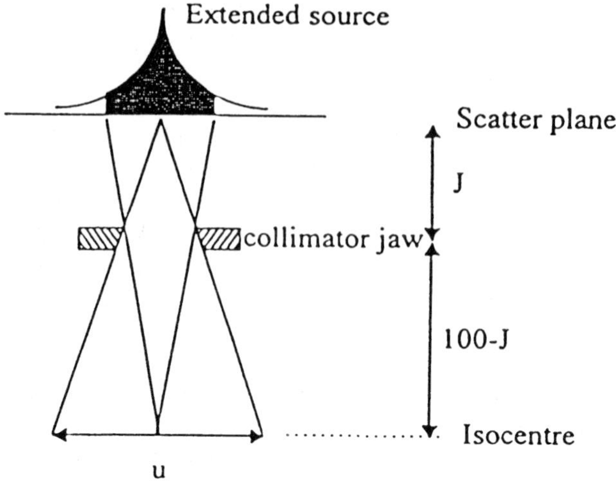

Figure 3.51. *A schematic diagram showing the relationship between the isocentric field size u and the area of an extended source of scattered radiation that is visible from the isocentre. A dose model for IMRT by the DMLC technique requires to account for the variation in scattered radiation as a function of source size. (From Hounsell and Wilkinson 1997b.)*

other authors make, given that interpreters generally operate on primary fluence only. Penumbra also must be modelled (Siochi 1999c, Kapur *et al* 1999).

Phillips *et al* (1999a) have developed a macro-pencil-beam model to calculate phantom scatter. Both head-scatter and phantom-scatter models use exponential functions fitted to a series of measurements to determine the model parameters. For two separate machines and three energies, comparisons have been made between calculated and measured dose distributions for a set of regular and irregular fields. For example, for a 20 step intensity-modulated field the accuracy was 3.4% compared with 18% with a conventional algorithm and the advantages of this model for IMRT were discussed.

Miller *et al* (1998) have also modelled the head-scatter factors of small field segments typical of those used in DMLC therapy. An extrafocal source model was used. Spirou and Chui (1998b) have shown that even when the extrafocal source was modelled, occasionally there were discrepancies between calculation and measurement for IMRT of the prostate, particularly when small-field-area segments were part of the treatment.

Arnfield *et al* (1999), Arnfield (1998) and Yu (1998b) have shown that the dose calculation for the DMLC technique must take account of the leaf transmission, rounded ends of the leaves, tongue-and-groove effect and scattering from the treatment head and MLC leaves. Yu and Sarfaez (1998a, b) have computed the output factors for field sizes of varying shapes and locations also using a scatter summation model with a virtual source created by the flattening filter. Once total fluence was computed, dose was computed by convolving a dose kernel with the TERMA. It was concluded that accounting for leaf and interleaf transmission changed the dose by 4.5%, accounting for tongue-and-groove effect changed the dose by 10.5% and accounting for the changes in output factor changed the dose by 10% relative to ignoring these effects. Siebers *et al* (2000) are developing Monte Carlo methods to compute the head scatter and MLC scatter and leakage through an MLC. At present, the study models the collimator in static mode but the concept can be extended to DMLC. It has been shown that MLC scatter cannot be ignored, particularly when the delivery efficiency is low and there is a large leakage component.

Backscatter into the monitor unit chamber also contributes to the recorded signal determining when the MUs will be deemed to have been delivered. In theory, this backscatter will vary depending on the locations of the leaves during the DMLC delivery of IMRT. This would call for a very complex separation of the effects of forward headscatter and backscatter in IMRT. Fortunately this proves to be unnecessary. Hounsell (1998b) has shown that, provided the anti-backscatter plate for the Elekta accelerator is in place, then the variation of backscatter with different delivered IMRT distributions was less than 0.3% and can be ignored in practice. This may not be the case for other accelerators.

Papatheodorou *et al* (1999) have also developed a primary scatter separation dose model to account for intensity-modulation generated by the dynamic MLC technique. Leaf penumbra was modelled by considering an extended source model with different collimation coefficients for the leaf tip and leaf side. Several test shapes were studied and measurements were made with radiographic film and ion chambers. Preliminary comparisons between calculations and measurements were satisfactory. The method uses a multiple-angular-segment technique.

Another issue is that of ensuring a sharp penumbra, first discussed by Sharpe (1997). Sharpe *et al* (2000) have created 'horny beams'. These are otherwise almost flat beams which have the additions of 'horns' at the periphery. This sharpens the 95–50% penumbra without compromising the 50–5% penumbra. Sharpe *et al* (2000) have shown that the addition of 50% fluence in a 10 mm rind is desirable. This has been implemented in the University of Washington planning system. They have delivered such beams by DMLC therapy and also by masks slightly smaller than the field shape. It has been shown that penumbra in lung has been reduced from 11 to 4 mm.

3.2.6.2. Monte-Carlo dosimetry

When the radiation passes through air cavities a discrepancy might be expected between the dose distributions computed from an inverse-planning system and the measured dose distributions. Ma *et al* (1998e) computed inverse plans using the NOMOS CORVUS system and then delivered them to a phantom constructed to simulate the patient planned. They then recomputed the dose distribution using Monte-Carlo methods based on the EGS4/BEAM code. The measurements were much closer to the Monte-Carlo calculations (about 1% different) than the CORVUS-computed distributions. This is not surprising since the Monte-Carlo calculation takes account of electron transport more accuractely. Ideally experiments should be correlated with these new calculation methods. Pawlicki *et al* (1999a) then showed that these Monte-Carlo calculated IMRT dose distributions agreed well with experiments close to tissue interfaces in phantoms whereas the dosimetry provided by the inverse-planning system, which did not have inhomogeneity corrections included, was significantly different in these regions. The experiments were done using the DMLC technique implemented on Varian 2100C and 2300D accelerators. Guerrero *et al* (1999) have reported further on the 28 patients treated with IMRT at Stanford since treatments commenced in October 1997. Many of these patients had tumours close to the spinal cord and it was shown that if the dose distributions were recalculated using the EGS4/MCDOSE Monte-Carlo system then the doses could be some 30% different from those generated by CORVUS in the regions of tissue–bone inhomogeneities. Ma *et al* (1999b) have given specific examples of the need for Monte-Carlo dosimetry in IMRT (figure 3.52). They computed a nine-field IMRT plan with CORVUS, created the leaf sequences for a Varian DMLC system and then replanned the dosimetry using the Monte-Carlo code, noting up to 10% dose discrepancies in the cord, attributed to the surrounding bone and the accumulation of small dose errors from each field when Monte-Carlo is not used (as in CORVUS).

Jiang *et al* (1999) have developed a two-stage treatment-planning process. First IMRT inverse planning was carried out with an analytical FSPB model. The fluence maps were then fed to a Monte-Carlo recalculation. If the dose distribution changed significantly (e.g. in the region of tissue inhomogeneities) then the fluences were adjusted; otherwise the results of the former calculation were accepted. They found that, for example, no changes were needed for prostate cases but changes were required for nasopharyngeal treatments.

Sixel (1999) has studied the dosimetric properties of a 3D treatment-planning system for forward planning of intensity-modulated fields. She has noted that standard tests used to evaluate a dose calculation algorithm and for conventional beam geometries would not reflect the complexity of these new techniques. Intensity-modulated fields are typically small with widths of 1–2 cm, typically off-axis and highly irregular. For accurate dose prediction, forward planning of the field segments is required. She studied the Theraplan Plus treatment-planning system from THERATRONICS under these conditions. The results showed that accuracy was highly dependent on the geometry. In particular, where electron transport issues prevailed, percentage differences in absolute dose of the order of 6% were observed. Typically, penumbra dose is also poorly predicted in regions of electronic disequilibrium. Sixel (1999) concluded that it was not possible to put these results into an iterative loop to redetermine the IMBs, because the nature of the problem leads to the conclusion that one just cannot always get the dose

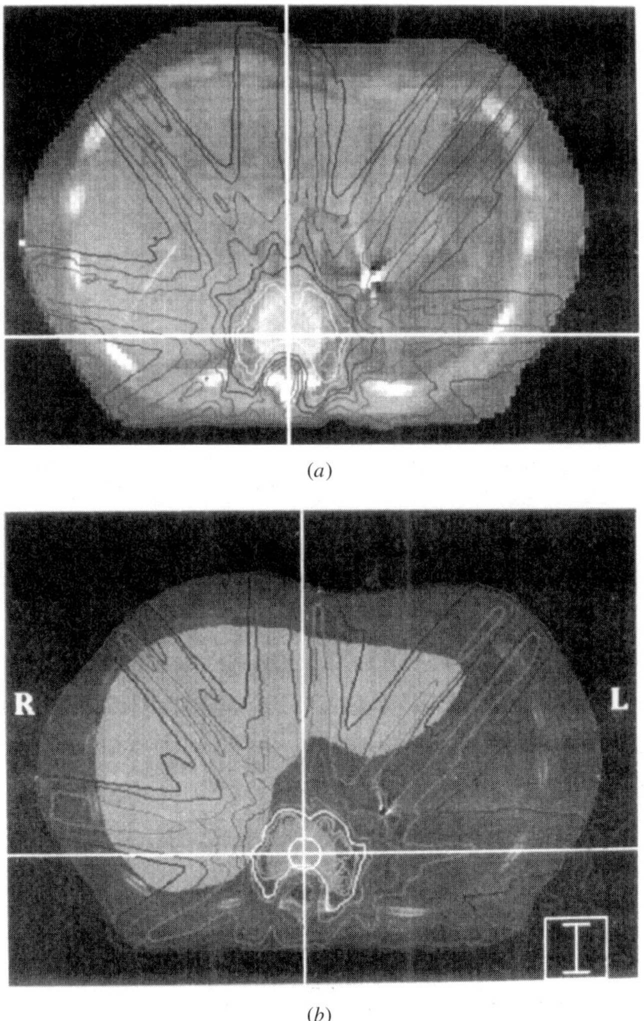

Figure 3.52. *Dose distributions for an IMRT case with nine intensity-modulated 6 MV photon fields calculated (a) by Monte-Carlo and (b) by the CORVUS inverse-planning system. The isodose lines given are 10%, 20%, 30%, 40%, 50%, 60%, 70%, 80% and 90% of the dose maximum. (From Ma et al 1999.)*

distribution that one would require. Calculations were made for a horseshoe shape with a 2 mm width and for off-axis rectangular shapes and for staircase-step shape, namely IMBs created by the superposition of multiple-static-field components.

Fujino *et al* (1999) have suggested that the PEREGRINE 3D Monte-Carlo code can be used to model the effect of the leaf motion on the DMLC dose delivery method. Specifically, they suggest using it to investigate tongue-and-groove underdose, the bullnose leaf-end effect, and dose due to scatter.

Fogg *et al* (2000) have used the MCNP4B code to create dose distributions from IMRT delivered both with multiple-static-fields and with the DMLC technique. They have shown good agreement between planes of dose and measurements on test irradiations.

3.2.7. Control of the Elekta DMLC

The Elekta protocol for controlling the DMLC technique moves the leaves and checks that at any time the leaves are not out of tolerance by more than 3 mm distance or the dose is not in error by more than 2%, whichever is the larger. If the system violates this then the pulse repetition frequency (PRF) is turned off until the criterion is satisfied. There is a special criterion for a 'move-only (zero-dose) step. Such 'interrupts' can be avoided by careful design of the interpreter. Early experience resulted in an unacceptably large number of PRF interrupts but Budgell *et al* (1997) have shown that these were reduced to zero provided the leaf velocity stayed less than 15 mm s^{-1}. The first release was called 1.1.1. For this, the labelling of the leaves is coincident with that of the jaws, i.e. X represents the direction normal to the direction of leaf travel and Y represents the leaf pairs. A negative leaf position means it is in overtravel and *vice versa*. This system is on a preclinical release to Consortium member sites. The first clinical release uses the 'step-and-shoot' approach.

The development of a commercial IMRT delivery system is a process of continuous evolution. IMRT system 3.1 corrected the fault in previous systems which led to PRF interrupts when leaves moved between a dynamic and a subsequent static position. This previous problem arose because the tolerance for leaves specified statically was tighter than for leaves in dynamic motion. Hence, as leaves moved from a dynamic situation to a static situation the tighter static tolerance would kick in to create an apparent error in leaf position and a subsequent PRF interrupt. This problem has vanished with system 3.1. System 3.1 also removes the necessity to specify where closed leaves are situated behind jaws and therefore the necessity to be concerned about whether such closed leaves are within tolerance, something which it is completely unnecessary to be concerned about. The system was successfully tested at the Royal Marsden NHS Trust and permitted a higher PRF and faster treatment times.

The new Elekta treatment delivery system is called the radiotherapy desk top (RTD). This comprises of a central patient database accessible from multiple consoles or workstations, no separate MLC interface, no error-prone re-entry of data. All the details are displayed and viewable. There are displays of the MLC leaves themselves. It is even possible to drag these leaves (with authorization) to amend a treatment field. There is a preview and then a live monitoring facility.

3.2.8. DMLC verification

This is an important issue. In the early days of IMRT development it was feared that the concept of moving components greatly jeopardized the safety of radiation therapy and hence many groups have developed ways to assure the quality of dynamic radiation therapy.

3.2.8.1. Using portal imaging devices

Partridge *et al* (1998b) have used a 2D high-sensitivity EPID to verify leaf position. This EPID (Mosleh-Shirazi *et al* 1998a, b) is capable of recording an image for each pulse (1/25 s) of an Elekta accelerator. It was shown that the 50% dose level on the measured penumbra was a good measure of the leaf position for all offsets and leaf speeds. The penumbra could also be suitably modelled by determining the fraction of the Gaussian source viewed both with static and with dynamic leaves. In fact, the EPID is not a good determiner of leaf velocity for this reason. Partridge *et al* (1998b) then created an image of the leaves set out in a comb arrangement (figure 3.53). A profile in the X-direction was sinusoidal and the peaks and troughs yielded the X-channels for each leaf pair. The position of the leaves at any instant was then found by searching for the 50% level along each of these lines. The motion of the dynamic leaves was formed into a video loop to show the quality of the movement (figure 3.54). Recent work has shown that the leaf motion can be tracked correctly even when the patient attenuation is also present, as is of course required for the method to be clinically useful. This method was used to show inaccuracies with the prototype installation and to correct them. Partridge *et al* (2000a, b) have developed a method to use the Theraview EPID to verify the DMLC leaf movement. A special dose display board was created to *image* the digital pulses at 1/64 of MU and so timelock the image to the delivered MUs. A novel louvred grid reduced optical scatter in the EPID.

James *et al* (1999, 2000) and Williams *et al* (1999a) have used the Elekta SRI-100 electronic portal imaging device to track the movement of Elekta MLC leaves during the DMLC technique. The method provides for geometric verification in real time and also *a posteriori* verification. A purpose-built data-capture device was developed to gate the video signal to monitor dose-rate signal from the linac and to trigger the EPID to produce snapshot images at equally spaced dose points during the irradiation. Real time qualitative verification of leaf position is provided during delivery by overlaying a template of the expected leaf positions onto the images as they were acquired. Subsequently, quantitative off-line verification is achieved using a maximum-gradient edge-detection algorithm applied to calibrated images to measure the individual leaf positions and compare them with the required

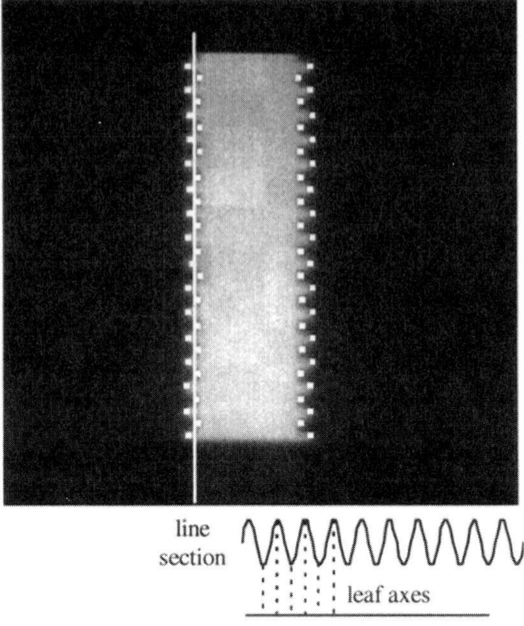

Figure 3.53. *Image of the leaves of a MLC set in a 'comb' arrangement, taken with a 2D portal imaging system. The white squares are the measured leaf positions. The vertical line shows how the leaf axes are defined. (From Partridge et al 1998b.)*

positions. The method has proved robust with respect to different dose rates, with respect to different attenuation thicknesses and with respect to the choice of maximum leaf velocity. James *et al* (2000) has provided detailed tables of the precise dependencies. Only certain combinations of image integration time and p.r.f. lead to successful verification. The off-line quantitative verification allows the measurement of leaf position to better than 1 mm even through the thickest portion of the anthropomorphic phantom and the *a posteriori* maximum intensity gradient technique works provided the maximum velocity does not exceed 8 mm s^{-1}, the pulse repetition frequency is greater than 100 min^{-1} and the image acquisition time is 140 ms. In 'dummy run' on-line verification the accelerator is operated at p.r.f. = 400 and the detector image integration time is 140 ms giving enough photons for submillimetre accuracy. In on-line clinical runs the accelerator is operated at p.r.f. = 100 and the detector integration time is 300 ms also giving submillimetre accuracy. The use of this monitoring technique increases the radiographer confidence in the DMLC IMRT technique. No clinical treatment required interruption because of discrepancies between images and templates.

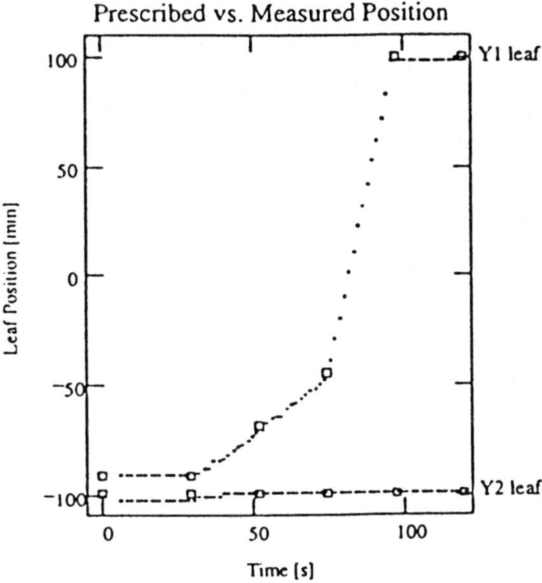

Figure 3.54. *Leaf positions for Y1 (upper/leading) and Y2 (lower/trailing) leaves measured with a 2D EPID when the leaves delivered a wedged radiation field. The positions were measured by the technique described in the text and each point shows a single measurement for a single leaf near the centre of the field. Squares show the position of the control points used to determine the delivery of a wedged distribution. (From Partridge et al 1998b.)*

Pasma *et al* (1999a, d, e) and van Esche (2000) have made a pre-treatment dosimetric verification of IMRT delivered using the DMLC technique by making use of a CCD camera-based fluoroscopic EPID system, the Elekta SRI-100. Without the patient present, EPID images are acquired for all beams which are produced with the DMLC and the images are then converted into two-dimensional dose distributions and compared with the predicted exit dose distributions. During this process, transmission through the leaves and collimator scatter are taken into account. The predicted dose profiles and the dose profiles measured with the EPID were compared at the centre of each leaf pair and predictions and measurements agreed within 2%. This verifies the leaf trajectory calculation, the correct transfer of the leaf-sequencing files to the treatment machine, and the mechanical and dosimetrical performance of the treatment unit. An example is shown in figure 3.55.

Fix *et al* (1999a, b) have used the Varian PortalVision EPID system to measure the position of the MLC leaves during DMLC delivery and also to integrate the total delivered dose. The method is similar to that used by Partridge *et al* .

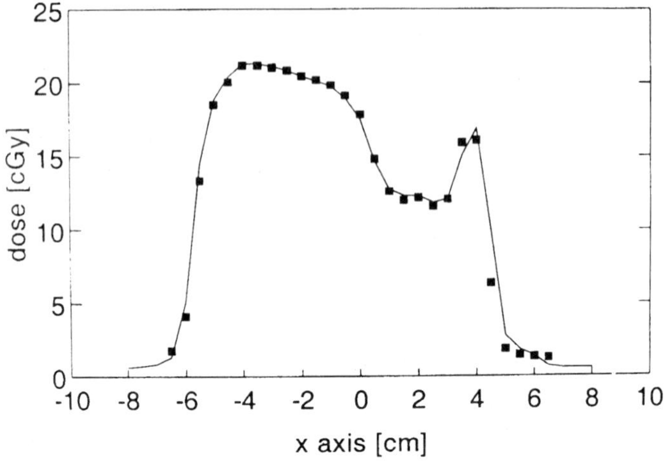

Figure 3.55. *Predicted (curves) and with the EPID measured (full squares) absolute dose profiles along a modulation line of a lateral 25 MV photon beam for a prostate cancer patient. (From Pasma et al 1999e.)*

Fix *et al* (1999b) observed that the Varian PortalVision EPID does not (unless calibrated) yield an image of fluence which precisely correlates with the TMS-computed dose distribution.

Murmann *et al* (1999) have also verified the DMLC IMRT technique using electronic portal imaging. The CADPLAN treatment-planning system was used to predict the two-dimensional intensity distribution at the plane of the EPID. To generate this reference image the total treatment field was segmented into strips of the same width as the MLC leaves and the calculation not only accounted for the movement of the leaf pairs but also took into account head scatter, leaf transmission and electron transport. Prior to patient treatment, a portal image was created of the modulated field and compared to the reference image. The technique was sufficiently accurate to verify that errors less than between 3% and 6% occurred.

Paul *et al* (1998, 1999, 2000a, b) have constructed an amorphous silicon EPID to make measurements of the fluence in the DMLC technique. The detector is 9.6×9.6 cm^2, and so is appropriate to a microMLC system. It has a spatial resolution of 2.6 mm. A key to its success is the relatively short readout time. Leaf edge positions are found by a Laplacian operator and compared with the interpreted prescription. Feedback to a gating switch (Paul *et al* 2000a, b) can account for leaf positioning errors.

Spies *et al* (1998, 2000b) and Hesse *et al* (1998a) have developed a method to deconvolve the scatter contribution to an EPID image before it is used to reconstruct MVCT images for verifying tomotherapy. Chang *et al* (1998c) have also used an EPID to verify the leaf position in the MSK DMLC application.

Convery *et al* (1998) have made use of the Wellhofer BIS system to create exit dose measurements. A similar approach is described by Chen *et al* (2000) as part of a full QA procedure for the Varian DMLC technique at Stanford.

Ploeger *et al* (2000) and Smitsmans *et al* (2000) have developed a technique to monitor leaf movement using a liquid-filled EPID. This requires monitoring the integrated MUs for each frame and taking account that the frame has a finite readout time during which the leaves move. Vetterli *et al* (2000) have used EPIDs to monitor integrated portal fluence in IMRT.

3.2.8.2. Using detectors in the blocking tray

Ma *et al* (1997) and Chen and Ma (1999) have also used an electronic imaging device to provide the quality assurance of DMLC delivery of IMBs. The technique is different from that of Partridge *et al* (1998b) and others described above in that the imager is situated in the blocking tray of the accelerator. It comprises a 0.6 mm thick gadolinium oxysulphide fluorescent phosphor screen preceded by and bonded to a 1 mm thick copper screen. A CCD video camera captures the image viewed through a 45° Mylar mirror. Images are fed into a frame grabber and the summed images are stored. This is another difference from the work of Partridge *et al* (1998b) which could record each image frame corresponding to a single pulse of radiation.

It was shown that the output of the system was linear with the number of MUs whatever the field size and so the detector acts as an integrator of beam fluence. Measurements have shown that the pixel size of the device was approximately 0.6 mm in both cartesian directions.

The device was used to record the integrated 2D fluence map resulting from the DMLC delivery of intensity-modulated radiation. Experiments determined that each MLC leaf width projected to 12.43 pixels at a distance for the detector of 74.5 cm from the source, agreeing with the mechanical calibration of 1 cm at 100 cm SSD for the Varian system. The measured integrated fluence $\Psi_k\,(x_i)$ for the kth leaf pair at position x_i along the leaf in the direction of travel was estimated by averaging the measured pixel value at x_i along a line central to the leaf projection image and the value corresponding to the same x_i but in the adjacent two lines either side.

This measured image $\Psi_k\,(x_i)$ was correlated to a theoretical image $F_k\,(x_i)$ computed from the known positions of the MLC leaves as a function of time and the correlation coefficient was used as a measure of how accurately the treatment had been delivered. The computed reference image was formed as follows.

(i) For each setting of each right and left pair labelled by k, at which an increment of fluence was delivered, the delivered incremental fluence was computed to be the convolution of the source function and the aperture function multiplied overall by an off-axis ratio. The source function was a double Gaussian as in the work of Partridge *et al* (1998b) and Convery and Webb (1997).

The parameters for the source function were chosen so that the computed penumbra function matched the measured penumbra data.

(ii) The incremental fluences were summed to give the total fluence at each x_i location in each leaf channel labelled by k.

Ma *et al* (1997) found that, for 26 experimental deliveries, the correlation between $\Psi_k(x_i)$ and $F_k(x_i)$ was always better than 95%. The quality assurance technique worked very well (figure 3.56). Xing *et al* (1999c) have reported further on the use of this entrance-portal detector to verify the leaf positions set up by the Varian DMLC technique. For nine fields created to deliver an IMRT distibution (see also section 3.3) the correlation coefficient varied between 0.96 and 0.99.

De Wagter *et al* (1997) has verified the Elekta DMLC delivery using film strapped to the accelerator head as an integrating dosemeter and also via the use of BANG gel (see section 4.4.3). Johnson *et al* (1999a) have also done this for clinical IMRT delivered with a Varian system as well as verifying dose distributions in a cubic vendor-supplied phantom.

3.2.8.3. *Other dosimetric QA techniques*

Other centres have suggested performing QA by: (i) a physical dummy run; (ii) delivering the IMBs to a water phantom and comparing with the corresponding plan (to the phantom not to the patient); and (iii) sampling a few dosepoints.

Boyer (1997) has constructed a cylindrical water phantom containing a Gadox screen viewed by a CCD camera for the purpose of verifying the positions of the leaves in DMLC therapy. Details have been reported by Boyer *et al* (1997b). The camera acquired 512 × 512 ten-bit images with an integration time of between 120 ms and 1 s. The images were processed by a frame grabber that could sum up to 2000 images acquired at a rate of eight images per second. Images of static fields and also of fields modulated by dynamic wedges were acquired over a range of dose rates. After calibration, the images had a spatial accuracy of 2 mm and agreed to within 5% of the measurements made in the interior of fields. The response of the imager has been found to agree well with the predictions of Monte-Carlo calculations of IMRT dose (Ma *et al* 1998b).

Beavis *et al* (1998b) have alternatively used a neural network to verify that simple IMBs, such as wedges, have been correctly delivered. The technique provides quality assurance for the enhanced dynamic wedge.

LoSasso *et al* (1998a) have developed an extensive QA programme for delivery of DMLC therapy in New York. In particular, they have highlighted the effect of small errors in the leaf position.

Several specific tests were described by Chui *et al* (1996) and applied to the Varian DMLC system in New York. These tests were as follows:

(i) Each pair of left and right leaves were swept across the field at a nominally constant equal speed but with a time delay on the starting time. Provided the speed is constant this generates a completely uniform intensity profile.

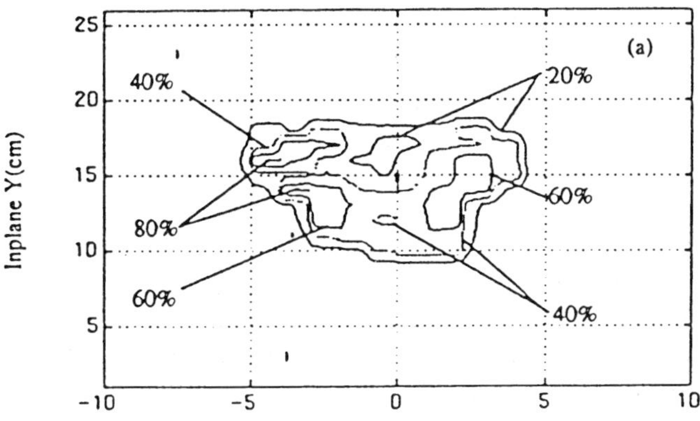

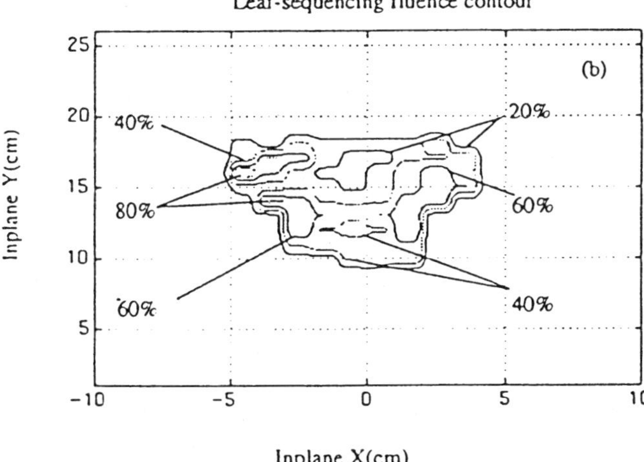

Figure 3.56. *The top part of the figure is the isovalue contour of a BIS-measured image. The lower part is the calculated reference image from the leaf-sequencing file. (From Ma et al 1997.)*

Different leaf pairs move at different speeds, as shown in figure 3.57, leading to the generation of different uniform intensities. Any departure from uniformity indicates an instability in the leaf speed. No such departure was observed for this system when profiles were recorded on film.

(ii) The effects of acceleration and deceleration of leaves were evaluated as follows. Once again, a leading and trailing leaf moved at the same speed and with a time delay nominally creating a uniform intensity. At some specified time, the beam is switched off and the leaves decelerated. Then, a time later the beam is restarted and the leaves accelerated. If the acceleration and deceleration were to be imperfect then hot or cold spots would arise. Again, for the equipment described, these were not seen, indicating no acceleration and deceleration effects.

(iii) The effects of leaf positional accuracy and round-end of leaves was assessed as follows. Again, a leading and trailing pair moved with the same speed and a time delay. At a specific location both pairs stopped with the beam on for a period of time before both recommenced. The nominal intensity profile is again uniform since nominally the difference between left and right leaf paths is constant despite this feature. In practice, for this equipment, hot spots occurred and were interpreted as due to the rounded leaf ends rather than positional inaccuracy. Chui *et al* (1996) recommended that leaf-end design be reconsidered (see also LoSasso *et al* 1998b).

(iv) Finally, an overall quick-look QA tool was designed. Leaf-pairs moved at the same speed but with a time delay and were then, as in test (iii), stopped for a period of beam-on time at specific locations before both were simultaneously restarted. However, a position delay was added of 1 mm between the stopping positions of left and right leaves. This leads to a deliberate hot spot. If all leaf pairs are set to behave the same way this leads to a hot stripe in the direction orthogonal to the direction of leaf motion. If, on the other hand, positional errors occur the location or thickness of the dark line on a film would move. Chui *et al* (1996) demonstrated this by deliberately introducing position errors but they did not see unwanted errors in normal practice.

LoSasso *et al* (1998b) have shown the importance of incorporating the through-MLC transmission, the transmission through the rounded leaf ends and the head scatter for small fields into the dosimetry for IMRT with the Varian DMLC system. They observed that the tolerance set on leaf position had a large effect on the treatment time but a very small effect on the dosimetry and therefore recommended a tolerance of 2 mm for which very few beam hold-offs occurred. Transmission averaged over leaf and interleaf was measured at about 2% and was factored into the dosimetry. Leaf-end transmission was measured and calculated and finally catered for by noting that the effect of the leaf-end transmission is equivalent to enlarging each subfield size by 1 mm border. Head scatter was modelled and again included in an iterative dose calculation. Chui and LoSasso (2000) have summarized the quality assurance tests for the DMLC technique.

Rhein *et al* (1999a, b) have experimentally verified the DMLC IMRT technique at DKFZ, Heidelberg using solid water phantoms for IMRT targets in the body and in the head-and-neck region. The phantoms contain films which can measure the delivered dose distributions. Calculated and measured dose distributions agreed to within 5% in most cases. But the disagreement increased to 10% when there were very spiky intensity-modulated fluences.

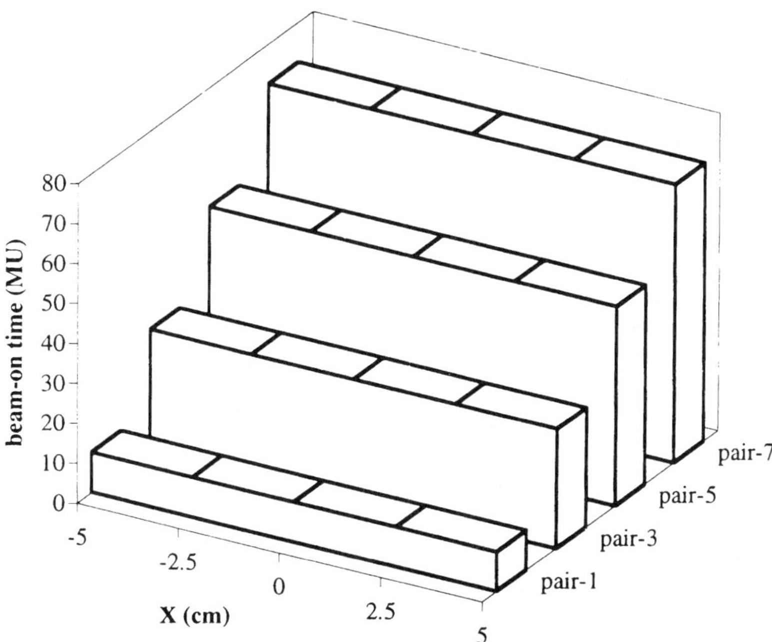

Figure 3.57. *The pattern of leaf motion to produce uniform intensity profiles. Only four pairs of leaves are shown. Leaf pair 1 moves at the lowest speed and generates the lowest uniform intensity. Leaf pair 7 moves at the highest speed and generates the highest uniform intensity. This enables the stability of the leaf movement to be investigated because any instability would lead to non-uniform intensity profiles. Also (not shown), by introducing deliberate periods in which the leaf pairs stop, either with the beam off or on, the effects of the leaf deceleration and acceleration, effect of leaf positioning accuracy and effect of leaf-end shape can be studied as described in the text. (From Chui et al 1996.)*

The dosimetric verification of IMRT at the DKFZ has been described in more detail by Rhein *et al* (1999c). When delivering IMRT using the SIMTEC option, either KONRAD or CORVUS are used to produce the intensity-modulated plan. For verification they then transfer the computed fields into a special phantom and recalculate the dose distribution for this. Recalculation has the advantage of having exactly the same conditions between calculation and verification by measurement. Special solid water phantoms for the head-and-neck region and the body have been developed with 1 mm slices between which films can be inserted.

Rhein *et al* (1999c) have analysed 32 IMRT plans delivered at the DKFZ and have shown that the deviation between the calculations and measurements evaluated for these 32 patients had a mean value of −0.9, a standard deviation of 2.17 and thus was within the ±5% tolerance specified. The number of patients rose to 77 by May 2000 (Rhein *et al* 2000). Additionally, dose profiles were measured by delivering the IMRT plan only once, but by inserting an LA48 (PTW) ion chamber array into a special IMRT verification phantom. Finally, a diamond detector was used to verify specific-point absolute dosimetry. Single-beam verification was also carried out, especially if there was some difficulty understanding the relationship between the calculated and the experimental IMRT distributions. An alternative verification technique using 'real' patient data has been described by Häring *et al* (1999b) for the DKFZ IMRT technique. The dose distribution delivered by IMRT becomes more independent of the patient contour as the number of beams increases. This well-known fact allows the transfer of IMBs created for a patient to the same beams reapplied to, for example, a cubic phantom for quality assurance. The technique of Häring *et al* (1999b) made use of a gelatin phantom built to fit a patient's specific scotch-cast head mask. When set, the gelatin phantom could accept an ion chamber and allow absolute dose calculations. Martens and de Wagter (2000) have also used the LA48 for IMRT verification.

Dirkx *et al* (1999a) have noted that when the beam intensity-modulation technique is applied to patients with lung cancer there may be dosimetric errors because tissue other than water-equivalent tissue is in the fields. An extensive dosimetric study was performed verifying the accuracy of the CADPLAN 3D treatment-planning system. Dose distributions were compared with film measurements performed in an anthropomorphic phantom and low-weight boost fields were superimposed on the most superior and most inferior part of an open field to sharpen the beam penumbra. It was found (Dirkx *et al* 1999b) necessary to increase the width and weight of these boost fields to account for the difference in density between water and lung.

Kurth *et al* (2000) have calculated IMRT doseplans with the Varian HELIOS system and delivered the modulated beams to a Rando phantom and BANG-gel. They noted agreement to within 3% between measurements, with film and gel, and calculations. Essers *et al* (2000) have performed a similar comparative study with the same conclusion regarding viability. Linthout *et al* (2000) have performed a comparative study of CORVUS-generated plans and measurements using the Elekta DMLC technique. Cozzi *et al* (2000a) have compared HELAX IMRT calculations with measurements using film, ion chamber, diamond and EPIDs for a Varian Clinac 2011 with 80 leaves.

3.2.9. *The relationship between the delivery of IMRT by the DMLC technique and by a compensator*

The DMLC technique, by which sets of leaf pairs move with differential velocity profiles, can deliver a 2D IMB. However, this same 2D IMB can be delivered by a metal compensator and indeed this is the technique which has been in practice for many decades. One might therefore reasonably ask why the new method is needed. Indeed, some (e.g. Sherouse 1997), make just this point, that the compensator delivers a 2D IMB with finer spatial resolution, without the need for QA of moving leaves, without constraints on leaf motions, without the need to interpolate the intensity-map to obtain the leaf patterns, without the need for complex dosimetry, without the need for EPI verification, without the 'tongue-and-groove' effect etc. However, conversely, the promise of the 'DMLC compensator' is that this 'software compensator' can be made quicker, can be checked more easily by a record-and-verify system, eliminates the need for blocking trays and associated mounting slots (wider access circle), requires less physical storage (computer files instead of lumps of metal), and can be changed at will if necessary between the treatment fractions as new information about the patient's tumour response becomes available. The argument of which is the better technique is set to rage for a long while. The technique of choice may well depend on local factors concerning skills and equipment availability.

The technical issue in question is whether a DMLC compensator can produce the same physical result as the metal compensator. There are also issues to explore concerning the dosimetric relationship between the techniques. As well as the work of Stein *et al* (1997a), reported earlier in sections 1.2 and 3.1, this subject has been fully explored by Geis *et al* (1996). They mimicked the modulation produced by a physical metal compensator using the Varian DMLC equipment. Film was used to integrate the dose delivered to a specified depth of compensation and a comparison of techniques was made showing the suitability of the DMLC technique. The example chosen was compensation of missing tissue in portions of a head-and-neck field.

The first step was the determination of the missing tissue values as a 2D map $L(x, y)$. This was done by taking Moiré camera images of a phantom. This is one of just many possible methods; others being the use of measuring rods, mechanical tracing machines, optical methods such as collimated light sheets, photogrammetry and the use of CT scans. The physical compensator was constructed by the 'usual' method taking into account the effective attenuation of the material and the geometric size reduction due to field divergence.

The DMLC compensator construction method is now described in more detail. Firstly, a compensation depth d_0 was chosen, greater than the deepest missing tissue deficit. Then, the residual tissue thickness $d(x, y) = d_0 - L(x, y)$ was computed for all points in the 2D field. At each point a fluence transmission factor, (FTF), was computed via

$$FTF(x, y) = TMR(S, d_0) / TMR[S, d(x, y)] \tag{3.55}$$

where S is the side of the square field equivalent to the treatment field and $TMR(S, d)$ is the tissue maximum ratio for a square field of length S and depth d. These FTF values were normalized to unity and then represented the fraction of relative fluence to be delivered along each rayline that passes through the point

(x, y) at which $L(x, y)$ was measured. The FTF of 1 corresponded to the rayline penetrating the most depth of tissue and the lowest value of FTF corresponded to where the missing tissue was greatest. The FTF values can be regarded as the 2D IMB fluence map which the DMLC technique is required to form.

The method of determining the leaf positions is by now well-known and not repeated here (Bortfeld *et al* 1994a, b) (see also review in Webb 1997d, chapter 2; see also figures 1.6 and 1.7). This creates multiple-static-fields, albeit delivered in 'DMLC mode'. For this work, the FTF was considered to be solely the difference between the time of uncovering (x, y) by the leading leaf and re-covering it by the trailing leaf. Full-depth leaf penetration was ignored. The 2D IMB was divided into fluence intervals of size $\Delta\phi$ with (NL) fluence intervals in total, each labelled by i. (NL) was the number of intervals required to deliver the 1D IMB 'track' in which the largest value of FTF occurred. Other tracks with lower intensity-maxima would require fewer fluence intervals. Each 1D IMB was delivered as a number N_s of static steps with elementary fluence $\Delta\phi$ delivered at each. The leading and trailing leaves were reset for each step to new positions with the radiation off between resettings. In this sense the delivery is a multiple-static-field delivery rather than a strict DMLC-with-beam-on delivery. It was also arranged that, of the many options possible (Webb 1998a, b), the 'leaf-sweep' option was selected. For each 1D track, the value of N_s depended on the complexity of the IMB in that track, specifically on the number of peaks and valleys. If N_i is the number of leaf-pair positions at fluence level i, then N_s is given by

$$N_s = \sum_{i=1}^{(NL)} N_i. \tag{3.56}$$

The largest value of N_s is designated $N_{s,max}$. For example, if figure 3.58 is intended to be the 1D IMB containing the maximum in the 2D field, from Geis *et al* (1996), we see that $(NL) = 8$ and $N_s = N_{s,max} = 10$. The component irradiations, each with its particular specified leaf position, were sent to the controller of the Varian MLC and the irradiation was recorded on film, which was subsequently digitized and corrected for film characteristics. Figure 3.59 shows the comparison of dose delivered by physical and DMLC compensator indicating the comparability of the two methods. Further details of the clinical use of the Varian DMLC technique at Stanford are discussed in section 4.1.2.

3.2.10. Delivery efficiency and absolute dosimetry for the DMLC technique

If a 2D IMB has any local minima the efficiency of delivering via the DMLC technique will always be less than that via a physical compensator, i.e. more MUs will be required. This is simply because the number of segments with uniform fluence interval will exceed the number of fluence intervals required to make up the maximum. Geis *et al* (1996) have defined the 'modulation scale factor' MSF as

$$MSF = N_{s,max}/(NL). \tag{3.57}$$

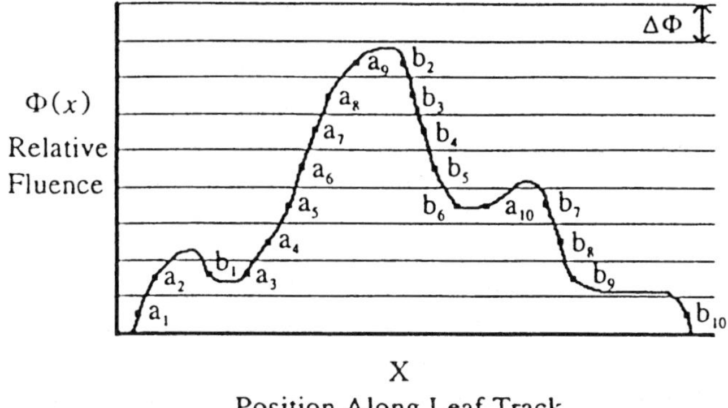

Figure 3.58. *A fluence profile divided into relative fluence increments. The full black curve represents the fluence profile as a function of position beneath a leaf pair. The dots show where the leading and trailing leaves are positioned in the leaf-setting sequence used to deliver this profile. This sequence has an efficiency of 80% compared with a physical compensator. (From Geis et al 1996.)*

This is the reciprocal of the efficiency E of the DMLC technique,

$$E = (NL)/N_{\mathrm{s,max}}. \tag{3.58}$$

MSF is also the ratio of the number of MUs required for the dynamic technique MU_{dyn} to that for the physical compensator $MU_{\mathrm{phys-comp}}$, i.e.

$$MSF = MU_{\mathrm{dyn}}/MU_{\mathrm{phys-comp}}. \tag{3.59}$$

Now, if MU_0 is the number of monitor units required to deliver a dose D_0 to the central axis of an open field of size S and at a depth d it follows that the corresponding number of monitor units for either the DMLC compensation or the physical compensation are linked through the transmission factors TF_{dyn} and $TF_{\mathrm{phys-com}}$ for the central ray in each technique via

$$MU_0 = MU_{\mathrm{dyn}}TF_{\mathrm{dyn}} = MU_{\mathrm{phys-comp}}TF_{\mathrm{phys-comp}} \tag{3.60}$$

where $TF_{\mathrm{phys-comp}}$ is simply the *FTF* though the central ray given by equation (3.55).

Hence, to determine MU_{dyn} one first determines $MU_{\mathrm{phys-comp}}$ from the second part of equation (3.60), knowing MU_0 and $TF_{\mathrm{phys-comp}}$. MU_{dyn} follows from equation (3.59) by using this value together with *MSF* from equation (3.57). Then, TF_{dyn} follows from the first part of equation (3.60).

The dose on the central axis from each technique is then $D_0 = D_{\mathrm{dyn}} = D_{\mathrm{phys-comp}}$ where (for the physical compensator)

$$D_{\mathrm{phys-comp}} = MU_{\mathrm{phys-comp}}TF_{\mathrm{phys-comp}}OF(S)FDD(S,d) \tag{3.61}$$

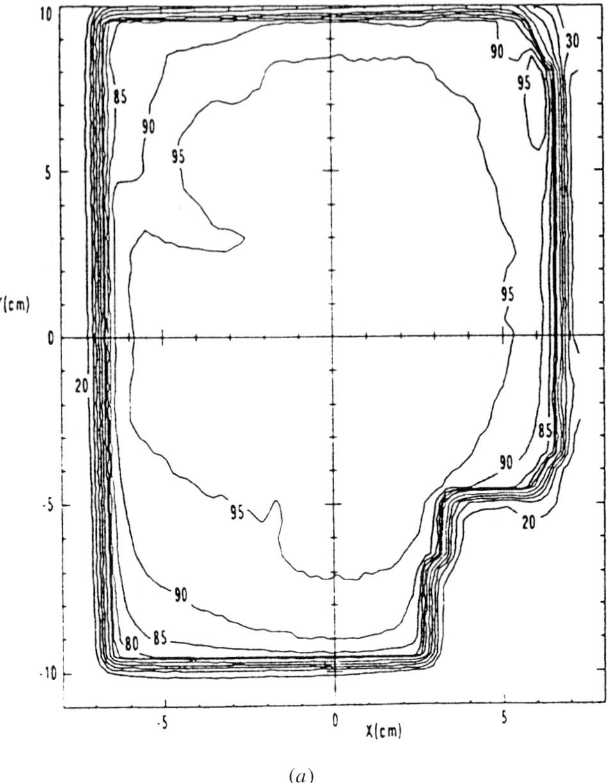

(*a*)

Figure 3.59. *Isodose lines from a film exposed beneath a wax head-phantom: (a) the beam was modulated by the DMLC technique. The dose variation is less than 10% over the whole field; (b) (page 169) the beam was modulated by a physical compensator. (From Geis et al 1996.)*

where *OF* is the output factor and *FDD* is the fractional depth dose; and (for the DMLC compensator)

$$D_{\mathrm{dyn}} = MU_{\mathrm{dyn}} TF_{\mathrm{dyn}} OF(S) FDD\,(S, d)\,. \tag{3.62}$$

Boyer *et al* (1999) have shown how to compute the absolute dosimetry for the form of the DMLC technique in which the field components are a large number of static fields delivered in quick succession. They describe the Varian implementation. As mentioned elsewhere this is strictly an MSF delivery. But because the number can be very large and the movement small between each component, the technique in the limit becomes not a step-and-shoot but a true dynamic treatment. They have addressed the question of how to compute the

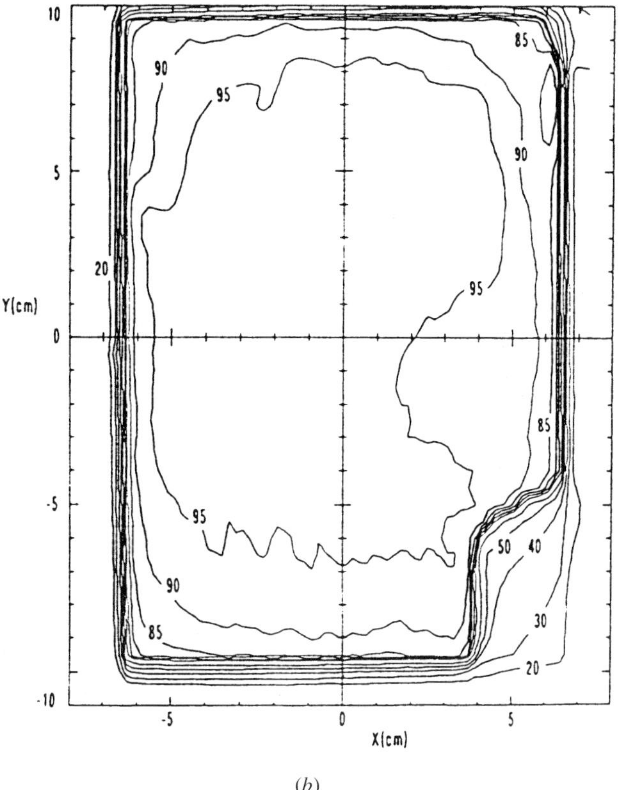

(*b*)

Figure 3.59. *(Continued.)*

absolute dose distributions, essentially a question of determining the appropriate monitor units for each IMB. A planning system, of course, returns dimensionless relative weights for each field component. They have described the method to convert from these weights to monitor units. This is a matter of choosing appropriate normalizations and determining the output factors which must be entered into the planning system (CORVUS in this case) for small field segments. The algebra, whilst appearing superficially different, embodies the same concepts as described in the appendix to Oldham and Webb (1997d). Boyer et al (1999) have also painstakingly reproduced the steps required to compute the leaf sequences via the technique of Bortfeld *et al* (1994a, b).

Boyer *et al* (1999) have considered three geometric phantoms each planned in three different ways. They have then compared the computed and measured doses at points where the gradients were small and found agreement to better than 4%. Xing *et al* (2000a) have continued this theme, presenting a closed-form equation for computing MUs in any IMRT technique in terms of the measured dose for any given open beamlet and the leaf or jaw transmission factor.

Xing *et al* (2000b, c) have amplified this theme. They have shown that the dose to a point can be decomposed into 'open' dose from open beamlets and transmission dose from closed beamlets, the former including both primary and scatter. Using the elementary beamlets as the basic building blocks they have given a closed formula for the monitor unit calculation. Kung and Chen (2000) have addressed the problem differently. They computed the CAX dose per MU. They exploited the fact that the scatter to the CAX from any bixel is dependent only on its radius from the CAX. Then, they replaced the bixels by circular annuli in which the intensity of each annulus is the circular integral of bixel intensities around the same radius. Then, they used a Clarkson integration over the annuli to obtain the required result. Huntzinger and Hunt (2000) have presented yet another method which agrees with measurements to within 3%.

3.3. DELIVERY OF IMRT BY THE MULTIPLE-STATIC-FIELD TECHNIQUE

Amongst the armamentarium of techniques for delivering IMRT is the multiple-static-field (MSF) technique whereby the IMB profile is built up from a series of multiple-static fields (Boyer and Bortfeld 1997). Between each field segment the radiation is temporarily turned off and the leaves move to new positions (see figure 1.6). Other tasks can be performed during this time, for example verification tasks. Strictly speaking, the MSF technique is not a *dynamic* method because the beam is off for the time during which the leaves move. However, if there are tens of field segments and these are automatically delivered (e.g. as in the Varian and Siemens methods) then it is convenient to consider this in the context of the DMLC method of delivering IMRT. On the other hand, the MSF method *with few segments* is clearly different and more correctly embodies the concept of the MSF method.

When originally proposed by Bortfeld *et al* (1994a, b) (see Webb 1997d, chapter 2) experimental deliveries took many hours. But recently, at least one vendor (Varian Associates) has developed the technology for rapid IMB delivery (Boyer and Bortfeld 1995) and Siemens and Elekta are also producing this capability (Bortfeld *et al* 1997b, Verhey 1997, 1998, Verhey *et al* 1998). Each field takes about 1 min and so a nine-field delivery of the kind pioneered could be completed in about 20 min. This turns out to be comparable with the time taken for a similar delivery by the NOMOS MIMiC tomotherapy device (Butler *et al* 1995). Both techniques are practicable in different ways (Carol *et al* 1992, Butler *et al* 1994, Grant *et al* 1994, Bortfeld *et al* 1994a, b) and each raises different issues. So there is a need for on-going investigation of both rather than any misunderstood competition, about which a debate has taken place (Boyer and Bortfeld 1995, Butler *et al* 1995).

3.3.1. Power-of-two fluence decomposition

The MSF method has the advantage of not requiring new FDA clearance. It is a simple extension of the basic use of the MLC. It may be less sensitive to patient movement (see also sections 4.2 and 4.4). Verhey *et al* (1997a, b, 1998) believe that simple modulation should be used where entirely adequate and that the MSF method has distinct advantages for starting work in this area of IMRT. Verhey (1997) is developing methods to improve the computational efficiency of delivery of MSFs through the use of 'power-of-two reduction' whereby the number of field components is the largest rounded off integer value of $\log_2(L)$ where there are L elemental fluence levels, e.g. fields with intensity values between 1 and 7 can be treated with three components of intensity, 4, 2 and 1 respectively (this is termed 'areal step-and-shoot') (Boyer 1998a, Xia and Verhey 1998). This turns out to be more efficient segment-wise than delivering uniform fluence intervals (the Bortfeld–Boyer method) or the method proposed by Galvin *et al* (1993). It can operate with an MLC allowing interdigitation (i.e. Varian) and also without (i.e. Siemens or Elekta). The philosophy behind the concept is that of the binary search. Verhey (1997) is also investigating whether small-area components can be discarded without seriously disturbing the dose distribution since these can have unwanted penumbral effects. The method may not however be entirely free of tongue-and-groove effects. The aim is towards using five intensity-modulated fields with ten levels in each. Indeed, Verhey *et al* (1998) have delivered IMRT by this method (SIMTEC) using as few as three to five fields with only three intensity levels.

The details of the areal step-and-shoot technique have been given by Xia and Verhey (1998). The algorithm is in two parts. In the first part, the largest intensity L is identified and the nearest rounded off integer is found

$$m = \mathrm{Int}\left[\log_2(L)\right]. \tag{3.63}$$

The 2D IMB $D(i, j)$ is then decomposed via N components $M_k(i, j)$ delivered with fluence d_k so that

$$D(i, j) = \sum_{k=1}^{N} d_k M_k(i, j). \tag{3.64}$$

The fluences d_k for the kth components ($k = 1, 2, \ldots, N$) comprise

$$2^{m-1}, 2^{m-2}, \ldots, (2^0 = 1). \tag{3.65}$$

This breaks up the delivery of any 2D IMB into N parts with the first part having roughly half the maximum fluence, the second roughly a quarter, and so on, until the last is just a unity fluence. Depending on the profile, some values of d_k are repeated with different mask patterns $M_k(i, j)$.

The second part of the algorithm is to maximize the area delivered at each component. Because the algorithm strips away the map and creates isolated islands

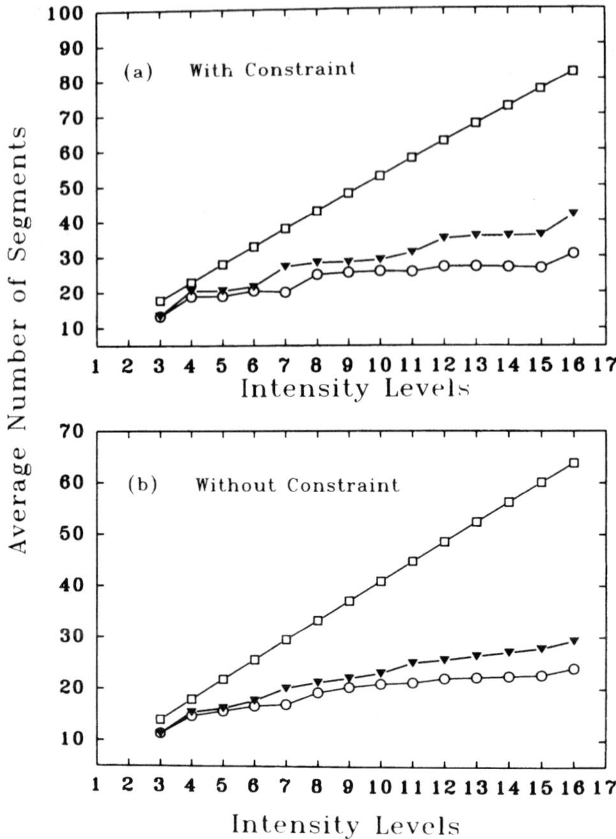

Figure 3.60. *(a) Average number of segments as a function of intensity levels for MLC systems with interleaf motion constraint, and (b) without interleaf motion constraint. The open circles represent the Xia and Verhey algorithm, the full triangles represent for the Galvin algorithm and the open squares represent for the Bortfeld–Boyer algorithm. (From Xia and Verhey 1998.)*

of residual fluence at each successive stage the number of resulting delivered components turns out to be larger than m. Also, the algorithm allows each level to be used more than once depending on the complexity of the 2D IMB. It also arises that the optimum number of fluence segments is a function of the collimator angle chosen in relation to the 2D IMB. Xia and Verhey (1998) have tested the new algorithm on 1000 random (15×15) arrays and also on several from the NOMOS CORVUS treatment-planning system. In general, the number of segments increased with the maximum intensity level but always 'won' over the number of segments required for the competing methods (figure 3.60).

However, the new algorithm was in general less MU-efficient than the Bortfeld–Boyer (1994a, b) method, the latter being linear with the maximum fluence level (figure 3.61). The goal of achieving the minimum number of segments is required because in the step-and-shoot method the treatment time is dominated by the operations which take place *between* segments.

Xia and Verhey (1998) have presented their results for the two cases of (i) allowed and (ii) disallowed interdigitation. Disallowing interdigitation increases the number of segments and decreases MU-efficiency but reduces the tongue-and-groove effect. Indeed, they prove mathematically that avoiding interdigitation is

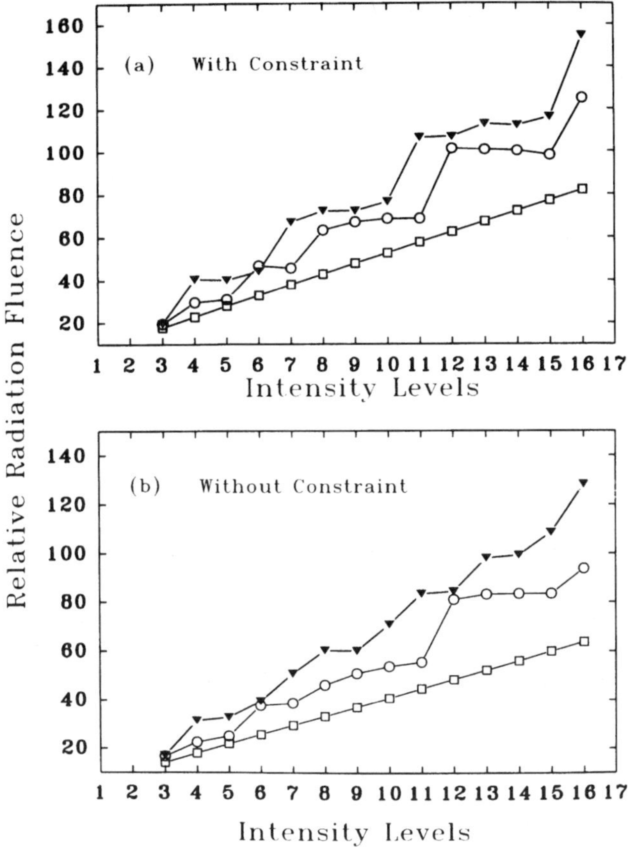

Figure 3.61. *(a) Relative energy fluences as a function of intensity levels for MLC systems with interleaf motion constraint and (b) without interleaf motion constraint. The open circles represent the Xia and Verhey algorithm, the full triangles represent the Galvin algorithm and the open squares represent the Bortfeld–Boyer algorithm. (From Xia and Verhey 1998.)*

a necessary but not sufficient condition to minimize the tongue-and-groove effect. Stein (1997) has also shown this (but that the reverse *is* true—eliminating the tongue-and-groove effect eliminates interdigitation (see section 3.2.2)). As we shall see later in this section, Webb (1998b) showed that achieving precisely zero tongue-and-groove effect may be impossible at maximum efficiency. Que (1999a) has also compared several agorithms for MLC field segmentation intended for step-and-shoot IMRT and has concluded that the method of Xia and Verhey (1998) was most efficient.

The methodology of Xia and Verhey (1998) and Que (1999a) has been extended by Que (1999b). He compared the performance of the algorithm of Xia and Verhey (1998) with that of Galvin *et al* (1993), Bortfeld *et al* (1994a) and the Siemens commercial sequencer IMFAST (see section 3.2.3). Que (1999b) also extended the Xia and Verhey algorithm to create four modifications of the algorithm in which the intensity level for the first step was varied from the nearest integer log to the base 2 value (given by equation 3.63). Firstly, it was shown that when fed with 1000 randomly generated 15×15 intensity matrices, the original algorithm of Xia and Verhey (1998) produced statistically the least number of segments. This observation may be added to that of Xia and Verhey (1998) themselves that their algorithm outpaced the algorithms of Bortfeld *et al* (1994a) and Galvin *et al* (1993) on a statistical average. Que (1999b) then went on to compare all eight algorithms for ten clinical treatment cases in which IMRT required a total of 41 treatment fields. Of the ten patients, three patients had brain tumours, three patients had prostate tumours and four patients had lung tumours. The interleaf restrictions for the Siemens' Primus MLC were applied to all algorithms. The observations, which follow, were striking.

(i) For a given clinical case, the difference in efficiency between different algorithms can be very substantial.
(ii) No single algorithm is the most efficient for all the cases, although that of Xia and Verhey (1998) most frequently worked best.
(iii) All eight algorithms have their turn to be the most efficient for a particular beam intensity map.
(iv) The algorithm of Bortfeld *et al* (1994a) had the least fluence for a treatment in all the cases.

It was thus concluded that it is desirable to have multiple algorithms available in a clinical treatment-planning system which will then search through these algorithms automatically and find the most efficient delivery sequence for a given treatment. It is clearly desirable to keep the number of MLC field segments to a minimum for many reasons. Minimum segments translate to minimum treatment time. More segments, in general, mean less monitor units per segment. Round-off errors can become significant when small MUs are delivered and the delivery accuracy may be less for small MU segments compared to larger MU segments (see section 3.3.3). In passing, it might be worth noting that it was observed that if the minimum intensity level in an intensity matrix is a positive integer, it is not always advisable to deliver this background intensity level first. In some circumstances, this can actually lead to an increase in the number of field segments. It is hard to find an exact algorithm to explain when this would be the case, but Que (1999b) has shown a specific example.

Boyer and Yu (1999) have given more examples of the areal step-and-shoot technique for delivering IMRT. An example is shown in figure 3.62. In this figure, the light grey areas represent the positions of the left leaves and the dark grey areas represent the position of the right leaves. The highest level in the pattern is five so the highest power-of-two contained in the levels is four. Each row in figure 3.62 is a 'step' in the first column and a 'shoot' in the second column. The first operation of the sequence moves the leaves so that they irradiate an area containing values of four or more. Then a fluence of four is delivered. The residual to be delivered is shown in the first row in the second column. In the second row there are still some areas of four or more to be treated so a second positioning of the leaves is required. In the second row a second fluence of four is delivered leaving the residual shown in the second column of the second row of figure 3.62. In the third row the next lower power-of-two, namely two, is delivered. A value of two must be delivered again in the fourth row. Boyer and Yu (1999) have shown that it takes three more patterns (rows) to irradiate all the areas containing values of one. The sequence was in fact shorter than the 1D profile step-and-shoot results for this particular intensity pattern.

In addition to the algorithm of Xia and Verhey (1998), Hamacher and Lenzen (2000) have presented a mixed integer programming approach to the problem of decomposing MLC fields for MSF IMRT avoiding interdigitation and the tongue-and-groove effect. This work is at an early stage and clinical problems have not yet been tackled.

3.3.2. Configuration options

One issue of practicability is the choice between the very large number of options for setting MSFs (figures 3.63 and 3.64). Webb (1998a) has derived the formula for computing this number for any arbitrary 1D IMB with any number of local minima and maxima, the constituents of 2D IMBs, under the arrangement whereby each field component has the same intensity interval of elemental fluence (the Bortfeld–Boyer (1994a, b) assumption) and the 1D IMB is delivered with the highest possible efficiency (given by equation (3.58)). For an IMB with M peaks of intensities $H_i, i = 1, 2, 3, \ldots, M$ and $(M - 1)$ local minima of intensities $P_i, i = 1, 2, 3, \ldots, (M - 1)$ we have that the number of physically-allowed combinations is

$$A = \frac{\Pi_{i=1}^{M} H_i!}{\Pi_{i=1}^{M-1} P_i!}. \tag{3.66}$$

'Leaf sweep' and 'close-in' are just two of many such options. Another very practicable option is the idea of delivering the first (widest) component first, with this verifiable by portal imaging, and then delivering the other component

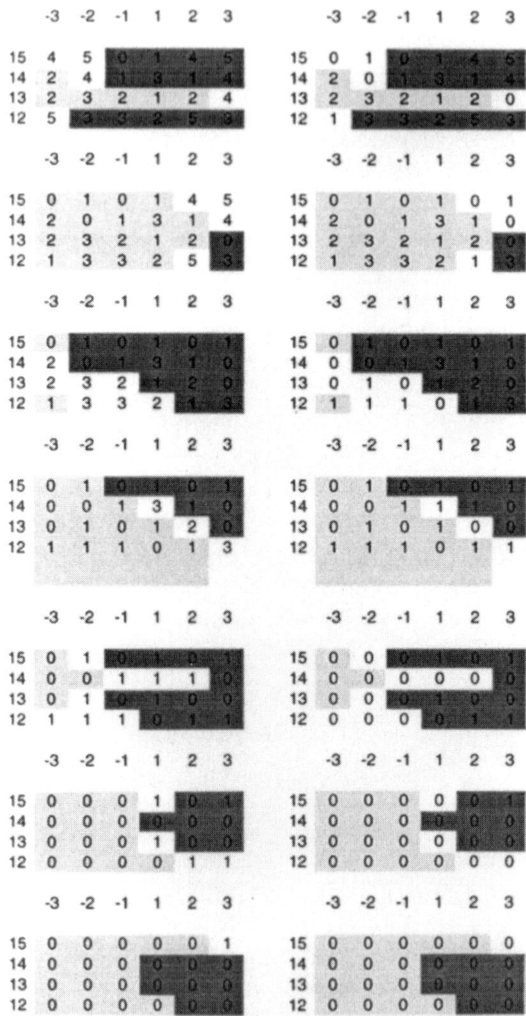

Figure 3.62. *The delivery of the intensity pattern by the areal step-and-shoot technique—see text for details. (From Boyer and Yu 1999.)*

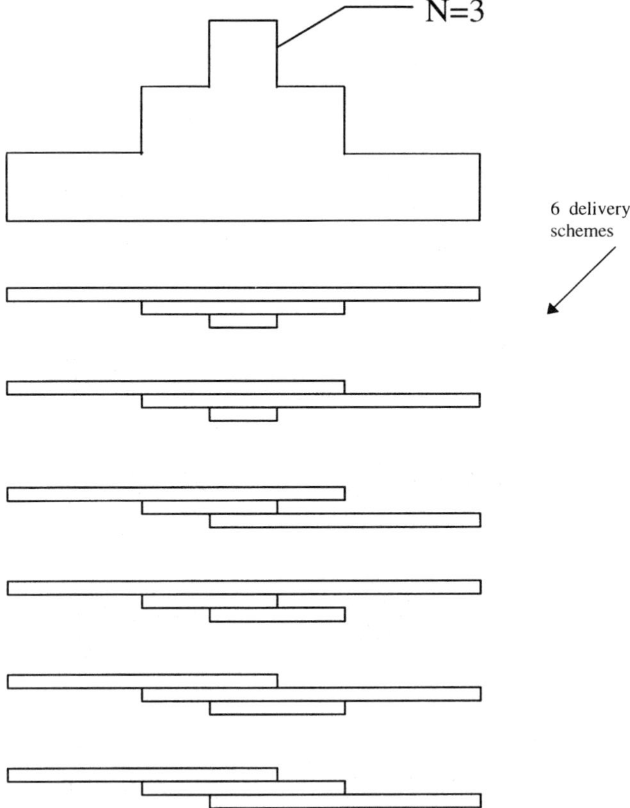

Figure 3.63. *A 1D IMB with a single peak and $N = 3$ levels. There are $3! = 6$ combinations of field-components, all with efficiency $\epsilon = 1$, which will yield this and these are shown below the profile. The uppermost combination is 'close-in' and the lowermost combination is 'leaf-sweep'.*

irradiations by a 'step-and-shoot' leaf-sweep arrangement (Webb 1998a). These configurations can sometimes be grouped together to give 2D IMBs with zero tongue-and-groove effect (Webb 1998b) (figure 3.65). This latter study showed that this is *sometimes* but not always possible. For example, 2D IMBs can be conceived which cannot be delivered with zero tongue-and-groove effect by conventional collimators with the highest possible efficiency. Indeed, the general search algorithm for all the possible permutations has also been shown to be unfeasible computationally (Webb 1998b). One particular configuration option that could be useful is to reverse the direction of leaf sequencing when applying the Bortfeld–Boyer technique. This could have advantages in removing tongue-and-groove underdose for certain IMBs. Yu (1998a) has commented on this problem

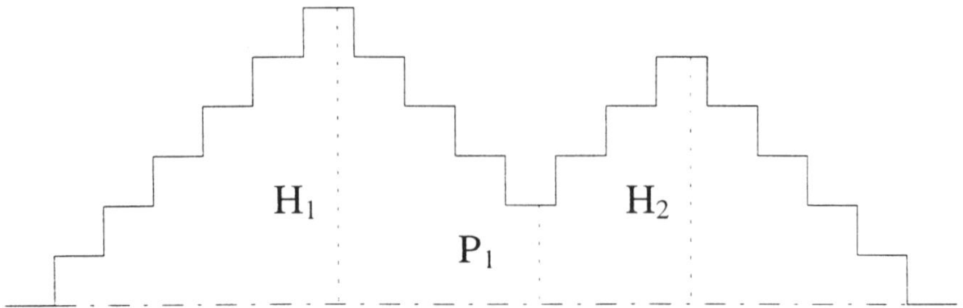

Figure 3.64. *Illustration showing a 1D IMB with two local maxima of intensity H_1 and H_2 with a local minimum of intensity P_1 between them. In this example the number of rising-fluence locations and falling-fluence locations is $N = 9$. Hence this 1D IMB can be constructed from nine 1D field-components. However there are not $9! = 362\,880$ possible configuration options because rising-fluence locations in the H_2 peak clearly cannot be paired with falling-fluence locations in the H_1 peak. There are actually $H_1!H_2!/P_1! = 6!5!/2! = 43\,200$ physically-allowed configuration options. The efficiency of delivery is $\epsilon = 6/9$.*

and proposed different leaf designs, as have Symonds-Tayler and Webb (1998). Siochi (1999a) has shown that the availability of several options to create IMBs in different ways leads to a flexibility which can be exploited, for example to minimize errors. In this work, the recommendation is that the number of IMB segments produced with the number of fields should not be more than 60.

Grosser *et al* (1999a) have developed an interpreter for MSFs which completely eliminates the tongue-and-groove effect but still leaves potential underdoses due to matching the ends of leaves (rather than the sides). They compensate for this by overrunning the leaf ends a little.

3.3.3. Delivery of small numbers of monitor units

Accelerators are usually calibrated and quality-assured for the delivery of large numbers of MUs (100 or more). The delivery of IMRT by the MSF method requires the delivery of small segments with small numbers of MUs. Hansen *et al* (1998a) have shown that the variation of output can be of the order 2% for a particular Elekta (Philips) SL25, 6MV accelerator, indicating that the machine can be used in this mode. One can see that 2% actually represents the theoretical minimum variation because each pulse delivers 1/60 MU at 400 pps and 400 MU min^{-1}. Since an *individual pulse* cannot be interrupted (since it has to be measured before a decision can be taken whether to allow it) this represents the minimum dose quantum and is roughly 2% of 1 MU. Flatness and symmetry of the fields were also within IPEM-recommended tolerance. Hansen *et al* (1998a) made the

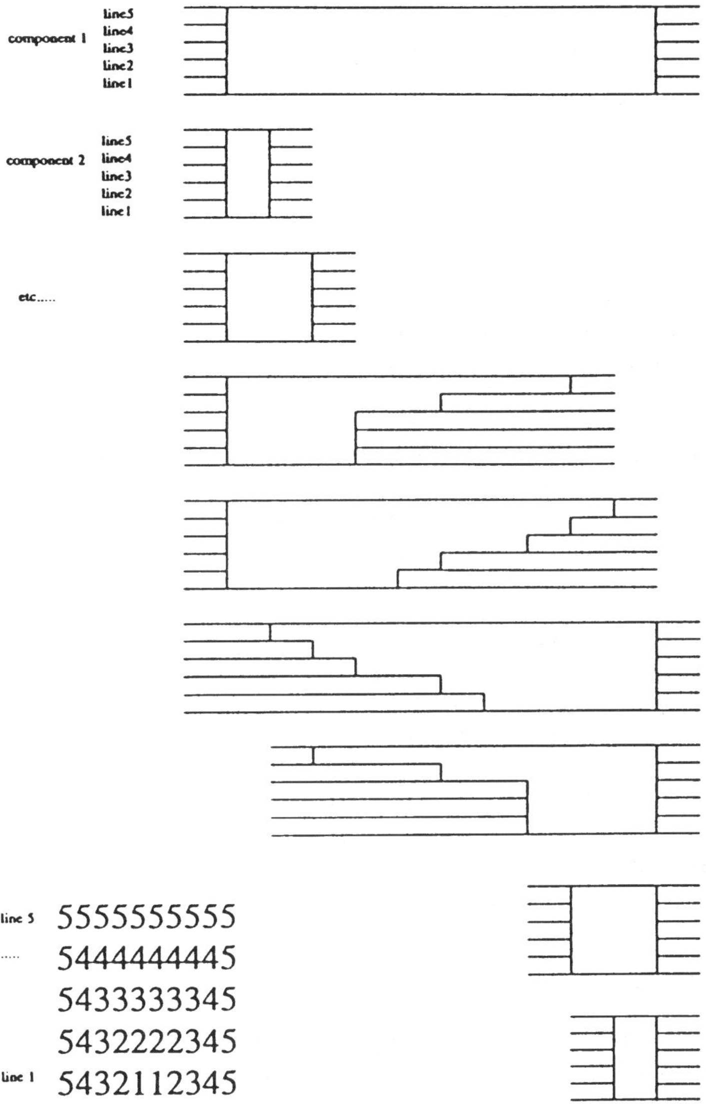

Figure 3.65. *This shows the nine component 2D IMBs required to give the inverted half-pyramid 2D fluence distribution shown at the lower left. The solution is a symmetrical 'forced baseline plus leaf sweep' solution. Each component shows the leaf positions. The shape of the field is the open portion contained by each leaf setting. There is no leaf collision and also no tongue-and-groove underdose.*

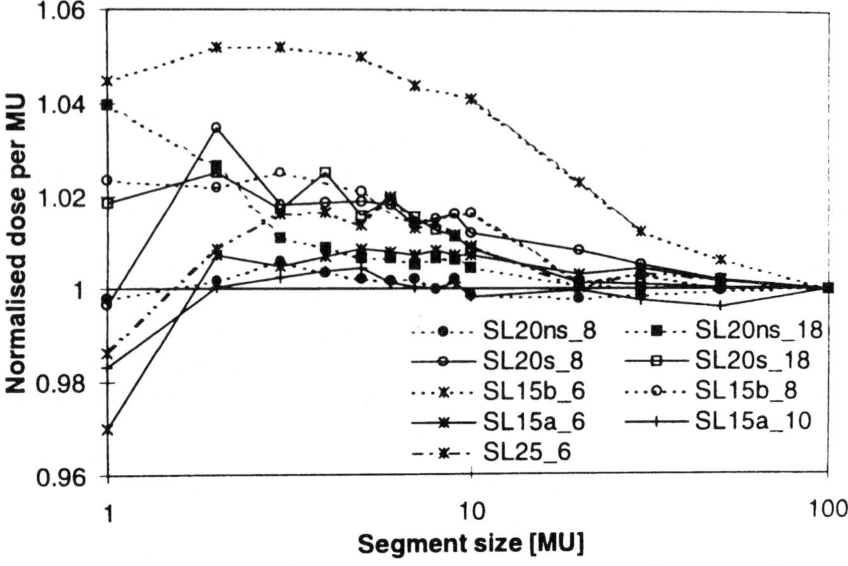

Figure 3.66. *The dose per MU as a function of the segment size, as measured using an ionization chamber. Data from five different accelerators and nine different beams are shown. The data are normalized to 1 at 100 MU. There is no correlation of different energies on the same machine, nor is there a clear correlation between the measurements of the same energy on different accelerators. The individual accelerators are identified in the paper from which this comes. (From Hansen et al 1998a.)*

same measurements on five accelerators at different energies. Four of the five had dose-per-MU accurate to within 2% but one had up to 5% variations in the dose-per-MU at low MUs (figure 3.66). They also found that the direction and magnitude of the effect was different when the pulse repetition frequency (p.r.f.) was lowered and they recommended that in some cases it may be advisable to use a low p.r.f. to obtain the more accurate delivery of small segments. The effects were stable with time and seem to have more to do with the design of particular tubes. The conclusion is that individual centres need to make these measurements on their particular accelerator and any one centre should perform the measurements approximately monthly to ensure reliability. Miller *et al* (1998) have reported similar findings with an older accelerator performing less well. Sontag *et al* (1998) have also shown that the delivery of 50 segments with 1 MU resulted in 40% greater dose than one segment of 50 MU due to the offset in the accelerator's ionization chamber which is linear with MU. Adjusting the offset reduced the error to 1%. Similar results to those of Hansen *et al* (1998a) have been observed by de Wagter *et al* (1999). Provided the elemental fluence increment was greater than 3 MU a precision of 1% was reached.

3.3.4. Incorporating MSF MLC constraints in inverse planning

The 'usual way' of invoking the MSF method is to regard the calculation of the 2D IMBs and their practical realization as separate computational problems. The realization becomes an 'interpretation' of the desired 2D IMBs and as such may introduce approximations. Gustafsson *et al* (1995, 1998) have coupled the two problems together so that the limitations on practical realization become embedded in the search for the optimum IMBs. Gustafsson *et al* (1998) have presented a technique to optimize IMRT delivered by segmented multileaf modulation. The technique uses a gradient-based algorithm which converges substantially faster than simulated annealing. Conventionally, segmented intensity modulation is obtained in a two-step process. Firstly, the IMBs are developed by inverse planning and this is followed by a translation of these beams into collimation segments. However, this second stage can compromise the modulation because constraints, such as the need to avoid overlap of opposing collimator leaves and other collimator limits, can constrain the modulation.

Hence, for a clinical treatment-planning system, Gustafsson *et al* (1998) have compared this two-stage process with a one-stage process in which the constraints on the MLC are built into the iterative calculation of the IMBs. In principle, this guarantees the realization of a nearly optimal solution to the optimization problem. They made use of ten intensity levels and the Bortfeld *et al* (1994a, b) algorithm for setting the leaves. They studied one case of optimizing IMRT of the prostate and performed this optimization both ways, i.e. either by the two-step process or by the single-step process. The results showed for this particular case that the method where MLC segments are generated in every iteration was not better than the conventional method where MLC segments were obtained after complete intensity-modulation optimization. In principle, there was no difference in delivery complexity between the two methods. In practice, the one-stage technique actually led to slightly worse dose homogeneity in the target in order to fulfil the dose-volume constraints in OARs. This technique is incorporated in the HELAX inverse-planning module. One might compare this methodology with that of Cho *et al* (1998b) who did a similar thing for the DMLC technique (see section 3.2.4).

Kolmonen *et al* (2000) have presented a direct technique to produce by inverse planning a series of MSFs to deliver with an MLC. The technique does not create an IMB followed by an interpreter, but the full physics of the component deliveries is included within the inverse-planning calculation. The results for one plan of a prostate patient were presented. There were five equiangled coplanar fields and initially three segments per field but when small subfields, which had too small an irradiation width were removed, this number decreased to nine. The inverse-planning technique was the gradient projection method.

The work of Keller-Reichenbecher *et al* (1999a) to incorporate the sequencer into the inverse-planning system KONRAD is discussed in section 4.1.6.

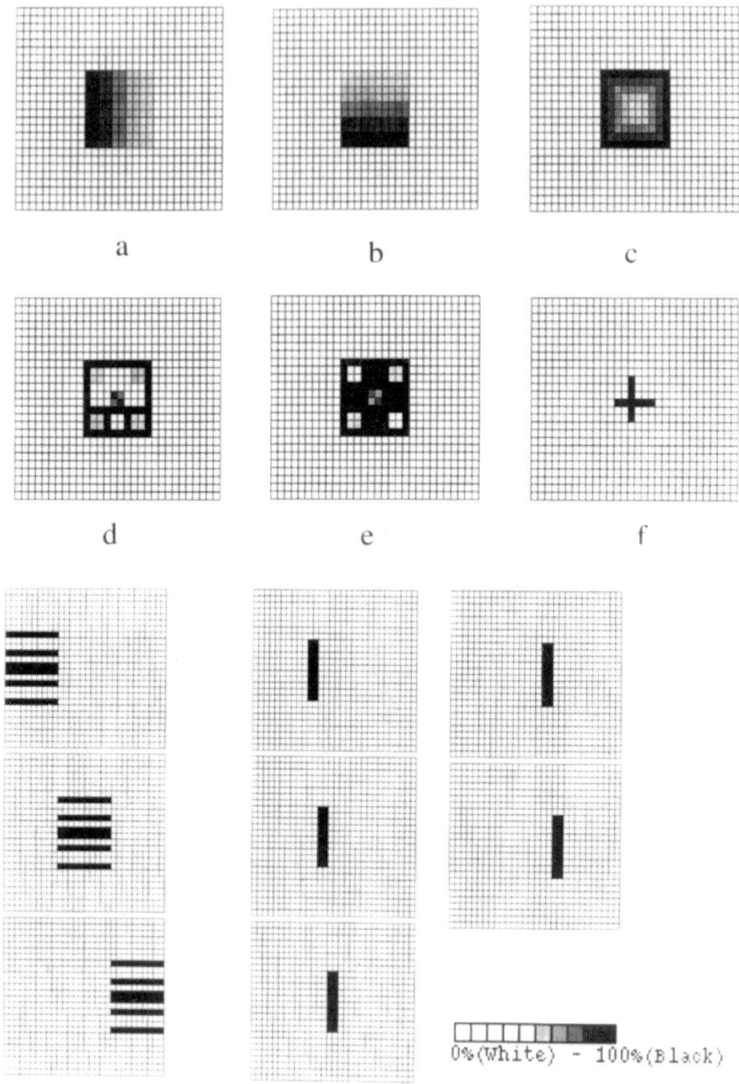

Figure 3.67. *Intensity-modulated beam patterns designed for testing the dose calculation algorithm in CORVUS and the dynamic MLC delivery of a Varian system. (From Xing et al 1999d.)*

3.3.5. Dosimetric studies

Xing *et al* (1998a) and Boyer *et al* (1998a) have shown that IMRT delivered by the MSF method delivers dose distributions which are in agreement with those computed by the NOMOS CORVUS treatment-planning system which can plan for multiple segmented fields. Xing *et al* (1999c) have reported the dosimetric verification of the use of the CORVUS inverse treatment-planning system with respect to delivering IMRT to phantoms using the DMLC method implemented on a Varian accelerator and thus delivering in 'step-and-shoot' mode. This study comprised a number of experiments as follows.

Firstly, the CORVUS system was used to plan a number of simple rectangular non-IMRT fields incident on a water phantom. The same fields (2×2 to 35×35 cm^2) were then delivered to a water phantom and measurements were compared with calculations. For percentage depth–dose curves and cross plane

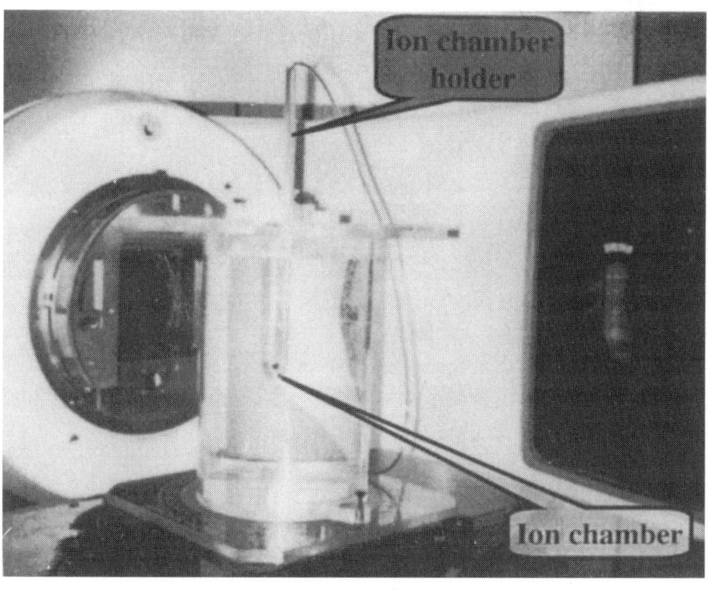

(*a*)

Figure 3.68. *(a) A cylindrical water phantom set up in front of a DMLC system on a Varian accelerator. The phantom can rotate in front of the stationary gantry to simulate gantry rotation. At each of nine 40° spaced angles, different CORVUS-computed IMBs were delivered by the DMLC technique. (b) (page 184) The nine-field plan on the central-axis plane of the cylindrical water phantom and doses at circular points which are spot measurements with an ion chamber. The discrepancy between the calculation and the measurement is shown in the figure as a %. (From Xing et al 1999d.)*

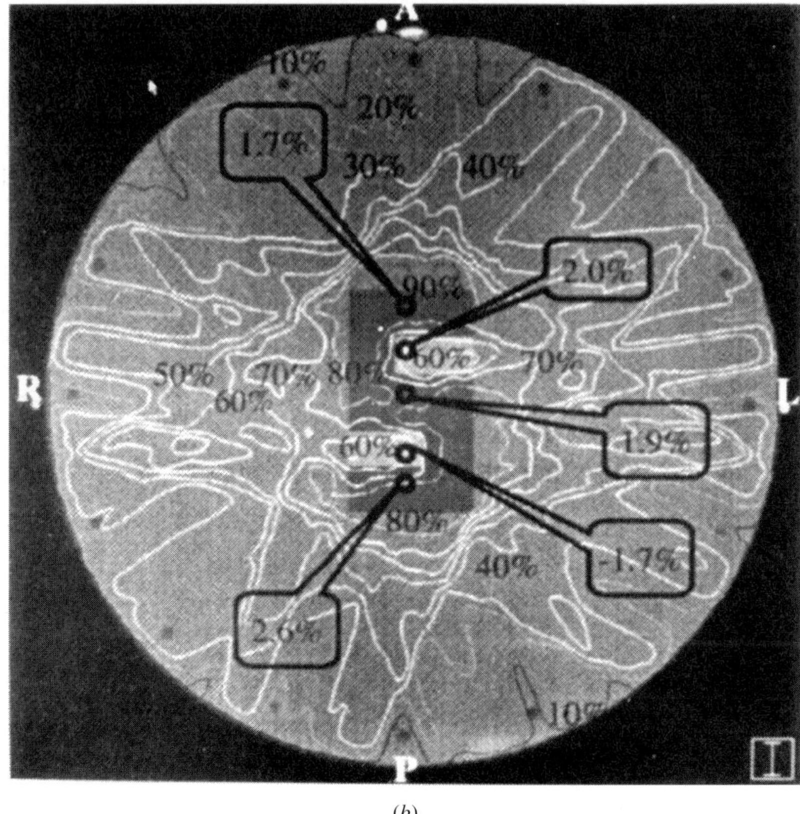

(*b*)

Figure 3.68. *(Continued.)*

plots the agreement was to within 3%, indicating the success of the finite pencil-beam algorithm in CORVUS. This experiment tested only the dose computation technique.

Secondly, Xing *et al* (1999c) created eight specially designed representative IMBs (figure 3.67) which were then delivered to a solid-water phantom containing radiographic film placed at a depth of $2d_{max}$. The dose delivered to the films was interpreted from calibrations and also recomputed by the CORVUS system. The agreement between measured and computed isodoses was less than 3 mm for all the beam patterns tested, except near regions of interleaf leakage.

Thirdly, Xing *et al* (1999c) used CORVUS to plan a variety of 3D PTVs of invaginated shapes enclosing OARs. These were an 'S' shape, a doughnut shape and a cylindrical target. Plans using one, two and nine coplanar fields were created and then the one-field and nine-field plans were delivered to a cylindrical water phantom on a turntable. The gantry was kept fixed, delivering a horizontal beam and the water phantom was rotated thus eliminating any possibility of error due to gantry sag (although this was estimated to be a negligible 1 mm). The dose was measured at a few points with an ionization chamber (figure 3.68). This procedure examined the optimization, the dose calculation algorithm *and* the delivery technique, providing the most thorough test of the IMRT system. Measurements and calculations agreed to within 4%. It was commented that this close agreement had much to do with the careful inclusion of leaf transmission in the computation of MUs. Finally, the same plans were delivered to a cubic phantom containing films. The dose distributions were recomputed for comparison with those determined from scanned films. The 90% isodose contours agreed to within 3 mm and the 60% to within 4 mm. It was thus concluded that the CORVUS inverse-planning system linked to the Varian DMLC technique could deliver accurate IMRT.

3.3.6. Flagpoles

De Neve (1999) has introduced the problem: 'What can we do about flagpoles?' Flagpoles are the name given to unwanted regions of open field irradiation when an MLC segment exists in just one quadrant of the beam's-eye view of the multileaf. In this case, for an Elekta MLC, it is not possible to bring the X-jaw over mid-line so as to completely collimate the field in addition to the multileaves. The problem is illustrated in figure 3.69 and solutions are indicated in figures 3.70, 3.71 and 3.72. Flagpoles maybe completely eliminated by making use of a combination of MLC leaves and backup diaphragms.

3.3.7. Very-few-segment MSF IMRT

Little has been said in this chapter about the most primitive form of MSF IMRT, namely that in which there are very few segments per field. As early as 1991, Webb (1991a) introduced the concept of two-weights-per-field applied such that any BEV of the target was divided into two parts: (i) that viewing just the PTV and (ii) that viewing PTV and OAR. The two weights were optimized by inverse-planning. Elsewhere (chapter 4) are reported recent implementations of this method at the University of California, San Francisco, and the method and its variants are also implemented at the University of Michigan, Ann Arbor, at the University of Ghent and at the Royal Marsden NHS Trust (chapter 4).

3.3.8. Multiple-static-field IMRT with jaws alone

So far, we have discussed the construction of 2D IMBs using a multileaf collimator. It goes without saying that one could of course deliver a 2D IMB using pairs of jaws alone and, until recently, this has been regarded as an impractical option. Two possibilities present themselves as obvious candidate techniques: either (i) 'pick off' each bixel individually and cycle through all the bixels, or (ii) deliver each line at a time using the MLC techniques described so far. Both are very inefficient.

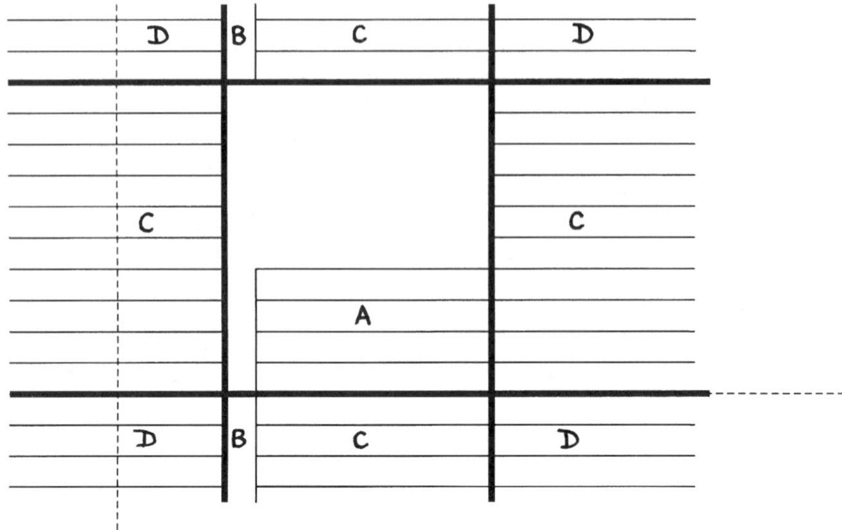

Figure 3.69. *In this figure we see an attempt to collimate a square of open-field irradiation using an ELEKTA MLC and backup diaphragms (jaws). The Y1 and Y2 jaws (vertical dark lines) can successfully collimate the left and right sides of the square. The X1 jaw (upper horizontal dark line) can successfully collimate one other side of the square. However, the X2 jaw (lower horizontal dark line) cannot cross the midline to complete the collimation (the dotted lines define the accelerator field midline). Hence, leaves have to be advanced to complete the collimation and, as can be seen, because of the Elekta constraint to not allow leaves to interdigitate, an open-field 'flagpole' is generated. Careful study of this figure will make it clear why the open-field flagpole arises. Key to figure: (A) region shielded by leaves only; (B) region collimated by an X-diaphragm only; (C) region collimated by leaves and 1 (Y) diaphragm; (D) region collimated by leaves and 2 (X and Y) diaphragms.*

Dai and Hu (1999) have presented two algorithms to deliver a 2D IMB using jaws alone and compared them with the MLC-MSF method of Xia and Verhey (1998). Perhaps the results are more encouraging than one might expect? There is no unique algorithm for delivery of a 2D IMB. Instead algorithms can be constructed around some desired constraints. The two algorithms were as follows. In both cases it was assumed that the matrices were square and the jaws moved parallel to one of the axes of the matrices.

The first algorithm searched the 2D IMB for the largest non-zero rectangular area as a segment and determined the four jaw settings. This segment, then delivered, was subtracted from the matrix to generate a new 2D matrix. These two steps were then repeated until the remaining intensity matrix was zero. This is purely heuristic.

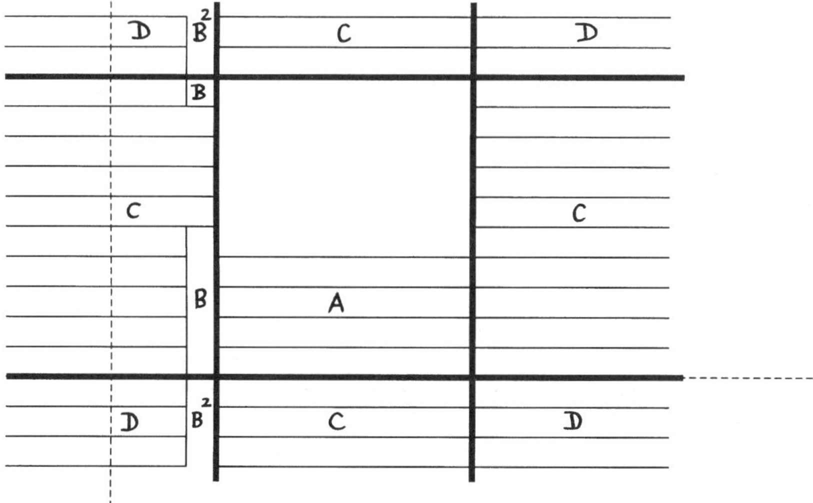

Figure 3.70. *This figure shows one possible solution to the 'flagpole problem'. Some of the leaves have been advanced a little further than is shown in figure 3.69 and their corresponding paired leaves have been retracted beneath a Y-diaphragm. However, instead of exposing an open-field 'flagpole' this now exposes a region shielded by diaphragm only. It also exposes a region shielded by the multiplicative combination of two diaphragms only (B^2). The key to this figure is the same as that for figure 3.69.*

The second algorithm searched the starting 2D IMB to determine the non-zero rectangular area (segment) that, once delivered and removed, minimized the complexity of the residual matrix. The complexity is the number of 'blocks' in the matrix, where a 'block' is defined as the largest rectangular area with elements of non-zero equal intensity level. The algorithm proceeds as follows. Firstly, the highest intensity level is found and the first segment is determined around this to minimize the complexity of the remaining matrix. This gives the first four jaw settings and fluence increment. This segment is subtracted and the new intensity matrix is searched in the same way. The process repeats until the matrix is emptied.

The outcome of both algorithms is a set of segments each labelled by four jaw coordinates and a fluence increment. A further outcome is the total number of segments and the total intensity to be delivered (the sum of the individual segment intensities). These segments are not, however, delivered in the order first found. Instead they are re-sequenced to minimize the total jaw-movement time. The total jaw-movement time is simply the sum of the maximum of the four times needed to move the four jaws between the ith and the $(i-1)$th segments. Of course, these times depend inversely on the velocity with which the jaws can move. The segments were re-sequenced using a simulated-annealing technique.

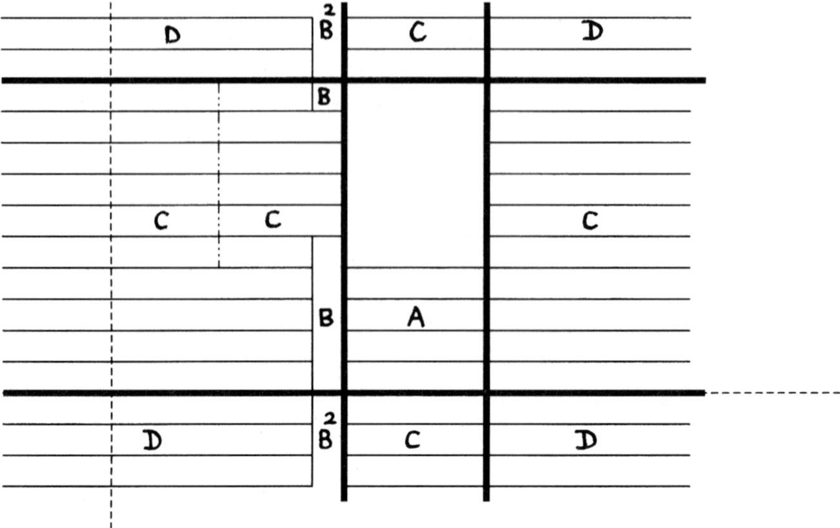

Figure 3.71. *An alternative solution using the same key. This is an attempt to irradiate the square field using two matched fields one of which is shown in figure 3.71 and the other of which is the logical inverse of this figure reflected around the central vertical axis of the square-field region. The key to this figure is the same as that for figure 3.69.*

Sections of segments were identified at each iteration and either reordered or placed elsewhere in the sequence. The new delivery time was computed. If this were an improvement, then the changed sequence was accepted. However, in order to avoid trapping in local minima some 'wrong-way' sequence changes were accepted according to the familiar rules of simulated annealing. This is analogous to the familiar travelling salesman problem.

The total beam-on time is simply the ratio of the total intensity and the machine fluence rate. The total treatment time is then the sum of the total beam-on time and the jaw-movement cumulative time. The efficiency of the technique is the ratio of the maximum intensity to the sum of the segment intensities. The reciprocal of this is the modulation scale factor. These two measures characterize the success of a technique for delivery.

Dai and Hu (1999) have given a simple numerical 4×4 example to illustrate the outcome of applying these two sequencing interpreters, and also that of Xia and Verhey (1998), which provide much insight on what arises. Firstly, it was observed that the algorithm generating the minimum number of segments does not at the same time also generate the minimum total fluence. Immediately this signals that the outcome of applying the algorithms will depend on the precise structure of the 2D IMBs (bixel values and their relative disposition) and that

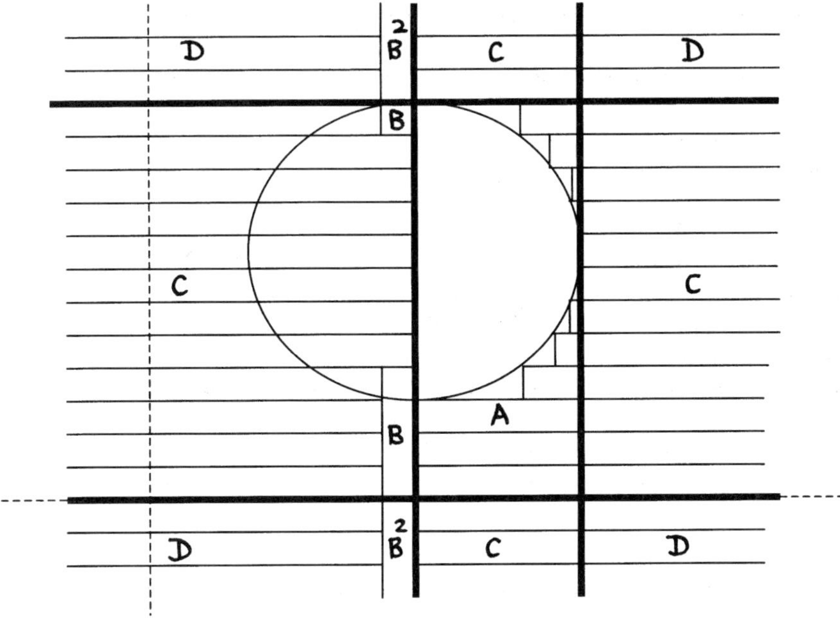

Figure 3.72. *It may be see that the method shown in figure 3.71 is the only way to collimate the field if the target area has an irregular shape, for example the circle shown. A quick inspection of this figure will show that a solution such as that shown in figure 3.70 would always leave flagpoles of open-field irradiation. The key to this figure is the same as that for figure 3.69.*

it is necessary to tabulate both the MU-efficiency (or modulation scale factor) and the total treatment time. Dai and Hu (1999) studied 1000 random sequences with matrix sizes of 5 × 5, 10 × 10 and 15 × 15, and for a varying maximum intensity level. The reader is referred to their paper for the detailed graphs of the outcome but, not surprisingly, it was found that the MLC-MSF algorithm had a much higher efficiency (lower modulation scale factor) than either of the two jaw algorithms by about a factor of four and that delivery times were much faster. It was also observed that the first jaw algorithm was always more efficient than the second (but not by much) provided the maximum fluence value was above two. For a given matrix size the number of segments increased monotonically with increasing number of intensity levels. For a given number of intensity levels the number of segments increased monotonically with increasing the matrix size. For a given matrix size the efficiency increased (modulation scale factor decreased) with increasing number of intensity levels.

Three clinical cases were analysed, one each for breast, prostate and nasopharynx. The dose rate was 250 MU min^{-1}, the dose 2 Gy and the leaf speed

1.5 cm s^{-1}. Compared with the performance for random sequences the average number of segments fell by about a factor of two and the efficiency increased by about a factor of four for the jaw algorithms and by two for the MLC-MSF algorithm for every case. This is because there is some correlation between adjacent bixels in clinical IMBs. The best (first) jaw algorithm was still some two to six times less efficient than the MLC-MSF algorithm.

The second jaw algorithm required the longest delivery time for all three cases. The first jaw algorithm was about was two to five times more *time*-inefficient than the MLC-MSF algorithm. The authors comment that this is perhaps not so bad as popular opinion may expect and they conclude that the first jaw setting algorithm is just about clinically feasible (times being about 20 m).

The jaw setting algorithm would improve by: (i) increasing machine dose rate; (ii) controlling reasonably the number of IMBs and their complexity; (iii) increasing the jaw speed; (iv) developing optimizing algorithms rather than these heuristic algorithms.

Finally Dai and Hu (1999) have pointed out four potential advantages of the jaw technique over other techniques: (i) they are less laborious than the use of compensators; (ii) matrix elements can be adjusted in size compared with the MLC-MSF method; (iii) there is the dosimetric advantage of no tongue-and-groove effect and no interleaf constraints; (iv) ease of control.

Perhaps this is a fitting place to recall that for all the complex technology of IMRT there may still be a place for simpler methods.

3.4. IMRT BY SCANNED BEAMS

Svensson *et al* (1997) have shown that, for concave dose distributions, a very high value of the probability of uncomplicated tumour control (P_+) could be achieved by combining a scanned photon beam with block or MLC. By reducing the source-to-skin distance (SSD) it was possible to reduce the photon beamwidth down to about 35 mm, not small enough to provide all the modulation by scanning alone. Blomquist *et al* (1997, 1998) have obtained good agreement between Monte-Carlo generated IMRT dose distributions and measurements made with the scanning beam of a MM50 accelerator. A significant problem with using the scanned beam of the MM50 to create intensity-modulated fields is the large FWHM of the elementary pencil beam quoted by Blomquist *et al* (1998) as 106 mm at 10 cm depth in water. Whilst this view may be criticized it could be argued that as of now this technique to generate IMBs does not rival the ones described previously in chapters 2 and 3.

Svensson *et al* (1999) have produced a proposal for a new compact treatment unit with an optimized bremsstrahlung target with which it is possible to generate a photon beam with a half width of about 3 cm at an isocentric distance of 100 cm and an initial electron energy of 50 MeV. If the isocentric distance is shortened even further, then it is possible to reduce the half width even further down to about 2 cm for photons. IMRT is delivered using such a scanning pencil beam.

Küster and Bortfeld (2000) have designed an elegant solution to reducing the width of the scanned beam. A collimator plate was put in the blocking tray. The plate contains a matrix of holes, of size and spacing which could be varied (figure 3.73). Monte-Carlo calculations using the GEANT 3.21 code showed that the width of the downstream penumbra could be reduced to 1 mm with no loss of fluence at maximum. The solid parts of the collimator block the unwanted wide wings of fluence from the scanned beam. The method is in its infancy but lends itself to the generation of IMRT by aiming the beam at different holes for different lengths of time proportional to the required relative fluence. The work won Best Paper Prize at the 13th ICCR, Heidelberg.

We conclude this chapter noting that the most exploited methods for generating IMRT are: (i) the compensator; (ii) tomotherapy; (iii) the DMLC technique; and (iv) the MSF method. No more will be said about the moving-bar method (Corletto *et al* 1998, Calandrino *et al* 1998) as it only generates 1D IMBs, nor the scanning beam method for the reason just described. In section 4.6, we shall introduce a seventh potential method (robotic IMRT). In the next chapter we proceed to review some of the issues associated with *clinical* implementation of IMRT.

3.5. COMBINED ELECTRON AND PHOTON IMRT

Korevaar *et al* (1999) have shown that when intensity-modulated *photon* and intensity-modulated *electron* beams are combined this can lead to dose distributions with desirable properties. The dose distributions have: (i) a steep depth-dose fall off at larger depths similar to pure electron beams; (ii) flat dose profiles and sharp penumbras for a large range of depths in tissue, as in photon beams; and (iii) a skin dose that is freely selectable in the range from photon to the electron skin doses. An example is shown in figure 3.74. This overcomes the disadvantage of pure electron beams of a broad penumbra at depth and relatively high skin dose. The appropriate mix of beams was computed by an inverse-planning algorithm. This was a two-stage process. In the first step an optimization routine selected the widths and fluences of a mix (actually two in the case shown) of electron beams and of the photon beam profile. A second stage optimized the desired 'mix ratio' (ratio of central axis maximum doses of electron and photon beams).

The method was tested experimentally by realizing the electron beams on a racetrack microtron with different settings of MLC and blocks. The intensity-modulated photon beams were created by the DMLC algorithm described by van Santvoort and Heijmen (1996) and Dirkx *et al* (1997b, c, 1998). The MM50 racetrack microtron allows the generation of the required modulation without intervention as it can store both static and dynamic fluence profiles. Similar concepts were developed by Li *et al* (1999a, b,2000), presented in section 4.1.14.

Jensen and Hebbinghaus (2000) have gone so far as to suggest that photon IMRT could be delivered by a *combination* of the use of MLC, compensators and high-fluence linacs.

3.6. SUMMARY

The development of the techniques for delivering IMRT with an MLC has flourished in the last decade and particularly in the last five years. Hence, necessarily, this has been a large chapter to encompass exhaustively all these developments. We have seen that many of the issues inter-relate and so it is quite hard to put down firm section headings when reviewing the subject. For example,

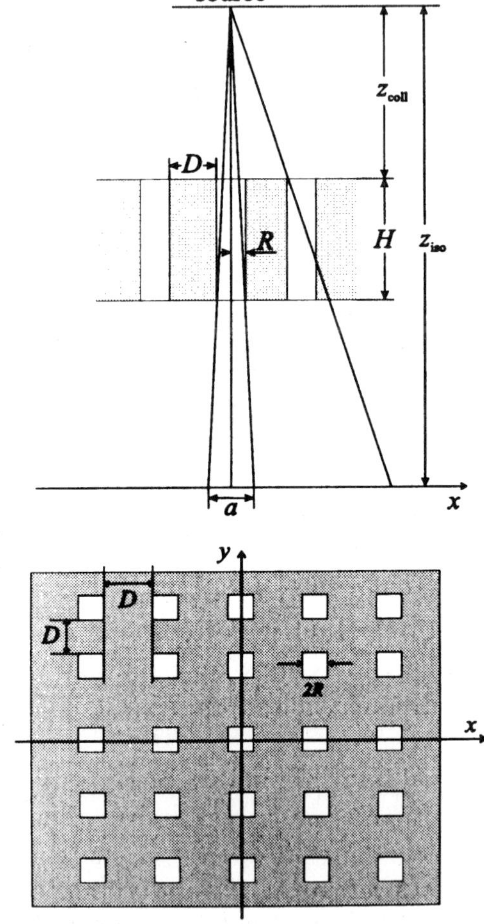

Figure 3.73. *Section and top view of the collimator plate with square holes. Source denotes the position of the bremsstrahlung target. In this study the distances were $z_{coll} = 15$ cm, $z_{iso} = 100$ cm and $H = 10$ cm. D and R could be varied. (From Küster and Bortfeld 2000.)*

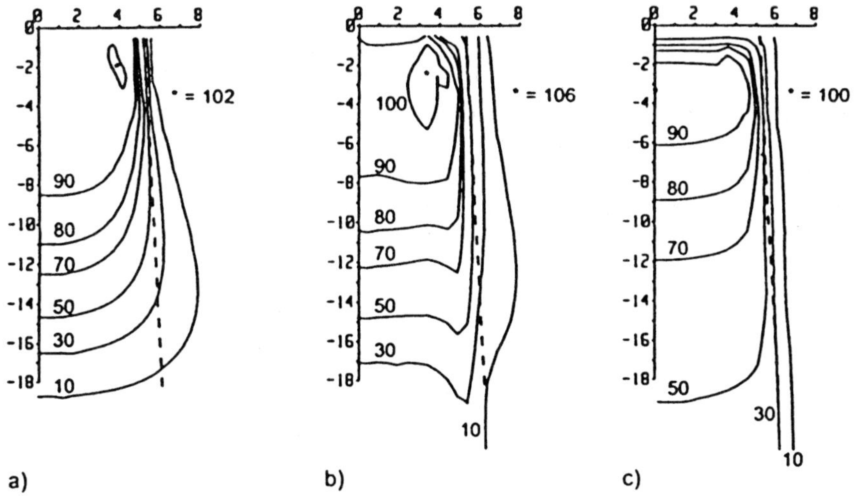

Figure 3.74. *Isodose plots of (a) a flat, 40 MeV electron beam, (b) an intensity-modulated, 40 MeV electron/25 MV photon mixed beam with an on-axis mix ratio of 0.8 to 0.2, and (c) a flat 25 MV photon beam. Dose values are percentages of the corresponding central-axis maximum doses. The field size is* 10.5×10.5 *cm*^{2} *at a source-to-surface distance of 87.5 cm. The dashed lines correspond to the effective field boundaries. (From Korevaar et al 1999.)*

the physics of the individual manufacturers' MLCs varies and influences the choice of interpreter and the way in which dosimetry is carried out. Also there is a difficulty with terminology. When the MLC MSF technique was first mooted (1994) this was the only way to use an MLC to realize a particular 2D IMB. It still is a perfectly valid way to create intensity modulation. Personally I would have preferred the term to be limited to the use of a few field components, as for example, used to create IMRT of the breast. However, I do not uniquely make the rules and the term has also been used to define those techniques in which the leaves take up many tens of static positions to create each field. When the movements between these positions are rapid, as is now possible with all three manufacturers' MLCs, the technique appears as if the leaves are in continuous movement. In that sense the technique appears to be a dynamic method. However, the term DMLC should strictly be (but is not) limited to that in which the leaves move with the radiation on. The 'step-and-shoot' DMLC technique is then really an MSF IMRT technique.

Because of this classification issue the interpreters have been discussed both in separate sections for interpreters for the true DMLC method and also in the section describing the MSF method. What the reader should be clear about is that, if the segments of a component delivery method (whether few or many components) are to be delivered with maximum MU efficiency and with equal fluence increment for each segment, then there are limitations on the number of options (see section 3.3). All the other leaf sequencers are methods to exploit the options with lower MU efficiency and in which there is, in principle, an infinity of options for the placement of the segments but the possibility to reduce the *number* of segments.

For the truly dynamic MLC delivery methods there is a distinction between the ideal interpreter, whose equations are simple to arrive at and standard, but which only applies to the generation of primary fluence for a 1D IMB with no constraints on IMB shape and leaf velocity, and the interpreter needed to cope with solving specific MLC constraints which apply to 2D IMBs. For the latter there is no 'correct' interpreter, the choice depending on application and goal. However, we can say with confidence that for the DMLC technique the tongue-and-groove problem is solved by sacrificing delivery (MU) efficiency.

There are a large number of available MLCs all with different properties. Notably there are at least three commercial macroscale MLCs (and many more in-house constructed ones) and at least five commercial microscale MLCs (and again in-house ones). Motorization is now considered a prerequisite design feature. There are also techniques to make a macroMLC behave like a microMLC.

Dose modelling and verification remain tricky but not insoluble issues. In general, there is a need to resort to models for the former and there are a variety of methods growing for verification. It has been necessary to study the delivery of small numbers of MUs.

As individual groups have developed IMRT with an MLC there has been significant but healthy competition. Many experiments have been carried out to 'test pathways'. In reporting development of IMRT in this chapter it has been impossible to avoid mentioning specific clinical activity but this has been kept to a minimum here. In chapter 4 the clinical issues are more thoroughly explored.

3A. APPENDIX: DETAILED MATHEMATICS OF THE SOLUTION OF THE TONGUE-AND-GROOVE EFFECT FOR THE DMLC TECHNIQUE

There now follows a general analysis of the 'tongue-and-groove' problem in dynamic MLC therapy including *all* partial transmissions. It will be shown that the solution known as synchronization (van Santvoort and Heijmen 1996) remains completely valid under these most general conditions and that there is also a 'partial synchronization solution'.

Let $M_{\mathrm{b,u}}$, $M_{\mathrm{a,u}}$, $M_{\mathrm{b,v}}$ and $M_{\mathrm{a,v}}$ describe the cumulative beam-on time (in MU) for the arrival of, respectively, the 'b' and the 'a' leaves, of the leaf pairs respectively labelled by 'u' and 'v'. The times are for arrival at some specific location x which will be implicitly understood throughout. Let the total time of irradiation for all leaf pairs be T.

Let α_{u} and α_{v} be the transmission through the stepped long edge of the two u-leaves and the two v-leaves respectively, these long edges capable of interlocking. The transmission of a full-depth leaf or fully-interlocking leaf pair is then, of course, $\alpha_{\mathrm{u}}\alpha_{\mathrm{v}}$.

There are three regions of interest in the vicinity of the interlock (figure 3A.1). The intensities in the beam's-eye view of the full-depth u-leaves and v-leaves are, respectively, I_{u} and I_{v}. The intensity in the thin-strip beam's-eye view of the interlocking edges is I_{s}.

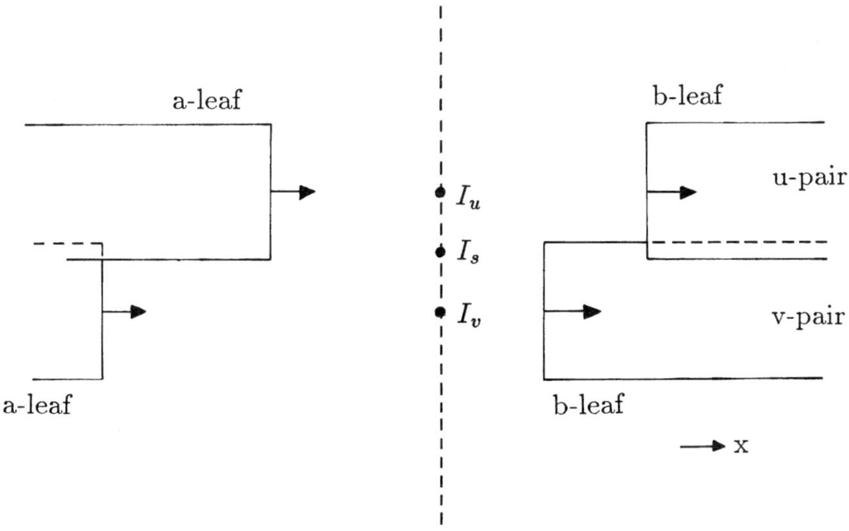

Figure 3A.1. *Showing the two leaf pairs and the three regions of interest for dose calculation.*

The problem is to ensure that I_s is never lower than the lower of I_u and I_v throughout the field.

For complete generality the expressions are derived assuming a spatially-varying upstream fluence represented by S. This could take care of 'head scatter' or, in principle, any other upstream modulation.

3A.1. General formalism for intensity in the beam's-eye view of the three regions

We assume throughout that the leaves pass in the order $[bbaa]$. First we have that

$$I_u = S\left(M_{a,u} - M_{b,u} + \alpha_u\alpha_v\left[T - \left(M_{a,u} - M_{b,u}\right)\right]\right) \tag{A.1}$$

and

$$I_v = S\left(M_{a,v} - M_{b,v} + \alpha_u\alpha_v\left[T - \left(M_{a,v} - M_{b,v}\right)\right]\right) \tag{A.2}$$

where the third term in each equation represents the total transmitted intensity through the full depth of the closed leaves. The times $M_{a,u}$, $M_{b,u}$, $M_{a,v}$ and $M_{b,v}$ are fitted to the 'reduced intensity' or 'pseudo-intensity' functions given by the inversions of equations (A.1) and (A.2), i.e.

$$M_{a,u} - M_{b,u} = i_u = \frac{(I_u/S) - \alpha_u\alpha_v T}{1 - \alpha_u\alpha_v} \tag{A.3}$$

and

$$M_{a,v} - M_{b,v} = i_v = \frac{(I_v/S) - \alpha_u\alpha_v T}{1 - \alpha_u\alpha_v}. \tag{A.4}$$

The leaf which is set to the maximum speed depends on the relation between

$$\frac{I_u\left(x_{i+1}\right)}{S\left(x_{i+1}\right)}$$

and

$$\frac{I_u\left(x_i\right)}{S\left(x_i\right)}.$$

In the beam's-eye view of the interlocking edges

$$\begin{aligned} I_s &= S \\ &\times\left[\min\left(M_{a,u}, M_{a,v}\right) - \max\left(M_{b,u}, M_{b,v}\right)\right] \\ &+\left[\max\left(M_{a,u}, M_{a,v}\right) - \min\left(M_{a,u}, M_{a,v}\right)\right]\left[\alpha_u H\left(M_{a,v} - M_{a,u}\right) + \alpha_v H\left(M_{a,u} - M_{a,v}\right)\right] \\ &+\left[\max\left(M_{b,u}, M_{b,v}\right) - \min\left(M_{b,u}, M_{b,v}\right)\right]\left[\alpha_u H\left(M_{b,u} - M_{b,v}\right) + \alpha_v H\left(M_{b,v} - M_{b,u}\right)\right] \\ &+\alpha_u\alpha_v\left[T - \left(\max\left(M_{a,u}, M_{a,v}\right) - \min\left(M_{b,u}, M_{b,v}\right)\right)\right] \end{aligned} \tag{A.5}$$

where the first term represents the intensity in the fully open field, the second term is the intensity through the trailing protruding leaf edge, the third term is the intensity through the leading protruding leaf edge, and the fourth term is the transmission through the full depth of the closed interlocking leaves. The expressions take account of all possible permutations of arrival times. H is the Heaviside step function which is defined by $H(x) = 1$ when $x > 0$ and $H(x) = 0$ when $x < 0$.

3A.2. Potential underdosage due to unsynchronized pairs—avoidance by 'partial synchronization'

Consider that the leaves arrive in the order $[bbaa]$ and suppose the 'u' b-leaf arrives first and that $I_u < I_v$. In this case the four times rank in the order $M_{b,u} < M_{b,v} < M_{a,u} < M_{a,v}$. From equation (A.5) we have that

$$I_s = S\Big(M_{a,u} - M_{b,v} + \alpha_u \left(M_{a,v} - M_{a,u}\right) + \alpha_v \left(M_{b,v} - M_{b,u}\right) + \alpha_u\alpha_v \left[T - \left(M_{a,v} - M_{b,u}\right)\right]\Big). \quad (A.6)$$

So

$$I_s = I_u + SM_{b,u} - SM_{b,v} + S\alpha_u \left(M_{a,v} - M_{a,u}\right) + S\alpha_v \left(M_{b,v} - M_{b,u}\right) + S\alpha_u\alpha_v \left[T - \left(M_{a,v} - M_{b,u}\right]\right) - S\alpha_u\alpha_v \left[T - \left(M_{a,u} - M_{b,u}\right]\right). \quad (A.7)$$

which, simplifying, gives

$$I_s = I_u + S\left(1 - \alpha_v\right)\left(M_{b,u} - M_{b,v}\right) + S\alpha_u\left(1 - \alpha_v\right)\left(M_{a,v} - M_{a,u}\right). \quad (A.8)$$

From equations (A.1) and (A.2) we have that

$$M_{a,v} - M_{a,u} = \frac{(I_v - I_u)}{S} + M_{b,v} - M_{b,u} + \alpha_u\alpha_v\left(M_{a,v} - M_{b,v} - M_{a,u} + M_{b,u}\right) \quad (A.9)$$

and from equations (A.3) and (A.4)

$$\left(M_{a,v} - M_{b,v} - M_{a,u} + M_{b,u}\right) = \frac{(I_v - I_u)}{S\left(1 - \alpha_u\alpha_v\right)}. \quad (A.10)$$

Putting equation (A.10) into (A.9)

$$M_{a,v} - M_{a,u} = \frac{(I_v - I_u)}{S\left(1 - \alpha_u\alpha_v\right)} + M_{b,v} - M_{b,u}. \quad (A.11)$$

Putting equation (A.11) into (A.8)

$$I_s = I_u + S\left(1 - \alpha_u\right)\left(1 - \alpha_v\right)\left(M_{b,u} - M_{b,v}\right) + \frac{\alpha_u\left(1 - \alpha_v\right)\left(I_v - I_u\right)}{\left(1 - \alpha_u\alpha_v\right)}. \quad (A.12)$$

The condition that $I_s \geq I_u$, is

$$\left(M_{b,v} - M_{b,u}\right) \leq \frac{\alpha_u\left(I_v - I_u\right)}{S\left(1 - \alpha_u\right)\left(1 - \alpha_u\alpha_v\right)} \quad (A.13)$$

but, if terms of the order α_u^2, α_v^2 and $\alpha_u\alpha_v$ are ignored in the equation (A.12) for I_s we obtain

$$\left(M_{b,v} - M_{b,u}\right) \leq \frac{\alpha_u\left(I_v - I_u\right)}{S\left(1 - \alpha_u - \alpha_v\right)}. \quad (A.14)$$

If we consider half-depth interlocking stair-steps, so that $\alpha_u = \alpha_v = \alpha$ then this reduces (with uniform upstream fluence i.e. $S = 1$) to

$$\left(M_{b,v} - M_{b,u}\right) \leq \frac{\alpha\left(I_v - I_u\right)}{(1 - 2\alpha)}. \tag{A.15}$$

Inequalities (A.14) and (A.15) show that 'partial synchronization' can ensure no underdose in this region.

Now, instead, suppose the 'v' b-leaf arrives first. In this case the four times rank in the order $M_{b,v} < M_{b,u} < M_{a,v} < M_{a,u}$. From equation (A.5) we have that

$$\begin{aligned} I_s = S\Big(&M_{a,v} - M_{b,u} + \alpha_v\left(M_{a,u} - M_{a,v}\right) + \alpha_u\left(M_{b,u} - M_{b,v}\right) \\ &+ \alpha_u\alpha_v\left[T - \left(M_{a,u} - M_{b,v}\right)\right]\Big). \end{aligned} \tag{A.16}$$

Proceeding through the same manipulations as led to equation (A.12) we find that

$$I_s = I_u + S\left(1 - \alpha_u\right)\left(1 - \alpha_v\right)\left(M_{a,v} - M_{a,u}\right) + \frac{\alpha_u\left(1 - \alpha_v\right)\left(I_v - I_u\right)}{\left(1 - \alpha_u\alpha_v\right)}. \tag{A.17}$$

The condition that $I_s \geq I_u$, is

$$\left(M_{a,u} - M_{a,v}\right) \leq \frac{\alpha_u\left(I_v - I_u\right)}{S\left(1 - \alpha_u\right)\left(1 - \alpha_u\alpha_v\right)} \tag{A.18}$$

and, if terms of the order α_u^2, α_v^2 and $\alpha_u\alpha_v$ are ignored in the equation (A.17) for I_s we obtain

$$\left(M_{a,u} - M_{a,v}\right) \leq \frac{\alpha_u\left(I_v - I_u\right)}{S\left(1 - \alpha_u - \alpha_v\right)}. \tag{A.19}$$

If we consider half-depth interlocking stair-steps, so that $\alpha_u = \alpha_v = \alpha$ then this reduces (with $S = 1$) to

$$\left(M_{a,u} - M_{a,v}\right) \leq \frac{\alpha\left(I_v - I_u\right)}{(1 - 2\alpha)}. \tag{A.20}$$

We have chosen to analyse all possible situations by considering that $I_u < I_v$ with either the 'u' b-leaf or the 'v' b-leaf arriving first. The second situation can alternatively be analysed by considering the 'u' b-leaf to always arrive first but with $I_v < I_u$ since this is no more than a reversal of the labelling of 'u' and 'v'.

3A.3. Generalized synchronization

For $I_u < I_v$ the goal of synchronization is to ensure that $I_s \geq I_u$. This is achieved if the trajectory of the leaf pair defining the minimum of the two adjacent intensities at x lies within that of the leaf pair defining the maximum of the two intensities at x, i.e. the trajectories are adjusted so that $M_{\mathrm{a,v}} \geq M_{\mathrm{a,u}}$ and $M_{\mathrm{b,u}} \geq M_{\mathrm{b,v}}$. From equation (A.5) we have that

$$I_{\mathrm{s}} = S\Big(M_{\mathrm{a,u}} - M_{\mathrm{b,u}} + \alpha_{\mathrm{u}}\left(M_{\mathrm{a,v}} - M_{\mathrm{a,u}}\right) + \alpha_{\mathrm{u}}\left(M_{\mathrm{b,u}} - M_{\mathrm{b,v}}\right) + \alpha_{\mathrm{u}}\alpha_{\mathrm{v}}\left[T - \left(M_{\mathrm{a,v}} - M_{\mathrm{b,v}}\right)\right]\Big). \quad \text{(A.21)}$$

So

$$I_{\mathrm{s}} = I_{\mathrm{u}} + S\alpha_{\mathrm{u}}\left(1 - \alpha_{\mathrm{v}}\right)\left(M_{\mathrm{a,v}} - M_{\mathrm{b,v}} - M_{\mathrm{a,u}} + M_{\mathrm{b,u}}\right). \quad \text{(A.22)}$$

Hence, from equation (A.10)

$$I_{\mathrm{s}} = I_{\mathrm{u}} + \frac{\alpha_{\mathrm{u}}\left(1 - \alpha_{\mathrm{v}}\right)\left(I_{\mathrm{v}} - I_{\mathrm{u}}\right)}{1 - \alpha_{\mathrm{u}}\alpha_{\mathrm{v}}}. \quad \text{(A.23)}$$

Since the second term on the r.h.s. is positive this ensures that $I_{\mathrm{u}} < I_{\mathrm{s}}$. By rearranging it is clear that $I_{\mathrm{s}} < I_{\mathrm{v}}$ and so $I_{\mathrm{u}} < I_{\mathrm{s}} < I_{\mathrm{v}}$ as required.

If instead $I_{\mathrm{u}} > I_{\mathrm{v}}$ and the leaves are synchronized so the v-gap is within the u-gap rather than *vice versa*, we obtain identical expressions but with the labels ‘u’ and ‘v’ reversed everywhere. This is no more than a change of labelling convention.

CHAPTER 4

IMRT: CLINICAL IMPLEMENTATION AND ASSOCIATED ISSUES

4.1. CLINICAL APPLICATIONS OF IMRT

In this chapter we look at the 'clinical implementation' of IMRT and associated issues. The term is considered broadly to include test IMRT irradiations of animals and planned IMRT patient treatments, whether actually delivered or not. Brief overviews have been provided by Chui (1999b) and Nutting *et al* (2000b).

It is difficult to know how best to categorize the review of clinical IMRT. This has been done partly by geographical centre, partly by clinical tumour site and partly by IMRT technique. It is recognized that, necessarily, (since these are not mutually exclusive categories), there is some overlap and arbitrary classification choices. Some discussion of clinical work has also been presented in chapters 2 and 3 where this illustrated the clinical use of specific technical developments being described.

4.1.1. IMRT of dogs at the University of Washington, Seattle

Sutlief *et al* (1997) have incorporated Yu's interpreter inside the PRISM treatment-planning system and have constructed an emulator to show both the leaf positions and the delivered dose at each stage through use of a convolution dosimetry technique. They use the Elekta DMLC technique on a Philips SL20 accelerator to deliver IMRT to dogs with head-and-neck tumours (Phillips 1997, Cho *et al* 1998a). The group have given a simple description of the issues in delivering IMRT with the Elekta equipment (Elekta 1998b). The treatment of dogs is worked up at a veterinary hospital in much the same way as that for patients. Dogs are CT scanned using a mobile CT scanner, and TLD chips held in catheters in the major airways record doses that agree well with dose calculations. The dogs are held in a Gill–Thomas bite-block frame and breathing is controlled externally (figure 4.1). Kippenes *et al* (1998) have shown how radiotherapy of dogs can be used not only as the source of therapy for companion animals but also as a model for IMRT of

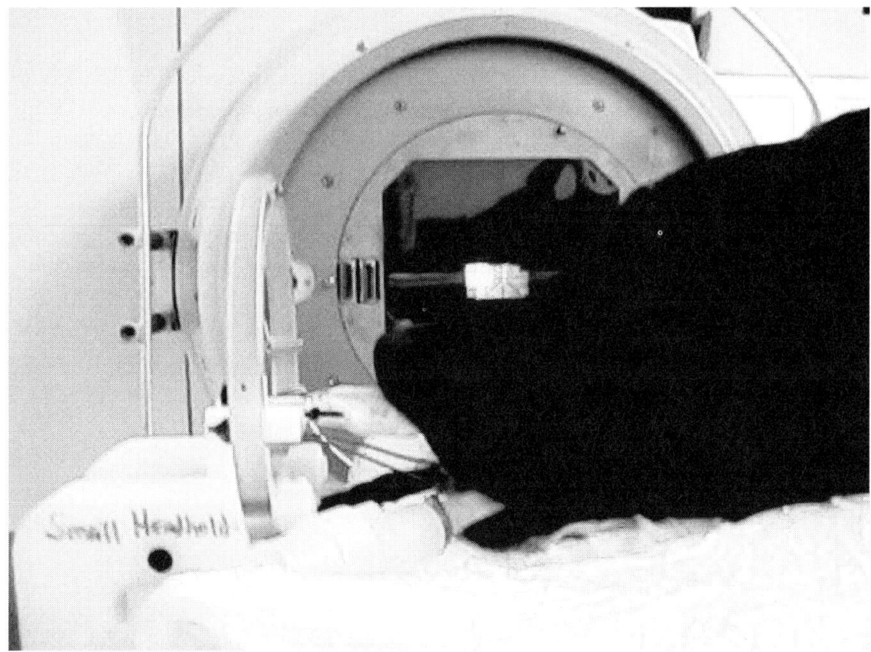

Figure 4.1. *Photograph of a dog positioned for treatment of a nasal tumor at Washington State University. A bite block is used to reliably reposition the dog with respect to a 'stereotactic' ring that contains radio-opaque fiducials. An identical set-up is used for obtaining the treatment-planning CT. If in vivo measurement of the dose is desired, a six French catheter containing 1 mm diameter TLDs can be inserted into the nasal or oral cavity. These TLDs have been used in conjunction with CT scans at the time of treatment to ascertain the dosimetric accuracy of the set-up and treatment. (Courtesy of Dr M Phillips.)*

humans. Specifically *in vivo* dosimetry has been carried out with precise 'patient' positioning in a stereotactic frame (Phillips *et al* 2000). The work irradiating dogs with IMRT has included gating respiration by a simple method and also comparing *in vivo* dosimetry (TLDs) with the predictions of a macro-pencil-beam model.

Phillips (1999) and Parsaei *et al* (2000) have given further details of the irradiation of dogs with head-and-neck tumours using IMRT. The work is carried out at two centres and is an example demonstrating that tele-radiology of IMRT works. Tumours are generally concave in shape and wrap around the spinal cord and are relatively small in size, particularly for purpose-bred beagels with para-spinal tumours. A full IMRT treatment-planning work-up with TLD catheter verification takes place. There is some indication that disruption of electron equilibrium in the sinuses is significant. Hence, the calculations referred to in section 3.2.6 are needed.

4.1.2. IMRT at Stanford University, California using the Varian DMLC technique

Boyer and Bortfeld (1997) have shown, via planning studies, that the delivery of IMRT by seven to nine fixed-gantry fields with modulated cone-beam delivery, the multiple-static-field (MSF) method, can give dose distributions that are competitive with those produced by arcing IMRT using the NOMOS MIMiC. Boyer *et al* (1997a) have used the capability of the Varian Oncology Systems prototype DMLC to deliver conformal IMRT to three clinical cases; they have shown this is an achievable goal. They postulate that development of a complete IMRT system is beyond the capability, not only of a particular university hospital, but also beyond that of any single manufacturer. The resources are simply not available. (The manufacturers may not agree with this!) Hence, their solution is to marry the IMRT products from two commercial vendors to create a working system. They have created IMRT fluence distributions using the NOMOS PEACOCK inverse-planning system, based on the simulated-annealing algorithm (Webb 1995b, 1997c). Planning was performed slice by slice, for nine static fields, corresponding to the part of the patient 'viewed' and irradiated by just one leaf pair. In this part of the process the bixel fluences were computed using just the in-slice dose prescription. Finally, a full 3D dose calculation was made by forward calculation, taking into account the contribution of each bank of bixels to the whole volume. Tissue inhomogeneities were however not included and they proposed that rectifying this, and also further developing the inverse-planning algorithm, is necessary (see also section 3.2.6). The second stage was to pass the fluence patterns into the Bortfeld–Boyer algorithm (Bortfeld *et al* 1994a, b) for generating the leaf-sequencing patterns. This was performed on a personal computer.

The leaf sequences were then fed to the DMLC capability of the Varian MLC controller. At the time of this study, this delivered the sequence of MSFs for just one gantry angle at a time, requiring operator intervention to change the gantry angle and re-feed the sequencer for a different orientation. However, the delivery of each 2D IMB proceeded automatically and required less than 2 min per orientation. Again, at the time of this study, the effective compensator transmission factor (modulation scale factor) (see section 3.2.10 and Geis *et al* 1996) was not included so the *absolute* dosimetry was not checkable. Mok *et al* (1999) have compared the IMRT delivered with the DMLC method and with a compensator and have concluded that the latter was needed for certain head-and-neck cases where fine spatial resolution was important.

Boyer *et al* (1997a) have planned conformal IMRT distributions for three patient cases. These were:

(i) a nasopharyngeal plan including cervical neck nodes, sparing the brainstem, spinal cord and parotid glands;
(ii) a unilateral vocal cord tumour sparing the contralateral cord;
(iii) an ethmoid tumour located between the eyes and anterior to the optic chiasm and the brain stem.

The study reported DVHs for the PTV and OARs and concluded that the aims of the conformal therapy were well achieved. Detailed tables of the volumes of tissues raised to tolerance doses or above were presented and isodose plots were shown in transaxial, sagittal and coronal planes (see e.g. figure 4.2).

The accuracy of the IMRT delivery was checked by exposing a film sandwiched between two slabs of polystyrene to each of the nine fields for case (i) (see figure 4.3). Up to 14 MLC leaves were required for each field. Additionally, a polystyrene head-and-neck phantom was prepared including a film sandwiched in one transaxial plane. This was exposed to the nine fields used for case (i) and, after the film was digitized, measured isodoses were qualitatively compared with the

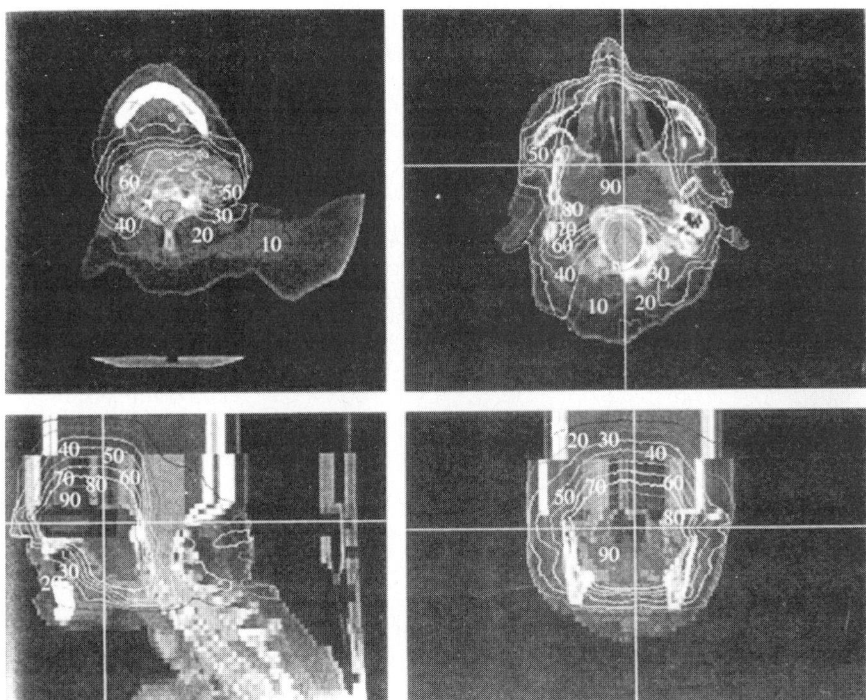

Figure 4.2. *Isodose curves computed by the NOMOS PEACOCK planning system for an intensity-modulated nasopharyngeal plan. Isodoses are shown in transaxial, sagittal and coronal planes. The shaded regions define a primary PTV and a secondary PTV. Also defined within the plan were the parotid glands, the spinal cord and the brainstem. The isodose curves are expressed as a percentage of the maximum dose of 90 Gy. The figure demonstrates qualitatively the conformity of the isodoses to the PTV and the reduction of planned dose to normal structures. (From Boyer et al 1997a; reprinted with permission from Elsevier Science.)*

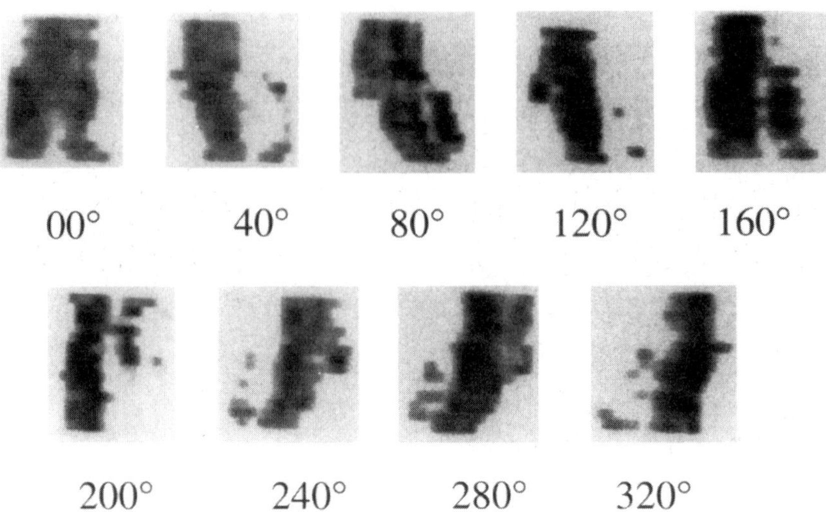

Figure 4.3. *Film measurements of the beam intensity modulation for the nine-field fixed gantry fields computed for the nasopharyngeal case shown in figure 4.2. The films were exposed using the computed leaf-setting sequences derived from the NOMOS PEACOCKPLAN using the Varian DMLC system. The films are oriented with the superior direction of the patient to the top of the figure. Gantry angles labelled 80° and 280° are 10° from true left-lateral and right-lateral views, respectively. (From Boyer et al 1997a; reprinted with permission from Elsevier Science.)*

corresponding isodoses in the planning slice (figure 4.4). A detailed quantitative comparison was not made at this stage because it was recognized that the planning algorithm was not yet completely satisfactory as discussed above. However, close agreement was observed, particularly in the high-value isodose lines.

A further update on a cohort of patients in this centre was provided by Tate *et al* (1998). CORVUS was used to plan head-and-neck treatments using nine intensity-modulated fields for the primary treatment and five intensity-modulated fields for a boost. Fields were delivered using the Varian 2300 C/D accelerator with the DMLC technique to a surrogate phantom containing film and ionization chambers. Improved dose distributions resulted, avoiding the use of electron boost for conventional photon fields.

4.1.3. IMRT at the Memorial Sloan Kettering Cancer Center, New York, using the Varian DMLC technique

Ling *et al* (1996, 1997, 1998), Burman *et al* (1997, 1999) and Hunt *et al* (1999) have reported on the use of the DMLC technique with the Varian MLC and accelerator

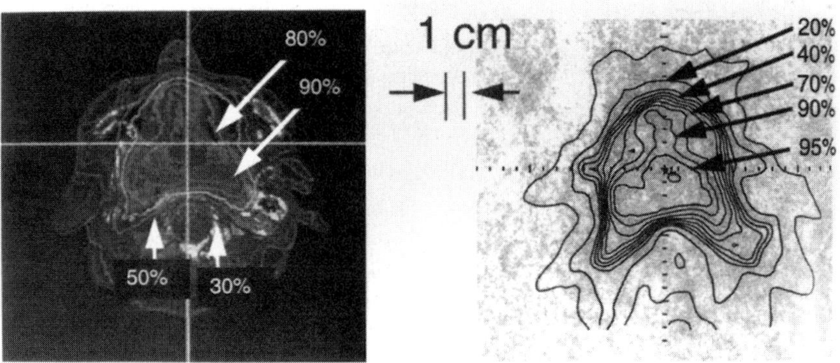

Figure 4.4. *The figure shows a comparison of isodose curves. On the right are isodose curves derived from film measurement. The isodose curves are calculated from relative optical density lines measured on a film located 0.5 cm superior to the isocentre. The isodose curves on the left were generated by the NOMOS PEACOCK treatment-planning system. (From Boyer et al 1997a; reprinted with permission from Elsevier Science.)*

at the Memorial Sloan Kettering (MSK) Cancer Center, where the method is used to treat prostate cancer. Six DMLC fields are constructed by the MSK inverse-planning technique. Initially, only the 9 Gy boost was delivered this way but since spring 1996 the full prescription dose of 81 Gy has been delivered by IMRT. Rectal NTCP was reduced from around 4% for a standard (non-IMRT) plan to around 2%. Happersett *et al* (1999) have shown that dose escalation in the prostate is possible with IMRT and 'dose painting'. Fuks (2000) has reported unequivocal evidence for clinical efficacy of IMRT.

The groundbreaking report of the use of the DMLC IMRT technique at this centre was from Ling *et al* (1996). At the time of this paper, just the 9 Gy boost field was delivered by IMRT and the group were gaining experience with the new technology. The Memorial inverse-planning system, originating with Bortfeld and improved by Mohan *et al* (see chapter 5), incorporated scatter into the inverse planning and used a pencil-beam convolution method of dose calculation. The same six-field angles were used for the IMRT-delivered boost as for the first phase of treatment. This implied a non-radical departure from conventional practice. In the region of overlap of prostate PTV and rectum OAR, priority was given to the rectum since limiting the probability of rectal injury was a paramount consideration. The important features of the composite plan were that the entire PTV was encompassed to 75 Gy and that a large portion (90%) of the PTV was raised to 81 Gy with the rectum outside the 81 Gy figure; 50% of the PTV was raised to 85 Gy, 30% of the rectal volume received less than 75.6 Gy. The calculated dose distribution for each field was compared with a measured dose distribution with the latter recorded on film at a depth of 5 cm. Comparisons with ionization chamber measurements were also made (Wang *et al* 1996). The mean disagreement between film measurements and calculations was reported to be 3%. Regarding the quality assurance of leaf movement, the leaf positions were assessed by redundant readouts every 55 ms and checked to a tolerance of 0.1 mm. If the position was not met within tolerance then the radiation was temporarily suspended until the condition was met. Each field was also verified initially by setting the leaves to their extreme starting and finishing locations and taking portal images. The set of six gantry angle deliveries was achieved without the need for anyone to enter the treatment room and delivery was complete in 12 min. Figure 4.5 shows typical intensity-modulated profiles.

Amols *et al* (1999) have given further details of the DMLC IMRT radiotherapy technique of the Memorial Sloan Kettering Cancer Center. Magnetic resonance spectroscopy imaging is used with strict dosimetric constraints for designing highly conformal treatment plans. The technique now delivers 91 Gy to the PTV and less than 30% of the rectal wall receives more than 76 Gy. Zelefsky *et al* (1999) have reported significantly decreased toxicity with high-dose IMRT of the prostate. Recently, the technique has been extended to deliver IMRT to nasopharyngeal tumours using seven coplanar equispaced portals from the posterior direction. It was found that the dose coverage could be improved whilst still sparing normal structures. The planning technique of Spirou and Chui (1998a) was used.

At a different centre, van Santvoort *et al* (1999) have developed the MSF IMRT technique to treat with post-operative radiotherapy patients with node-positive cervical cancer. The kidneys and rectum are dose-limiting organs and they investigated whether the application of inverse planning could contribute to a solution to this problem. They used the HELIOS programme which is part of the Varian CADPLAN software. This is an implementation of the optimization algorithm that was developed in the Memorial Sloan Kettering Cancer Center. A conformal four-field box technique was compared to a four-field inverse plan with appropriate dose-volume constraints and all inverse plans resulted in a significantly lower dose to the small bowel and rectum.

4.1.4. Boosted intensity margins for various tumour sites

An issue of concern in delivering IMRT with MLC-shaped fields which are coplanar is the possibility of underdose due to superposed penumbras in the superior and inferior margins of the PTV. One way to avoid this is to add margins but this may irradiate too much unwanted normal structure. Dirkx *et al* (1997a, b, 1998) have developed an alternative technique. They reduced the field margins from 1.5 cm to an average of 0.7 cm and added a boosted intensity to the most-

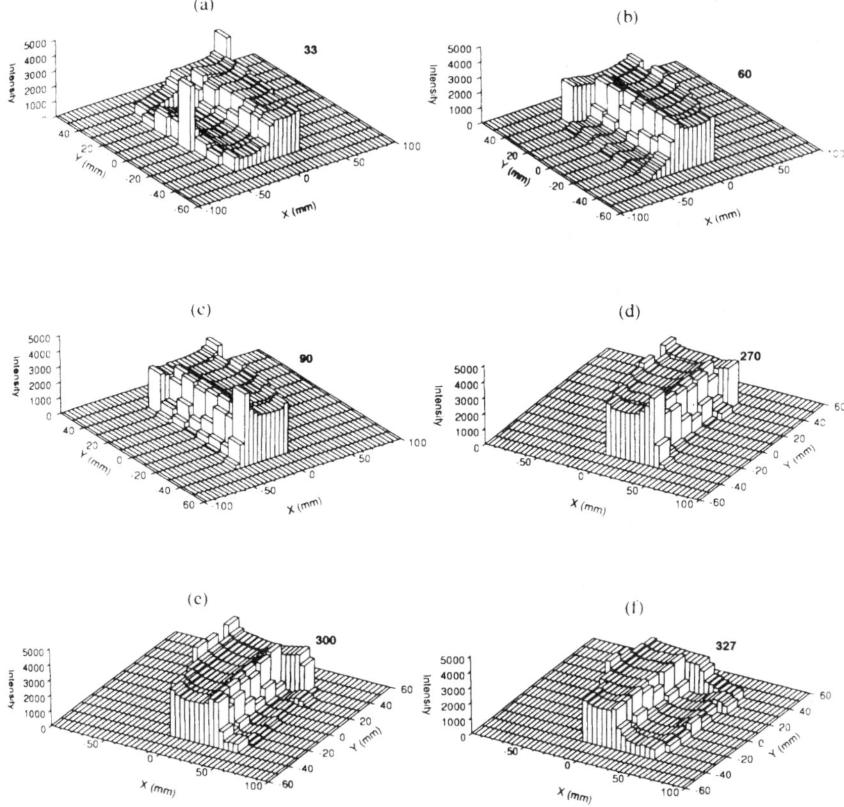

Figure 4.5. *The intensity profiles in the isocentric plane of the six fields derived from inverse planning: (a), (b) right posterior oblique at gantry angles of 33° and 60°; (c) right lateral; (e), (f) left posterior oblique at gantry angles of 300° and 327°; (d) left lateral. Positive Y is in the cephalad direction. At 90°, positive X is towards the anterior and at 270°, positive X is towards the posterior. (From Ling et al 1996; reprinted with permission from Elsevier Science.)*

superior and most-inferior leaf position. They showed that this achieved three things. It simultaneously sharpened the 50–90% penumbra to the value it had been with the large margins, it reduced the dose inhomogeneity within the PTV and it reduced the volume of rectum and bladder irradiation for prostate treatment. Ten patients were irradiated with the Scanditronix racetrack microtron.

Following on from the use of the method to improve prostate dosimetry the same group have applied it to improve the dosimetry for lung cancer (Dirkx *et al* 1999a, b, Dirkx and Heijmen 1999b). Although it is recognized that lung

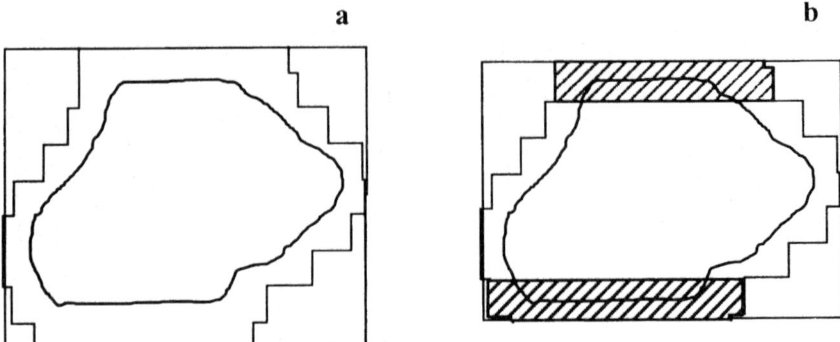

Figure 4.6. *Posterior field for a patient in a study to improve the dosimetry of lung cancer using beam intensity modulation: (a) the standard technique with large field margins and no modulation; (b) the BIM technique. Boost fields (shaded) are superimposed on the superior and inferior parts of the treatment field. (From Dirkx et al 1999b; reprinted with permission from Elsevier Science.)*

cancer treatment requires adjuvent therapy, any improvement in the radiotherapy will benefit all concomitant strategies. Figure 4.6 shows the beam intensity-modulation method of reducing the superior and inferior field margins from 1.5 to 0.8 cm and adding a boost 'strip field' of higher intensity. Note that this can be achieved by forward planning and does not require inverse planning. The boost fields were delivered simultaneously by closing leaf pairs in the central part of the beam. Figure 4.7 shows a typical result from the study. Plan comparison showed that (averaged over 12 patients) the volume of healthy lung tissue receiving a normalized total dose of over 20 Gy reduced by 9.7% and the calculated NTCP reduced from 10.7% to 7.6% on overage. A potential for a 6–7 Gy dose escalation was recorded. However, the clinical priority was to reduce complications without dose escalation and this is already being routinely applied in the clinic (Dirkx *et al* 1999a, b).

The same idea of reducing the field margins and adding a 'rind' of high intensity to the border, an idea attributed to Mohan *et al* (1996), has been investigated experimentally by Brugmans *et al* (1999). Experiments were done to evaluate the reduction in field size and the border dose increment required to maintain dose homogeneity within a cork insert to a phantom representing the irradiation of lung.

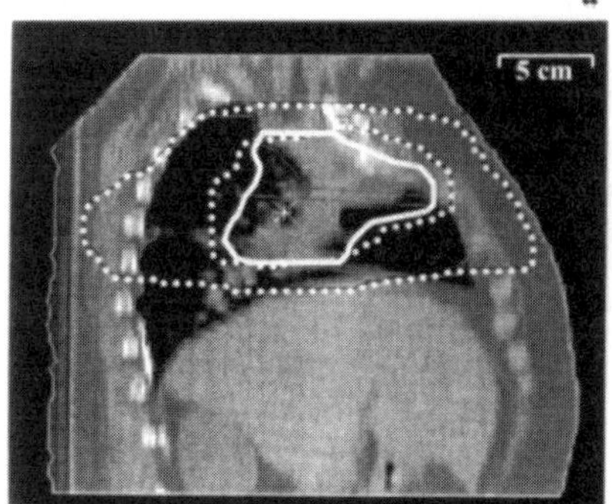

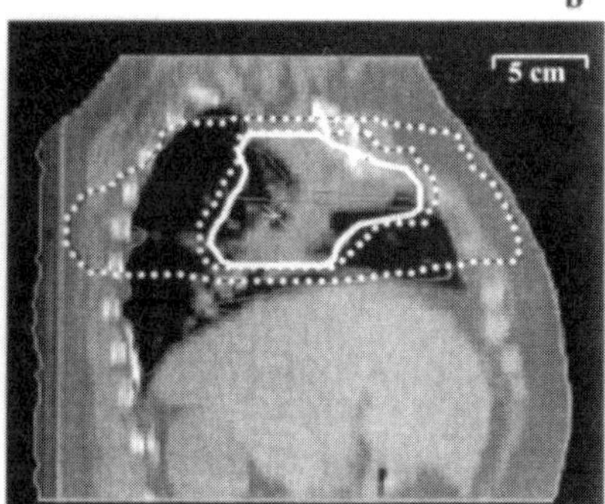

Figure 4.7. *Dose distributions in the sagittal plane through the isocentre for a patient in the trial to improve dosimetry for lung cancer: (a) for the standard plan; (b) for the BIM technique. Depicted are the PTV (full curve) and the 50% and 95% isodose curves. (From Dirkx et al 1999b; reprinted with permission from Elsevier Science.)*

4.1.5. *IMRT at the University of Ghent by segmented field delivery with the Elekta MLC*

De Neve *et al* (1999b) have discussed specific clinical situations in which IMRT is preferable to conventional radiotherapy. Having begun by noting that concave PTVs are a clear indication for IMRT, they have pointed out that the butterfly-shaped PTV, first postulated by Brahme and co-workers, representing advanced cervical cancer, is rarely likely to appear in affluent countries where early cervical cancer is treated surgically. Perhaps the prime indication for IMRT is the treatment of nasopharyngeal tumours and also those tumours which are in the location of the spinal cord. They have argued that the treatment of the prostate by IMRT is indicated not so much because it is a concave volume tumour, but because it is important to deliver a differential inhomogeneous dose distribution to the prostate so as to deliver a lower dose region to the overlap volume between the planning target volume and rectum. They have called this an 'intentionally inhomogeneous dose distribution'. The same situation arises in ethmoid cancer where tumour edges are adjacent to one or more of numerous OARs and IMRT allows the irradiation with an inhomogenous dose distribution so that at each fraction the PTV portion that intersects with, or is close to, the OARs is slightly underdosed.

De Neve and de Wagter (1997) have used the Elekta MLC to deliver DMLC therapy using a series of individually optimized multisegments (see also Elekta 1999c). They have concentrated on two types of tumours, pharyngeal relapses and non-small-cell lung cancer. For the former, tumours of the nasopharynx, oropharynx and hypopharynx were studied and it was shown that IMRT achieved good coverage of the postero-lateral wall, the deep jugular chain, decreased dose to mandibular bone, parotid glands and spinal cord. Concave targets around the spinal cord were conformed. The early treatments took place within a clinical waiver agreement, using a preclinical version of the Elekta system (de Neve *et al* 1996, 1998a, b, c, Elekta 1999c). The lung tumours were a group of unresectable and/or inoperable T4 tumours, mostly centrally located in the mediastinum. Cord dose and radiation pneumonitis of the lung were dose limiting and it was shown for ten patients that IMRT kept dose to normal structures below tolerance, allowing dose escalation (de Gersem *et al* 1997, Derycke *et al* 1997, 1998). The expected dose distributions were checked by irradiating films and TLD chips contained in anthropomorphic phantoms (de Wagter *et al* 1997). By combining the knowledge of the relative dose from film and the absolute dose at a few points from TLD, the experimental confirmation yielded a map of the absolute dose distribution. Eisbruch (1998) has also shown that static multisegment IMRT of head-and-neck tumours can spare dose to the parotids. He has actually measured salivary flow following conformal radiotherapy and shown it to be much more preserved than with conventional therapy. Van Dieren *et al* (1998) have also shown that the use of IMRT delivered by the DMLC technique can spare the salivary glands and prevent xerostomia. A 2–6 field IMRT method was compared with the conventional parallel-opposed method. The work of the Ghent group has been reported in semi-lay terms in the July 1999 Issue (vol 3, no 2) of the Elekta journal *Wavelength* along with the dosimetry insight provided by BANG gel dosimetry (see section 4.4.3).

De Neve *et al* (1998a, b, c) have argued that the use of class solutions respects existing departmental procedures and obscures the difference between IMRT and 3D planning, thus overcoming potential resistance. There is also little increase in planning and treatment time, although Williams (1998) has argued that studies of cost-effectiveness can backfire if the statistics involved are small. Cost-effectiveness of IMRT should not be the sole criterion for accepting IMRT treatments. De Meerleer *et al* (1998, 1999) from the Ghent group have shown that IMRT of the prostate by the DMLC technique can increase the TCP for fixed NTCP compared with conventional 3D planning.

De Gersem *et al* (1998) have used IMRT by the DMLC technique to treat relapsed pharyngeal cancer. Vakaet *et al* (1998) have implemented a computer-graphic technique to avoid selecting non-coplanar beam directions which would cause collision.

De Neve *et al* (1998a, 1999a) and de Gersem *et al* (2000a, b) have given details of the Ghent step-and-shoot IMRT method. Class solutions of simple segments are determined and the beamweights optimized. A programme called CRASH (Combine, Reorder And Step-and-sHoot) then creates the field shapes delivered by the Elekta MLCi DMLC technique. Perhaps these should be better regarded as MSFs for this implementation. As well as presenting the plans for two interesting patients, these papers have diagrams of the actual leaf sequences which aid the appreciation of the method (figure 4.8).

The starting point of the technique is to note that many relapsing nasopharyngeal cancers sweep backwards into the para-pharyngeal space and

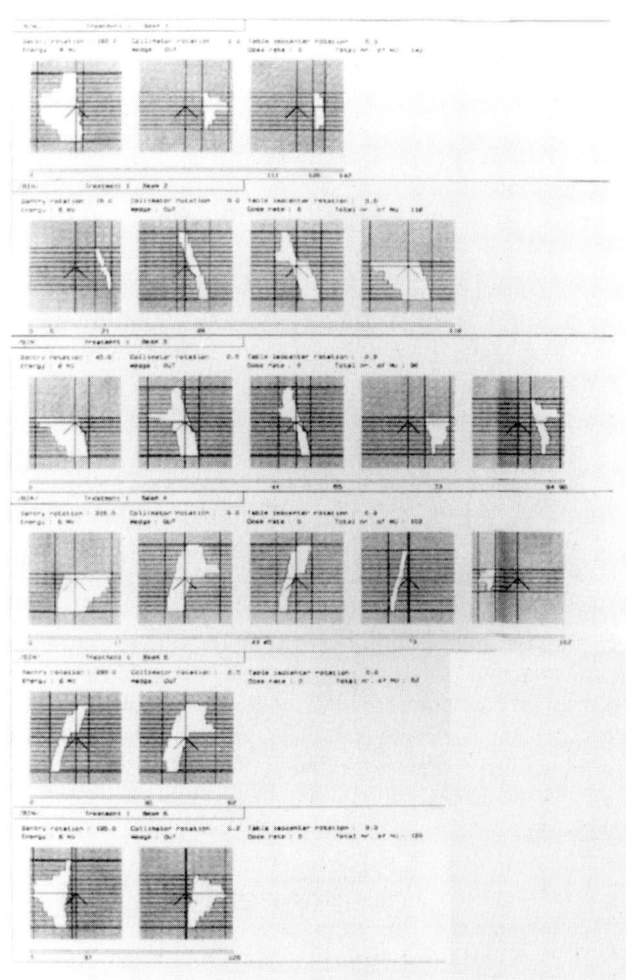

Figure 4.8. *Graphical printout of the sorted (CRASH) sequence of beamparts for each IMB for a patient with a large transglottic larynx carcinoma. From left to right, the MLC configurations of the consecutive beamparts to be delivered during an IMB. The numbered bar below each row of beamparts represents the cumulative monitor count reached at the end of each beampart. From upper to lower, the sequence of IMBs as ordered by CRASH. (From de Neve et al 1998a.)*

create a horseshoe-shaped PTV requiring IMRT. These cancers are generally inoperable (de Neve *et al* 1999a). Planning is based on a class solution with fixed beam directions and the weights of static beam segments were determined using an optimization scheme within GRATIS. De Neve *et al* (1999a) have minimized the function '$-BF$' where B is a biological factor given by

$$B = (TCP)^{2.5} \times (1 - NTCP_{\text{cord}})^{1.0} \times (1 - NTCP_{\text{brainstem}})^{1.0} \tag{4.1}$$

and F is a physical factor given by

$$F = \left(\frac{dose_{\text{min}}}{dose_{\text{max}}}\right)^{0.8} \times \left(\frac{dose_{\text{mean}}}{\sigma_{\text{dose}}}\right)^{1.0} \tag{4.2}$$

where the powers were determined empirically during development of the class solution. The segment weights were optimized on a DEC ALPHA 255/300 workstation and took about 1 h CPU time when 1000 dose points were sampled for optimization. The intensity modulation is built up from physically-overlapping beamparts. The segments were ordered by the CRASH routine.

CRASH works as follows. The process begins with a number of n machine states equal to the number of beamparts (de Neve *et al* 1999a) with a specific number of MUs assigned to each state. The aim of CRASH optimization is to reorder the n machine states (minus those with zero weight) into a logical sequence which minimizes the treatment time. Firstly, the beamparts are sorted by machine parameters (e.g. couch position, gantry angle etc) to create m beamparts with identical beam incidences, collimator rotation, tray status and wedge status. To force the delivery into 'step-and-shoot' mode, each of the m beamparts was duplicated to construct $2m$ control points. Zero weights were assigned such that movement between geometric states (control points) occurred with the radiation switched off. A so-called reverse prescription directory was also created so that in the event of an abnormal termination the unfinished beam could be subsequently executed (figure 4.9).

The security of delivering the beam components was assured by strapping a film, sandwiched between 3 cm thick perspex, to the treatment head so that this records the integration of the beamparts. Figure 4.10 shows an example. Moreover, BANG gel dosimetry enabled a 3D verification of the delivered dose. A RANDO phantom was replanned using the clinical fields and then these fields were delivered to a PVC copy of the RANDO containing BANG gel (see also section 4.4.3). In this way, the whole chain of therapy processes through imaging, planning, optimization, segmentation, treatment delivery were quality-controlled. Figure 4.11 shows an example of the resulting dose distributions arising from the application of the fields shown in figure 4.10. These clearly show the desired concave dose distribution sparing the spinal cord. Altogether, de Neve *et al* (1999a) have shown that the delivery of IMRT does not need to radically depart from many conventional processes.

De Gersem *et al* (1999) have performed inverse planning for ten patients with stage-three non-small-cell lung cancer, either with conventional 3D planning

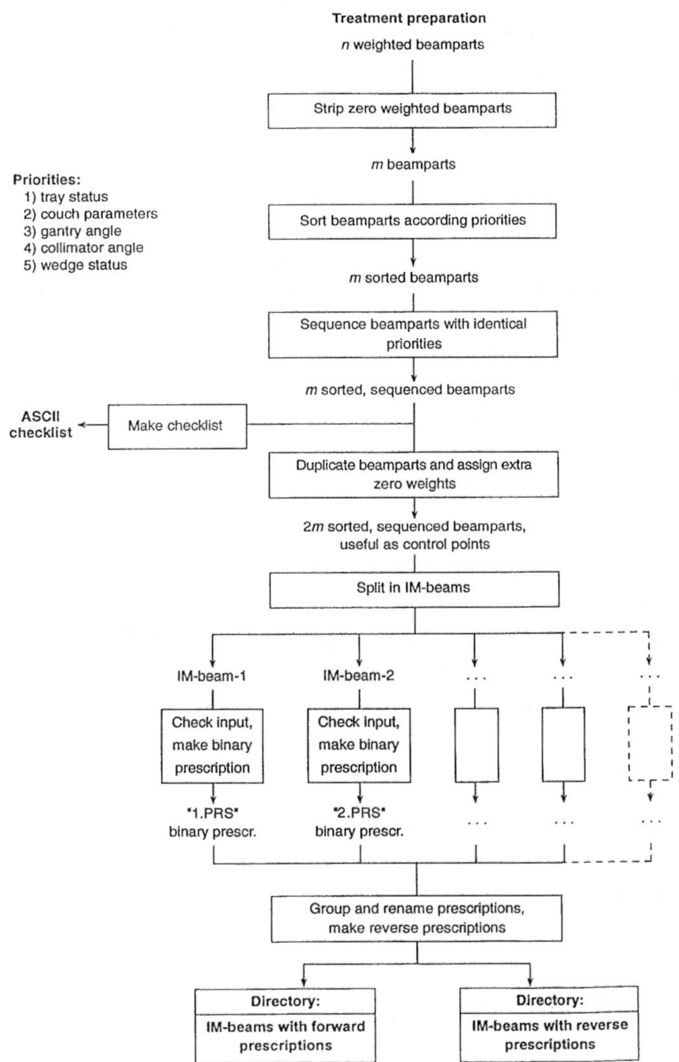

Figure 4.9. *Illustrating the procedure known as CRASH—see text for details. (From de Neve et al 1999a; reprinted with permission from Elsevier Science.)*

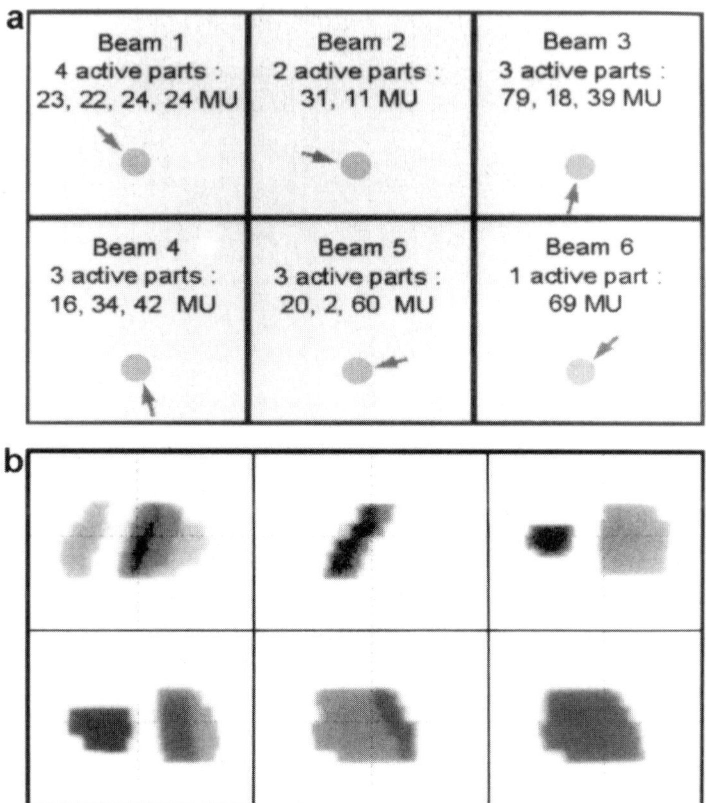

Figure 4.10. *IMRT plan of a patient with a relapsed nasopharyngeal tumour as applied to a RANDO phantom for dosimetric verification of the IMB. Panel a: incidence and beampart intensities per beam. The beampart intensities are those specifically optimized for the RANDO outline (outline different from the actual patient). Panel b: resulting beam-intensity patterns recorded on film during a test run. Beams 1, 3 and 4 are split around the brainstem. In the beam's-eye view, beam 2 passes tangentially to the right, while beams 5 and 6 pass tangentially to the left of the brainstem. (From de Neve et al 1999a; reprinted with permission from Elsevier Science.)*

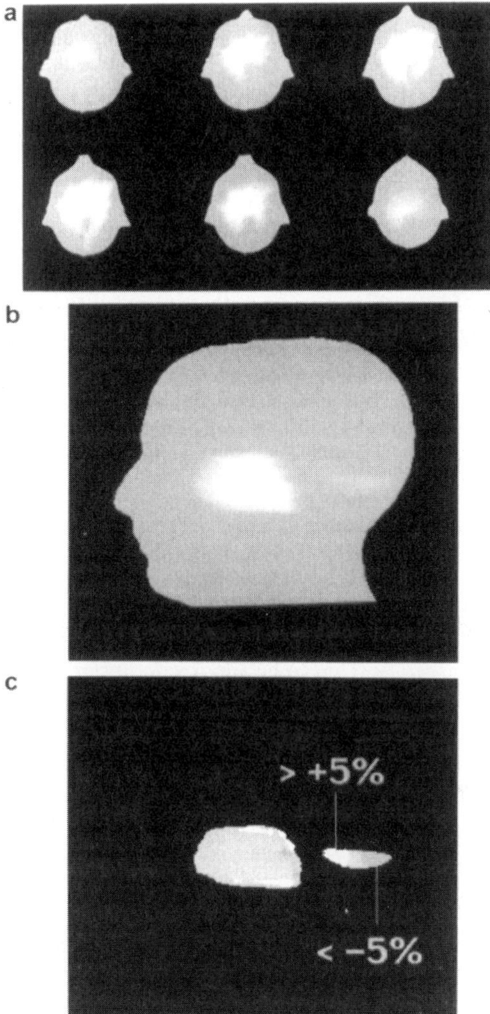

Figure 4.11. *MRI dosimetry on a polymer-gel phantom for verifying IMRT. The outline of the phantom is similar to that of the RANDO phantom. Panel a: dose distribution measured in six different slices (10 mm thickness and 10 mm separation) ranging from cranial to caudal. Pixel size is 0.98 mm. Panel b: dose distribution measured in the sagittal mid-plane (10 mm slice thickness). Pixel size is 1.2 mm. Panel c: dose differences higher than 5% normalized dose are mainly confined to a secondary hot spot. (From de Neve et al 1999a; reprinted with permission from Elsevier Science.)*

or with beam-intensity modulation. They have made use of either a biological cost function

$$B = (TCP_{\mathrm{PTV1}})^{2.5} \times (TCP_{\mathrm{PTV2}}) \times \left(1 - NTCP_{\mathrm{lung}}\right) \times (1 - NTCP_{\mathrm{heart}}) \times (1 - NTCP_{\mathrm{cord}}) \quad (4.3)$$

or a biophysical cost function BF where

$$F = \left(\frac{D_{\mathrm{min}}}{D_{\mathrm{max}}}\right)^{1.0} \times \left(\frac{D_{\mathrm{mean}}}{\sigma}\right)^{0.8} \quad (4.4)$$

where D_{max}, D_{min}, D_{mean} and σ are the maximum, minimum, mean and standard deviation of dose in the PTV_1. PTV_1 refers to the target volume of high clonogenic cell density, PTV_2 refers to a second target volume encompassing PTV_1 and containing a lower clonogenic cell density in the non-overlapping region. The NTCP values are calculated for the OARs indicated. Optimization was by minimizing either $-B$ or $-BF$ respectively. Note that the powers in equation (4.4) can be adjusted and are different from those used for the prostate planning study reported earlier by de Neve *et al* (1999). The results of the planning studies are complex and can only really be appreciated by studying the tables provided by de Gersem *et al* (1999). However, it can be stated that for the plans with beam-intensity modulation:

(i) biophysical optimization led to better dose homogeneity;
(ii) biological optimization led to the converse and also led to higher maximum and median tumour doses;
(iii) biological optimization led to a higher combined TCP and a higher uncomplicated local control probability;
(iv) because biological optimization led to much larger dose ranges the spatial locations of these required checking and safety concerns were raised.

De Neve *et al* (2000a, b) have given some figures for treated clinical cases. Over the period of three years up to September 1999, 31 patients received IMRT for head-and-neck cancer. Over the 18 months up to December 1999, 31 patients received IMRT for prostate cancer. Patient numbers in 2000 were expected to be about 100. The 'do-it-ourselves' approach was commented on as very expensive and more turn-key IMRT would be welcomed.

4.1.6. IMRT at DKFZ, Heidelberg using the Siemens DMLC technique and compensators

Grosser *et al* (1999b) have discussed the implementation of the DMLC IMRT technique at DKFZ, Heidelberg. The technique is implemented on a Siemens Primus linear accelerator operating at 6 and 15 MV. IMRT plans are produced using KONRAD and CORVUS treatment-planning systems and a sequencer tool was developed to minimize match lines and number of fields. The number of field components was limited to less than 80 per fraction to limit the treatment time and, following the commencement of patient treatments in 1998, 12 patients with cancers in the head-and-neck had been treated by April 1999.

The Siemens integrated IMRT package is called IMART. Intensity-modulated beams can be created, for example by CORVUS or KONRAD. A module known as IMFAST then preconditions the accelerator for the shortest treatment time. This module allows a variety of interpreter options. The resulting solution is then fed to the delivery module which is known as SIMTEC (see also section 3.2.3).

IMRT is achieved at the DKFZ using either the dynamic multileaf collimator technique or compensators. At the time of the first IMRT Winter School in Heidelberg, 40 patients had been treated, 37 with a segmented multileaf step-and-shoot technique and three with the MCP96 compensators. Also, 29 plans had been optimized with KONRAD and 11 plans with CORVUS. The segmented multileaf step-and-shoot technique utilized the Siemens delivery system and details have been provided by Grosser *et al* (1999c). These authors pointed out that two particular problems arise with this technique. One is the potential for tongue-and-groove underdose and the other is an along-the-leaf matchline effect created because of non-complete compensation of penumbra. The two effects are similar in magnitude and due to the double-focusing of the MLC are nearly independent of off-axis position. It can be shown that by choosing suitable superpositions of field components the tongue-and-groove effect can be removed. However, as a consequence, along-the-leaf matchline defects arise. Grosser *et al* (1999c) have described that they removed these effects by small field overlapping of the two adjacent field components by 2 mm. This matchline can be fixed independent of the off-axis position because of the double-focusing MLC (see also section 3.3.2).

One difficulty with the Siemens multisegment approach is the deadtime between segments in which other tasks are performed. This is of course reduced to a minimum by the sequencer (see section 3.2.3.2). However, this still leads to lengthy treatment times and it is not anticipated that improved sequencers will give less segments. Grosser (2000) has given a neat solution to the problem. This is as follows. Firstly, each 2D IMB is doubled. Then, it is ‘sliced’ by the usual Bortfeld–Boyer technique. The resulting 2D IMB is then divided into two parts, one called even and the other odd. The ‘even’ 2D IMB comprises for the first MLC channel all the even-numbered field components, and for the second MLC channel all the odd-numbered field components, and so on sequentially. The ‘odd’ 2D IMB is the reverse of this, i.e. the first MLC channel contains all the odd-numbered field components and the second MLC channel contains all the even-numbered field components and so on sequentially. Thus, on average each of the even and odd 2D IMBs delivers about the same maximum dose but has about half the number of field components. The even 2D IMB is delivered one day; the odd the next, and so on alternately, during a changed fractionation scheme. The summed relative dose distributions and absolute dose distributions for both a conventional and two consecutive days of a changed fractionation/irradiation scheme are of course the same. The only remaining problems are (i) lack of ability to do this automatically with current software and (ii) residual concerns about potential radiobiological changes.

Bortfeld and Oelfke (1999) have also described how to use compensators for IMRT as implemented at the DKFZ. The compensators have a minimal transmission, smaller than 5%, and a spatial resolution of approximately 6 mm. They are made of the material MCP96 poured into a special moulding material called Obomodulan. The milling of each compensator is performed in a series of levels with a commercial milling machine with a maximum of 17 mm milling depth. The individual compensators take account of beam divergence and are stacked in the blocking tray of the accelerator. Each level is a slightly magnified version of the level above it. The compensators can be designed using the KONRAD inverse-planning system which allows a bixel resolution higher than the usual 10 mm square achievable with a regular size MLC. The advantages of a compensator are many, namely: high spatial resolution is achievable; high precision is achievable; one can have an infinite number of intensity levels; there are no matchline problems. Delivery can be relatively fast but there is a problem generating very low intensities and manufacturing time is very lengthy.

Salz *et al* (2000) have developed compensators made of a tin granulate wax mixture with a density of approximately 4.8 g per cubic cm. Compensated plans have been prepared using the HELAX TMS and the transmitted fluences have been compared with experimental measurements and shown to agree within 1.4%.

Keller-Reichenbecher *et al* (1999a) have noted that, conventionally, inverse planning does not take account of the method of delivery. For example, an inverse plan prepared by the KONRAD inverse-treatment-planning system would be followed by a calculation of the leaf positions required to deliver the IMB using the DMLC technique. However, new work presented by Keller-Reichenbecher *et al* (1999a) has incorporated the sequencing process into the inverse-planning process itself to take account of the number of intensity levels, the necessity for monitor units to be integer, leaf transmission, tongue-and-groove underdose and its solution and leaf collision avoidance. Computations were compared with experiments and it was observed that, when the sequencer was included in the inverse-planning process, the experimental values more closely matched the predicted dose distribution than if the sequencer was applied *a posteriori*.

4.1.7. IMRT at Thomas Jefferson University

Galvin *et al* (1998) have used the DMLC method to deliver 2D IMBs which comprise wedges and compensators derived *without* the use of an inverse-planning algorithm. This is a novel technique but they show it generates acceptable dose distributions for prostate patients. Galvin (1999a) has shown that forward planning can be used to create PTVs with invaginations provided dose uniformity constraints are lifted. The technique makes use of nine fields, initially shaped with a beam's-eye view of the target volume. Further field segments are then added using a rule-based forward-planning system to create field segments. The field segments combine to create a PTV with invaginations. It is claimed that the distributions are little worse than those created using the NOMOS CORVUS inverse-planning system (Galvin 1999c). My contention is that this conclusion must clearly depend on the shape of the PTV and its disposition with respect to OARs. This technique was used for treatment of nodal regions and boost volumes for head-and-neck tumours that surround critical normal structures. Furthermore, Xiao (1999a, b) and Xiao *et al* (1999) at the same centre have then optimized the weights of the subfields using the Cimmino algorithm. The technique is very reminiscent of that developed by Webb (1991a) in which each field was divided into two segments. Xiao (1999b) has used the method for optimizing treatment of the prostate. Each field was divided into two segments viewing prostate-only or prostate-plus-rectum. They showed that, as the weight applied to reducing the dose in the rectum was increased, so the conformality around the rectum indeed improved but at the expense of a worsening PTV dose homogeneity. Bednarz *et al* (1999) created concave dose distributions for treating head-and-neck tumours using forward-planning methods and compared them with those from CORVUS inverse planning. They claimed the former were just as good and also had the advantage of including significant open fields.

4.1.8. IMRT at the Netherlands Cancer Institute, Amsterdam, using the Elekta MLC

The Netherlands Cancer Institute (NKI), Amsterdam perform IMRT of the prostate using the Elekta DMLC as follows. Five fixed field directions are segmented into two or three subfields viewing the PTV-only or PTV-plus-rectum, an idea reminiscent of that of Webb (1991a). The weights of the subfields are computed by simulated annealing, implemented with a dose-volume cost function on UMPLAN (figure 4.12). Very-small-weight fields are eliminated. The NKI five-field technique for irradiation of the prostate with IMRT is verified by replanning the IMBs on a cylindrical phantom and then delivering them to such a phantom containing film. This validates the coronal and sagittal dose distributions. This team also verified that small numbers of MUs could be delivered with the Elekta MLC (Wittkämper *et al* 1998, Brugmans *et al* 1998, Hansen *et al* 1998a).

Bos *et al* (1999) have optimized segmented intensity-modulated radiotherapy of prostate cancer at the Netherlands Cancer Institute. The method is based on the University of Michigan treatment-planning system incorporating a simulated-annealing process to optimize beamweight configuration. It was shown that a simultaneous treatment of the prostate with 68 Gy plus a boost of 10 Gy is advantageous compared with the serial irradiation of both treatments with respect to the total dose to the rectal wall.

De Boer (1999) has given more details of the NKI segmented-treatment technique for radiotherapy of the prostate. The number of beams is currently limited to five with a maximum of three segments per beam, the reason being that

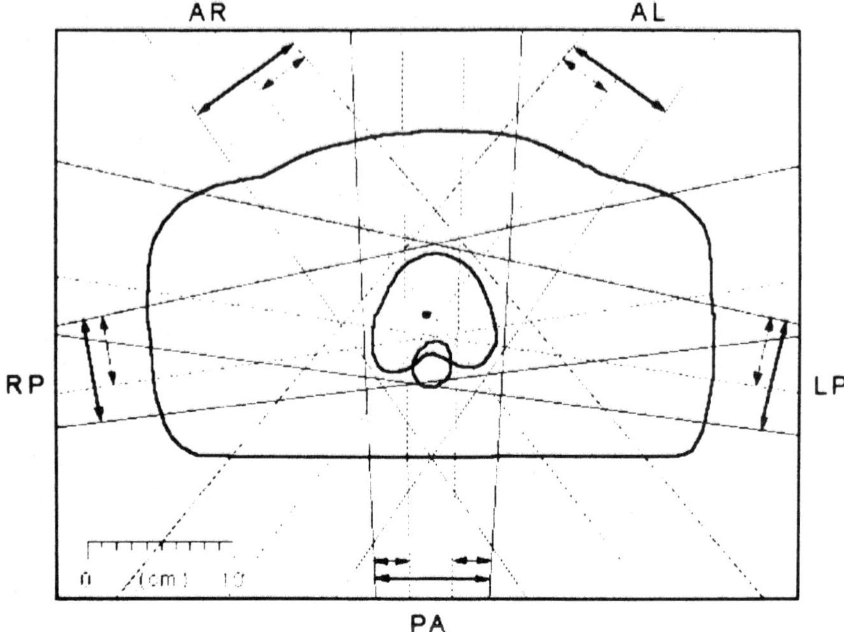

Figure 4.12. *Transverse slice illustrating the segmented five-field technique. The external contour, the outer rectal wall and the PTV are delineated. The beam edges of the open segments are indicated with full curves, while the edges of the rectum-shielding segments are represented as dashed curves. Note that 11 segments are defined in all and their weights optimized. (From Wittkämper et al 1998.)*

the NKI wish to operate the ELEKTA accelerator in 'clinical mode' rather than in 'local service mode'. The reasons for this are that to operate in local service mode it is necessary to switch modes, there is no networking so you have to type in all the data from scratch, there is no 'finished beam' facility, the maximum dose rate is 100 MUs min^{-1} and it would be necessary to have a physicist present at each treatment. Conversely, using a small number of beams the clinical practicalities are currently acceptable.

Rasch (1999) at NKI has developed IMRT for advanced bilateral head-and-neck cancers, e.g. T4 base-of-tongue carcinoma extending over the mid-line in such a way as to spare the dose to the parotid gland. Comparisons have been made between using three to six non-opposing fields with two to four segments per field and a full, KONRAD-generated, unlimited number of segments and a simple conventional two-lateral-field treatment. The latter led to 36% of the parotid receiving more than 60% of the tumour dose. The KONRAD technique led to 21% of the parotid receiving more than 60% of the tumour dose and, in between these, the multiple-segment technique led to 24.5% of the parotid receiving more than 60% of the tumour dose. The segmentation took place so as to divide fields into segments which see either the PTV and OARs overlying or underlying or alternatively PTV alone. The inverse-planning technique used simulated annealing with penalty functions in the OARs. In addition, the dose to the eyes, optic nerves and other proximal structures was spared. Ten patients have been treated so far.

4.1.9. IMRT at the University of Michigan, Ann Arbor, using the multisegment technique

Fraass *et al* (1998, 1999a, b, c), Marsh *et al* (1996), Martel *et al* (1998a, b) and Eisbruch *et al* (1996, 1998, 1999) have also developed a multisegment approach which uses MLC-equipped treatment machines, but not the DMLC technique. They reported in 1998 that 340 patients had been treated with a variety of sites included. Optimization used relatively simple tools to achieve an effective dose distribution. Fraass *et al* (1999a) have developed segmental IMRT in the clinic for more than five years. Some of these treatments for patients on dose escalation involved prescription doses of 90 Gy or more for brain, lung and liver tumours. Fraass *et al* (1999a, b, c) have compared forward and inverse-planning approaches for optimizing conformal treatment plans and found that there were advantages and limitations in segmental IMRT compared with other forms of IMRT.

Fraass *et al* (1999a) have described in detail how segmental IMRT treatments have been used in the Department of Radiation Oncology of the University of Michigan since January 1994. Multisegment IMRT makes use of fixed-field shaping to create intensity-modulated fields by adding the dose from several different shaped beam portals that have the same beam direction. Depending on the complexity of the intensity modulation, the number of segments can be upwards from two to twenty. An example of the three-segment lateral field use for treatment of the prostate is shown in figure 4.13. Segmental IMRT is available using both gantries of the MM50 racetrack microtron and several different Varian Clinac accelerators. Segments can be delivered automatically under computer control for the racetrack microtron but require manual entry for the Varian Clinac system. Segmental IMRT is planned using the 3D planning system UMPLAN running on a Dec Alpha workstation. The optimization makes use of the simulated-annealing algorithm based on a cost function which is a combination of many 'costlets'. Segmental IMRT has been used to treat nearly all clinical sites including prostate, liver, abdomen, lung, breast, chest wall, head-and-neck and brain. Figure 4.14 shows the number of patients treated in each category and the number of intensity-modulated segments per plan for the various clinical sites.

An important part of the treatment-planning process has been an attempt to minimize the number of intensity-modulated segments that are used and it was determined for the first 475 clinical plans that more than four additional segments were necessary only in a handful of cases. However, it is expected that, as planning protocols become more complex, the number of segments will increase. For

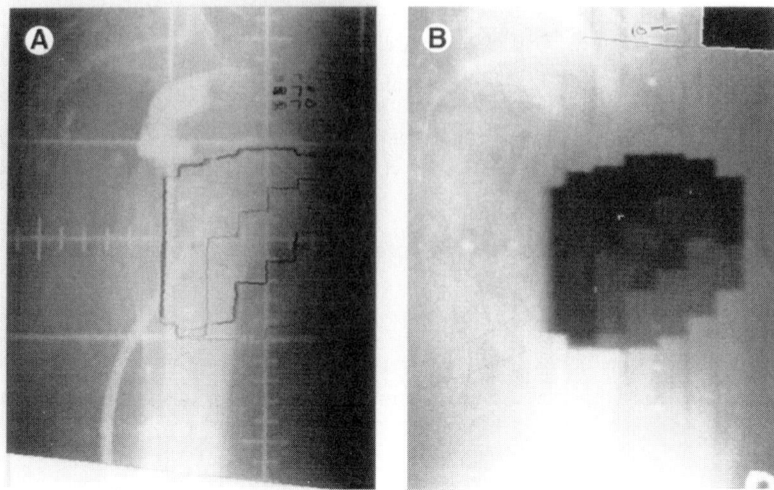

Figure 4.13. *(a) A simulator film image of three-segment lateral field used for treatment of the prostate. (b) Portal image for multisegment prostate field (lateral). (From Fraass et al 1999a.)*

prostate treatments it was found that two segments were generally all that were required. For the brain, analysis of brain recurrences showed that virtually all failures were inside the original target volumes. Three PTVs were defined, each of which were designed to receive a different radiation dose given sequentially before IMRT became available. However, the differential dose to the planning target volumes could also be achieved using an IMRT plan which treated all the volumes concurrently using segmented IMRT, now the preferred technique. After some experimentation, a standard template-driven plan was determined, consisting of five fields which typically used nine segments in total. Segmental IMRT plans were developed for parotid-sparing head-and-neck treatments which make use of non-coplanar beam geometries and shaped electron beams as well as intensity-modulated multisegment fields. The complexities of head-and-neck segmental IMRT planning led to wide variations in the number of intensity-modulated segments used. Fraass (1999a) have explained that multisegment IMRT and its optimization is a smaller paradigm shift from conventional treatment planning and therefore can be more readily adopted by clinics who do not have to change completely the planning or delivery paradigm with which they are most familiar. At the University of Michigan, human interactive planning is used to optimize the placement, number and weight of treatment segments. Vineberg *et al* (1999) have reported how, at this centre, the dose to glioma could be escalated to 130 Gy without exceeding normal brain toxicity by careful use of and optimization of non-coplanar beamlets.

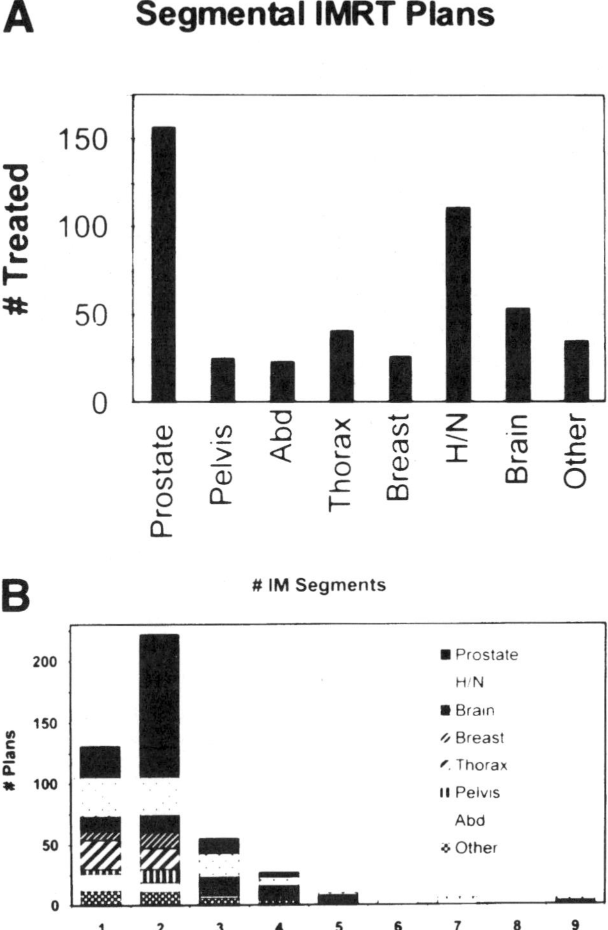

Figure 4.14. *(a) Multisegment IMRT plans for each clinical site treated at the University of Michigan. (b) Number of intensity-modulated segments per plan for various clinical sites—see text for details. (From Fraass et al 1999a.)*

The goals of IMRT can be many and varied. IMRT can be used to deliver a concurrent boost treatment; it can be used to improve target uniformity; it can be used to avoid normal tissues better; it can be used to minimize the number of beam directions or to compensate for axial-only beam directions. Finally, the classic example, it can be used to create concave isodose surfaces or low-dose areas surrounded by high dose. Fraass *et al* (1999a) have shown that the full automatic IMRT optimization based on the use of costlets can, for a five-field head-and-neck treatment, produce results which are somewhat improved over segmental IMRT. However, the converse may be stated that the segmental IMRT is still able to generate a very acceptable dose distribution.

4.1.10. IMRT at UCSF, California

Pickett *et al* (1999) have attempted to improve the dosimetry to the prostate using static-field IMRT (ST-IMRT). Specifically, they aim to treat the bulk of the prostate to more than 70 Gy whilst simultaneously treating a dominant intraprostatic lesion (DIL) to 90 Gy without increasing complications. The DIL was defined using magnetic resonance spectroscopic imaging. Seven fields were defined and divided into two or three segments whose weights were determined using the UMPLAN treatment-planning system. Pickett *et al* (1999) have shown detailed pictures of the field segments for a typical patient from the set of 22 studied so far. It was found that 73.8 Gy could be delivered to the bulk of the prostate. Multifocal disease in general in other organs would seem to be a candidate for this technique which was deemed advantageous compared with DMLC therapy in view of its simplicity, ease of verification and ease of delivery with (for example) the Siemens SIMTEC control system. The authors believe that long-term follow-up is required to prove that the presently observed lack of complication continues and also that Monte-Carlo treatment planning will be needed for precise dosimetry.

Verhey (1999a) has also used segmented-field IMRT for treatment of both head-and-neck tumours and the prostate. He, however, chooses to call this technique 3D conformal radiotherapy because it can be planned with forward treatment planning. The method derives from the two-weight-per-field technique invented by Webb (1991a). Figure 4.15 shows an example of the three-weight-per-field technique applied to radiotherapy of the prostate.

Verhey (1999a) has studied three clinical problems comparing the delivery of radiotherapy using three techniques: (i) 3D conformal radiotherapy, including simple IMRT; (ii) inverse-planned static-field IMRT; and (iii) the DMLC technique with the NOMOS PEACOCK MIMiC. The three patients planned were: (i) a patient with carcinoma of the prostate; (ii) a patient with a para-spinal sarcoma that is adjacent to the spinal cord and close to the kidney; and (iii) a patient with nasopharyngeal carcinoma adjacent to the brain stem, spinal cord, optic chiasm and parotid. Plans were compared using dose-volume histograms. For the prostate patient, the PEACOCK MIMiC plan showed the lowest dose to the rectum and the most uniform dose to the prostate. The MSF IMRT technique showed a slightly higher dose to the rectum than the 3D conformal radiotherapy plan. All three plans yielded an acceptably uniform dose to the prostate and rectal doses well below tolerance. The plans ranked in the same order for the head-and-neck tumour patient, and for the patient with a para-spinal lesion the MSF technique also defeated the 3D conformal radiotherapy technique. The paper by Verhey (1999a) gives very useful comparative dose-volume histograms for the three techniques in the three clinical applications.

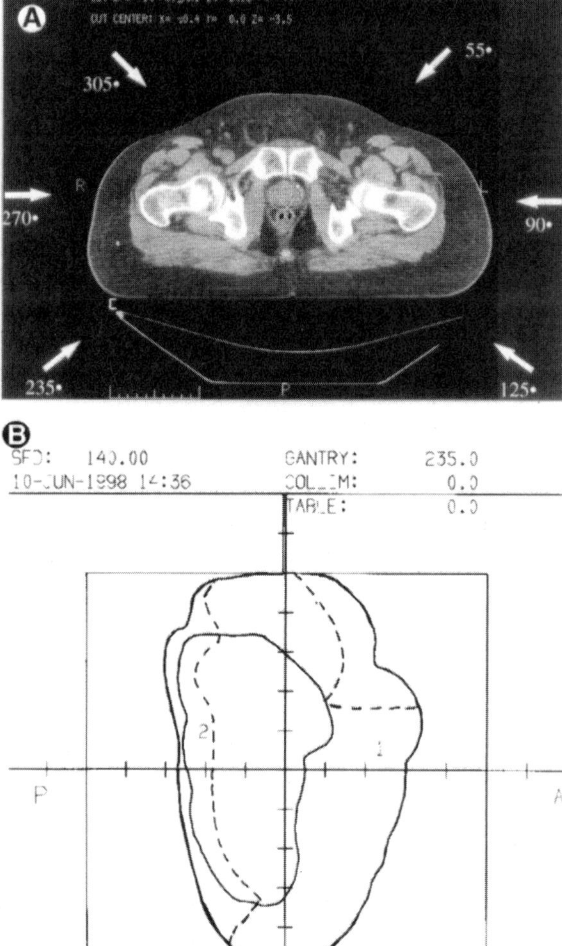

Figure 4.15. *(a) An axial CT section shown with the six axial beam directions and the prostate and dominant intra-prostatic lesion contours indicated. (b) A beam's-eye view projection for the right posterior/beam showing three separate apertures: one treating the entire prostate; one (shown dashed) indicating the areas shadowed by the rectum and bladder, which are excluded for a portion of the treatment; and one indicating the treatment of the dominant intra-prostatic lesion with 8 mm margins. (From Verhey 1999a.)*

Ma *et al* (1999c) have compared the use of the gamma knife with the IMRT MIMiC fan-beam technique for treating intracranial lesions. They decided that for very small lesions the gamma knife would be clearly chosen and for large lesions the MIMiC technique would be chosen. Hence, they have only investigated three model and three patient tumours of intermediate size where the choice is less obvious. The planning study showed that the gamma knife spared more normal brain tissue but did not give so homogeneous or conformal a dose to the PTV as did the fan-beam MIMiC technique.

4.1.11. IMRT with the NOMOS MIMiC

Whenever a new treatment modality becomes developed to the point of clinical utility assorted quotations for patient numbers treated appear in various sources, e.g. the PARTICLES newsletter updates the worldwide proton totals every six months. At the time of writing, it is believed that the most *clinical* IMRT treatments have taken place with the NOMOS MIMiC equipment. Any numbers quoted must be out of date by publication so one can only quote past milestones. However, even these usefully document the speed of clinical progress in the manner of milestones. In August 1998 there were 31 centres using the MIMiC clinically, 1753 completed treatments and 162 patients on treatment (Bleier 1998). Engler *et al* (1999) noted, at this later date, that over 3000 patients have been treated worldwide using the PEACOCK/CORVUS system in conjunction with a MIMiC collimator. Over 350 of these treatments have been at the New England Medical Center. The BAT ultrasound device is used for prostatic target localization on a daily basis and a complex process of quality control ensures positional and dosimetric accuracy. It was noted that the MIMiC vanes bounce slightly which leads to a departure of measured dose from calculated dose. This is particularly problematic in fast switching dynamic radiotherapy. In January 2000 there were over 65 centres and over 6000 completed treatments (Bleier 2000). Grant and Butler (2000) have reported that patient numbers had reached 1200 at Baylor by June 2000.

Woo *et al* (1997) have reported on the early clinical experience with the NOMOS MIMiC at the first centre where treatment has taken place, Baylor College of Medicine, Houston, Texas. By the time of this report (March 1996), 22 patients had been treated with benign CNS tumours (six pituitary adenomas, twelve meningiomas, two neuromas, one optic glioma, one neurocytoma) and seven patients with head-and-neck tumours (recurrent nasopharyngeal carcinoma, recurrent pharyngeal carcinoma, glomus tumour, nasopharyngeal angiofibroma, maxillary sinus carcinoma, soft palate carcinoma and left tonsillar carcinoma). Invasive fixation (the TALON, which secures the skull to the couch) was used but it was recorded that no severe complications arose. At one to three months after treatment, three patients had complete clinical response, three had greater than 50% decrease in size of tumour and 21 had between greater than 25% decrease and no change. It was reported that several patients regained optical or sensory function. It was also established that acute and subacute morbidity was decreased by the IMRT treatment. Woo *et al* (1996) had previously reported that superior dose distributions could be obtained for brain tumours with large irregular shape although the method was not particularly superior to conventional radiosurgery plans for smaller regularly-shaped targets. Carol *et al* (1999) have shown that the choice of whether to use fixed field or arc therapy is somewhat patient dependent.

Grant and Woo (1999) have updated the situation regarding the use of the NOMOS PEACOCK system at Baylor College of Medicine/Methodist Hospital, Houston, Texas. Following the first ever IMRT treatment in March 1994, at the time of writing 40% of all patients are treated with IMRT using two linear accelerators dedicated totally to IMRT. The diseased sites treated are cranial, head-and-neck and prostate. Figures 4.16 and 4.17 show the great power of the PEACOCK IMRT system to deliver high dose to complex target volumes whilst sparing normal tissues. Butler and Grant (2000) have shown reduced xerostomia when irradiating head-and-neck using IMRT.

Grant and Woo (1999) have shown that the Medicare-allowable charges for conventional radiation therapy are higher than those for Peacock IMRT of the head-and-neck but lower than those of PEACOCK IMRT for the prostate. Regarding clinical outcome in head-and-neck cancer, PEACOCK IMRT has kept the volume of the parotid glands receiving greater than 30 Gy to below 10% and xerostomia has been almost universally avoided using the PEACOCK technique.

Woo *et al* (1998) have treated 23 paediatric patients with brain or head-and-neck tumours using the NOMOS MIMiC system, 15 of which had active invasive immobilization. No acute symptoms were reported but longer term follow up was advised. Woo *et al* (2000) reported that Baylor had treated 51 paediatric patients with brain tumours between September 1994 and December 1998 with few complications.

Kuppersmith *et al* (1999) have reviewed the first 28 clinical cases with head-and-neck tumours at the Methodist Hospital, Houston, treated with IMRT delivered with the NOMOS MIMiC. The first treatment was on 21 March 1994, since when ten patients have been treated who had a history of previous (non-IMRT) radiotherapy and 18 patients have been treated *ab initio* with IMRT. Specifically, the acute toxicity was graded according to RTOG criteria and found to be minimal for the ear, eye, salivary glands and skin. Some grade-three mucosal complications were observed when patients were treated to a full dose that covered a large volume of the mucosal membrane within the oral cavity and oropharynx in an accelerated fractionation scheme. Analysis of dose distributions showed that the parotid glands received an exceptional degree of tissue sparing. Kuppersmith *et al* (2000) have performed IMRT of patients with recurrent juvenile angiofibroma.

Augspurger *et al* (1999) have improved the radiotherapy of optic nerve sheath meningioma using the PEACOCK MIMiC system. This was shown to be preferable to surgery in terms of sparing vision.

Mueller (1996) has written a popular account of how the NOMOS system has been clinically implemented at the Mercy Cancer Center, Oklahoma. Here, the primary focus has been on the treatment of head-and-neck disease and brain

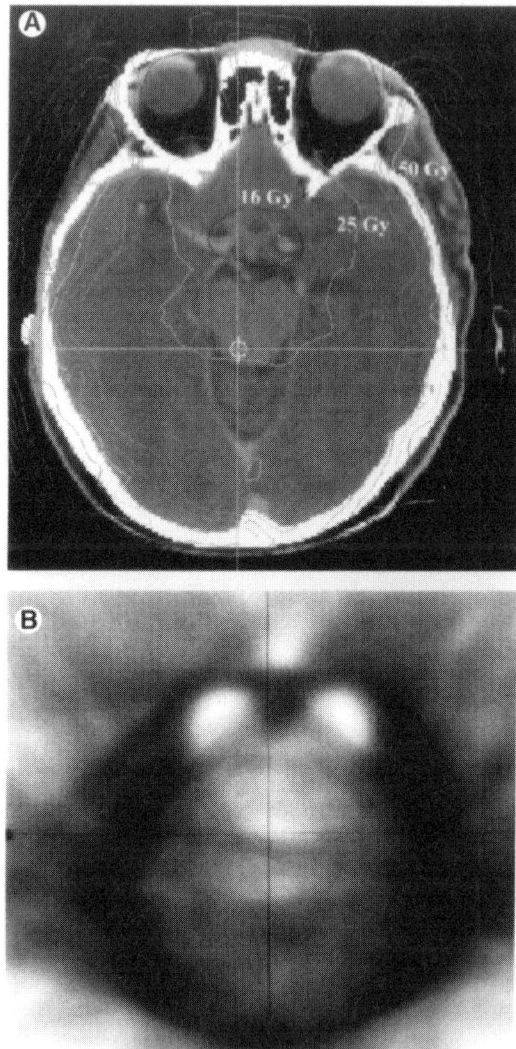

Figure 4.16. *(a) The treatment planned for a melanoma of the scalp. (b) The quality control film for the delivery showing the enhanced sparing of the optic chiasm and orbits available from the use of the NOMOS MIMiC treatment IMRT technique. (From Grant and Woo 1999.)*

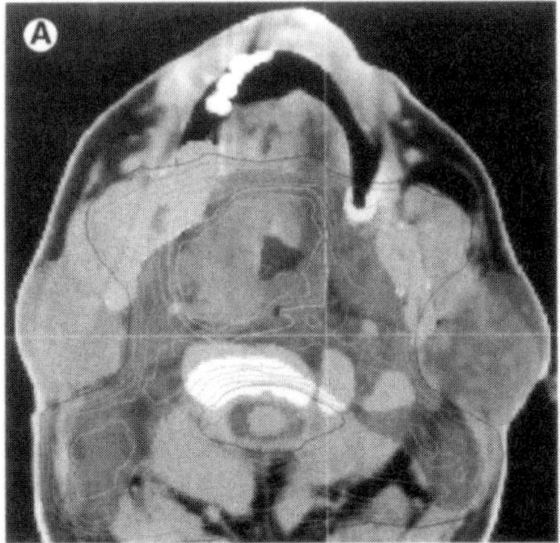

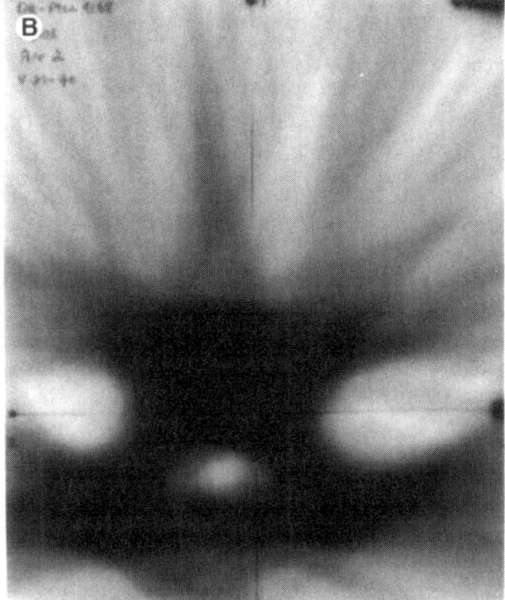

Figure 4.17. *(a) The treatment planned for a head-and-neck cancer patient showing the primary target receiving 2.4 Gy per fraction and the nodal targets, receiving 2 Gy per fraction, which are treated simultaneously. (b) The quality control film shows the distribution and the concomitant sparing of dose to the parotid glands and the spinal cord consequent on use of the NOMOS MIMiC IMRT treatment technique. The quality control of tomotherapy is made using software from a vendor called Phantom Plan which calculates the resulting dose distribution if a patient's plan is delivered to any arbitrary phantom containing a film. (From Grant and Woo 1999.)*

tumours. Lee *et al* (1996) have demonstrated the case study of an improved treatment for optic nerve sheath meningioma treated with the NOMOS system.

The NOMOS MIMiC is capable of IMRT with multiple couch positions, so-called dual-dynamic IMRT. Pagnini *et al* (1999) have shown that the conformality improved for certain brain lesions as more IMRT arcs are included. The effect of including the non-coplanar arcs is striking for the first few and then reaches a point of diminishing returns. When the target is located in the right middle cranial fossa there was little advantage to increasing beyond four arcs. When the tumour was in the cerebello–pontine angle it was advantageous to use seven arcs.

IMRT was observed to have four major advantages for radiation oncology: conformal avoidance of normal tissue, computerized inverse planning, a reduction in the overall treatment time and an emphasis on partial-organ tolerance. Case studies presented have shown the following.

(i) IMRT allowed physicians to re-treat previously irradiated patients with minimal doses to adjacent normal structures (see also Lomax 2000).
(ii) IMRT can be used to trace nerves to the base of the skull and different doses can be delivered to the nerve paths and the primary site.
(iii) With the simultaneous modulated accelerated radiation therapy (SMART) boost technique, different doses can be delivered to a primary tumour in the head-and-neck and to the neck itself which makes possible once-a-day radiotherapy that can be completed in a shorter time (Butler *et al* 1999).
(iv) Multiple targets can be treated while minimizing dose to adjacent normal structures.

Teh *et al* (1998, 2000f) have shown that IMRT can significantly reduce acute toxicity compared with conventional and six-field conformal radiotherapy of the prostate. A balloon rectal immobilization device was used. Uhl *et al* (1999) and Greco *et al* (2000) have created a phantom to model this treatment technique and have shown that the dose to the rectal wall was indeed reduced due to the presence of the balloon whilst the dose to the prostate remained much the same. Teh *et al* (2000a, e) have extended the method to deliver IMRT to post-prostatectomy patients with acceptable complications. Teh *et al* (2000c) have shown that SMART prostate IMRT can compensate for coldspots created by technically difficult seed implants.

Low *et al* (2000) have described a most novel use of IMRT using serial tomotherapy. They propose that high-dose-rate brachytherapy could be replaced by external-beam IMRT. The key to their suggestion is that a rigid applicator substitute is put into the vagina, so localizing the tissues to be treated to a solid object which can also be viewed at CT-planning stage. Hence the high-dose volume can be accurately targeted to the tumour. They show that the DVH of the PTV has then the characteristic sharp profile of external-beam IMRT rather than the characteristic curved profile of brachytherapy. Dose to OARs was also reduced.

4.1.12. IMRT at the Royal Marsden NHS Trust, London

Nutting *et al* (1999) have compared IMRT with conventional therapy for the thyroid and for the parotid gland. IMRT treatments were planned using the CORVUS inverse-planning system. For thyroid cancer the goal was to deliver 60 Gy to the thyroid bed and upper deep cervical lymph nodes and conventional treatment achieved a minimum target dose of 48 Gy with spinal cord dose of 44 Gy. On the other hand, IMRT reduced the spinal cord dose to 30 Gy. In delivering conventional treatment of 60 Gy to malignant parotid tumours, the mean doses to the cochlea, contralateral parotid gland, oral cavity and spinal cord were 48 Gy, 2 Gy, 20 Gy and 20 Gy respectively, whereas the corresponding figures for IMRT were 27 Gy, 2 Gy, 16 Gy and 20 Gy leading to the conclusion that IMRT spared dose to normal structures. Figures 4.18 and 4.19 show typical dose distributions obtained for these two categories of patients indicating the highly conformal IMRT distributions. Wu *et al* (1999c) have also demonstrated that IMRT of head-and-neck tumours leads to better dose distributions and that IMRT planning with a facility in a commercial system was not particularly complicated. Maes *et al* (2000) have presented results of an equispaced five-field parotid sparing treatment technique.

The above observations were based on a comparison of a nine-field equispaced set of beams with a conventional conformal (wedged pair) arrangement. Rowbottom *et al* (2000b, c) have shown that a better arrangement can be obtained with just three IMBs with their orientations carefully selected by a double-loop algorithm. Orientations are selected from an acceptable set which avoid machine collisions, passage through tissues where CT data have not been gathered and direct passage through the contralateral parotid gland. Then, the IMBs for this chosen set were optimized with a modification of the Bortfeld algorithm. A conformality index—a kind of *a posteriori* cost function—was then computed. Then, sequentially, new random choices of orientation were made, the IMBs reformed, the conformality index recomputed and the process continued to maximize the conformality. Rowbottom *et al* (2000b) have shown that the dose to the whole contralateral parotid could be reduced to below 2 Gy. At the same time the advantages for the dose to the cochlea of nine-field IMRT were maintained. This observation seems to contradict some conventional wisdom (see section 1.2) that nine fields are needed. But it was pointed out that there is in fact no contradiction since that view was established largely from studies of cancer of the prostate where the OAR (rectum) was adjacent to the PTV, which is not the case for the parotids which are well separated. Pugachev *et al* (1999, 2000) have also observed improved conformality with optimizing beam directions for IMRT even with nine beams for some tumour sites. Braunstein and Levine (2000) have provided a very complicated mathematical analysis, based on the analogy with X-ray CT theory, of why, for small beam numbers, preferred directions can be found.

Nutting *et al* (2000a) have compared geometrically-shaped two-phase, five-field conformal radiotherapy of the oesophagus optimized to a spinal cord dose

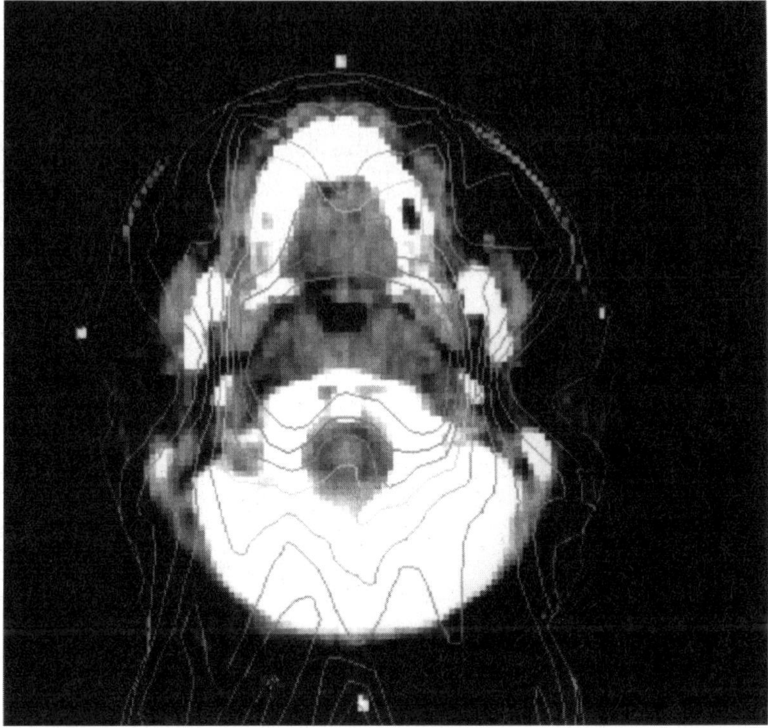

Figure 4.18. *Showing a highly conformal dose distribution with a surface concavity. An IMRT dose distribution is produced by inverse planning to treat a patient with carcinoma of the nasopharynx and allow parotid and brainstem sparing. (Courtesy of Dr C Nutting.)*

of 45 Gy (Guzel *et al* 1998, Bedford *et al* 2000b) with a same-five-field-direction intensity-modulated therapy. The results, averaged over five patients, have shown that the five-field IMRT led to a more homogeneous PTV dose distribution with a lower mean lung dose and lower volume of high-dose to lung than the conventional conformal radiotherapy, leading to a significantly lower lung NTCP. Once again, a comparison of the five-field and nine-field IMRT has shown that the former was superior given that the lung is a parallel organ in which NTCP is mainly reflected by mean dose.

Nutting *et al* (2000c, e) have compared optimized conventional three-field, conformal 3D radiotherapy and coplanar IMRT for irradiation of the prostate with pelvic lymph node involvement. This condition generates a horseshoe-shaped PTV at the inferior border of the sacroiliac joint in the cusp of which lies the small and large bowel. Plans were prepared with the CORVUS system. With conventional radiotherapy the mean percentage of small bowel and colon receiving greater than

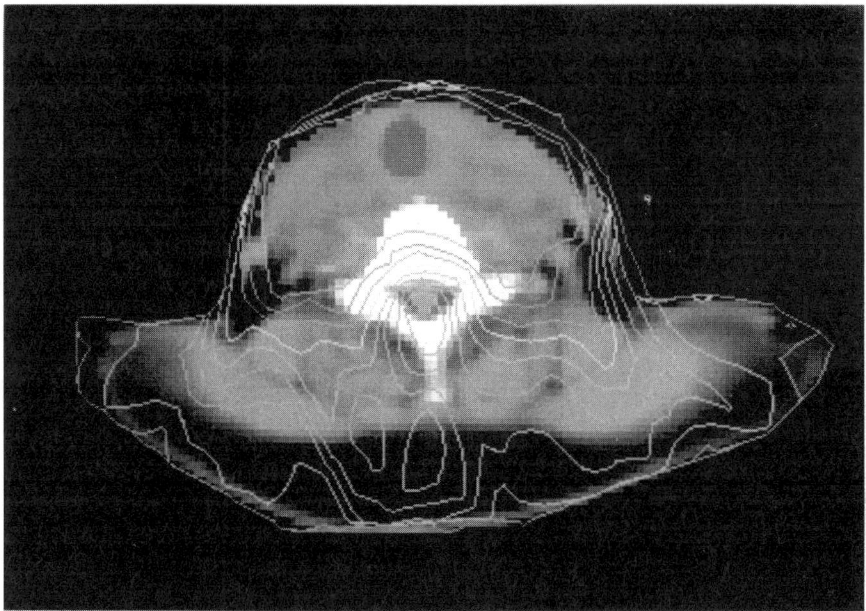

Figure 4.19. *Showing a highly conformal dose distribution with a surface concavity. An IMRT dose distribution is produced by inverse planning to treat the thyroid bed and adjacent lymph nodes while sparing the spinal cord in a patient with thyroid carcinoma. (Courtesy of Dr C Nutting.)*

45 Gy (90% PTV dose) was 21.4%. For conformal therapy it was 18.3% and for nine-field IMRT it was 18.3%. The rectal volume irradiated to more than 45 Gy was 89.4%, 50.5% and 5.8% for the three configurations, respectively. The bladder volume greater than 45 Gy was 75.9%, 52.2% and 7%, respectively. The PTV dose homogeneity was maintained with the IMRT technique. Nutting *et al* (2000c) importantly noted that whilst seven-field IMRT did not reduce these advantages greatly, the advantages slipped away with five- and three-field (equispaced) IMRT. These figures were means for a study of ten patients. It was concluded that dose escalation to the prostate or to parts of the prostate should be possible with IMRT.

Nutting *et al* (2000d) have compared conventional radiotherapy, a conformal two-phase radiotherapy technique and IMRT with nine, seven, five or three beams for treating carcinoma of the thyroid (mean of six patients). The best possible (mean of six patients) with conformal radiotherapy was a mean PTV dose of 51.8 Gy and mean spinal cord dose of 46 Gy (tolerance). IMRT with nine equispaced fields changed these to 59.7 Gy (PTV) and 42.1 Gy (OAR), respectively. The advantages of IMRT held down to the five-beam technique but the three-beam technique was poorer.

Khoo *et al* (1999a) have compared IMRT with stereotactic radiotherapy for convex brain tumours concluding that there was little advantage for the latter, as one may expect. Similar work has been reported by Cardinale *et al* (1998b) and by Kramer *et al* (1998). Ma *et al* (1999c) have compared MIMiC-based IMRT with the gamma knife concluding, for intermediate-size lesions, that the former gave more conformal and homogeneous PTV distributions but irradiated more normal brain.

Breast IMRT at the Royal Marsden NHS Trust is discussed in the next section.

4.1.13. IMRT of the breast

The technique for IMRT of the breast at the Royal Marsden NHS Trust makes use of measuring the breast thickness using an EPID and designing compensation to be added to the tangential wedged pair to reduce the dose inhomogeneity in the breast. The technique has been described in detail by Evans *et al* (1995) and reviewed by Webb (1997d; chapter 4).

Donovan *et al* (1999) have shown that IMRT of the breast can improve the dose homogeneity. Twenty patients were planned with simple wedged tangential fields and replanned using IMRT. The latter decreased the breast volume receiving over 100% dose from 7.3% to 1.3%. Target volume coverage between 95% and 105% increased from 88.8% to 90.4%. It was concluded that, at present, it was too early to decide whether increased dose uniformity correlated with improved late complications. Evans *et al* (1999) have discussed a randomized trial currently assessing the efficacy of IMRT of the breast delivered with compensators.

Partridge *et al* (1999) have shown how the hand-cut lead-fabricated compensators can reliably be replaced by milled compensator moulds filled with a stainless steel granulate. Polypropylene foam blocks were milled to a tolerance of 0.5 mm using an ACD-3 milling machine (Par Scientific A/S, DK-5250 Odense SV, Denmark). A close agreement was shown between profiles recorded on film of test irradiations using the two methods with and without the use of a wedge in the beam. The quality assurance measurements were also presented.

Evans *et al* (1999), Wong (1999) and Sharpe *et al* (2000) have all developed techniques for improving the homogeneity of dose delivery to the breast using breast-compensation techniques based around the MSF method. Typically 85–90% of the dose is delivered using an EPID-verifiable open field topped-up with three to four segments delivering the remaining 10–15% of the dose to fields with different shapes. In particular, Wong (1999) and Sharpe (2000) have combined this technique with active breathing control of patients receiving treatment for breast cancer (see section 4.2). It was observed that the diaphragm changes position during treatment but the outer contour of the breast does not change much.

Hansen *et al* (1997), Bleakley *et al* (2000), Donovan *et al* (2000), Reise *et al* (2000) and Evans *et al* (2000a, b, c) have further described how the top-up fields required for IMRT of the breast can be constructed using the MSF technique. They have shown that the technique is dosimetrically sound, achievable in reasonable clinical times and verifiable. The method begins with the pair of tangential fields whose modulation has been created from portal imaging measurements. Each of the two fields is then decomposed into a wedged pair and three top-up fields. Since the wedged pair comprises weighted open and universal wedged fields this comprises two fields. The limitation to three top-up fields is from the requirement of the planning system used (TARGET-2) to have five fields per direction only. The three top-up fields are created by the 'close-in' technique. A filtering operation ensures that the raw 2D IMB is decomposed into subfields in which the superior–inferior direction is scaled to a leaf width of 1 cm, which can be physically delivered, which do not violate leaf-collision constraints and which do not have a tongue-and-groove underdose. This filtering operation also takes care of the noise in the 2D IMBs resulting from the measurement technique which creates them. The resulting filtered beams can be delivered in either 'clinical mode' in which the beams are described separately or in 'dynamic mode' which uses the Elekta local service mode and in which the three top-up fields are specified as one beam created by control points. Two different methods were tested for dealing with the 'flash region' (the region inside the largest rectangular wedged field but outside the breast contour). Either the minimum intensity was extended everywhere or the intensities were extended from the edge of the breast in the anterior direction. The MLC leaves move in the transaxial direction since the top-up fields have no wedge.

The principal findings were that the physical compensator and the two methods of MSF delivery reduced the volume of breast outside 95% to 105% by an average of 9%. This is expected to have significant clinical impact and will be evaluated in future clinical trials. Clearly, the compensator has to be best because it has no spatial resolution limitation in the direction orthogonal to leaf travel. Dosimetrically, measurements and planning calculations corresponded to within 2%. The argument in favour of the MSF technique (and the second flash technique described above was preferred) is that, in principle, fields can be redesigned to take account of the change in breast volume with progress of treatment (Hector *et al* 2000a, b, c).

Hector *et al* (2000a) have shown that, whilst movement reduces the conformality of the breast IMRT technique, the degraded technique is still more conformal than the non-IMRT method. Hector *et al* (2000b, c) have compared five methods of IMRT of the breast with respect to patient movement. They have: (i) designed a compensator based on the images taken on day one and reused for all subsequent occasions; (ii) designed a compensator for each imaging occasion; (iii) designed multiple-static MLC-shaped fields using just four fluence increments for each imaging occasion; (iv) designed multiple-static MLC-shaped fields using fluence increments equivalent to the compensator for each imaging occasion; and (v) designed a DMLC technique for each imaging occasion. They have found that technique (ii) gave significantly better stability with respect to dose to the breast than (i), indicating that redesigning compensators would be dosimetrically useful (if impractical). However, techniques (iii) and (iv) gave no statistically significant advantage over technique (i) due to the poorer spatial resolution. The fifth technique, the DMLC technique, did lead to a more stable conformal distribution than technique (i) and so could be used to mimic the redesign of compensators.

Ma *et al* (2000a) have shown that a four-field intensity-modulated technique can improve the treatment planning of intact breast radiotherapy when the supraclavicular lymph nodes also required to be treated. The dose-volume prescription was 45 Gy for at least 80% coverage of the intact breast and 50 Gy for at least 80% coverage of the lymph nodes. With a standard conformal plan, only approximately 40% of the lymph node volume received 50 Gy whereas 95% received this for the IMRT treatment. At the same time, IMRT provided more uniform dose coverage for the lymph nodes and dramatically reduced the hot spots in heart and lung.

Tyburski *et al* (1999) have created IMRT of the breast as follows. For each tangential field direction, open field calculations were performed to establish preliminary isodose contours. Then, for each tangent, MLC segments were entered to conform to the isodose lines in 5% increments. No more than five segments were used corresponding to the 110% to 90% isodose lines. Two additional rectangular tangent segments were defined with lung blocks to minimize hot spots in the medial and lateral periphery of the breast. Then, a series of points were defined throughout the breast volume and the inverse-planning capability of the ADAC PINNACLE system was used to optimize the segment weights to make the distribution of dose as homogeneous as possible. No less than 5 MUs were permitted per segment and so if the number dropped less than this, the segment was removed and the optimization restarted. For most patients three segments were adequate and uniformity was of the order ±5%.

The technique of using MSFs to improve homogeneity of breast dose and to reduce dose to lung has also been developed by van Asselen *et al* (2000). They based the technique on the use of CT data and firstly generated maps of the breast thickness. Then, they designed just four MLC-shaped fields to correspond to regions of approximately constant equivalent path length. This accounted for the effect of lung. This was done by eye. The field shapes were transferred to the Nucletron PLATO RTS version 2.2 and field segments were manually weighted by a forward planning process. Approximately 90% of the fluence was delivered by the conventional open field covering the whole breast. This technique, unlike that at the Royal Marsden NHS Trust, does *not* use wedges in the field because the leaves are set in the direction of the transaxial slices. They showed differential DVHs indicating a much more homogeneous breast dose and also noted that the maximum dose moved out of the lung.

Hong *et al* (1999) have studied the planning of ten patients with breast cancer (five left, five right) with conventional wedged tangential fields compared with intensity-modulation created through an inverse-planning process. The goal was to assess whether the use of intensity modulation could improve dose homogeneity to the intact breast, reduce contralateral breast dose, reduce dose to lung tissue and reduce dose to the heart (for left-sided treatments). With conventional treatment there is a known increased risk of myocardial infarction, particularly as more young women are receiving radiotherapy to the intact breast. It is also known that conventional treatment leads to an increased risk of contralateral breast tumours and to lung damage causing breathlessness and cough.

Forty CT images were collected for all patients at 0.5 to 1.0 cm spacing, ranging from 6 cm inferior to the border of the breast to 6 cm superior to the sternal notch. Target and OAR volumes were outlined (a not insignificant task in itself). 'Standard plans' were generated using the usual procedure at Memorial Sloan Kettering Cancer Center whilst, without changing the beam geometry, intensity-modulated plans were created with inverse-planning software using a quadratic dose-based cost function with penalties applied to PTV and OAR doses which violate set constraints. The standard plans were recalculated using inhomogeneity corrections based on CT data, whereas these were automatically included in the inverse plans based on the use of CT data. Beams were extended 2 cm beyond the skin margin in both cases. Further details of the precise PTV determination have been given by Hong *et al* (1999).

Typical intensity profiles for one patient are shown in figure 4.20. These are wedged shaped and fairly smooth. The main wedge shape appears in the section overlying breast tissue, the slope varying in the superior and inferior directions. In the posterior section overlying lung tissue, the intensity decreases to decrease the dose to lung. In the area near the field edge, the intensity increases to sharpen the penumbra, which is a feature of this inverse-planning technique. It was found that the IMRT plans (compared with the standard plans):

(i) improved PTV uniformity—typically the maximum dose was reduced from 120% to 112%;
(ii) decreased the lung dose—typically the lung volume receiving the prescription dose reduced from 10.2% to 6.6%;
(iii) reduced the dose to coronary arteries—the mean dose was reduced from 21.3 Gy to 14.8 Gy and the dose encompassing 20% of the volume decreased from 36.1 Gy to 26.7 Gy;
(iv) reduced the dose to contralateral breast by 35%;
(v) reduced dose to other soft tissues.

Hong *et al* (1999) have commented that: (i) the DMLC technique is better than the use of a compensator because the latter provides too much scatter dose; (ii) the Varian DMLC technique is limited by leaf constraints at present; (iii) the gains described above can be achieved without significant increase in planning complexity.

Lo *et al* (2000) have developed a 'field-within-a-field' technique for IMRT of the breast. They noted that when tangential wedged fields were applied to the breast, even with optimized wedge angles and weights, the dose homogeneity to the breast was some 10% to 25% depending on the patient breast size and that hot spots tended to reside in lung and heart (for left breast). Their technique collimated

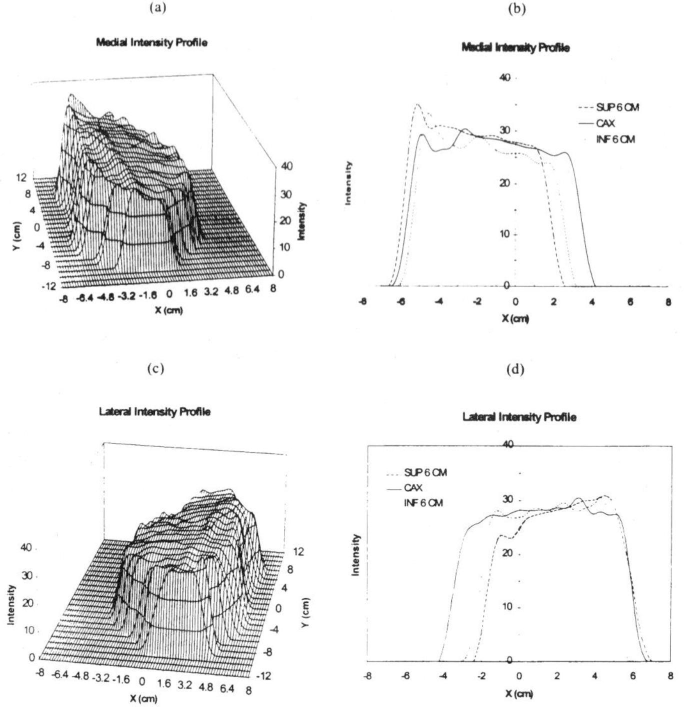

Figure 4.20. *Intensity profiles for the medial and lateral tangential fields for a left-breast patient (intensity in arbitrary units). Field size was* 11 × 20 *cm. The corresponding standard plan had a 30° wedge in both the medial and lateral fields: (a) medial field intensity profile, with positive X direction toward anterior and positive Y direction toward superior; (b) medial field intensity profile through CAX, 6 cm superior and 6 cm inferior planes; (c) lateral field intensity profile, with positive X direction toward posterior and positive Y direction toward superior; (d) lateral field intensity profile through CAX, 6 cm superior and 6 cm inferior planes. (From Hong et al 1999; reprinted with permission from Elsevier Science.)*

the geometric border by use of an MLC to avoid as much lung as possible and to include the internal mammary nodes. The planning was based on the use of CT data and performed with a CMS FOCUS planning system. To improve the outcome, the dose to these fields was reduced by about 10% to reduce the dose to OARs but also causing underdose in the breast. Then a second, smaller, MLC-shaped field was added to bring the breast dose back to prescription. The effect of this dual-field treatment was to reduce the breast inhomogeneity to 7–15% and to place the residual hot spots in the breast rather than in normal tissue. Also, the nodes receive a more homogeneous dose. The results of 20 patient cases have been presented. We may note that the breast homogeneity is not as great as that created by the Royal Marsden technique but Lo *et al* (2000) do comment that further subfields could be added.

Chang *et al* (1999b) have made a detailed measurement study of the dosimetry consequent on the use of eight different intensity-modulation techniques for treating the breast. These included the use of conventional wedged pair tangential fields, an LM wedge only, 2D intensity modulation delivered by a compensator either to LM only or both tangential fields, 2D intensity modulation delivered by the MLC-MSF technique, either LM only or both tangential fields. The multileaf modulation method used the Siemens LANTIS delivery system and five levels of intensity per field which led to up to 14 field segments to be delivered. The planning was done with the PLanUNC treatment-planning system with fields sequenced by IMFAST. The compensators were tin-filled milling-machine-cut devices. The goal was to determine the treated breast homogeneity, the dose to the contralateral breast and the overall treatment time. The planning system PlanUNC was compared with the NOMOS CORVUS system by Svatos *et al* (2000).

Chang *et al* (1999b) have found that the techniques in which both tangential fields were modulated created the best dose homogeneity to the treated breast and the least contralateral breast dose. There were small differences due to the finite spatial resolution of the MLC-MSF technique and the actual magnitudes depended on the geometry. They also scored the eight techniques according to the time required for their execution. Not surprisingly, the multileaf modulation technique (especially when applied to both tangential fields) scored low given the (current) large inter-segment treatment time delay of the Siemens system. Chang *et al* (1999b) decided not to recommend a single technique but to score them according to the priorities of the user. In some circumstances the use of just a single tangential modulation may be appropriate.

Iori *et al* (2000) have compared a variety of techniques varying the beam number, energy and locations. They concluded that the optimum set-up depended on the irradiated geometry, specifically whether nodes were included in the PTV. Teh *et al* (2000b) have shown that the SMART boost technique and IMRT can lead to improved dose homogeneity in the breast for treatments of pectus excavatum.

4.1.14. IMRT at Medical College of Virginia

Wu *et al* (2000) have made a planning study of the potential for sparing the parotids and escalating the dose to targets in treating head-and-neck carcinoma. They compared CFRT conventional 3D plans with IMRT with equally-spaced beams and varied the number from fifteen, nine, seven to five. Plans were optimized using an in-house constructed planning system and delivered with the 'step-and-shoot' technique. Inverse planning made use of either a dose-based cost function, essentially a quadratic cost with penalties for exceeding specified limits, or a dose-volume constraint set. The goal was to keep the dose to 25% of the parotid volume less than 30 Gy. It was found that certainly 50% of the volume could be kept this low using the fifteen or nine-field IMRT technique. There was no perceptible improvement in reducing from fifteen to nine but the IMRT suffered from reduction to seven and five fields. For this reason it was concluded to use nine fields for clinical practice. Wu *et al* (2000) pointed out that, provided a sufficient number of beams were used, there was little need (in their opinion) to optimize beam directions, a difficult task. However, they acknowledged that even fewer beams could have been used if such orientation optimization could be carried out (as in fact achieved by Rowbottom *et al* (2000b); see section 4.1.12).

4.1.15. IMRT with combined modalities

Li *et al* (1999a) have combined an electron field with four intensity-modulated fields for IMRT of the breast. Two of the photon fields were in the usual tangential directions whilst the other two were angled by 25° with respect to these. It was shown that this arrangement reduced the dose to the contralateral breast and to the heart, and also was less susceptible to patient breathing motion.

Li *et al* (2000) have described clinical applications of combining photon IMRT created with the DMLC technique with electron beams. The idea is that when electron beams are combined with unmodulated photon beams, the penumbra of the two modalities do not match, leading to hot and cold regions by the side of a junction line. However, when electron beams are combined with intensity-modulated photon beams, the profile shows no such regions of over and under dose. This is because the intensity modulation of the photons effectively broadens the photon penumbra and makes it complementary to the electron penumbra at a specified depth. At the same time, this technique makes the combined modality delivery less dependent on positioning errors (see also section 4.2.1).

Li *et al* (1999b) have given details of the calculation technique. The photon and electron fields are overlapped by 2 cm. Then, an optimization routine (Xing *et al* 1997a, 1998b) is called to minimize a quadratic dose cost function between the prescription and the sum of the photon and electron doses. This leads to a modulation of the photon field in the overlap region. This modulation was fed to an interpreter (Ma *et al* 1998a) and the combined fields were then delivered to an experimental phantom. It was shown that the measured dose profile became more uniform than when combining electron and unmodulated photon fields and that measurements matched closely the calculated profile. The technique was applied to head-and-neck tumours and also to a follicular lymphoma decreasing the dose inhomogeneity to the target.

Ma *et al* (2000b) and Lee *et al* (2000) have developed an electron multileaf collimator consisting of 30 steel leaves, each leaf being 0.476 cm wide, to deliver intensity-modulated electron fields for treating breast cancer. They have shown that intensity-modulated fields improve the homogeneity of dose to the breast and reduced the volume of lung treated to high dose.

4.1.16. Other IMRT studies

Meeks *et al* (1998b) have conducted a planning study to show when IMRT plans, computed via PEACOCKPLAN, are superior to those created by BEV shaping alone or, worse still, 'conventional' 2D planning. Plans were prepared for several patients with brain tumours, head-and-neck cancers, breast cancer, prostate cancer and lung cancer. The plans were scored dosimetrically and also in terms of the overall probability of uncomplicated tumour control. They have shown that radiotherapy for convex brain tumours was no better performed by IMRT than by stereotactic multiple-arc therapy. Also, breast radiotherapy was not improved by arcing IMRT. For the breast, a couple of fixed fields are best and these could indeed be modulated with consequent improvements. Radiotherapy of the prostate, lung and head-and-neck tumours was always better with arcing IMRT than with simple geometrical field-shaping.

Wijers *et al* (1998) have demonstrated reduced xerostomia when using IMRT, delivered by the DMLC technique, for cancers of the oropharynx and larynx, including the elective neck. Levendag *et al* (1998, 2000) have studied whether the DMLC technique of delivering IMRT can result in improving the dose distributions of head-and-neck tumours. Thirty patients with midline tumours were studied. Conventional plans had led to 46–70 Gy to the salivary glands with residual salivary flow less than 10% of baseline. With 3D target definition of the primary and elective neck and IMRT, between 25–72% and 43–85% of the parotid volume and 0–15% and 3–20% of the submandibular volume received a dose less than 40 Gy for soft palate/tonsillar fossa and supraglottic larynx tumours respectively (see also Lagerwaard *et al* 2000). Nowak *et al* (2000) have carried out a multi-institutional study (Amsterdam, Ghent, Brussels, Ann Arbor and Rotterdam) on one patient demonstrating a clear advantage to IMRT in terms of reduced xerostomia.

Kachnic *et al* (1998) have shown the improvement of treating oesophageal cancer by IMRT using a variable number of fields. Williams *et al* (1998) have shown that multiple field arc IMRT of the breast has dosimetric advantages.

Broggi *et al* (1999) have compared the use of IMRT for treatment of prostate cancer with a conventional prostate cancer technique. The IMRT plans were generated using just five equispaced beams and the one-dimensional single absorber modulation technique previously reported by this group. The IMRT technique generated concave-shaped treatment volumes. It was shown that the average gain in tumour control probability with a normal tissue complication probability set to 5% when considering IMRT versus non-IMRT plans was 7.3%. Similar increases were noted in other biological indices.

Yi *et al* (1999b) have investigated whether it is better to perform static-gantry IMRT or dynamic-arc IMRT. They investigated this question for five sites: brain, head-and-neck, breast, lung and pelvis. No single technique proved superior—the outcome depended on the site. At least for the cases investigated it was found that dynamic IMRT was more suitable for the brain, static-gantry IMRT was more suitable for the head-and-neck, lung and breast, and the prostate was equally well treated with either.

Lagendijk and Hofman (1999) and van der Heide *et al* (2000) have shown that a micro-boost to the solid tumour centre can improve the local control of solid tumours and that this micro-boost can be integrated concomitantly with the last fractions of normal dose delivery using intensity-modulated techniques.

Chao *et al* (1999) have also compared conventional therapy with MIMiC tomotherapy and with static-field step-and-shoot IMRT for nasopharyngeal cancer. They concluded that parotid sparing was improved with IMRT and that the static-field technique was better than tomotherapy for sparing normal tissue but at the expense of homogeneity in the target.

Sultanem *et al* (1999) have observed almost no cases of locoregional failure in IMRT of nasopharyngeal cancers but Dawson *et al* (1999) made a different observation that several in-field recurrences did occur and that there was a need for dose escalation. The outcomes depended on the clinical history.

Bues *et al* (1999a, b) have linked the KONRAD inverse-planning system to the Radionics microMLC and have shown that IMRT created by the DMLC technique is superior to large MLC IMRT. The segmentation algorithm takes into effect the finite leaf transmission and also penumbra effects.

Parker *et al* (1999) have combined IMRT fields to the spinal cord with other fields delivered with the asymmetric jaws to improve the dosimetry of whole CNS irradiation. A simple step-and-shoot DMLC technique was used.

Klein *et al* (1999b) have developed a technique to deliver a different dose distribution to the prostate alone as to the prostate plus seminal vesicles. The method uses seven fields, each with two segments appropriately weighted but each delivered by the DMLC technique. The calculations were performed by forward planning. Experimental phantom measurements agreed to 1% with the calculations.

Portelance *et al* (1999) have shown that IMRT delivered via either four, seven, or nine fields can significantly improve the dosimetry of pelvic lymph nodes compared with more conventional radiotherapy. The amount of small bowel, rectal volume and bladder included in the high-field region decreased with IMRT. Planning was performed with CORVUS. Robertson *et al* (1999) have shown that radiotherapy of rectal carcinoma would also benefit from IMRT, sparing more small bowel when compared with conventional therapy. Kavanagh *et al* (1999) have shown that IMRT used to create a butterfly-shaped PTV for treating cervical cancer can significantly spare rectal tissue. *In vivo* dosimetry confirmed the accuracy of the technique.

Schmidt *et al* (2000) have interfaced the NOMOS CORVUS planning system to a Siemens DMLC system and within three months of installation had treated seven patients for different tumour sites. They reported almost a turn-key system.

Olch and Lavey (2000) have pointed out that in irradiating children's cancers it is important to reduce doses to growing OARs below the levels tolerated by adults. They used the PLATO system to show that nine-beam and five-beam IMRT of neuroblastoma led to improved conformality.

Mott *et al* (2000b) have studied the improved conformality with IMRT for treating the maxillary antrum studying the effect of varying the beam number and orientation.

Conway *et al* (2000) have made a study of the improved conformality of IMRT for head-and-neck patients. They used the HELIOS planning system and a cohort of 31 patients. However, Cozzi *et al* (2000b) have questioned the use of IMRT which, whilst improving OAR sparing, led to reduced target coverage in their study.

4.2. IMRT AND MOVEMENT

4.2.1. Movement studies and models for IMRT: 'smoother' IMBs

Yu *et al* (1997a, b, 1998) have studied the effect of intra-treatment movement during the delivery of IMRT via the sliding-window DMLC technique. In this technique, if we ignore radiation transmission through the collimator for the purposes of studying movement effects, a point receives radiation only if it is in the 'temporal tunnel' between the arrival of the leading leaf and that of the trailing leaf (figure 4.21). If the patient moves during the irradiation, points can move into and out of the temporal tunnel and the irradiation will then depend on the speed of the leaves and on the amplitude and the phase of the movement. It also depends on the width of the leaf-opening aperture. Yu *et al* (1997a, b, 1998) have created a model of this problem and have shown that the phenomenon could lead to 100% errors in the delivered dose. It can be easily appreciated that the issue is one of 'phasing' or 'beating' between movement of the leaves and leaf-gap and movement of the target. The same problem arises in tomotherapy (see below). It is for this reason that techniques such as active breathing control have been developed (Lebesque *et al* 1998) (see also later this section).

Highly conformal dose distributions can be generated by intensity-modulated radiotherapy. Intensity-modulated beams (IMBs) are generally determined by inverse-planning techniques designed to maximize conformality. Usually, such techniques apply no constraints on the form of the IMBs which may then develop fine-scale modulation. Webb *et al* (1998, 1999) have presented a technique for generating smoother IMBs, which yielded a dose distribution almost identical to that without the constraint on the form of the IMBs (figures 4.22–4.24). The method applied various filters successively at intervals throughout the iterative inverse planning. It was shown that the IMBs, so determined using a simple median window filter, have desirable properties in terms of increasing the efficiency of delivery by the DMLC method and may be 'more like conventional beams' than unconstrained, highly-modulated IMBs. Given that they are 'more blocky' than unconstrained IMBs they also should be less sensitive to movement. This idea of median window filtering (MWF) has been developed further by Kessen *et al* (2000) and incorporated into the KONRAD system. Filtering took place after each KONRAD iteration. They showed that a 3×3 MWF was not successful and

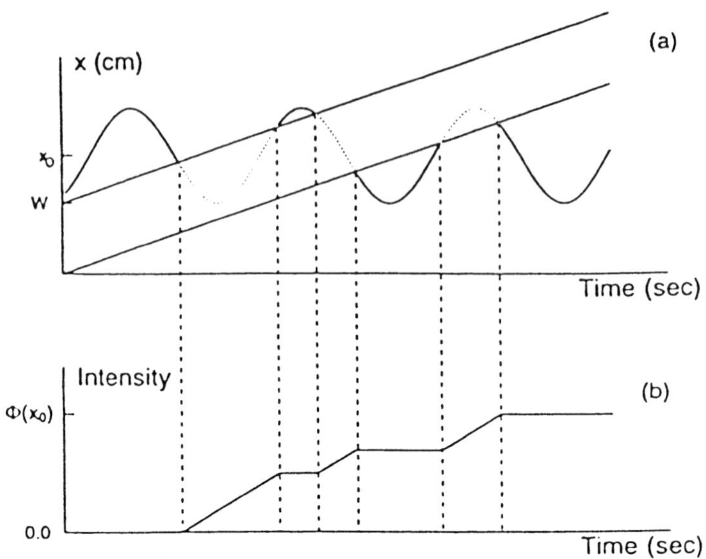

Figure 4.21. *A one-dimensional model for analysing the effects of intra-fractional organ motion. The target point at x_0 is making a sinusoidal motion while a slit beam scans across it. The cosine wave in the top diagram (a) represents the position of the point as a function of time and the two lines represent the position of the two beam edges. Only when the point is within the beam does the point receive radiation. The accumulated beam intensity received by the point as a function of time is shown in (b). (From Yu et al 1998.)*

instead continued to use line-by-line filtering but operating on data with high resolution along the leaf-travel direction. Llacer (1998) has also constrained the inverse-planning problem with Bayesian smoothing. This disallows large variations of fluence from adjacent spatial locations in the IMB. It was shown that smoother beam profiles could be generated without any substantial loss of conformality. Shepard *et al* (2000) have also commented on this, pointing out that it is a consequence of the ill-conditioning of the inverse problem and the degeneracy of certain cost functions.

Alber and Nüsslin (2000) and Alber *et al* (2000) have developed a similar idea in which the IMBs along the direction of leaf movement are smoothed by incorporating a minimization of the second derivative of the fluence profile. This models the IMB as a rubber membrane with a given tension whose curvature is to be minimized. Laub *et al* (2000a) have compared the planning code Hyperion which incorporates this smoothing with KONRAD for IMRT of colorectal cancer, preferring the former because it generates significantly less segments with all the attendant advantages this has.

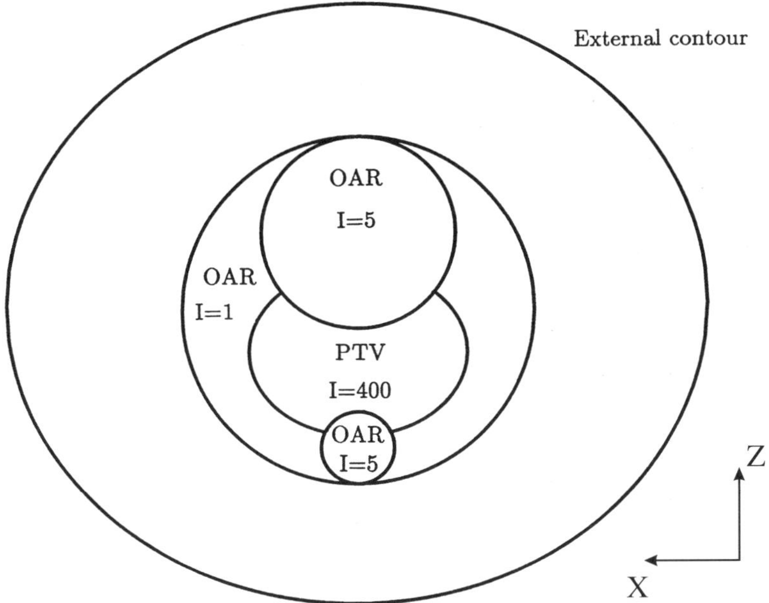

Figure 4.22. *A model geometry representing a prostate PTV with a concave outline with adjoining OARs, bladder and rectum. The aim is to create a high-dose region conforming to the shape of the prostate whilst sparing the OARs. The 'I' values are importance factors assigned during the inverse planning.*

Another reason why creating IMBs which are as smooth as possible is a good concept is that strong gradients lead to small sliding gaps in the DMLC technique and small gaps are themselves associated with potentially larger errors due to leaf imprecision, due to the delivery of small numbers of MUs and due to the difficulty in establishing an output factor (see section 3.2.6).

In the MLC-MSF technique smooth beams lead to the need for fewer segments and thus reduced deadtime between segments and shorter overall treatment times.

Yang *et al* (1997) have made a thorough and detailed study of the matchline issue in spiral tomotherapy and MIMiC-device delivered (non-spiral) tomotherapy. The study included a series of measurements which determined the importance of breathing motion and also indicated some desirable parameters for gantry rotation speed (in spiral tomotherapy) and slit beam size in relation to the amplitude and periodicity of breathing. The motion of importance is superior–inferior motion because the patient would be translated longitudinally in both techniques.

Yang *et al* (1997) have constructed a phantom-positioning device for simulating these two methods of delivery, shown in figure 4.25. The phantom could

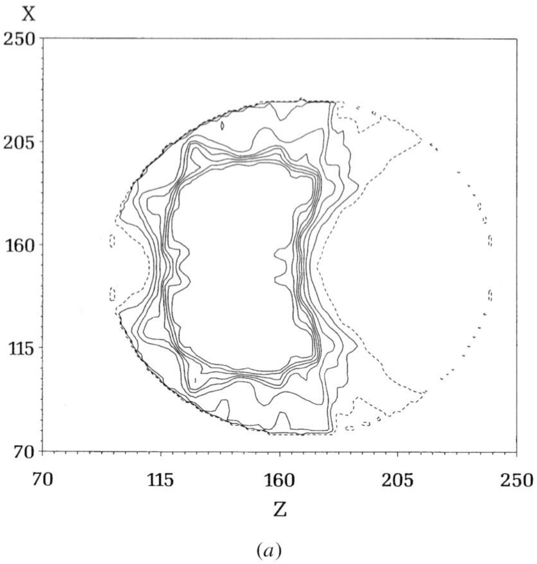

(*a*)

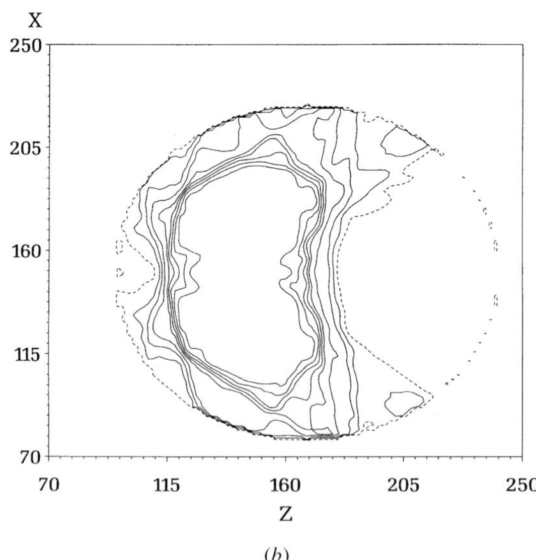

(*b*)

Figure 4.23. *The result of (a) unconstrained inverse planning and (b) constrained inverse planning, showing the 2D dose distributions and the relative conformality developed. The highest (inner) isodose line is 90% of the maximum and the isodoses are shown at 90%, 85%, 80%, 75%, 70%, 60%, 50%, 40%, 30%.*

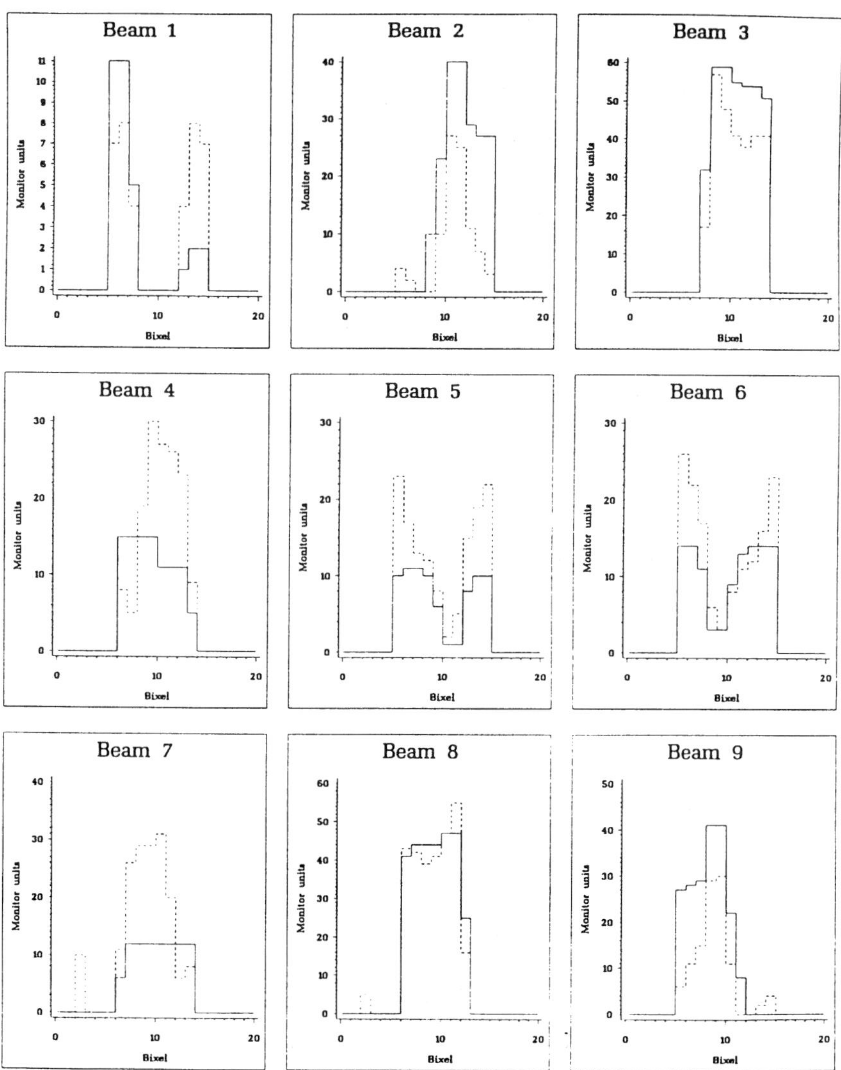

Figure 4.24. *Shows the nine IMBs (at 40° intervals of gantry orientation around the slice) resulting from the computation, with unsmoothed beams represented by dotted lines and smoothed beams by full lines.*

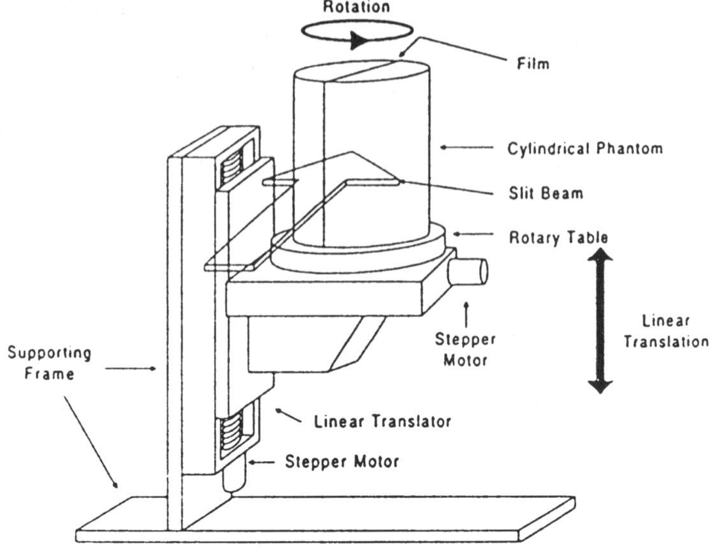

Figure 4.25. *A schematic drawing of the phantom positioning device used to simulate helical tomotherapy beam delivery and sequential rotational beam (MIMiC-device-like) tomotherapy. (From Yang et al 1997.)*

be rotated on a turntable and the whole turntable platform could be translated either continuously during the rotation (simulating spiral tomotherapy) or in discrete steps between each discrete rotation (simulating MIMiC-device delivery). A cylindrical water phantom, composed of two hemicylinders enclosing between them a film, was mounted on the platform for making measurements of the dose distribution primarily in the longitudinal direction. The gantry rotated at 4 r.p.m. and, for spiral tomotherapy, the translation was arranged as a pitch of 1, i.e. the translation was by one field width (generally 2 cm) during each rotation. Measurements were made with the radiation beam static and also oscillating at between 5–20 cycles min^{-1} with amplitudes between 0.5–2.0 cm.

Using these techniques the following conclusions were reached for helical tomotherapy.

If the beam were *static* simulating treatment of a *motionless patient*:

(i) The measured central longitudinal dose profile for one rotation of the helical beam was an approximately triangle function of width 4 cm (simply the convolution of the square beam profile of width 2 cm with itself).
(ii) The longitudinal dose profile along the phantom axis for continuous multiple-rotation helical delivery was approximately flat.
(iii) The relative flatness of the dose profile was independent of changes in the

beamwidth but the absolute value of the peak changed, e.g. if the beamwidth were 1.9 or 2.1 cm instead of 2.0 cm, the height of the plateau region changed by −5% and +5% respectively. This indicated that spiral tomotherapy is insensitive to these changes. The insensitivity of the profile shape is, as mentioned above, due to the 'forgiving' nature of the superposition of the triangle functions. The height change could be accommodated by adjusting the MUs.

If the beam simulated treatment of a *moving patient*:

(i) With a frequency of simulated target motion of 8 cycles min^{-1} and oscillations of magnitude 0.5 cm and 1.0 cm, up to ±5% hot and cold spots in dose were delivered (figure 4.26).

(ii) With an amplitude of 1.0 cm, and the motion varied as 5, 10 and 20 cycles min^{-1}, the magnitude of the hot and cold regions varied enormously, being as large as ±15% for 5 cycles min^{-1} but almost unmeasurable for 20 cycles min^{-1} (figure 4.27). This is due to the phasing of simulated motion with respect to the rotation frequency. When the rotation frequency and the simulated breathing frequency were very close, large variations in dose arise, simply because the moving target either tracks the movement of the beam ('beating') or 'anti-tracks' it. The measurements suggest the advantage to rapid breathing in spiral tomotherapy. The amplitude is of less importance.

(iii) When the simulated motion is at 10 cycles min^{-1} and an amplitude of 1.0 cm, the variation of dose falls with decreasing beamwidth, 0.5 cm being best. However, even with a beamwidth of 2.0 cm the variation in dose was only an acceptable 2%.

Conversely, when the delivery was by discrete-step discrete-rotation (MIMiC) then, if the field width slightly mismatched the translation step (e.g. 2.1 cm or 1.9 cm instead of the exactly required 2.0 cm), hot or cold spots would arise of magnitude about 15% *without patient movement*. However, introducing a patient movement of 10 cycles min^{-1} and an amplitude of 1.0 cm, the variation of dose falls back to about 7%, i.e. *movement actually improves this kind of non-spiral tomotherapy*.

This is an important study which gives insight into the relative merits of the two methods of tomotherapy. It shows primarily that spiral tomotherapy is little affected by motion provided this motion is rapid with respect to the gantry rotation motion; this method is also very insensitive to the exact beamwidth. Conversely, the rotate-then-translate method is more sensitive to precise beamwidth and motion seems to actually improve the dose distribution when such mismatches occur. An unstudied effect is that of the randomization on a fraction-by-fraction basis of the movement. The study presented (and also that of Yu *et al* 1997a, b, 1998 above) was for movement at a particular fraction. Chui (1999a) has also studied the effect of movement in breast radiotherapy using the MIMiC technique and concluded that, provided the magnitude of motion was less than 3 mm, the effects

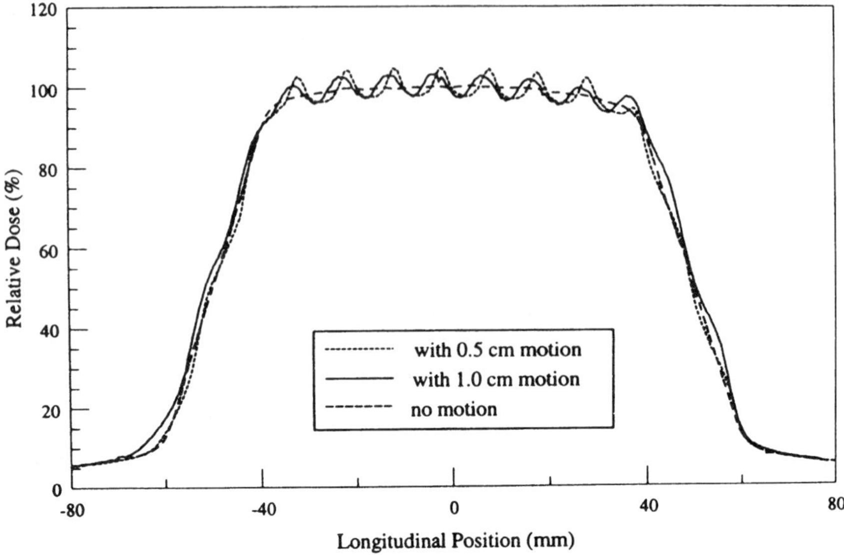

Figure 4.26. *The measured longitudinal dose profiles for helical beam delivery along the phantom axis of symmetry with simulated longitudinal target motion. The target motion frequency was 8 cycles* min^{-1} *with magnitudes of 0, 0.5 and 1.0 cm. The beam size was* 8.0 × 2.0 cm^2. *The helical beam was rotated at 4 r.p.m. and linearly translated at 2.0 cm per rotation. (From Yang et al 1997.)*

were minimal. Fitchard *et al* (1998a, b) have created an algorithm that adjusts the delivery sinogram for the Wisconsin tomotherapy device to account for patient movement between radiotherapy fractions. Fitchard *et al* (1999) have shown the experimental verification at the University of Wisconsin tomotherapy workbench.

The first few hundred IMRT treatments with the NOMOS MIMiC have been carried out for tumours in the head-and-neck, partly because this site represents a real problem for achieving conformal radiotherapy and partly because immobilization of the head-and-neck is more straightforward. The NOMOS system uses a device known as the TALON which is attached with bone screws to the inner table of the skull (see figure 2.4). The device is removable and can be re-attached reproducibly for each fraction. The TALON is attached to the NOMOGrip and CT pointer at the time of 3D CT imaging allowing the image coordinate system to be established accurately in relation to the treatment system (Curran 1997).

Low *et al* (1997b) have studied the effect of systematic gantry and collimator angular rotation errors on the dose delivered by IMRT with fixed portals. The way they did this was to take some 2D IMRT maps created by inverse planning and then, after disturbing the fluence maps by a 2° angular error in either the gantry

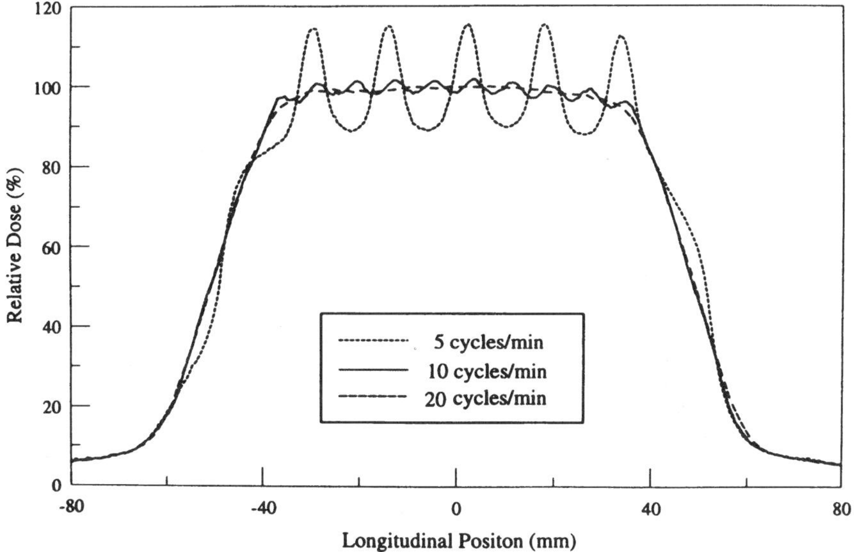

Figure 4.27. *The measured longitudinal dose profiles for helical beam delivery along the phantom axis of symmetry with simulated longitudinal target motion. The target motion frequency was 5, 10 or 20 cycles min*$^{-1}$ *with magnitudes of 1.0 cm. The beam size was* 8.0×2.0 *cm*^{2}. *The helical beam was rotated at 4 r.p.m. and linearly translated at 2.0 cm per rotation. (From Yang et al 1997.)*

angle or the collimator rotation angle, convolve the fluence distributions with the appropriate dose kernels for 6 MV and for 50 MV radiation. For 6 MV beams the magnitude of the dose error was in general numerically much smaller than for 50 MV beams and the area encompassed by these errors was much smaller.

Xia *et al* (2000) have concluded from a modelling study of the effect of patient rotation within a mask that the outcome of IMRT would not be degraded beneath the planning constraints. They studied five patients with nasopharyngeal tumours.

Xing *et al* (2000d) have also studied the dosimetric consequences of gantry, collimator and couch rotation errors and also longitudinal displacement errors for the NOMOS MIMiC technique for IMRT delivery. To do this they used the NOMOS CORVUS treatment-planning system and made 'optimized plans' for two clinical cases with no errors in the equipment placement. Then, for each planning case, they replanned making a deliberate error ΔX_i in some quantity Q to determine the discrete partial derivative $(\delta Q/\delta X_i)$. For example, in each plan (which had nine fixed gantry orientations) they varied just one orientation to find the partial derivative with respect to change in that orientation. The same was done for the variations with respect to position. Changes were observed in dosimetric quantities (mean, min and max) pertaining to PTV and OAR. It was noted that the effect of position changes were much greater than the variations with respect to orientation. Using these data and the principle of linearity, they were able to state that the total change in quantity Q due to several simultaneous changes was $\Delta Q = \sum_i (\delta Q/\delta X_i)\, \Delta X_i$. This linearity was verified for selected sets of combined changes.

Kung and Chen (1998) have also studied the effect of misregistration of the patient when treated with the NOMOS MIMiC system. They first computed an IMRT dose distribution for the patient in the desired position. Then the patient data were translated incrementally in steps of varying length along the three orthogonal coordinate axes and the dose distribution recalculated. It was concluded that a 1 cm translation could lead to an underdose or overdose of some 25% reducing to about 10% for a 5 mm error.

Xing *et al* (1997b) have shown that if an intensity modulation is added to the edge of otherwise unmodulated fields and then two such fields (with opposing modulation, e.g. a ramp-up and ramp-down) are overlapped then the dose homogeneity in a PTV becomes less sensitive to motion as the degree of overlap is increased. The IMRT can be delivered by the DMLC method. Li *et al* (1999a, b, 2000) have also shown that, when modulated photon beams are abutted to unmodulated electron beams, the dosimetry is both more uniform and less sensitive to movement error (see section 4.1.15)

Löf *et al* (1998) have made a theoretical study of movement with respect to IMRT. It is still an open issue which IMRT treatment delivery technique is least sensitive to tissue movement. One can certainly appreciate that 'beating' must be avoided. Either the patient must breath fast or not at all (!) with respect to equipment movements, or some method to gate the treatment is required (Kubo 1998). MSF delivery and the compensator may avoid these problems.

Samuelsson *et al* (2000) have studied the effects of set-up mistakes of 0.5 cm in six directions for a seven-field IMRT plan of the head and neck. They concluded that errors of these size had a negligible effect due to the margin already added to create the PTV. However, they noted the important point that the dose normalization point must be carefully chosen. Because PTV dose distributions are usually inhomogeneous, prescribing to the isocentre is inappropriate. They recommended prescribing to the mean dose in the PTV.

Jones and Hoban (2000) have shown that both single set-up errors and random movements each fraction led to changes in the equivalent uniform dose (EUD) for PTV and OARs.

The studies by Hector *et al* (2000a, b, c) of the effect of movement on IMRT of the breast have already been described in section 4.1.13.

4.2.2. Movement control in IMRT

One of the major potential difficulties with IMRT is the possibility (probability?) that movement during the therapy will compromise the advantages of the method. Kubo (1998) has reviewed the methods which have been attempted to gate out movement. Some of the first attempts were made by those engaged in proton radiotherapy (Ohara *et al* 1989). Kubo and Hill (1996) have reviewed the technical feasibility of respiration-gated therapy and details of the UC Davis system were given by Kubo *et al* (2000). Sontag (1999) pointed out that the point of maximum exhalation is probably the most stable point on which to gate a radiation treatment. There are several types of apparatus to measure changes in the abdomen (e.g. impedance plethysmography and pneumotachometry) and some radiation systems can be gated.

Mageras (1999, 2000) has classified respiratory-motion-induced treatment control into two types of technique. In the first, the linear accelerator is respiration gated while the patient breathes freely and the second attempts to control patient breathing, so that the radiation is delivered at certain intervals in the patient's respiratory cycle. Mageras (1999) has measured that the tumour position for lung tumours varies by 1.0–2.5 cm and for liver and kidney tumours by 1.5–3 cm due to breathing. Strategies for overcoming the unwanted effects on dosimetry include gated radiotherapy, voluntary or forced breathholding and hybrid combination. At the Memorial Sloan Kettering Cancer Center a technique has been developed asking patients to deeply inspire, then deeply expire prior to voluntary breath-hold. Lung NTCP decreased allowing dose escalation.

Wong *et al* (1997a, b) have developed active breathing control (ABC) (figure 4.28) (Sharpe *et al* 1999). MRI movies have shown the lung location to move dramatically (up to 2 cm) during breathing, not surprisingly. The ABC technique customizes the length of breath-hold and the radiation is only turned on during the active breath-hold. The length of this period can be customized to the patient depending on what can be comfortably tolerated (figure 4.29). It has been argued strongly that the use of single CT scans is dangerous for dose-escalated therapy. Wong *et al* (1998a) have applied the technique to irradiation of the liver.

Wong *et al* (1999) have stated that in the absence of ABC the total beam aperture expansion to take account of breathing, set-up variations and beam penumbra can be as large as 2.5 cm, giving rise to significant risk to normal structures. The William Beaumont Hospital ABC device potentially overcomes these problems. The prototype was a modified Siemens 'servo' ventilator model 900C. The flow monitors and scissors valves are interfaced to a personal computer which can select a period of breath-hold of varying length, at different flow directions and different lung volumes or different functional residual capacity. This has been tested with twelve patients, four with Hodgkin's disease, four with lung cancer and four with liver cancer. During normal breathing in training sessions patients preferred breath-hold either at deep inspiration or at the beginning of expiration, the opposite of that proposed by Kubo and Hill (1996). All liver and

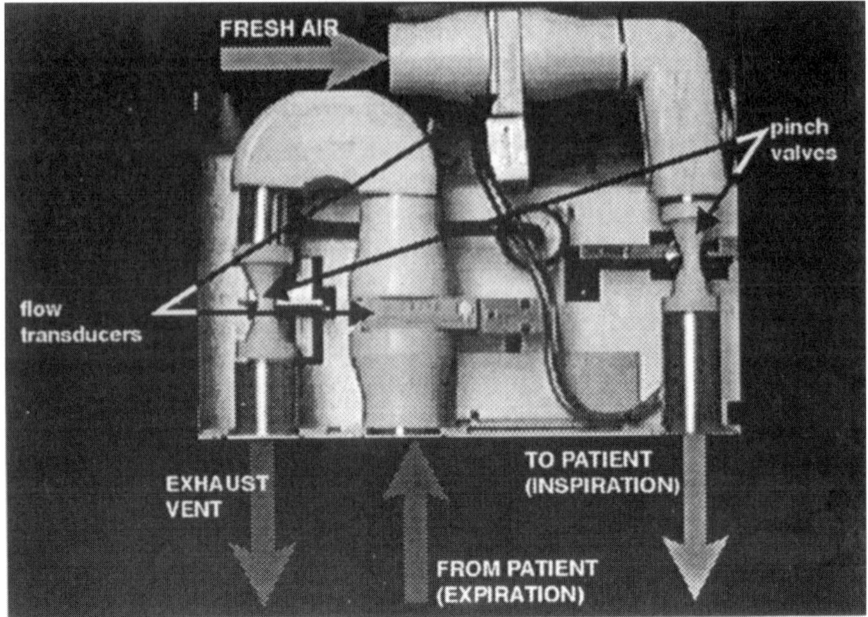

Figure 4.28. *This figure shows the arrangement of the flow monitors and scissors valves of a modified ventilator for active breathing control. Figure 4.29 shows the real time display of the air flow and lung volume for a normal subject during normal breathing. The ABC level is also shown. Figure 4.29 shows the display where the ABC was activated to hold the subject breathing for 5 s. (From Wong et al 1997a.)*

Hodgkin's patients could tolerate a breath-hold period of 20 s, some much longer. Lung patients could tolerate 15 s.

Figure 4.30 shows an example set of CT slices showing that the ABC device effectively generates a very similar set of CT images in the upper thorax whereas the non-breath-hold images show considerable artefacts due to lung motion. Analysis showed that lung edges in the two ABC CT profiles were within 3 mm of each other. It was established that projection images were *not* suitable for evaluating changes in lung profile because of the non-uniform nature of 3D organ motion. This somewhat argues against the concepts developed for the University of Wisconsin tomotherapy machine. ABC-controlled repeat CT scans were contoured and used to determine changes in tissue volumes. Not surprisingly, the liver volume changed very little and excursions of the liver 'centre-of-target' were also no more than 1 mm. However, lung volume changed by some 6% due to air compressibility but still corresponded to no more than 3 mm change in centre-of-target (Wong *et al* 1999). The ABC technique also allows the study of the variation in organ position

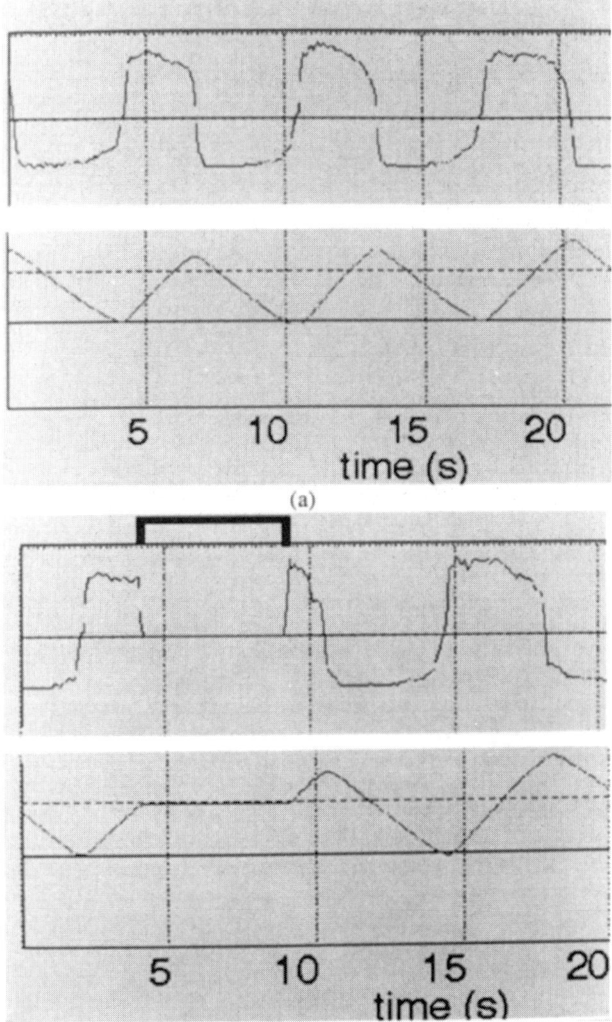

Figure 4.29. *(a) Normal breathing trace, (b) ABC applied for 5 s. Top: air flow; bottom: lung volume. (From Wong et al 1997a.)*

during the breathing cycle by making movie-loops from CT datasets recorded at different phases of inspiration. The technique is adaptable to the Elekta linear accelerator even though it does not have instantaneous beam-on capability because a temporal offset can be programmed in. Finally, it is expected that the ABC method will greatly assist the planning and delivery of intensity-modulated radiotherapy which would otherwise be compromised by breathing motions.

Sontag and Burnham (1998) have also constructed apparatus for respiratory

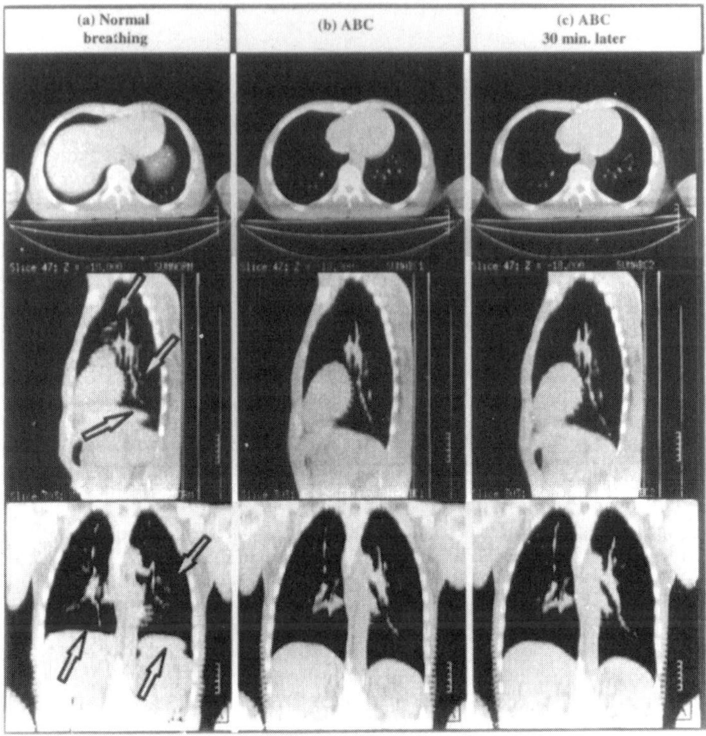

Figure 4.30. *Transversal, sagittal and coronal views of CT scans for a Hodgkin's patient acquired: (a) during normal breathing; (b) with ABC during deep breathing; and (c) with ABC at the same deep respiratory phase, 30 min later. The arrows highlight motion artefacts in the normal breathing scan which are not seen in the ABC scans. (From Wong et al 1999; reprinted with permission from Elsevier Science.)*

gated radiotherapy. They have measured the CO_2 level in the patient's mouth. This level turns sharply negative at the start of inhalation and sharply positive at the start of exhalation. The position of organs is imaged by continuous radiography and the treatment is gated (Siemens PRIMUS accelerator) to full exhalation which was found to be the most reproducible part of the respiratory cycle.

How can we know that breathing-synchronized radiotherapy has achieved its goal? This requires images of the patient recorded during the breath-hold period and this in turn requires a fast imager. Kubo *et al* (1999, 2000) have used the Varian prototype si-EPID based on amorphous silicon technology in two modes, fast and normal; the latter taking about twice as long to acquire and read out an image. Data are given for the imaging times, which depend on a number of factors but typically 512 × 512 images with pixels of size 508μ can be made in 1–2 s, much shorter than the typical breath-hold period. Kubo *et al* (1999) have shown that images with some kind of active breathing control show considerably less tissue movement than images taken on the same patients the day after with normal breathing. They took images at the beginning, middle and end of the 22 s breath-hold period. They propose to use the system to take a series of images during the step-and-shoot DMLC technique hopefully combined with synchronizing the breath-hold to the 'shoot' intervals.

MacKay *et al* (1999) have developed an imaging tool to show by animation the movement of tissues and to compute the biological consequences in terms of TCP and NTCP. Patterns of motion were applied to simulate inter-fraction motion and set-up error. It was shown that the consequence of inadequately set margins was a reduction in TCP and increase in NTCP. It was shown that if the margins were very small then it would be necessary to intervene to correct for movements which had been observed using portal imaging.

Shirato *et al* (1999) have developed a technique whereby the movement of the organ being irradiated is tracked using four X-ray TV systems which view an implanted gold seed. The radiation can be gated to be on only when the target is in the correct location. The system has a rapid automatic response.

Ramsey *et al* (1999) have shown that the use of gating on a Varian 2100C/D accelerator does not greatly affect the physical dosimetry. They created a square field with and without a carefully controlled set of gated deliveries of different burst length and different time interval between bursts. They showed that the maximum output deviation (0.8%), flatness deviation (1.9%) and symmetry deviation (0.8%) occurred when the beam windows consisted of 2 MU sequences for a total of 10 MU. These were deemed clinically acceptable, showing that dosimetric consistency had been improved compared with similar measurements a decade or so ago. Interestingly, these measurements were made using the DMLC capability of the accelerator in which the beam was gated off for leaf movement and moving just one leaf which was not in the open field but under a jaw.

Solberg *et al* (2000) have investigated the dosimetry of gated intensity-modulated radiotherapy. They operated the UCLA NOVALIS linear accelerator with an integrated microMLC and delivered a variety of intensity-modulated two-dimensional fields. These included wedged fields as well as full intensity modulations. The integrated portal fluence was measured with an amorphous silicon detector. Experiments were performed gating the accelerator between 0.2 and 1 Hz and for monitor units varying between 25 and 200 MU. They compared the integrated fluences for both gated and ungated fields. Differences of up to 18% were observed for dynamic wedged fields when a low number of MUs were delivered. At 100 MU and above, no variations exceeded 3.5%. It was thus concluded that gating would not destroy the required intensity modulation and could be applied to control breathing effects on the movement of target volumes.

4.3. INDUCED CANCERS FROM IMRT?

Followill *et al* (1997) have worked out the whole-body dose-equivalent (WBDE) in mSv received from different treatment techniques at three energies 6, 18 and 25 MV. They have compared (i) conventional unwedged therapy, (ii) wedged therapy, (iii) Varian MLC IMRT and (iv) NOMOS MIMiC IMRT. The WBDE was evaluated by first computing the X-ray and neutron scatter and leakage dose per unit isocentre dose at the appropriate energy. Then these figures were multiplied by the dose delivered by each technique (70 Gy) and in turn multiplied by the factor representing the increased number of MUs required to deliver therapies in which some or all of the beams were attenuated. Then they used the BEIR estimates for induced secondary cancers and leukemias of 5.0×10^{-2} Sv^{-1} and 5.0×10^{-3} Sv^{-1} respectively.

They reported that at 6 MV the lifetime risk for induced secondary cancers rose from 0.4% for conventional unwedged therapy to 1.0% for MLC MSF radiotherapy to 2.8% for NOMOS MIMiC tomotherapy. There is controversy over the exact figures, and also the comparisons at higher energies are not needed because it has already been shown that MIMiC IMRT does not require higher energies. Verellen and Linthout (1999) and Verellen and Vanhavere (1999) have worked out that the MIMiC technique led to an increased risk of induced cancers by a factor of eight compared with conventional head-and-neck therapy but were willing to accept this for the benefits of IMRT. IMRT required 1969 mSv whereas conventional therapy to give the same target dose required 242 mSv.

Mutic and Low (1998a, 2000) have also reported that the NOMOS MIMiC IMRT technique led to higher superficial doses than the corresponding plan delivered conventionally. Superficial doses also differed slightly from calculated doses and they cautioned that care should be taken if a target was within 1 cm of the surface. Mutic and Low (1998b) reported that the dose at 10 cm from the target volume due to internal scatter and leakage was 2.5% of the target dose. The corresponding dose at 30 cm reduced to 0.5% of the target dose. They concluded that this is relatively uniform throughout the measurement phantom and may lead to a possible slight increase in radiation-induced fatal malignancies.

4.4. VERIFICATION OF IMRT

4.4.1. *EPI and MVCT*

Megavoltage electronic portal imaging (EPI) has become a standard method to verify the beam's-eye view of MLC-shaped treatment fields (Mubata *et al* 1998). It has also been suggested as a way of verifying IMRT. Systems for megavoltage imaging were reviewed by Webb (1993). Some studies using an EPID to verify the DMLC IMRT technique have already been reviewed in section 3.2.8. Here, we concentrate more on the developments in MVCT. Partridge *et al* (1997a, b, 1998a) used an area scintillating EPID to study the output variations from an

Elekta (Philips) SL25 linear accelerator. They have shown that: (i) pixel-by-pixel quadratic exponential calibration was required in order to convert individual pixel values on the EPID to a measure of tissue attenuation or thickness; (ii) the beam does not stabilize until 8 s after the start signal; (iii) there was a pulse-by-pulse output variation of about 0.7% on every pulse (i.e. every 0.04 s frame); (iv) random noise of $\simeq$6% was seen per pixel on each pulse; (v) larger variations of about 4% occurred randomly and suddenly over longer time intervals (minutes rather than seconds) and these were caused by shifts in the electron focal spot on the steeply sloped flattening filter; (vi) 9% sinusoidal variations with gantry angle were systematically observed. Consequences for electronic planar portal imaging and for megavoltage CT (MVCT) have been discussed.

Megavoltage CT has been proposed for many years as a method of verifying the position of the patient during conformal radiotherapy (see review in Webb 1993, chapter 6). A secondary use for MVCT images could be the direct performance of treatment planning using the X-ray linear attenuation coefficients for the treatment energy, directly measured by MVCT. Many systems have been designed from as early as about 1982, some using purpose-built detectors either for single-slice or multi-slice cone-beam MVCT and others using commercial EPIDs. However, a major limitation is the difficulty of obtaining a sufficiently large number of projections in a realistic time. Generally the maximum speed of rotation of the linac leads to an acquisition time of the order of 1 min which is too long. Nakagawa (1998) has reported further use of a Japanese copy of the Swindell MVCT system for monitoring patient positioning.

Hesse *et al* (1998a) have also developed a MVCT system for verification of intensity-modulated conformal radiotherapy. This uses an electronic portal imaging device with a copper-plus-gadox screen viewed by a CCD camera. The system is the BIS710 manufactured by Wellhöfer Dosimetrie. Hesse *et al* (1998a) first showed, by comparison with measurement of projections using a diamond detector, that the EPID was measuring integrated transmitted fluence. Then, they took 120 projections of a number of phantoms demonstrating that a contrast resolution of 9% could be obtained for a circular object of diameter 3 cm. Additionally, small 100% contrast objects of 3 mm diameter could also be resolved. Images were also shown of the Alderson Rando phantom. It was intended to continue this work by studying the effects of deconvolving the scattered radiation from the EPID with a view to removing beam-hardening effects in MVCT and also eventually improving the accuracy of transit dosimetry (Hansen and Evans 1998b). The background of the work is the intention to develop a tomotherapy IMRT system based around a conventional linac (Hesse *et al* 1997a,b). Hesse *et al* (1998b) have argued that there is no need for a special CT attachment to a tomotherapy machine since the megavoltage imager can provide adequate information (figures 4.31 and 4.32).

Guan and Zhu (1998) have investigated whether the Philips SRI-100 EPID can be used for MVCT. Specifically they proposed a new reconstruction algorithm called a multilevel scheme algebraic reconstruction technique (MLS-ART) which

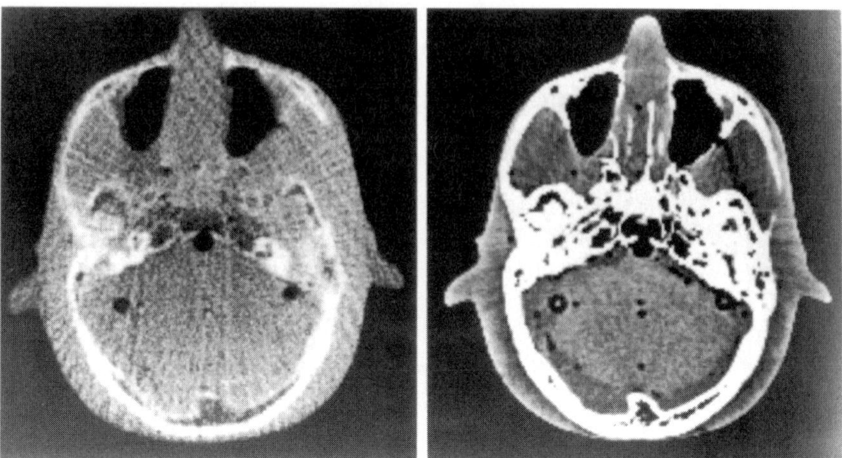

Figure 4.31. *Reconstructed slice of the Rando Alderson head at 6 MV beam quality (left) and 120 kV CT image of a similar slice (right). (From Hesse et al 1998b.)*

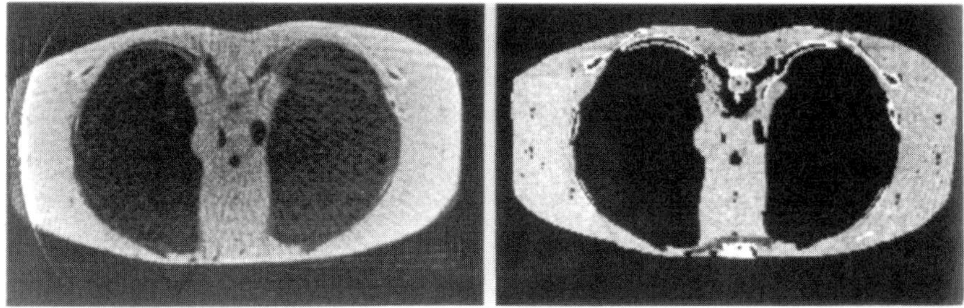

Figure 4.32. *Reconstructed slice of the Rando Alderson thorax at 6 MV beam quality (left) and 120 kV CT image of a similar slice (right). (From Hesse et al 1998b.)*

has the property to select maximally orthogonal projection sampling order so that just one iteration of the algorithm is adequate. The beam was collimated to a 25×1 cm^2 slit to render scatter negligible. Projections were obtained by manually rotating the linac, simulating the action of a third-generation CT scanner. They used the well-known relationship (Barrett and Swindell 1981) between dose, spatial resolution, slice width, signal-to-noise ratio, mass–energy absorption coefficient, detector DQE, X-ray linear attenuation coefficient and beam energy to predict that with a DQE of 1% and a spatial frequency of 0.4 lp mm^{-1}, a slice-width of 1 cm and an object radius of 12.5 cm, an SNR of 100 (i.e. 1%

contrast resolution) could be obtained using a dose of 0.9 Gy. This is a high value, 20 times higher than for diagnostic CT, comparable to the dose per fraction and so could not be the basis of regular MVCT fraction-by-fraction. The authors recognized that this is a rough prediction because the conditions under which the Barrett–Swindell equation were derived were somewhat idealized.

Guan and Zhu (1998) corrected images for a variety of detector problems including shift of the slit centre with gantry rotation and slit tilt. Data were smoothed by a median window filter to remove artificial spikes. Experiments were performed on three phantoms: (i) a model with air, bone and lung; (ii) a contrast-detail phantom; (iii) a Rando head phantom. For the first two phantoms, 5 MUs were delivered for each of 40 projections, a factor of ten undersampling in angle. Contrast detectability was determined to be $\simeq$2.5% for a 5 $\times$ 5 mm^2 square and the FWHM of the line-spread function was $\simeq$2.5 mm using the MLS-ART reconstruction method. For the third phantom, 2 MUs were delivered for each of 99 projections (figure 4.33). In all cases, reconstruction by convolution and backprojection or 'original ART' were found to be inferior to reconstruction by MLS-ART. Possibly this work has been somewhat eclipsed by the cone-beam MVCT with very low MUs per projection acquisition.

We have already seen (section 4.2) how Kubo *et al* (1999) propose to use an amorphous silicon imager to verify IMRT by the DMLC method. Groh *et al* (1999, 2000) have also constructed a 6 MV MVCT system using an amorphous silicon flat-panel detector. The detector comprises pixels of size 0.75 mm^2 and uses a standard gadox phosphor. MVCT images of an Alderson phantom have been constructed using a dose of less than 5 cGy. Spatial resolution of the MVCT scanner system was determined to be about 1 mm and the DQE is 1.1%. From the reconstructed images a contrast resolution of about 2% was derived. The results are promising for MVCT but the system still requires a scatter correction to be developed. Yin *et al* (2000) have also developed MVCT.

Libby *et al* (1999) have computed the exit dose distribution in IMRT using a Monte-Carlo method and then correlated this with experimentally measured dose distributions created by amorphous silicon technology. Paul *et al* (1999, 2000a, b) have used an amorphous silicon detector to create time-stamped images of moving leaves in the DMLC delivery technique for a microMLC. The method of determining the leaf positions was similar to that of Partridge *et al* (1998b).

4.4.2. Non-invasive patient positioning

Lappe *et al* (1997) have shown how a stereophotogrammetry system could be used to follow the small movements of the head of a patient in real time with feedback to the couch controlling the position of the patient, with respect to that of the small beam used in stereotactic radiotherapy. Two cameras observed the location of four highly-reflecting dots on a plate joined rigidly to a dento-maxillary fixation onto the teeth. The repositioning accuracy was almost as good as that of a stereotactic frame. This could also form the basis for ensuring high position accuracy in IMRT.

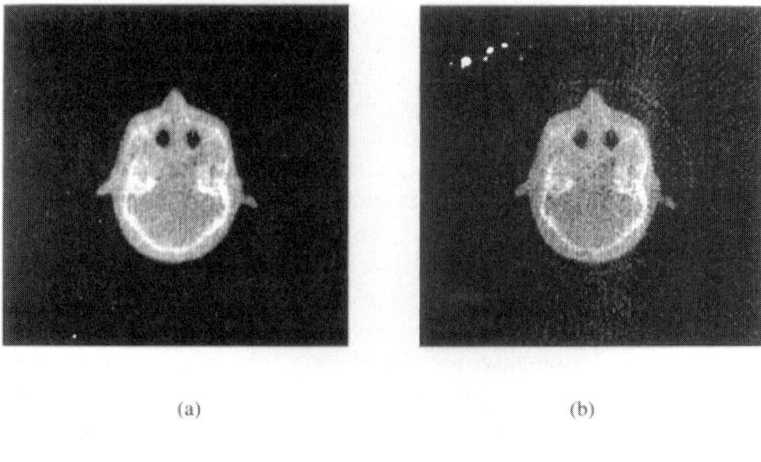

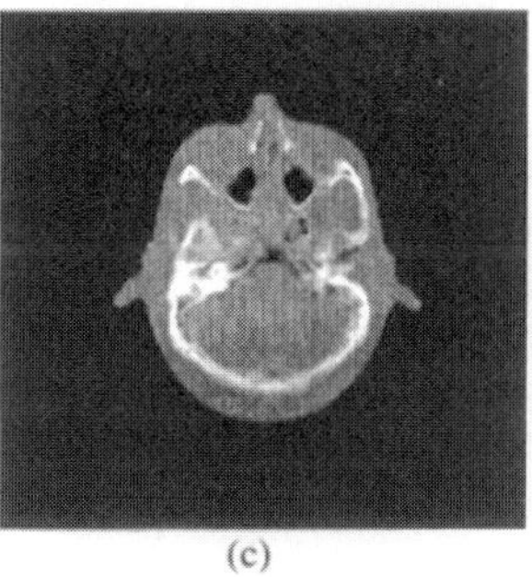

Figure 4.33. *(a) A reconstruction by the MLS-ART technique of a slice across the nose and ears of the Rando phantom; 2 MUs were delivered for each of 99 projections. (b) The same slice reconstructed by the CBP algorithm. (c) The corresponding CT slice scanned with a diagnostic CT scanner. (From Guan and Zhu 1998.)*

Schlegel *et al* (2000) have summarized the possibilities to adapt immobilization techniques such as these for extracranial targets.

4.4.3. Verification of 3D dose distributions—developments in BANG gel radiation dosimetry post-1996

4.4.3.1. Theory

Polyacrylamide gel dosimetry has largely replaced Fricke gel dosimetry although studies with the latter for IMRT verification are still being reported (Belanger *et al* 2000). The group based at Yale University and at MGS Research Incorporated have continued to develop the physics and chemistry of BANG gel dosimetry. Strictly, BANG is the name of the particular gel made by this group and the generic name is PAG (polyacrylamide gel) since these gels are polymer gels made of bis, acrylamide, nitrogen and gelatine. The so-called BANG-2 gel replaced acrylamide with acrylic acid. They are almost water-equivalent (Keall and Baldock 1998, 1999). Keall and Baldock (1999) computed that the electron density of PAG_1 and PAG_2 gels were 1% and 2% different respectively from that of water. The attenuation coefficient, absorption coefficient, radiative stopping power and scattering power were all within 2% of that of water provided the photon energy was above 100 keV. Monte-Carlo depth-dose curves at 6 MV were also equivalent to those in water to within 1%.

PAG gels change both their optical attenuation (Shahnazi *et al* 1999) and nuclear magnetic resonance (NMR) relaxivity $R_2 = 1/T_2$ proportional to radiation dose. They can thus be used to 'store' dose for later readout. This readout may be via magnetic resonance imaging (MRI) or optical tomography (Ranade *et al* 1998, Jordan 1999). Thus they integrate dose as required for a verification dosimeter for IMRT (McJury *et al* 2000). In principle, the accuracy can be of the order 2–4% over a useful dose range of 10 Gy with a spatial sampling on a scale of 1 mm (Maryanski 1997). Knisley *et al* (1997) later reported the dynamic range to be 20 Gy. The method is a volumetric dosimeter. There are certain potential toxic hazards associated with the manufacture of PAG gels and so precautions are taken. However, the risk has been assessed as 'low' (Baldock and Watson 1999). Polymer gel dosimetry has now reached the stage of development where the first international conference dedicated to this subject alone has been organized (Lexington, KT, USA, July 1999). The application to IMRT verification is expected to grow.

Maryanski *et al* (1997) have shown the dependence of the relaxivity (R_2) dose sensitivity on: (i) the composition of the cross-linker and on (ii) the temperature at which the NMR measurements of R_2 were made. The BANG gels all comprised gelatine (5% w/w), acrylamide (6% w/w total co-monomers) dissolved in water (89% w/w) at 60°C. The bis weight fraction (%C) was varied from 17% through 33%, 50%, 67% to 83%. Gelled samples were equilibrated at 21°C room temperature and irradiated with 250 kV X-rays filtered through 2 mm Al with a dose rate of 3.5 Gy min^{-1}. (High-energy performance had previously been shown to be identical to low-energy performance.) They observed the following.

(i) The R_2 sensitivity (defined as the initial R_2 increment per unit dose), at any specified temperature, initially increases with increasing %C peaking at 50%C and thereafter decreases with further increases in %C (figure 4.34).

(ii) The R_2 sensitivity, for each composition by %C, increases with decreasing temperature; for each temperature investigated the peak was at 50%C (figure 4.35).

(iii) The greatest increase in dose sensitivity with temperature change occurs at the 50%C level.

(iv) Each R_2 versus dose curve reaches a plateau at $R_2^{\max}$ and the value of $R_2^{\max}$ increases with %C cross-linking and is greatest for the lowest temperatures.

It is therefore necessary to make a compromise between the dynamic range and the dose sensitivity. The results indicate just how important it is to allow the whole sample to equilibrate to a fixed known temperature before the NMR measurements are made. The work also indicates the future potential for changing the composition of the BANG gels.

Djennaoui (1997) has shown that with a BANG gel composition of 1.5% acrylamide, 1-5% methylenebisacrylamide, 0.5–1.5% agarose, the dosimetry was only accurate to about 10% at high dose and 20% at low dose suggesting that BANG gels are better suited to high-dose measurements and even then require further improvements. Bonnett *et al* (1997) found a better accuracy of between 2–4% for verifying five-field intensity-modulated plans. Audet *et al* (1997) found that the BANG gel response varied throughout the volume from 0.30–0.34 s^{-1} Gy^{-1}. Ertl *et al* (1997) have compared the use of GafChromic film and BANG gel dosimetry for measuring stereotactic radiation 3D distributions with multiple hot spots produced by a gamma knife. Both techniques were found to agree with computer-generated isodose lines to within 10%.

Maryanski and Barry (1998) have developed a new BANG gel dosimeter which employs only aqueous gelatine and methacrylic acid monomer with no cross-linking agent at all. This showed a substantial increase in MR sensitivity (4 s^{-1} Gy^{-1}) and optical sensitivity (0.8 cm^{-1} Gy^{-1}) at 543 nm.

Pappas *et al* (1999) have presented the dose sensitivity of a new MRI-readable polymer gel called VIPAR. VIPAR is N-vinylpyrrolidone argon, contains the monomer N-vinylpyrrolidone (4% w/w), the cross-linker N, N′-methylene-bisacrylamide (bis) (4% w/w) and type A gelatine (300 bloom) (5% w/w) in a water environment (87% w/w). Argon was used to extract the air. Otherwise the gels were prepared in the same way as for BANG gels. Again it was important to allow the gels after irradiation to equilibrate to a constant temperature. Vials were irradiated or partially irradiated and exhibited diffusionless opacity reponse. The irradiated vials were then imaged along with calibration vials of known T_2. The dose response (figure 4.36) was linear according to

$$R_2\left(\mathrm{s}^{-1}\right) = 0.095D\,(\mathrm{Gy}) + 0.974 \tag{4.5}$$

with a correlation coefficient of 0.998. More importantly it was found that this relationship *did not change with time*. The gels were imaged at 4, 5 and 15 days after irradiation and showed the same response. Also, previously irradiated gels were further irradiated and even then the dose response stayed the same. The response was found to be reproducible albeit about a factor of three lower than that for BANG gels.

Murphy *et al* (2000b) have shown that the consumption of monomers can be estimated spectroscopically and also related to dose, thus giving a different method to measure dose. Murphy *et al* (2000c) have shown that substituting

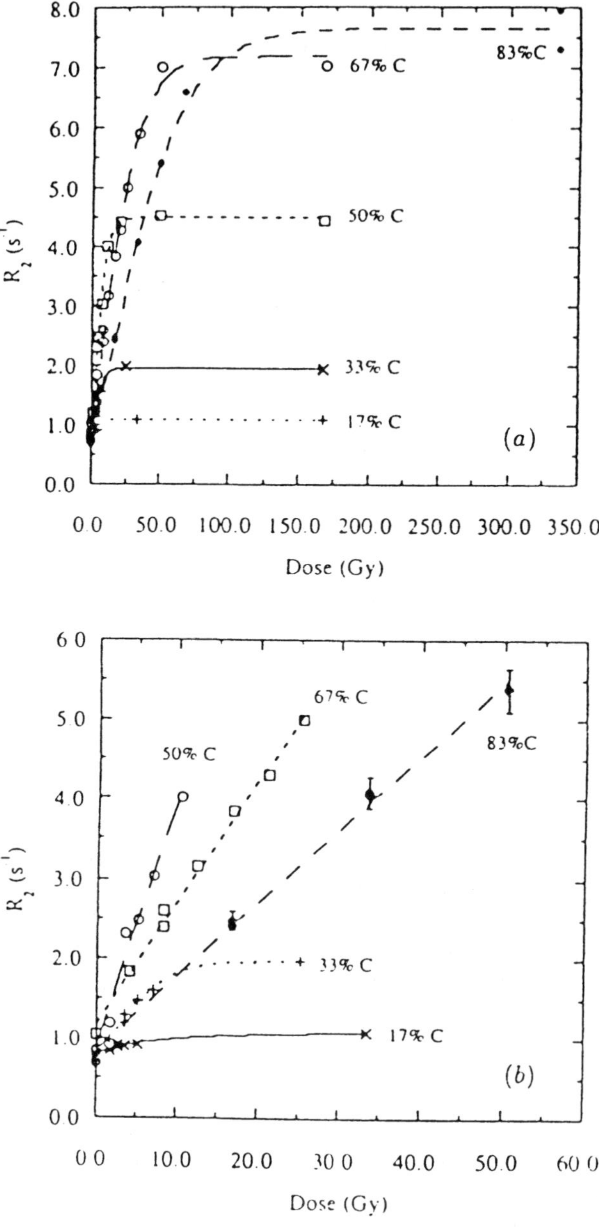

Figure 4.34. *(a) Dose-response curves obtained from R_2 measured at 20°C in gels with different weight fraction of cross-linker per total co-monomer (%C). (b) Lower dose region of the data from (a). (From Maryanski et al 1997.)*

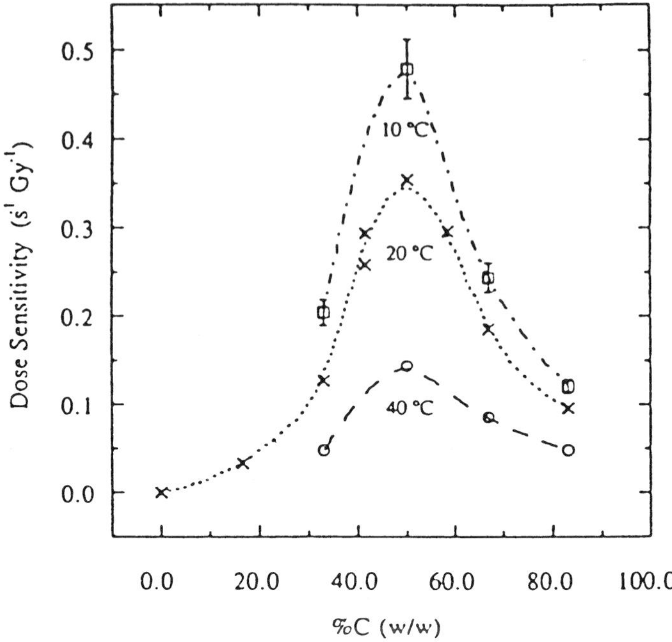

Figure 4.35. *Dependence of the initial dose-response sensitivity on the cross-linker content for three different temperatures at which R_2 was measured. (From Maryanski et al 1997.)*

sodium methacrylate for acrylamide also gives a gel with a comparable dose-response to BANG-1 (although the baseline [zero dose] value and slope depend on pH). This new gel has the additional advantage that, as well as being less toxic, it gives a methyl spectrosopic peak well separated from the peak due to bis (MBA) and so the consumption of the two types of monomers with increasing dose can be separately monitored. It was found that the MBA was consumed more efficiently than the methacrylate.

Bonnett (1999) has reviewed the use of polymer gels for 3D radiation dosimetry. It was concluded that optical measurements are likely to develop as an alternative to MRI. Gore *et al* (1996) have shown how to construct a prototype optical tomography system for reading out the optical density of irradiated BANG gels and, since optical density is proportional to the delivered dose, have thus created a new, MRI-free, method of obtaining the 3D radiation dose distribution (figure 4.37). The concept is very simple, taking its inspiration from the first-generation X-ray CT scanners. A laser beam of 632 nm is taken through a beam splitter onto a mirror which can be translated with respect to a sample. The split beam also arrives at a reference detector to calibrate any drift. The scanning mirror

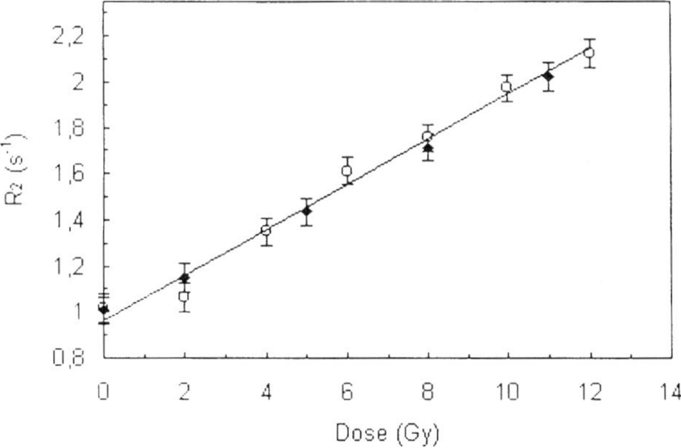

Figure 4.36. *Dose response of VIPAR gels. Measurements for two batches of gels are shown differentiated by symbol. A linear relationship between absorbed dose D (Gy) and the reciprocal* R_2 (s^{-1}) *of the spin–spin relaxation time is observed. The full line fits to both sets of data. (From Pappas et al 1999.)*

is in tandem with a second scanning mirror which receives the attenuated beam after passage through the sample. This passes the laser beam through a lens and onto a single large area (1 cm^2) photodiode detector. The pair of mirrors can be driven up to a maximum speed of 20 cm s^{-1} with a maximum scan length of 23 cm. A 16 bit A/D card is used for data acquisition so that, with sample averaging, each projection effectively contains about 70 macroscopic sampling points recorded in 3 s. The gel is then rotated by between 2° and 3° in the range 0–180° before the translation stage is repeated, and so on in sequence. The projection data were used to reconstruct 60 × 60 pixel images (total acquisition time about 6 min) using the filtered backprojection technique and either a Ram–Lak filter or a Shepp–Logan filter. Maryanski *et al* (1996) have shown that the projection data were entirely resultant from scattering and not absorption so that Beer's law held good. In order to minimize the effects of refraction at interfaces, the scanned portion was confined to some 90% of the diameter of the sample. The gel was contained within a perspex circular container which was itself placed within a parallel-walled tank and the laser beam was arranged to fall at a small angle from normal to prevent direct back reflection. A computer model was made to predict the angular and spatial deflection of the beam but it was found that by matching the refraction of the outer chamber solution these deflections could be made negligibly small.

The spatial resolution depended on: (i) the diameter of the laser beam; (ii) the accuracy of centring the projection data on the axis of rotation (the usual common problem in CT); (iii) the backlash in the scanning mechanism; (iv) the number of projections and filter parameters. In practice, careful experimental arrangement

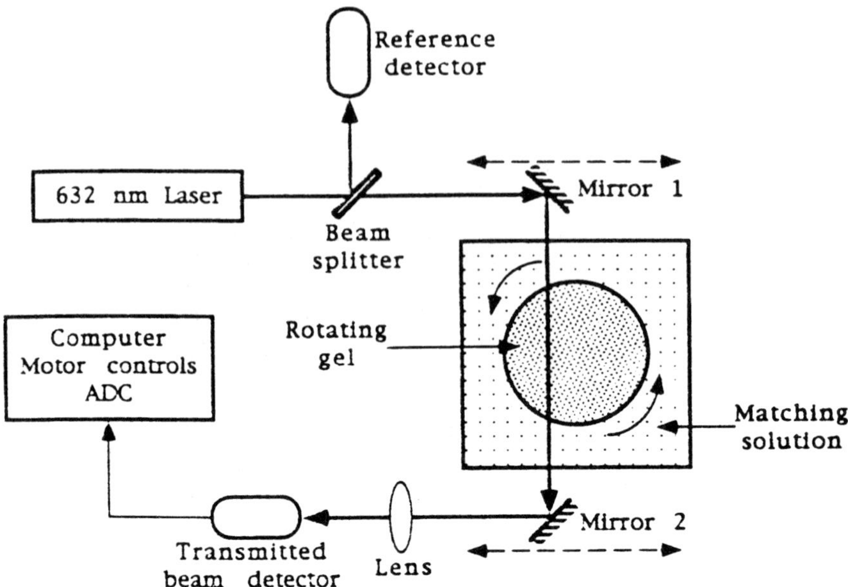

Figure 4.37. *A schematic diagram of the prototype optical CT scanner for reading a BANG gel. The mirrors translate left-to-right to obtain projections of the gel optical irradiation. Between each translation the gel is rotated by a second stepper motor. (From Gore et al 1996.)*

eliminated problem (ii). Problem (iii), though potentially a determinant of resolution, was not, since reconstructions with 'one-way' translation were shown to be identical to those from alternate-direction scanning. The main determinants of the spatial resolution remained (i) and (iv) and they found the resolution was some 2.0 mm FWHM with the Ram–Lak filter and some 2.4 mm FWHM with the Shepp–Logan filter. The signal-to-noise ratio was *not* dependent on the detector noise since this was negligible. The main determinant of signal-to-noise was the limited data sampling and the properties of the reconstruction algorithm. Large timescale drifts in the laser output were corrected by use of the reference detector. Reconstructed optical attenuation coefficients had a standard deviation of 3.5% of the mean corresponding to a dose of some 0.05 Gy and thus indicating that dose increments of the order of 0.1 Gy could be reliably detected at that resolution. From a reconstruction of a sample (figure 4.38) with a series of uniform dose patches, the optical parameter was established as 0.63 mm^{-1} Gy^{-1} for the composition of gel studied.

The main conclusion of these preliminary studies was that the optical technique could in principle replace or at least complement the MRI imaging method of dose-readout and that the areas where attention would be focused

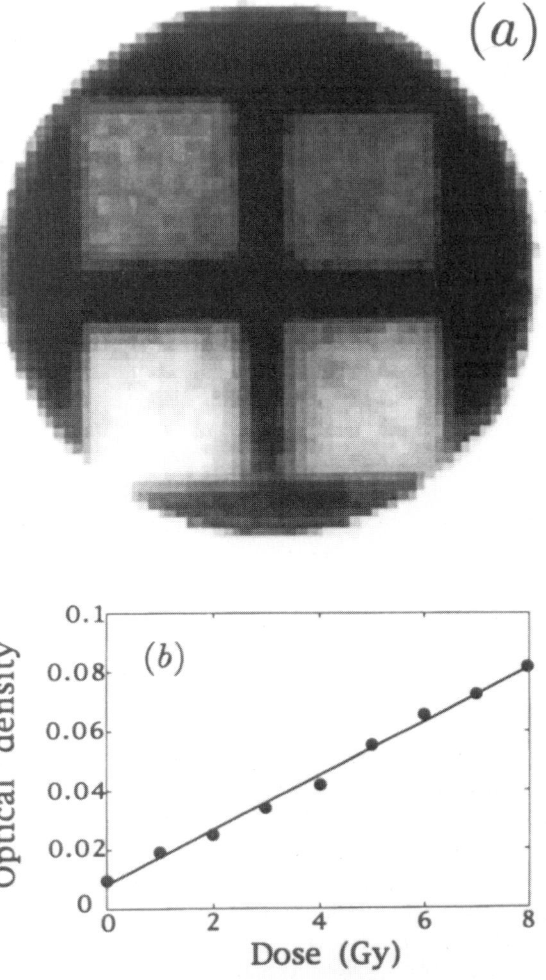

Figure 4.38. *Optically scanned 2D dose distributions of irradiated polymer gels: (a) the calculated dose map of a cylindrical sample of radius 10 cm in which four rectangular fields of different doses were placed; (b) the relationship of optical attenuation per pixel to dose. (From Gore et al 1996.)*

for improving the optical method were: (i) the accuracy of the centre-of-rotation determination; (ii) the laser cross-section size; (iii) the parameters of reconstruction; and (iv) the linear and angular sampling intervals. The paper by Maryanski *et al* (1996) is a detailed study of the reasons for the optical scattering. There is also scope for studying the dependence of the tomography method on the composition of the gels. Maryanski *et al* (1998) have also developed a BANG gel optical imaging system with a 100 micron spatial resolution capability. It also uses a rotate–translate technology. This has been used so far to read out the 3D dose distributions created by endovascular brachytherapy. Fryer *et al* (1997) have developed a similar method of optical tomography. They surrounded the sample by water thus limiting the deflection of the beam at the detector to less than 5°. This allowed use of a collimated detector accepting radiation from within only 2°. The sampling density per projection was not constant and so a backproject-then-filter reconstruction method was developed. Images were formed of both photon and electron irradiation.

Maryanski *et al* (1999) have developed a new optical BANG gel CT device in which the sample is not contained in a liquid of similar refractive index and in which the laser beam makes the scanning with respect to a stationary phantom. This so-called ‘dry scanner’ eliminates problems due to bubbles of gas in the surrounding liquid. Filtered back projection was used for reconstruction. Shahnazi *et al* (1999) have used optical imaging techniques to show that the polymerization was similar at different dose rates.

Jordan (1999) has reviewed the advantageous properties of gels for optical tomographic readout. Specifically, at the energies of visible light the refractive index varies from one to two, and so both refraction and reflection of light take place. This allows mirrors and lenses to be used to control the optical CT. In X-ray CT radiation dose limits the performance. In optical CT it is the temperature rise due to local absorbtion. Jordan (1999) has shown the range of optical CT scanners that have been developed (figure 4.39). Bero *et al* (1999) have developed a *white-light* cone-beam optical CT scanner for imaging ferrous xylenol gels. White light is passed through a filter to select an imaging wavelength, expanded by a lens, and the image is collected on a CCD camera.

Oldham *et al* (2000) have developed a laboratory optical CT scanner for reading BANG3 gels. A rotate–translate device allowed the collection of line projections every 1.8° with 1 mm steps prior to reconstructing the optical map by inverse Radon transform. Diluted green antifreeze provided an optical coupling medium and the phantom rotated in a stationary water bath to avoid ripples. The optical properties of the irradiated gels were calibrated by scanning three irradiated flasks and measuring the attenuation of HeNe laser light. A flask irradiated with the prescription for a complex stereotactic irradiation was read by both MR and optical methods, the former showing less noise and being much faster.

Recently, it has been discovered that BANG gel is also polymerized by 100 kHz ultrasound in proportion to deposited dose, raising the interesting possibility to image therapy ultrasound distributions (Maryanski 1999).

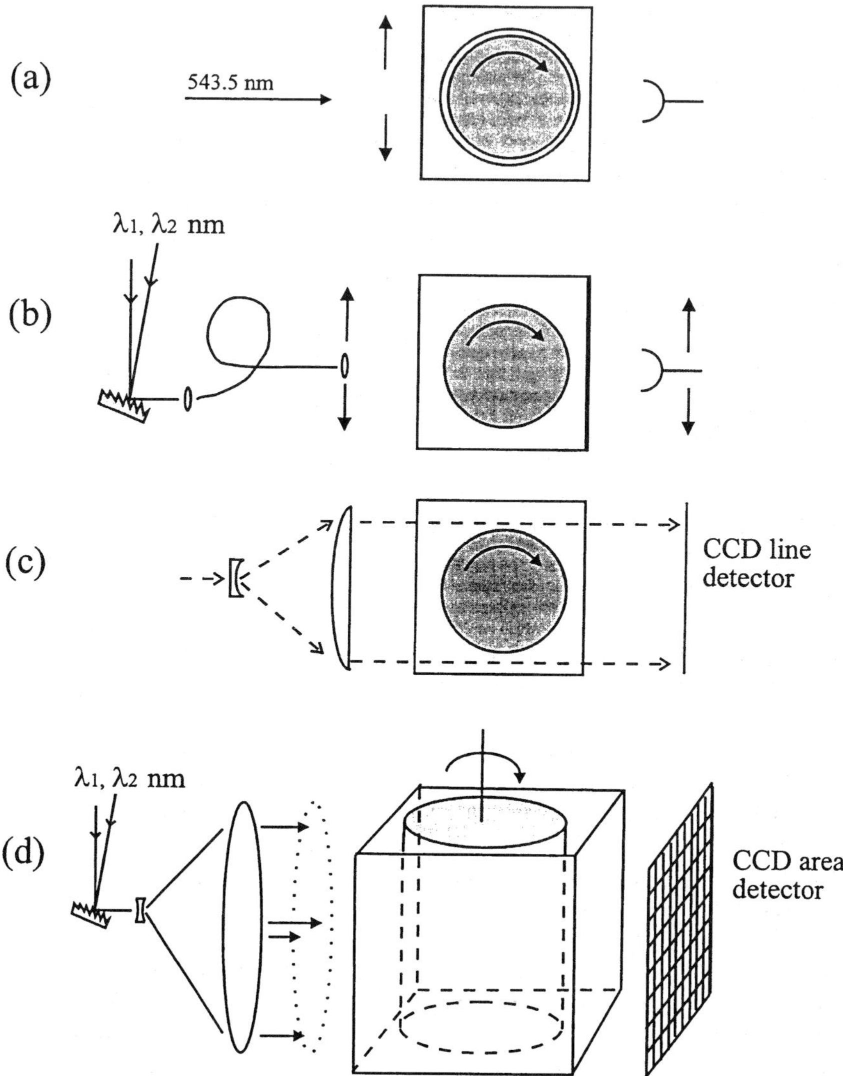

Figure 4.39. *Schematic diagrams of optical CT scanner approaches for BANG gel dosimetry: (a) first generation; (b) differential absorption; (c) planar scanner (fast); (d) broad-beam scanner (very fast). (From Jordan 1999.)*

4.4.3.2. Clinical and phantom applications

Oldham and Webb (1997a, b, c, d) and Oldham (1997) have measured, with film, IMRT dose distributions delivered with the NOMOS MIMiC and compared them to the results of the component-delivery (CD) mode calculation, which takes into account the penumbral characteristics of the actual IMB (see chapter 2). The 90% isodose distribution was consistently in spatial agreement to within 3 mm and at the 50% isodose level there was consistent agreement, again to 3 mm (figure 4.40). Volumetric dosimetry of IMRT with the NOMOS MIMiC has been performed using BANG gels by Oldham *et al* (1997a, b, 1998a) and Oldham and Webb (1997c) for nine-field static tomotherapy (figure 4.41). Low *et al* (1998b) have also used BANG gel to validate plans created by the NOMOS PEACOCK inverse-planning system.

Ibbott *et al* (1997) have used BANG gel dosimetry to measure 3D dose distributions from a gamma knife radiotherapy unit delivering radiation to multiple isocentres. They reviewed the known properties of the BANG gel with respect to their utility for verifying conformal therapy. These include the following.

(i) The BANG gel is a volumetric dosimeter. Film is planar and diodes, TLD and ion chambers are point measurement devices.
(ii) They are tissue-equivalent.
(iii) They have a linear response over a clinically useful range of doses.
(iv) They are stable for a sufficiently long period before irradiation that they can be used as a mailed dosimeter.
(v) They are stable after irradiation and so repeat readouts can be made.
(vi) They are uniformly sensitive and calibration between batches may be used.
(vii) The dose response is independent of both dose rate and beam energy allowing a calibration from a linac beam to be used for converting a distribution created by a cobalt unit (see also Shahnazi *et al* 1999).

A gel was prepared in a 16 cm diameter glass flask fitted to a radiosurgical head frame. Eight individual 'shots' of radiation were delivered by an Elekta gamma knife to create a concave volume localized about a glass rod within the phantom. Calibration tubes were also irradiated to obtain an almost linear map of R_2 versus dose. After temperature stabilization the 3D dose map was read out using a MRI scanning sequence known as 3D RAGE. Using the glass rod as a registration tool, the 3D dose map was overlaid on the MRI images and also on the 3D planned dose distribution. Ibbott *et al* (1997) concluded that the delivered distribution showed a systematic shift of 1–3 mm with respect to the planned distribution and that this shift was not an artefact of the experiment due to imprecise registration. Also calibration is a difficult issue as discussed by Oldham *et al* (1998b).

Gluckman *et al* (1999) have compared planned dose distributions with BANG gel MRI measured distributions for stereotactic irradiation of a spherical flask containing gel housed in a BRW head-frame. Three irradiations were carried out. The first was a five-arc single isocentre irradiation; the second was a ten-arc

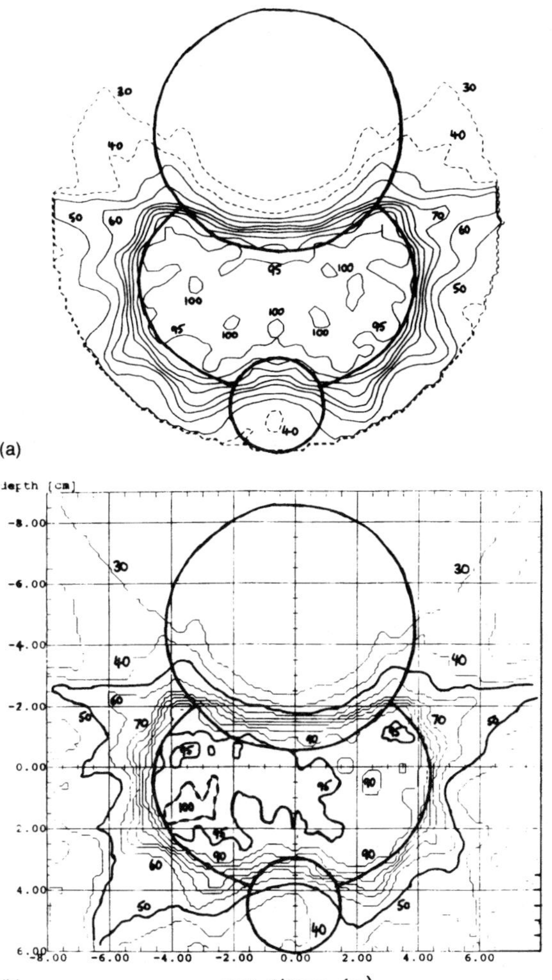

Figure 4.40. *Isodoses of the computer-calculated and film-measured dose distributions delivered by nine fields, each shaped with the static NOMOS MIMiC. The planning geometry is shown in figure 4.22. Isodose lines on both plots are (starting from outer lines and moving inwards) 30%, 40%, 50%, 60%, 70%, 75%, 80%, 85%, 90%, 95% and 100% of the maximum dose. A scale drawing of the PTV and OARs has been superposed onto the isodose lines: (a) the predicted dose distribution as found by the 'CD' mode calculation; (b) the delivered dose distribution measured by verification film. (From Oldham and Webb 1997d.)*

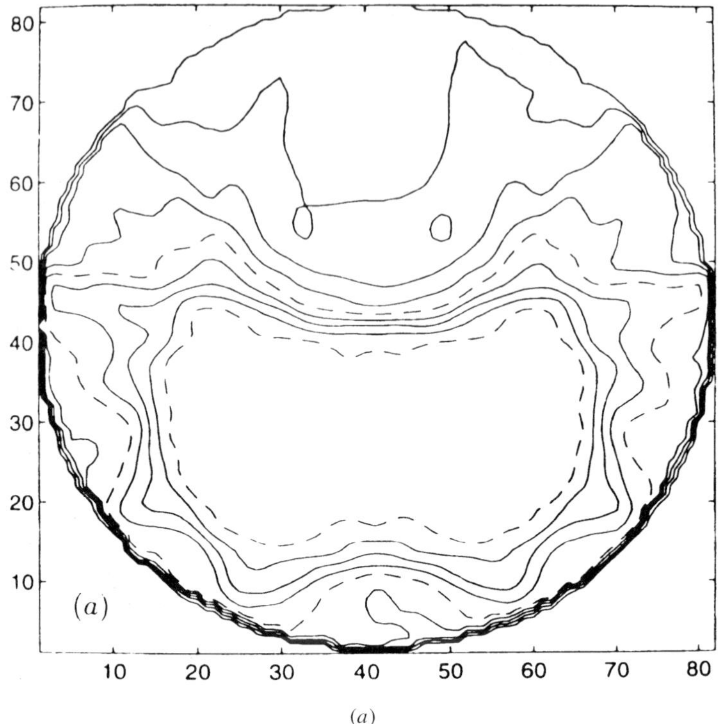

(*a*)

Figure 4.41. *(a) Transaxial view of the predicted dose distribution by the 'CD' mode technique for a circle escribing the regions of interest shown in figure 4.22. Normalization is to the maximum dose and isodoses from 90–10% in 10% intervals are shown with the 90% and 50% dashed. Axis units are in pixel/2. This circle corresponds to the region occupied by a gel insert into the phantom representing figure 4.22. (b) (page 275) The gel-measured dose-distribution for the same slice—a transaxial slice with 2 mm thickness, positioned centrally in the tomotherapy delivery slice. Isodoses are shown from 90–10% in 10% intervals with the 90% and 50% dashed. (From Oldham et al 1998a.)*

two isocentre irradiation creating a peanut-shaped dose distribution. The third was a six-field, single-isocentre irradiation. A special image-comparison package was developed to enable CT-planned dose distributions to be carefully correlated with MRI measured BANG gel distributions. It was concluded that the calculated and measured isodose surfaces agreed to within 2 mm but that absolute doses did not necessarily agree.

Gluckman *et al* (2000) have developed automatic DICOM-compatible software for comparing BANG gel measured dose distributions with predicted

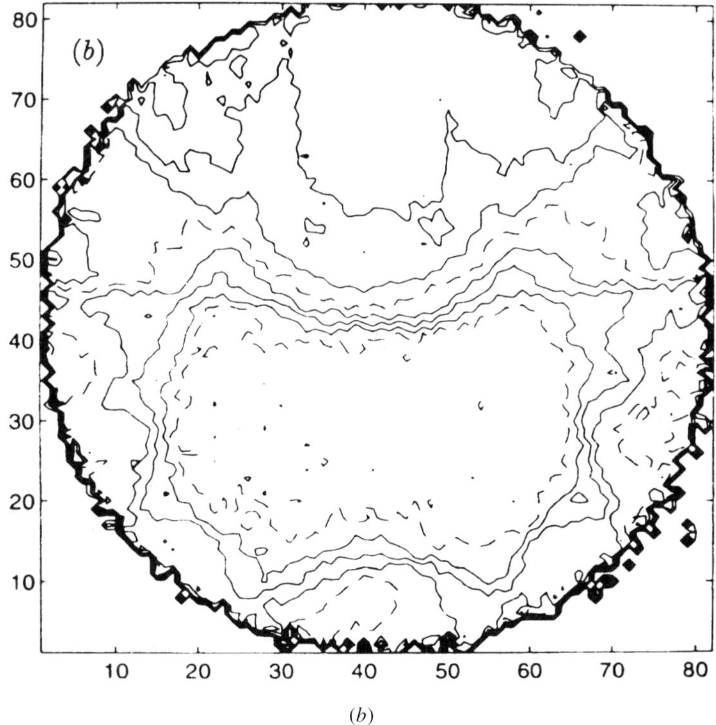

(b)

Figure 4.41. *(Continued.)*

dose distributions. Results for a six-field single-isocentre non-IMRT plan showed an agreement within 5%.

Pfaender *et al* (1999) have delivered dose distributions using the Brainlab microMLC and have measured the distributions with BANG gel. They noted agreement between computed distributions (BrainScan stereotactic planning system) and measured distributions of 1 mm spatial accuracy.

One difficulty with using BANG gels is always the necessity to spatially correlate a planned with a measured distribution. In some of the above studies this is just done by eye. Meeks *et al* (1998a) have instead used landmark-based image correlation to align the MRI determined dose images with those computed by a planning computer. Four planned dose maps of varying complexity were delivered to a BANG gel contained in a spherical pyrex flask (Meeks *et al* 1999b). The flask was fitted with three fiducial markers containing a proprietary hydrogel that is visible in both MRI and CT scans. The flask was held with suction cups to a stereotactic system which was localized either with a frame or with an optical position sensor reading light-emitting diodes attached to the phantom. All the plans delivered 15 Gy to the 70% isodose surface. The gels were imaged using a Hahn spin-echo sequence in a 1.5 Tesla Siemens Signa MR scanner. The R_2 values were related to dose by linearly scaling the value corresponding to zero dose and the maximum (21.4 Gy) dose. Hence the dose maps generated are *relative* not absolute dose maps. The dose maps were transferred to an image correlation program called ImMerge and fused to the treatment-planning CT and planned dose contours. The technique used was the well-known rigid body transformation based on singular-value decomposition to create the transformation matrix. Figure 4.42 shows the comparison of calculated and measured relative dose distributions. Because there are large dose gradients it made more sense to compute the spatial discrepancies between isodose lines rather than the dose discrepancy at a point, a technique adopted by most workers comparing BANG gel dose distributions and calculated ones. Meeks *et al* (1999b) found the discrepancies were no more than 2 mm. The figure provides more detail. It was concluded that the spatial resolution of the MR scans provided the ultimate limitation but perhaps this is because absolute dosimetry comparisons were not attempted. The motivation for the work was to provide a technique to 'close the quality-assurance chain' for precision radiotherapy including intensity-modulated radiotherapy.

De Deene *et al* (1998a, b) have verified some of the complex IMRT deliveries using BANG gel phantoms. They optimized the echo time spacing to minimize stochastic noise. The technique has also been used by Haraldsson *et al* (1998) to verify stereotactic radiosurgery.

The gel dosimetry work of de Deene *et al* (1998b) has verified the irradiation of complex head-and-neck tumours using the Ghent MSF technique (see section 4.1.5). Careful calibrations were made between gel, film, TLD and a diamond detector. The temperature dependence of the dose-relaxivity curves was determined. Figure 4.43 shows the dose distributions measured by both film and BANG gel. De Deene *et al* (1998b) computed the percentage absolute value of the difference (*PAVD*)

$$PAVD\,(x,y) = \frac{200\times \mid D_{\mathrm{gel}}\,(x,y) - D_{\mathrm{film}}\,(x,y) \mid}{D_{\mathrm{gel}}\,(x,y) + D_{\mathrm{film}}\,(x,y)} \tag{4.6}$$

between dose maps obtained by film and gel dosimetry respectively and concluded that the mean *PAVD*, excluding border regions where wall inhibition of polymerization had occurred, was about 8%. This discrepancy was attributed to a combination of: (i) systematic excess of 4% gel dose over film dose; (ii) random noise; (iii) geometrical mismatch of the images; (iv) inaccuracies in film calibration; (v) partial-volume MR effect; (vi) the Rando phantom was inhomogeneous whereas the gel phantom was uniform.

De Deene *et al* (1999) have used BANG gel dosimetry to model the clinical radiotherapy of a mediastinal carcinoma. A model of the chest region of the Rando phantom was constructed with watertight PVC. A cylindrical bore of 10 cm diameter and 30 cm in height was then able to accept either a cylindrical flask filled with BANG gel or a cylindrical pile of PVC in which film had been sandwiched. Six beams were then delivered each with several segments for both 6 MV and 25 MV

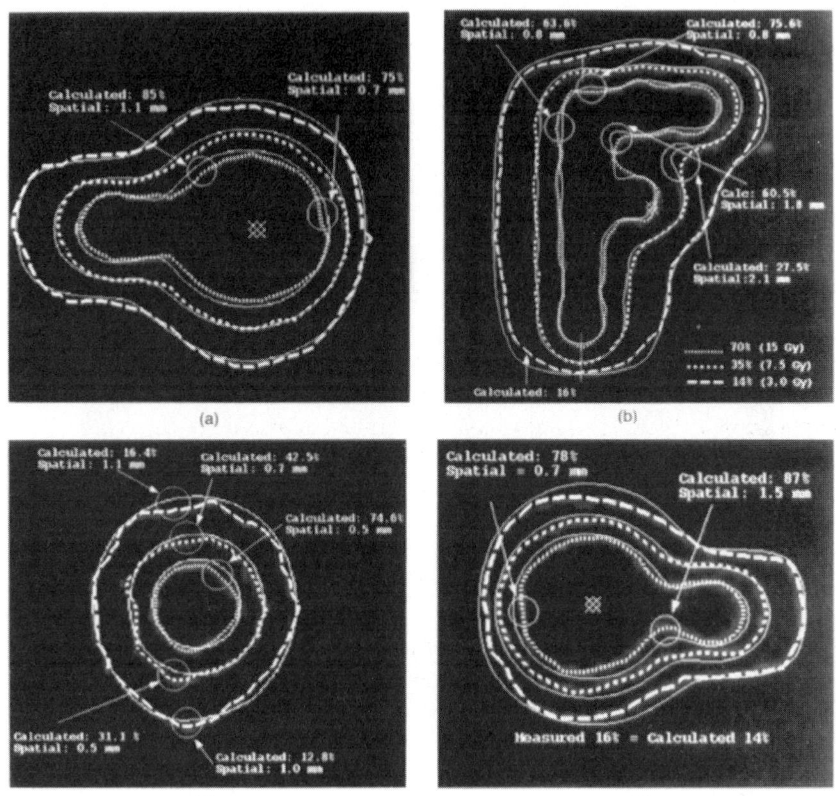

Figure 4.42. *Quantitative comparison of the measured and calculated isodose surfaces for four plans delivered to a spherical phantom with stereotactic localization. The calculated 15 Gy (70% of maximum), 7.5 Gy (35%), and 3 Gy (14%) isodose surfaces are drawn as full lines, and the corresponding measured isodose lines are shown as dashed lines. The regions of largest spatial disagreement are circled, and the magnitude of dose disagreement is circled in the figure. The 'calculated' label indicates the calculated dose point (% of maximum) along the measured isodose curve, and the 'spatial' label indicates the vector distance between the measured and calculated isodose surfaces. (From Meeks et al 1999b; reprinted with permission from Elsevier Science.)*

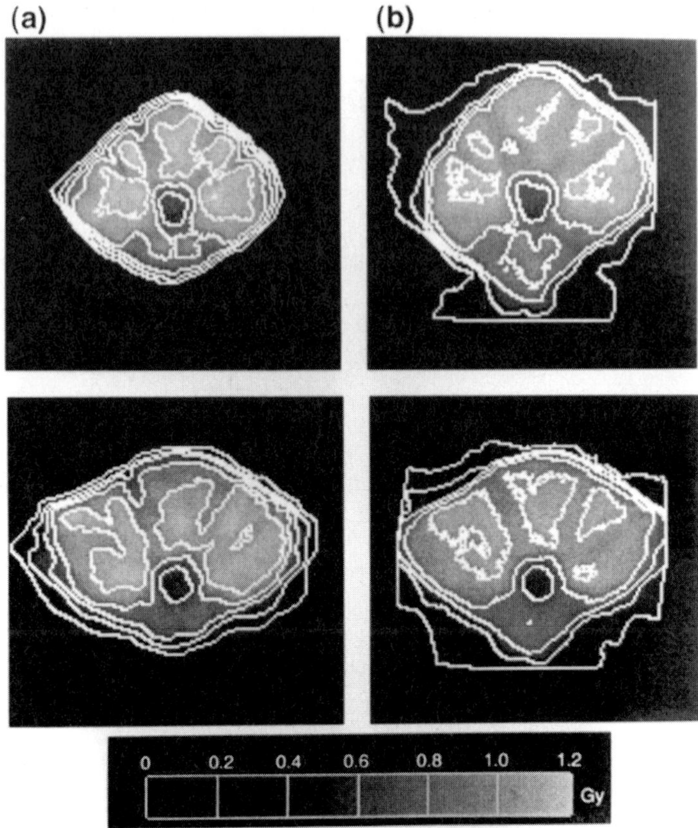

Figure 4.43. *Dose distributions obtained in two cross-sections of the Rando phantom (upper and lower rows) obtained by two methods (right and left columns). (a) Dose distributions are obtained by gel dosimetry. (b) Dose distributions are obtained by film dosimetry. Isodoses are drawn for 0.2, 0.4, 0.6, 0.8 and 1 Gy. (From de Deene et al 1998b; reprinted with permission from Elsevier Science.)*

beams. The dose distribution as recorded by the gel was compared with that as recorded by film and both in turn were compared with the planning dose distibution. The average root-mean-square (r.m.s.) deviation between corresponding maps obtained by the two different dosimetry techniques amounted to less than 3% for maps located at the site of the irradiation target. The average root-mean-square deviation between gel and planning amounted to less than 4% for the same sites. De Deene *et al* (1999) concluded that BANG gel dosimetry was a time consuming but useful process.

De Deene and de Wagter (1999) have provided a detailed discussion of the errors which can arise in gel dosimetry as well as further details of the use of BANG gel to quantify CFRT delivered by IMRT. They emphasize that r.m.s. deviations of voxel-by-voxel dose measurements may not be completely appropriate as they are strongly dependent on correctly aligning the volumes to be compared. They introduce the concept of creating a histogram of the shortest distance of any isodose point obtained by one dosimetry technique to the isodose surface obtained by another dosimetry technique. De Deene *et al* (2000a) have shown the 3% average r.m.s. deviation between gel and film dose maps for multi-segment IMRT of a Rando-type phantom containing BANG gel or films.

Low *et al* (1999a) have evaluated the use of polymer gel as a dosimeter for verifying IMRT delivered dose distributions. Four treatment plans were created and delivered to phantoms using a single arc of the NOMOS MIMiC system. The phantoms were cylinders of BANG gel contained in a box structure. The same box phantom was also irradiated containing an ion chamber, TLDs and film for comparative measurements, specifically for quantitative comparisons. The dose-R_2 response of the gels was calibrated using seven tubes of gel by the technique described by Oldham *et al* (1998b). MRI measurements of both the calibration set and the phantoms were made over a period of four weeks to investigate ageing effects. Care was taken with alignment for comparative measurements and temperature was stable. A Wiener filter was applied to the gel phantom data to smooth measurement noise. It was found that a third-order polynomial was required to fit the dose response; a linear dose response was not observed and had it been used would have led to large (7.5%) errors.

It was observed that the dose response changed with time and that the sensitivity gradually decreased. Even the unirradiated gel R_2 increased with time. Ageing has also been observed by McJury *et al* (1999a) but de Deene et al (2000b, c) have observed ageing a factor of 10 less. De Deene *et al* (2000b, c) have observed that the gradient of the dose response stabilizes at about 12 hours whereas the zero-dose R_2 intercept continues to change for up to 30 days. They have provided detailed measurements to understand the chemistry. Low *et al* (1999a) therefore recommended reading gels within three days of irradiation. Regions of high-dose, low dose-gradient were identified and the ratio of the gel measured to the ion-chamber measured dose was obtained for all four phantoms and for all four measurement times. It was found that just after (72 h) irradiation this ratio ranged from 0.2–6.5%. However, the ratio increased with time to as much as 22% by the end of week four. From this it was concluded that, not only were the gel doses larger than the ion-chamber reference doses, but that the gels of different sizes (e.g. phantom and calibration vials) aged differently so that the ratio did not stay the same with time. Also, even with the same volume, the gel history differed. Hence, Low *et al* (1999a) calibrated each measurement to reference using the appropriate ratio so that gel dose maps could be compared with film and TLD measured and calculated dose maps. On doing this the gel measurements and the dose calculations agreed to within 3% (points in regions of low dose-gradient) or

3 mm (in regions of high dose-gradient) for the most part. Using the same criterion more than 90% of the TLD measurements agreed with the gel measurements. The same was true for film comparisons with gel. A graphical plan evaluation tool was used to depict the regions where agreement fell outside required limits.

One technique used to verify IMRT is a tool in the inverse-planning system (e.g. CORVUS) which allows the intensity-modulated beams computed for a real patient case to be reapplied to a phantom (e.g. the box phantom) which has totally different geometry. This has been described on page 68. An alternative has been developed by Häring *et al* (1999a) for verifying head-and-neck plans. The thermoplastic mask was reused as the mould for a gelatine model of the patient (note *not* BANG gel) to be made and point dosimeters were placed at different positions within the set gel. After insertion of the dosimeter the gel 'closes' behind the insertion probe. This has been used to develop confidence in IMRT (see also section 3.2.8.)

Kaurin *et al* (1999) have irradiated an anthropomorphic phantom containing BANG gel in the position representing the prostate to compare measured dose distributions with calculated. The fields were non-coplanar but unmodulated. The relative dosimetry was good to 3mm for the 90% isodose line but the absolute measurements were about 5–8% too high indicating that as yet absolute dosimetry is unreliable. Gustavsson *et al* (1999) have applied two unmodulated fields to a BAREX phantom containing BANG gel and constructed around the shape of an anthropomorphic Alderson Rando phantom. Relative measurements and calculations performed on a HELAX treatment-planning system agreed well.

Cosgrove *et al* (2000a, c) have verified DMLC delivered intensity-modulated radiotherapy distributions using BANG gel dosimetry. A special phantom was CT scanned, images transferred to the CORVUS planning system and a horseshoe-shaped PTV defined in a set of transaxial slices. Five coplanar fields were then set up at 72° intervals and the sets of IMBs were created. It was noted that, despite the cylindrical symmetry of the arrangement, different IMB sets were calculated for different slices, each corresponding to a representation of the required dose distribution. These IMBs were then passed to the interpreter for the Elekta accelerator described by Convery and Webb (1998) and a set of DMLC leaf sequences was computed. These sequences were then delivered to the phantom which contained BANG gel. The phantom was imaged by MRI and dose maps were computed using calibrated data for dose versus R_2. The measured dose distributions in the central plane and in planes ± 2 mm from this plane were compared with the CORVUS calculation in the central slice. It was found that the shape of the relative isodoses above 90% agreed very well but that the gel measurement was systematically higher for lower isodoses (figure 4.44). Also, the measurements of absolute dose were systematically about 15% higher than the calculations. Cosgrove *et al* (2000b) have studied the stability of polymer gel dosimetry.

Love *et al* (1999, 2000) have compared dose distributions (i) predicted from a treatment-planning system, (ii) computed by Monte-Carlo methods and (iii)

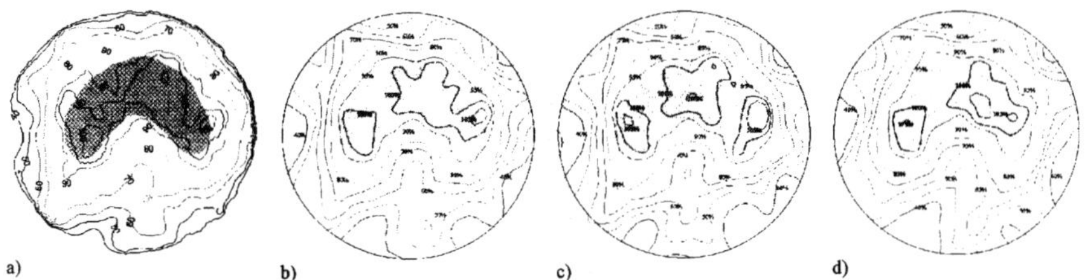

Figure 4.44. *Dose distributions in the transverse plane: (a) gel measurement, (b) CORVUS calculation in a central plane, (c) calculation in a plane 2 mm inferior to central, (d) calculation in a plane 2 mm superior to (b). (From Cosgrove et al 2000a.)*

measured by BANG gel in a BAREX phantom. They noted agreement between the three methods for homogeneous phantoms but (not surprisingly) disagreement with the planning system for inhomogeneous phantoms.

Haas *et al* (1997a, b) have optimized the beam orientation of IMRT using a genetic algorithm as well as developing the method to compute compensators from the 2D IMB profiles taking into account in-compensator scatter. Concave treatment volumes have been created and measured using film and polymer-gel dosimetry.

Hepworth *et al* (1998) have also developed a BANG gel dosimetry system for measuring the dose build-up and build-down near the interfaces of different tissues (e.g. soft-tissue/bone). They have commented on some difficulties associated with gel manufacture and noted the need to mix chemicals in certain order, at the correct temperature and in oxygen-free conditions. This work is at an early stage.

Farajollahi *et al* (1999) have used BANG gel to measure brachytherapy dose distributions. Firstly, they confirmed that (i) the gels were water equivalent to within 1%, (ii) dose response curves were reproducible in time to within 4%, (iii) diffusion in time was absent and (iv) the gel response was energy independent from 0.3–8 MV. They have shown that the dose response increased from 0.28 ± 0.01 s^{-1} Gy^{-1} to 0.50 ± 0.02 s^{-1} Gy^{-1} with increasing the total monomer concentration from 6–9%. The deposited dose around a line ^{137}Cs source was then measured with BANG gel, TLDs and also calculated, and the relative dose distributions agreed to within 4%. However, the absolute dose distributions showed significant disagreement. McJury *et al* (1999b) have also studied the distribution of dose around a HDR brachytherapy line source using BANG gel with good relative agreement.

4.4.4. Verification of plans—computation of exit dose distributions

The liquid-filled matrix ionization chamber was reviewed by Webb (1993). Here, we consider applications. Essers *et al* (1996) have studied the use of the liquid-filled matrix ionization chamber EPID as a portal dosimeter. There were two main aims. The first was to determine whether the EPID could precisely substitute for the point ion-chamber dosimeter. If it could, then clearly instead of making just point measurements of transmitted radiation, a full 2D map of such measurements would be available. This would be particularly useful for measuring the transmitted dose through blocks or wedges or even the transmitted intensity in IMRT. The spatial resolution of the EPID is better also than that of the point dosimeter giving an extra advantage. Since the matrix ionization chamber takes about 5.6 s to collect an image it is best suited to measuring the transmitted dose (fluence) rate rather than the transmitted dose (fluence).

The second aim was to find out the relationship between the transmission dose rate and the exit dose rate. The transmission dose rate is measured by the EPID at some distance from the exit contour of the patient but was corrected back to that position by the inverse-square law. Conversely the exit dose rate is measured by a point ionization chamber dosimeter placed at the exit surface or just 1.6 cm beyond it. A basic problem is that the transmission measurement, if made very distant from the patient, contains little scatter, whereas the exit dose measurement contains a large (of the order 30%) scatter.

The 256 × 256 pixel EPID had a sensitive area of 32 × 32 cm^2. The relationship between the dose rate $\mathrm{d}D/\mathrm{d}t$ and the pixel value I at the detector had been previously established as

$$I = G\left(\frac{\mathrm{d}D}{\mathrm{d}t}\right) = a \times \frac{\mathrm{d}D}{\mathrm{d}t}^{1/2} + b \times \frac{\mathrm{d}D}{\mathrm{d}t} \tag{4.7}$$

where a and b are constants. In practice, it was necessary to correct for the varying sensitivity $S_{i,j}$ of each pixel (i, j) in the matrix chamber via

$$\frac{\mathrm{d}D}{\mathrm{d}t} = G^{-1}\left(S_{i,j} I_{i,j}\right). \tag{4.8}$$

Measurements were made at 6 MV with and without an additional 5 mm of polystyrene build-up on the EPID obtaining $a = 1.155 \times 10^3$ (Gy/min)$^{-1/2}$ and $b = 76$ (Gy/min)$^{-1}$ (with build-up) and $a = 1.139 \times 10^3$ (Gy/min)$^{-1/2}$ and $b = 72$ (Gy/min)$^{-1}$ (without build-up). The ‘gold standard’ transmission measurements were made with an ion chamber in either a ‘mini-phantom’ providing just build-up or in a slab of polystyrene of size 30 × 30 × 2.5 cm^3, this latter providing lateral scatter into the ion chamber equivalent to the lateral scatter expected in the EPID.

It was established that the 5 mm build-up was needed on the EPID at this energy and that it should be calibrated with this in place. The ratio R_{ch} between the EPID measurement and that of the ion chamber was found to decrease monotonically and by a maximum of 5% with decreasing field size *if the mini-phantom was used*. However, R_{ch} was more or less constant with field size if the ion chamber was conversely in the slab of polystyrene. This observation was taken to indicate that the field size dependence of the former ratio was due to scatter in the EPID. A method to parameterize and then to deconvolve this scatter contribution was formulated.

Measurements were also made of the ratio R_{ch} below a volume of scattering material for varying thicknesses of material, varying field sizes and varying phantom-to-detector distances. It was observed that *again if the ion chamber were in the mini-phantom and the EPID did not have its build-up applied* the ratio increased with increasing phantom thickness, increased with increasing field size and decreased with increasing phantom-to-detector distance. Again, this was interpreted as due to scatter detected in the EPID, since the variation with phantom-to-detector distance and phantom thickness went away if the build-up were applied and the variation with field size then became the same, whether the phantom were present or not. In summary, it was determined that, if the EPID were used with an extra 5 mm of build-up and the scatter contribution within the detector were deconvolved, then the EPID agreed with the ion chamber dose rate measured in air to within 1/2%. This established the success of the first investigation, namely that the liquid-filled EPID could be used as a 2D measure of transmission dose rate.

Regarding the second aim, the ratio between the transmission and exit dose rate decreased rapidly with increasing field size, increasing phantom distance and increasing phantom-to-detector distance. The discrepancy was as large as 30% for the largest parameters studied. The reason is that at large distances between the phantom and the detector only a small fraction of the secondary photons created in the patient are present and so the EPID detector detects almost only primary radiation. It is emphasized that the ratio was formed *after the EPID measurement was corrected back to the output surface where the ion chamber exit measurement was made*. Hence this is not a distance effect as such but an effect of the loss of scatter at distance. Essers *et al* (1996) confirmed this by some simple calculations of the loss of scatter. However, it does mean that provided the EPID is a large distance from the source and output surface, the ratio is almost constant.

Essers *et al* (1996) measured the off-axis ratio of both exit dose and transmission dose, when the detector was at various distances from homogeneous and inhomogeneous phantoms. For the homogeneous phantom there was quite close agreement in *relative* dosimetry provided the in-detector scatter deconvolution was performed. However, for the inhomogeneous phantoms and large phantom-to-detector distances the agreement was poorer. If the EPID were directly behind the patient the agreement of absolute dosimetry would be good but more sophisticated methods were needed for large patient-to-detector distances to accommodate the loss of scattered photons. These were later developed by Boellaard *et al* (1997a, b).

One may comment that it is important to separate in mind the two types of scatter referred to here. It is the in-EPID scatter which may be deconvolved, solving the problem of equating the EPID measurement with a point ion chamber measurement *at the same position*. The discrepancy between transmission measurements by an EPID some distance from the patient and an exit dose-rate measurement is due to loss of scattered photons *produced in the patient*.

Boellaard *et al* (1996) have provided the theory to explain the dose-rate dependence of the EPID response, specifically the two components. The square-root term arises from calculation of the free-ion pair concentration with time when there is no polarizing high voltage. The production of ions was shown to be in equilibrium with the recombination even for pulsed 400 Hz radiation. The linear term arises when the high-voltage readout is applied because the ion concentration near the electrodes is determined by the average delay between switching on the high voltage and reading the electrometers. At low dose rate the square-root dependence dominates.

Boellaard *et al* (1996) have described three different ways to read out the electrometer: (i) 'normal mode', reading one line at a time and taking 5.12 s (20 ms per line); (ii) 'half-resolution mode', reading two rows simultaneously; and (iii) 'smoothed mode', reading rows 1, 2, 3 then 2, 3, 4 then 3, 4, 5 etc. They studied the dose rate response for varying photon energy, pulse-repetition frequency, gantry angle and image acquisition mode. At a fixed p.r.f. the value of 'a' increased and 'b' decreased with increasing mean photon energy. 'a' increased and 'b' decreased with increasing p.r.f. Only small changes resulted from different gantry angles. Large differences resulted from changed readout mode. The values also change for different EPIDs of the same nominal design and the take-home message is that each centre has to calibrate its own EPID. There are also differences between the measurements presented by Essers *et al* (1996) and Boellaard *et al* (1996) illustrating this point. Boellaard *et al* (1996) have also demonstrated that the thickness of build-up required at the entrance face of the EPID increases with beam energy. This build-up is needed so the ratio of EPID measurement to dose measurement in a mini-phantom stays constant with changing phantom-to-detector distance. Essentially the EPID with build-up measures the true scatter from the patient.

Boellaard *et al* (1997a) have developed a technique to compute the distribution of exit dose from a patient making use of the information collected by a liquid-filled EPID. The technique uses a convolution to compute the contribution from scatter. The importance of the development is that unlike the work of Hansen *et al* (1996) there is no requirement for CT data and a full 3D convolution.

The first three stages of the computation are common with those of Hansen *et al* (1996), namely: (i) the transmission dose is measured by an EPID with a large ($\geq$50 cm) air gap between the patient and the EPID; (ii) an estimation of scatter at the EPID plane is made (for example, knowing that the distribution is almost flat and proportional to the scattering volume) and this is subtracted from the total signal to give the primary dose at the EPID plane; (iii) the primary dose is backprojected to the exit surface of the patient, correcting for beam divergence.

Then comes the new part of the calculation. The scattered exit dose $S_{\mathrm{calc},i,j}$ as a function of position (i, j) is computed from

$$S_{\mathrm{calc},i,j} = \left[P_{i,j} \times NSPR\left(T_{i,j}\right)\right] \otimes EDSF\,(r) \tag{4.9}$$

where $P_{i,j}$ is the primary exit dose at pixel (i, j) derived from images made with the $\geq$50 cm air gap and the three steps above, $NSPR\left(T_{i,j}\right)$ is a normalized scatter-to-primary ratio and *EDSF* is the exit dose spread function. Equation (4.9) is a convolution expression. The *EDSF* is the distribution of scattered dose in a phantom resulting from a pencil beam and is given by

$$EDSF\,(r) = \text{norm} \times \exp\left(\mu r\right)/r^{q} \qquad \text{for} \qquad r > \text{pixsize} \tag{4.10}$$

and

$$EDSF\,(r) = \text{norm} \qquad \text{for} \qquad r < \text{pixsize} \tag{4.11}$$

and it was determined experimentally that $\mu = 0.041$ cm $^{-1}$ and $q = 1.2$. The parameter norm is the normalization constant for a reference thickness of 20 cm. $NSPR\left(T_{i,j}\right)$ takes care of the change in normalization for different attenuating thicknesses and can be evaluated on a pixel-by-pixel basis (taking account of tissue inhomogeneity) as

$$NSPR\left(T_{i,j}\right) = SPR\left(T_{i,j}\right)/SPR\left(T_{20}\right) \tag{4.12}$$

where T_{20} is the transmission through 20 cm of polystyrene and $T_{i,j}$ is the transmission at pixel (i, j). This model still has limitations but Boellaard *et al* (1997a) have shown that it leads to a computation of the scattered component of exit dose to about 5% accuracy. The total exit dose is then

$$E_{\text{calc},i,j} = P_{\text{calc},i,j} + S_{\text{calc},i,j}. \tag{4.13}$$

Since scatter is about 33% of the total dose, the total dose is thus estimated to an accuracy of about 2%. The 'transmission' in the above computations reflects radiological thickness and the model was shown to work very well for inhomogeneous phantoms when comparisons were made with ion chamber measurements.

There is perhaps some concern that the magnitude of the scatter at the EPID plane as specified by Boellaard *et al* (1997a) is somewhat smaller than computed by Swindell and Evans (1996) and is perhaps flatter than their computations. Spies *et al* (2000a, b) have further considered this issue.

Boellaard *et al* (1997b) have slightly modified the method to account for the changed scattering conditions with low-energy photons and large inhomogeneities. They found that implementation of their basic model for computing exit dose distributions was otherwise in error when applied in these conditions. They also made a large series of measurements confirming the utility and practicality of the (extended) method.

The extension was to introduce an additional geometry factor $G\left(Z_{i,j}/Th_{i,j}\right)$ into equation (4.9) to give instead

$$S_{\text{calc},i,j} = \left[P_{i,j} \times NSPR\left(T_{i,j}\right) \times G\left(Z_{i,j}/Th_{i,j}\right)\right] \otimes EDSF\,(r) \tag{4.14}$$

where G is the ratio of the *SPR*s at the exit side, measured in a phantom with a symmetrically placed inhomogeneity and thickness Th and that measured in a homogeneous phantom with the same radiological path length Z. G was again assumed independent of field size. The reason for G is that in the presence of large inhomogeneities (e.g. air gaps or lung tissue) a considerable part of the radiation that is scattered in tissues in front of the inhomogeneity does not reach the exit surface behind the inhomogeneity. Also, lower energy radiation is more isotropically scattered. In all other respects the work of Boellaard *et al* (1997b) followed the exact same model as that of Boellaard *et al* (1997a).

Boellaard *et al* (1997b) compared exit dose measurements at points made by an ionization chamber with the predicted calculated exit dose distributions using the measured data on the Varian PortalVision EPID together with the model for converting to exit dose. This was done for a variety of phantom arrangements, some purely geometrical, others anthropomorphic and also for some preliminary patient measurements (exit doses instead measured by diodes). They came to the following conclusions:

(i) For irradiation of homogeneous phantoms with open beams the accuracy was 1.2%.
(ii) For irradiation of homogeneous phantoms with wedged beams the accuracy was 1.7%. In both cases (i) and (ii) and for low-energy irradiation, inclusion of the geometry factor made no difference.
(iii) For irradiation of inhomogeneous phantoms the accuracy was better than 2% provided, for low-energy irradiations, the geometry factor was included. (Without it the error was as large as 8%.) *With* the factor the error was minimized when the inhomogeneity was central in the phantom.
(iv) For irradiation of a head-and-neck anthropomorphic phantom the error was less than 1.5%. For the irradiation of a lung anthropomorphic phantom the error crept up to about 2.5%.
(v) For irradiation of a patient the error was within 1.1%.

A disadvantage of the modified technique is that a knowledge of the patient contour is required. However, it is clear that Boellaard *et al* (1997b) have demonstrated that the Varian EPID can be used to create exit dose distributions using the data correction techniques (calibration and in-detector scatter deconvolution) and the model described to include patient scatter.

Boellaard *et al* (1998a) have tackled the problem of computing the mid-plane dose distribution using data from the EPID. The strength of their method is that once again it does not need CT data and, being based on EPID data, generates a distribution of mid-plane dose across an entire plane rather than just a series of point measurements. The method is relatively simple. It begins with the measured exit dose. This is multiplied by (i) the inverse-square law to correct back to the mid-plane distance from the source, (ii) the inverse of the exponential attenuation between the exit position and the mid-plane, (iii) a factor determining the rescaling of the scattered dose between the exit position and the mid-plane. This arithmetic is performed on a pixel-by-pixel basis to generate an image of mid-plane dose.

Boellaard *et al* (1998a) then compared this with three other extant methods: (i) the arithmetic mean of entrance and exit dose; (ii) the geometric mean of the same; (iii) the so-called Rizzotti method. They compared the four methods with measurements made in a large number of phantoms with different geometries representing homogeneous bodies, symmetrically-inhomogeneous bodies irradiated by just one beam and parallel-opposed irradiation of bodies with asymmetric inhomogeneities, both with regular field shapes and fields shaped with an MLC.

In most instances the new method scored at least as well as the Rizzotti method and defeated the other two. Accuracy was to 2%. However it was noted that the new method did not do well for irradiation of phantoms with large inhomogeneities particularly with small fields, and also in the narrow parts of the irregular-field irradiation. The reason was due to a lack of electron equilibrium. In the thorax, for example, the method is inapplicable.

In a companion paper Boellaard *et al* (1998b) applied these methods to clinical cases. They predicted the mid-plane dose distribution and compared it with the calculation delivered by the UMPlan treatment planning system. Five patients with larynx cancer, two patients with breast cancer, five patients with lung cancer and ten patients with prostate or bladder cancer were studied. It was found that the mid-plane dose derived from the portal dose measurements agreed generally within 2.5% with the treatment-planning data for most parts of the field for all patient treatments. However exceptions occurred when (i) large inhomogeneities occurred such as the presence of lung, (ii) the patient anatomy had changed with respect to the anatomy at the simulation and planning stage (the object of the study of course is to permit this observation), (iii) the presence of rectal gas disturbed the anatomy, (iv) the regions concerned were in electron disequilibrium, e.g. near the patient contour or field edges.

Curtin-Savard and Podgorsak (1998) have used a scanning liquid-filled ionization chamber (SLIC) EPID to record the integrated exit fluence in DMLC therapy. They were able to do this because they performed DMLC therapy in step-and-shoot mode, hence images from individual field segments were separately recorded and summed *a posteriori*.

The SLIC EPID measures dose-rate not dose and therefore is not an integrating dosimeter. Hence, the only way to use it to record an integrated dose map from a 2D IMB is to separately record each of the contributions from each static field component and then add them together. Curtin-Savard and Podgorsak (1999) have shown that the SLIC EPID has a very good long term dosimetric stability. However, it is relatively slow in returning to its original state after irradiation due to the low charge mobility in the iso-octane layer. It was shown that if insufficient time is allowed between reirradiation the normalized response to a given stimulus rises. For example if ten successive images are recorded with only 6 s between each, then the tenth image will be 7% too high compared with its correct value. It was recommended that a full 60 s be taken between successive imaging events for which rest interval there was essentially no 'memory' effect. A second precaution taken was to calibrate each image to reinstate the effect of the beam off-axis horns since these are taken out by the manufacturer's software.

Curtin Savard and Podgorsak (1999) created two 2D IMBs, the first a mathematical circularly symmetric function which was delivered in 88 segments (hence the experiment took well over an hour) and a field representing the compensation for non-uniform contours in the neck region which was delivered in 18 segments. The EPID measurements were compared with both the CADPLAN calculation and also with the output of a beam profiler. It was determined that the EPID and beam profiler measurements agreed to within 2% or ± 2 mm between corresponding isodose lines in regions of steep dose gradient. However, the CADPLAN calculation was not so accurate especially in the regions of high dose-gradient. Figure 4.45 shows the result for the latter case.

4.4.5. *Portal dose measurements*

Kroonwijk *et al* (1997) have further developed the method for computing portal dose measurements using planning CT as input. The method has been clinically implemented and despite large gradients in 'tissue density' (phantom) the difference between measured and predicted transmission was better than 1%. The method to determine absolute portal dose images from measured EPID images has been described by Pasma *et al* (1997). EPID dosimetry has also been developed by Conte *et al* (1997).

Pasma *et al* (1998, 1999d) have shown that the Philips SRI-100 EPID can be used to measure the absolute portal dose in IMRT delivered by a DMLC system to better than 1% accuracy when compared with an ionization chamber.

Pasma *et al* (1999b) have used an electronic portal imaging device to measure the thickness of compensators constructed for treatment of head-and-neck cancer patients. EPID measurements are converted into a measure of the thickness of the compensator and it was determined that this could be found with an accuracy of 1/2 mm.

The method of converting EPID images obtained with the Philips SRI-100 to measurements of thickness of steel-granulate compensator has been described by Pasma *et al* (1999c). It rested on accounting for cross talk in the detector, for the scatter component and converting from radiological path length to path length in steel. It was tested on four compensators: a simple slab, a spherical compensator, a wedge and a clinical compensator whose design was based on an inverse-planning system. The 0.5 mm accuracy achieved corresponded to a 1% change in transmitted dose and was considered acceptable for clinical use.

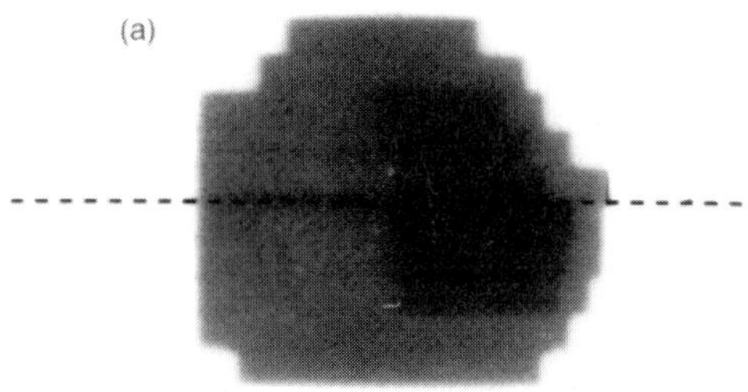

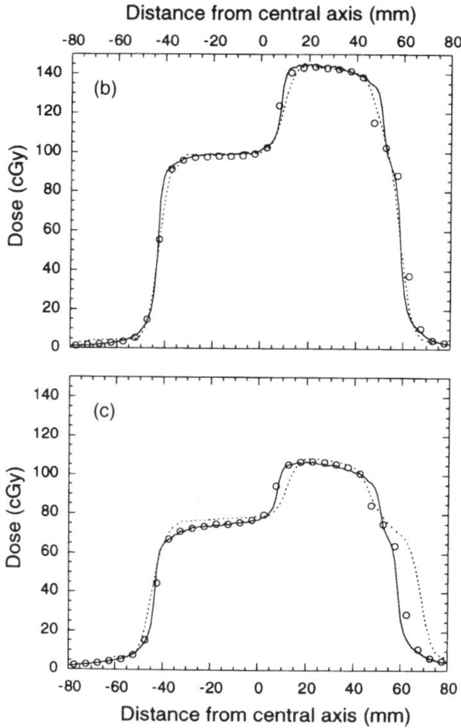

Figure 4.45. *Dose distribution at 6 MV photon energy for one lateral field of a clinical case measured at a depth* d_{max} *for a SAD set-up: (a) the weighted sum of 18 calibrated EPID images; (b) a dose profile along the dashed line of (a) showing excellent agreement between the EPID measurement (dotted curve), the measurement with the beam profiler (open circles), and the calculation of the CADPLAN treatment-planning system (full curve). (c) represents the same dose profile as in (b) except that it is given for a 10 cm depth in phantom in a SAD set-up. (From Curtin-Savard and Podgorsak 1999.)*

4.5. POTENTIAL LIMITATIONS FOR CFRT WITH IMRT

As a broad statement it has been estimated that at diagnosis about six in ten tumours are truly localized, three in ten have clinically metastasized and one in ten is occult metastatic (Cionini 1998). Of these without metastases at diagnosis, four in six can be cured and two in six cannot be cured. Of those with metastases at diagnosis, half in four (one in eight) are cured. It is estimated that of the three and a half in four (seven in eight) with metastases not cured, one in eight is because of persistence of disease at the primary site.

The fraction of deaths due to local failure depends strongly on the tumour site as follows: brain (95%), prostate (61%), uterine cervix (60%), oesophagus (59%), bladder (54%), head-and-neck (41%), breast (14%) and lung (10%). Local failure of radiotherapy is difficult to manage because subsequent salvage treatments have a low probability of cure, high risk of side effects, can be more toxic and are psychologically distressing for the patient.

Conformal radiotherapy and specifically IMRT aims to improve local control. There are conflicting views whether increasing local control decreases the likelihood of distant metastases. Some observations support this view whilst others indicate that metastatic spread has already occurred before diagnosis and treatment. Hence, determining the probability that the tumour extent is entirely local is an issue since it affects the treatment objective.

Why does radiotherapy fail to achieve a cure in localized tumours? There are many possibilities: (i) there may be radioresistive cell clones; (ii) the dose to nearby OARs may compromise the tumour dose; (iii) the diagnostic modality may have failed to yield the true extent of tumour; (iv) there may be inaccuracies in dose planning and delivery. Issue (i) may be insuperable. Issue (ii) is a challenge to treatment planning and largely governed by the physics of photon-tissue interactions. Issue (iii) demands the use of multimodality imaging, collaboration between diagnostic radiologists and radiotherapists and training in reading images. Issue (iv) demands good immobilization, position verification and quality assurance.

4.6. A LOOK TO THE FUTURE—ROBOTIC IMRT?

Which technique of delivering IMRT will 'win'? Everyone would like to know which horse to back but the competition is not a fair one. The result may not be determined by science alone. Certainly we have seen differences between the sensitivity of techniques to movement, differences in practicality, in time-consumption, in technical development and complexity... and so on. A major determinant is the scale of the vendor support. For example, Varian have specifically engineered to deliver the MSF method with minimal operator intervention. Since there is large number of Varian accelerators in the USA this will be a factor which may establish this method faster than it would otherwise (Verhey 1997, 1998). Siemens have an automated technique as well. The Elekta

DMLC technique is gaining ground at least in the UK. The MIMiC development has the support of a major manufacturer in the US. Each technique may well favour one clinical application rather than another. Given the distribution of effort in developing this field of radiotherapy by IMRT it would seem most probable that all the techniques will continue to thrive together rather than that one will 'win'.

We close this section with a glimpse at a seventh candidate for the delivery of IMRT (see section 1.1.4 for list of the other six). It has been proposed (Webb 1999a, b, 2000a, b) that IMRT could be delivered optimally by a short-length linac mounted on a robotic arm at (ϕ, θ) polar coordinate relative to an (x, y, z) Cartesian frame. The robot would allow the linac to 'plant' narrow pencils of photon radiation with any orientation (excluding collision zones) relative to the planning target volume (PTV). The treatment is specified by the trajectory of the robot and by the number of monitor units (MUs) delivered at each robotic orientation. An inverse-planning method to determine the optimum robotic trajectory has been developed. It was shown that for complex PTV volumes, specifically those with concavities in their outline, the treatment was improved by the use of a complex trajectory in comparison with a less complex constrained trajectory and this improvement was quantified. It was concluded that robotic-linac delivery would lead to a great flexibility in those clinical IMRT treatments requiring very exotic dose distributions with complex 3D shapes. However, even using very fast computers, the goal of determining whether robotic-linac delivery is the ultimate IMRT cannot be conclusively reached at present.

One of the patient models studied is shown in figure 4.46. This is a complex geometry in which the PTV is that part of a sphere centred at the isocentre 'O' and of radius 4 cm minus the volume represented by two OARs: (i) a sphere of radius 4.5 cm centred at $(x, y, z) = (0, 2, -5)$ cm and (ii) a cylindrical tube of radius 2 cm passing through the point $(x, y, z) = (0, 0, 3)$ cm and tilted at an angle 20° away from the y-axis in the $(y\text{–}z)$ plane. Figure 4.47 shows the dose distributions and the DVHs for the case of 36 equispaced ϕ angles about the y-axis equispaced in the $(x\text{–}z)$-plane and nine θ values at 50°, 60°, 70°, 80°, 90°, 100°, 110°, 120°, 130° to the y-axis. The excellent conformality is apparent.

The philosophy behind this study should be explained. It is emphasized that the proposal of this new technique is not intended to divert attention from those alternative methods currently being developed to achieve practical IMRT in the clinic with today's existing technology. The aim was to work towards establishing the ultimate limits to delivering IMRT (see also the discussion in chapter 5).

This study did *not* attempt to discuss the practicalities of constructing a method of robotic-linac delivery but this is known to be feasible (Delpy *et al* 1998, Shiomi *et al* 1998). Once the robotic trajectory has been established, an *a posteriori* sorting would rearrange the orientations to minimize the time moving between them. Issues such as the stability and reproducibility of orientation would have to be addressed along with issues of delivering small numbers of MUs. Since this method of IMRT treatment would be quite unlike any other, its susceptibility to patient movement would have to be investigated. However, there are already

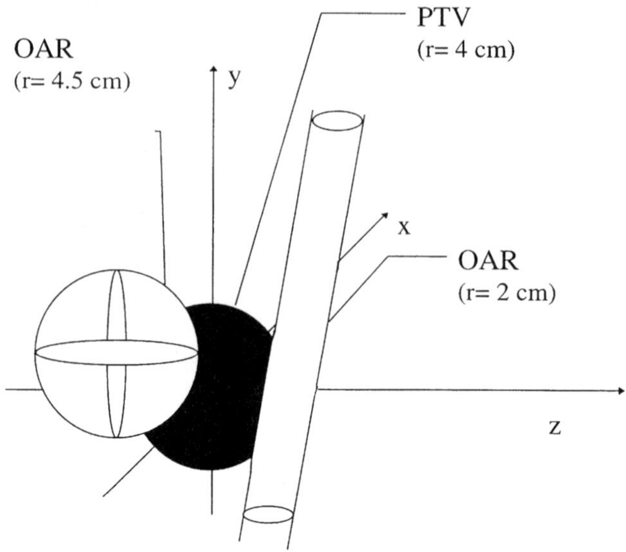

Figure 4.46. *The geometry of the IMRT planning problem—see text for details. The dark volume is the PTV and the open volumes are the OARs. (From Webb 1999a.)*

known feedback techniques to correct for patient movement in stereotactic delivery which could be adapted (see section 4.4.2). For example, Phillips *et al* (1999b) have developed a system using a bite block viewed by two wall-mounted stereoscopic infra-red cameras to locate the patient during IMRT. The system is non-invasive and could in principle be used to reposition an out-of-alignment patient in robot IMRT. Sharma *et al* (1999) have also developed a form of gating for robotic IMRT. They have used a geometrically calibrated video camera to observe the movement of a light-emitting diode relative to which the PTV is known. Then, gating of the robotic source—the Cyberknife (see page 294)—was performed in one of two ways. Either a 'human-in-the-loop' approach was taken in which an observer controlled the gating via a key press or a 'computer-in-the-loop' approach was used in which the radiation was gated on only when the target was within a specified distance of the expected static position. Test irradiations were performed for film sandwiched in a phantom. It was found that the therapeutic ratio improved with both forms of gating compared with no gating. Bel *et al* (2000) have described a computer-controlled couch for repositioning.

Apparatus for performing robotic radiosurgery has indeed been constructed (Tombropoulos *et al* 1995, 1996, 1998). Initially this was called the Neurotron-1000 and is now called the CyberKnife. It is manufactured by Accuray Inc. in conjunction with the Stanford Medical Center (Adler and Cox 1995,

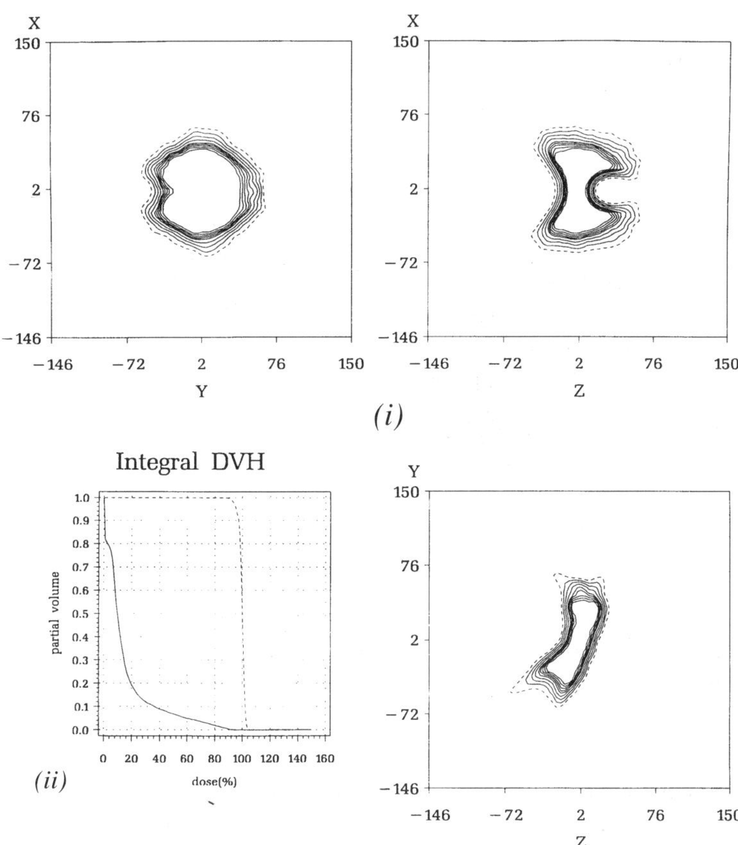

Figure 4.47. *(i) Three orthogonal dose planes passing close to the origin of coordinates and (ii) the PTV and OARs DVHs for the problem modelled in figure 4.46 when there are 36 equispaced ϕ angles and nine θ values equispaced at 10° intervals about $\theta = 90°$. Isodose contours are 90%, 85%, 80%, 75%, 70%, 60%, 50%, 40%, 30%. (From Webb 1999a.)*

Adler *et al* 1997, 1998, 1999). Schweikard *et al* (1994a, b, 1995, 1996) have created a robot-controlled 6MV X-band linac with six degrees-of-freedom (figure 4.48). The linac weighing only 285 pounds (Chang *et al* 1999a) is mounted on a robotic arm manufactured by GMFanuc, Auburn Hills, MI, USA. This can position the beam with an accuracy of 0.3 mm. Specifically, the robotic-linac had attached a (small) rectangular or circular collimator and planning techniques were developed to create the desired shape of dose distribution by combining robotic 'sweeps' across the beam's-eye view of the target with rotations of the robot. A feature was that the pencil beams did not have to point at a single isocentre but

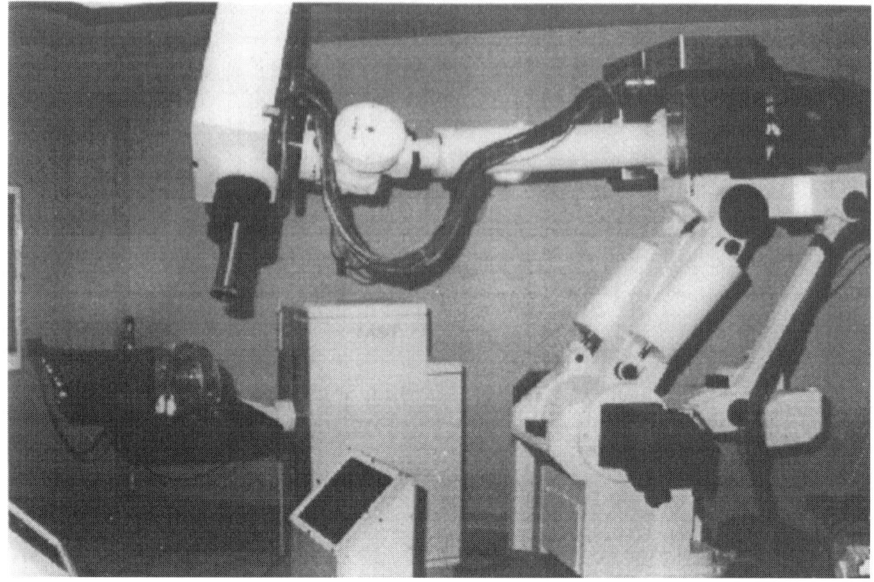

Figure 4.48. *The robotic radiosurgery system known as the Neurotron-1000. It is also known as the 'Cyberknife' system. (From the Schweikard web-site.)*

could effectively take up any direction in space subject to collision constraints. A consideration was the need to minimize the radiation delivery time and to also minimize radiation penumbra. Preliminary experiments indicated that concave dose distributions were deliverable (figure 4.49) and that the distributions were preferable with the rectangular collimator to those with a circular collimator. A most important feature of this development was that the patient did not have to wear an invasive frame for localization because two cameras were able to track (once per second) any real time patient movement and to adjust the radiation delivery accordingly (Murphy and Cox 1996, Adler *et al* 1999, Chang *et al* 1999a). This can be achieved with a measured accuracy of better than 1 mm. Essentially the Cyberknife operates in step-and-shoot mode with the radiation off between movements. A maximum of 300 'nodes' can be activated with the source located at these node positions. However, the linac does not point to a single isocentre but can point anywhere within a sphere about the isocentric origin. Guerrero *et al* (2000) have used the EGS4/MCDOSE code to compute the radial profiles, depth dose characteristics and collimator output factors for the beam and compared these with measurements, with close agreement.

Tombropoulos *et al* (1999) have presented a detailed account of the treatment planning which has been developed for the Accuray Cyberknife. They call this system CARABEAMER. The goal of the treatment planning is to satisfy a set of dose constraints in which the dose is forced to lie inside a small range for the

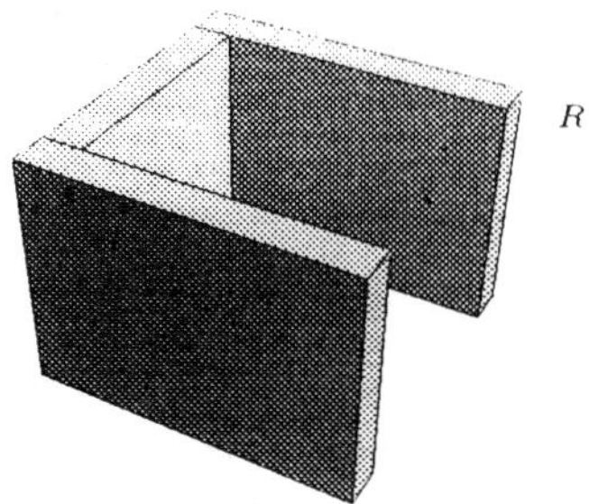

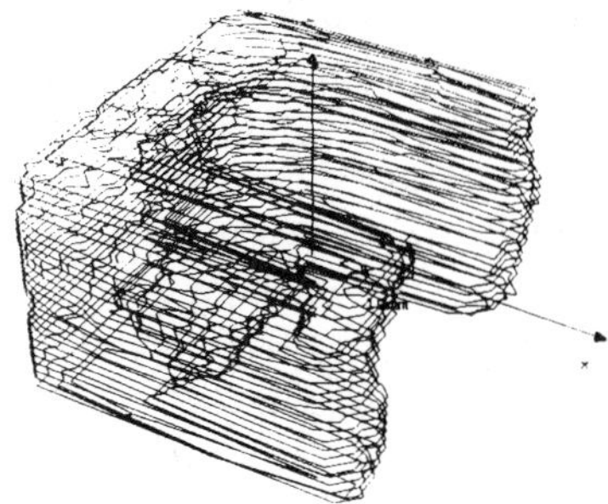

Figure 4.49. *(a) Desired extruded u-shape high-dose volume. (b) 80% isodose surface generated with jaw collimator and automatic beam weighting using a robotic-linac fitted with a small rectangular collimator and executing translations and rotations. (From Schweikard et al 1995.)*

PTV and below some maximum for each OAR. The goal is also to do this with the smallest number of beams since this determines the overall treatment time. Hence, this is an entirely different ambition to that of the work of Webb (1999a, b, 2000a, b) who was simply trying to establish the ultimate IMRT arrangement, however impractical. If CARABEAMER can find a solution it provides it; if not it provides a solution with the minimum departure from that required by the constraints. CARABEAMER begins by specifying a maximum number (usually quite small 100–600) of beam orientations for the robot. It does this by arranging that the beam point at a set of points distributed on the surface of the PTV but entering from totally random directions. They all have unity weight. Forward calculations are performed with an extremely elementary dose model (uniform dose in the cylinder centred on the beam axis. Then a second stage weights the beams. A third stage then removes those beams which have very low weights and replaces them by the same number of beams from different locations and pointing at different surface points. This process iteratively cycles. The optimization invokes a linear simplex algorithm. Finally, the process concludes by substituting a more realistic beam model. Options exist also to start with a very large number of beams and come down to a more realistic number. Options also exist to make use of collimators of different radii. Tombropoulos *et al* (1999) have shown the results of applying the planning technique to a range of twelve clinical cases of brain tumours and then to 36 synthetic cases to stretch the limits of performance. They found that the tighter the dose constraint to the PTV, the fewer the beams allowed and the more complex the PTV–OAR geometry, then the more time it took to find a solution. Finally they proposed that the Cyberknife could be used to treat prostate cancer. Murphy *et al* (2000a) have described the use of the Cyberknife for treating lesions in the thorax and abdomen. The robot was guided by feedback from images which monitored the positions of implanted markers. Also some respiratory gating was employed.

Following the paper by Adler *et al* (1999) there were three published critiques. Whilst welcoming this exciting technology there were concerns expressed that: (i) paediatric patients may not be able to lie still even within 1 cm and would be better accommodated in an invasive frame; (ii) the relatively small number of beam directions used may compromise conformality of normal structures; (iii) the actual use of the system had been low (22 patients by June 1998) and so clinical experience was lacking (a familiar 'catch-22'); (iv) the technique was based on CT and not MRI, whereas MRI can provide the basis for much conventional stereotactic radiosurgery; (v) the method was unproven in comparative trials with, for example, the use of a dynamically shaped microMLC.

Fantini *et al* (1999) have attached a 7 MeV intra-operative radiotherapy electron accelerator to a six-axis robotic arm (Piermattei *et al* 1999). The accelerator is equipped with an external tungsten target to generate photons which are then collimated by a small circular aperture so as to create a pencil-beam of radiation. The diameter of the collimator is 2.5 mm at the collimator. This equipment allows the delivery of radiation using a painting process. Fantini *et al* (1999) have already delivered uniform radiation to structures, including a bow-tie shaped structure and a circular structure containing a region to be protected, and shown using film dosimetry that this very-low-efficiency mode technique can yield uniform dose distributions. Moreover, they can provide fluence modulation by changing the electron current. So, in principle, this equipment has all the capability to perform robotic IMRT using the technique described by Webb (1999a, b, 2000a, b). At present, the accelerator operates with a very low current (approximately 1.5 milliamps) but a high-current accelerating structure is under construction.

It has been suggested (Wong 2000) that there is no reason to restrict to just one robot. Several robots could operate simultaneously working on subsets of the required trajectory and decreasing the treatment inefficiency.

4.7. SUMMARY

The clinical introduction of IMRT techniques has been a patchy progression. In attempting to review the developments completely it is not possible to tell a story with a simple single-tracked theme. Firstly, whilst IMRT has grown explosively in the last five years, we still must recognize that, compared with the use of conventional radiotherapy (whatever we take this to mean), the number of IMRT treatments worldwide is very very few so far and they are distributed in a small number of key centres throughout the world. As a result, whilst there have been some structured collaborations, the main *modus vivendi* of progress has been that of parallel individual developments. Of course we live in an age of good communication but, nevertheless, many of the developments have been repeat attempts to solve the same problem and one cannot report that the developments have been in a serial progression, one following on neatly from the accrued wisdom of the others. The developments have been instead in parallel.

We have seen that there are a number of competing IMRT delivery technologies and even for a single class of technologies there are several interested commercial vendors. To some extent centres have become linked to one or more particular vendor, perhaps because of historical links or because of identifying a manufacturer as the lead 'national' manufacturer. Perhaps there have even been external pressures from health authorities. It is certainly very difficult to line up competing manufacturers and pick a certain lead contender. Hence the history of IMRT, its development and progression, to some extent depend on which centre becomes the focus of attention. This is reflected in the lengthy clinical review in section 4.1 in which sometimes the classification of research and development can be by centre and sometimes by technique and/or manufacturer. According to Boyer *et al* (2000b) at the time of finishing this book about 8000 patients have been treated with the NOMOS MIMiC at more than 70 centres. There are 30 MLC centres performing clinical IMRT with 1500 completed patient treatments. The balance may change.

There is no clinical consensus which technology best suits particular tumour sites. This is one of the few statements all workers would agree with. It is also not possible for a dispassionate observer to conclude whether the technology adopted by a particular centre (both technique and manufacturer) is genuinely because it is considered head-and-shoulders above its competitors or whether the issue has been blurred by the constraints above.

This somewhat unhelpful comment will displease the reader who is hoping to have an answer to the question 'what exactly should I buy?' Perhaps I have been over-academic in my reticence to publicly back a winner. I guess I am trained to be objective and so I have to genuinely report that, like many choices in life, there are pluses and minuses attached to each potential purchase. For this reason I have concentrated on providing more facts than opinion and I leave the reader to make a decision based on local circumstances. Maybe it will always be like this. I have a feeling that ten years on in 2010 the situation will not have changed much and that choice will be as great or maybe even greater a decade on.

However, what everyone hopes will change is that IMRT will become routinely established for clinical work and there is every expectation that this *will* happen. As we have seen in this chapter, the clinical implementation will depend on the ability to control the effects of the live, breathing, moving patient (section 4.2), to accurately verify the treatment (section 4.4), to overcome any concerns about the radiation dose (section 4.3), and will also feed off improved knowledge of what actually causes cancers to fail to respond to aggressive radiotherapy (section 4.5).

Finally, in this chapter (section 4.6), I have hinted at the issue of what constitutes the ultimate IMRT tool and the question of how good do we want our IMRT techniques to be? This leads naturally on to the subject of the next chapter, treatment planning for IMRT.

CHAPTER 5

3D PLANNING FOR CFRT INCLUDING IMRT

It would be a mistake to imagine that the only way to improve CFRT is to implement intensity modulation. IMRT is required when the shape of the target volume demands it or the juxtaposition of target and OARs makes it impossible to obtain an acceptable plan by other methods. Many treatments will continue to require beams which are not intensity modulated and for this reason we shall discuss such situations and what can be done to optimize them. This will drive into focus even better the specific role of intensity modulation.

5.1. A PHILOSOPHICAL NOTE ON THE OPTIMIZATION (CUSTOMIZATION?) OF TREATMENTS

The term ‘optimization’ has recently been discussed and criticized at length (Rowbottom *et al* 1999a). The argument is as follows. There are several areas of concern in attempting to improve treatment planning. The first involves the compromise between the PTV and the OAR doses. The trade off is achieved via the use of importance factors which specify the relative importance that each volume in the patient is ascribed for achieving conformality, whether this be conformal therapy or conformal avoidance. For each patient, the human planner may set these differently and the optimization algorithm then finds the best solution for the compromise sought. Some cost function to be minimized is specified which may be based on dose-volume constraints with these importance parameters. Authors then forcefully refer to these plans as *the* optimum whereas really they are the best outcome for the constraints applied (see also Xing *et al* 1999a). The term ‘optimum’ should strictly be reserved for the one—probably unachievable—plan which leads to unity TCP with zero NTCP.

A second area of concern is that optimization in radiotherapy is a multiparameter problem, requiring choice of treatment modality, energy, number of beams, orientation of beams, variation of beamweights and intensity (in IMRT). These parameters should ideally be optimized simultaneously since they depend on each other. This is impractical, hence most of the parameters are fixed and the optimum value of the remaining parameters is sought. In that sense the resulting plans are *improvements* on standard plans and are not *the optimum*. One cannot qualify optimum by adjectives of degree and so it would be better to refer to these plans as 'constraint-customized plans' and the process of seeking plan improvements as 'constrained customization' (figure 5.1) (see also section 5.11).

What is the goal of customization? As well as the physical objective of creating a treatment plan with an improved therapeutic ratio, implementing 'customization' has a secondary goal which has more to do with human dynamics than physics. At first sight, it might be thought that a computer customizing a treatment plan would save the time of a human planner who might be possibly overworked, possibly harassed by time constraints, possibly bored through tedious repetition of the same task. Such a view might seem reasonable, along the lines of domestic automation. However, those whose task it is to create and subsequently tune customization algorithms quickly find that these tasks are themselves not inconsiderable. The more computer power that is brought to bear seems to generate ever increasing opportunities to try more alternative options. The more flexible the algorithms become, the more complex become the possibilities for implementing them. Thus, as is commonly observed, the use of a computer to perform a task previously attempted by a human does not remove work from that human. If anything, it may even create *more* work. The key point to observe is that the type and quality of such work changes for the better. The work becomes more cerebral, more intuitive. It gives the planner a greater sense of control over the outcome. It has more variety. It allows the possibility of even further development. It is quite simply more open ended, less closed.

Would it be a good thing if a genuinely automated customization technique could be created? At first sight the answer might seem to be affirmative. However, after a while, the skills of human judgement would cease to propagate. Planners might even forget what controls the goodness of outcome. The planning task could become turnkey. It could become dangerous. Hence I would argue that complete automation is *not* a desirable objective.

It is not hard to visualize everyday analogies which emphasize these points. Aeroplanes today can fly without pilots, so are inherently safer. However, by turning over a large part of the control to computers this frees the human pilot to take on a more cerebral role. He can monitor behaviour, observe for departure from normality, consider how to fine tune the flight beyond what is strictly necessary (e.g. seeking calmer air, balancing the plane). Most importantly, he is still there for the occasional emergency for which he certainly needs to retain manual flying skills. So the treatment planner should have the tedium removed but still understand the methodology and be given more time to concentrate on fine tuning it. Using a different analogy, but the reversed time argument, it might be argued that driving an old, less automated car is safer since the driver must be constantly alert to what is controlling safe progress (figure 5.2). A more fully automated vehicle might allow the (to be discouraged) simultaneous telephoning

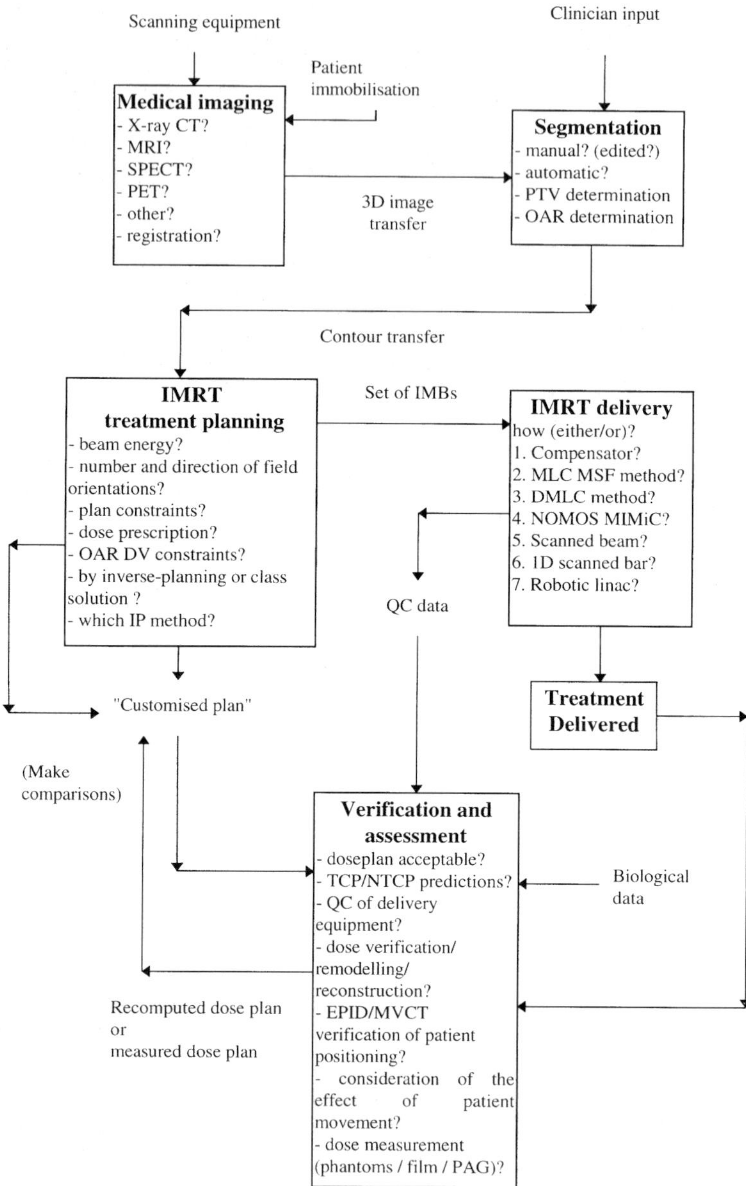

Figure 5.1. *Showing the component 'processes' of IMRT and the several quality assurance and comparative loops that characterize the radiotherapy technique. The issue of 'customization' versus 'optimization' relates to the 'IMRT treatment-planning' component. The other component processes and the loops are discussed elsewhere in this book.*

Figure 5.2. *This photograph is here to illustrate the days when a driver perhaps needed to know more about the car in order to make it run. The analogy with radiotherapy planning is that in our quest to optimize planning and introduce more automation we must not lose sight of the basic principles which govern good quality practice. The figure actually has a radiological link since it shows Dr Twining, the pioneer of classical X-ray tomography in the UK in the 1930s (Twining 1937). Classical tomography, like later CT, was used to underpin radiotherapy treatment planning (Webb 1990).*

of friends and listening to entertainment but may be less safe because the driver is almost unaware of what governs safety and may indeed be concentrating on non-driving tasks (figure 5.3). The advantage of modern cars should be to remove the difficulties of driving but not remove 'contact with the road conditions' and knowledge of how the car works. So, treatment plan automation should not become an excuse to take less time in planning but the reason to use the saved time more productively.

There is no evidence that computer customization is freeing up valuable time for treatment planners. What it does do is allow better use of their time, the opportunity to individually influence each plan in a customized manner whilst preserving the human experience, indeed expanding it, and to retain the full professionalism required for the task, especially when a complex and challenging task presents itself.

Figure 5.3. *Driving these cars (the 1957 Austin A35 car on the road was driven by the author for 22 years) involves much more thought than driving a modern automated one, and so could be safer since the driver is only thinking about driving and not something else. Similarly, automation of treatment planning should not remove the human from key decision making.*

It has sometimes been argued that clinicians, and maybe patients, need to understand inverse planning in an intuitive way. The claim is 'why is inverse planning such a mystery when forward planning is not (sic)?' To me this is an unsatisfactory, possibly arrogant, position taking the moral high ground. Inverse planning is eminently understandable. Indeed it can be argued that the best results from inverse planning will feed forward planning. One must not interpret the quest for ensuring safety in the delivery of IMRT as an indicator towards abandoning the highly complex methods to improve dose distributions. After all, one can drive a car without understanding the complex details of the engine and other mechanics, but understanding can equally help to optimize performance and ensure safe progress.

5.2. PTV DETERMINATION, FUZZY LOGIC AND ADAPTIVE RADIATION THERAPY

The issue of how to determine the PTV from the gross tumour volume (GTV) is still subject to debate. According to ICRU-50 the GTV should be expanded to a clinical target volume (CTV) to account for microscopic spread of disease and then a margin should be added to the CTV when forming the PTV to account for the possibility of patient movement, target movement and set-up uncertainty. Traditionally this has been done by adding a 2D margin in transaxial slices. The technique is in principle simple: a circle of fixed and specified radius is rolled around the 2D contour of the GTV in each transaxial plane and the envelope of circles becomes the PTV contour in that plane. Bedford and Shentall (1998) and Bedford *et al* (1999a) have developed an alternative 3D margin-growing technique. This is a two stage process. Firstly transaxial CT slices are interpolated every 1 mm between measured slices. A sphere is then rolled around the GTV contour within each transaxial plane including these extra planes. The intersection of the spheres with the transaxial planes is again circular and the union of these circular intersections provides the contour of the PTV in each and every transaxial plane. Thus, the grown margin takes into account the form of the GTV superior and inferior to each transaxial plane (figure 5.4).

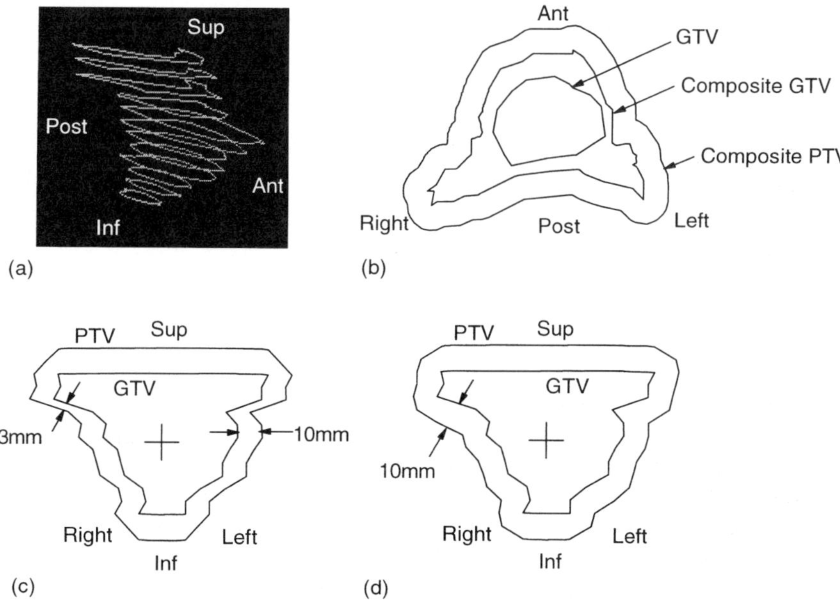

Figure 5.4. *(a) A stacked contour perspective of the GTV for a stage T3 prostate tumour. (b) A transaxial slice through the isocentre, showing the GTV, the composite GTV and a composite PTV which is the composite GTV plus a 10 mm margin. (c) An anterior beam's-eye view of the GTV with a slice-wise 2D margin of 10 mm added. The true margin is generally around 10 mm but just inferior to the seminal vesicles becomes as little as 3 mm due to the obliquity of the GTV. (d) The same anterior view, but with a genuine 3D (10 mm isotropic) margin added to the GTV. The margin is now correct in all regions. (From Bedford and Shentall 1998.)*

How much does this matter? Khoo *et al* (1998) have compared the outcomes of the two methods (2D and 3D) for a cohort of ten prostate patients with just the prostate outlined and a similar cohort when both prostate and seminal vesicles were outlined. In the former case, the difference between the 3D and 2D determination of the GTV was 17%; for the latter case, it was 20%, the 3D GTVs always being greater of course. The 2D method is most in error when there is a large angle between the normal to the GTV surface and the transaxial plane. Khoo *et al* (1998) carefully quantify the differences by location within the GTV and by orientation within the patient. Also evident from the study by Khoo *et al* (1998) is the observation that when a 6 mm 3D margin is added to the prostate (or prostate plus seminal vesicles) the PTV becomes between three and four times as large as the GTV, emphasizing the urgent need to develop conformal techniques.

Recently, the problem of determining the margins to add to the CTV to create the PTV has been separated into the addition of an 'internal margin' (IM) to create the 'internal target volume' (ITV) to account for tissue movement, and a separate 'set-up margin' (SM) to account for set-up error (Aaltonen-Brahme 1998, Dobbs 1998). A new ICRU report 62 (supplement to report 50) has been issued (Purdy 1999). Wambersie (1999) has discussed the improvements in reporting dose distributions, embodied in ICRU report number 62.

Landberg *et al* (1999) have presented the latest ICRU approaches to margin definition and external-beam treatment planning. The new recommendation is that the GTV and the CTV are defined in the same way as before but that an IM should be added, leading to an ITV and an SM, which accounts for patient and beam positioning and the treatment unit geometry, should be added to create a PTV. The PTV is not just the sum of the CTV plus the IM plus the SM but must account for the probabilities of tissues lying in certain geometries. It is argued that these changes are necessary because we now have more information about the geometry of patient organs at the time of treatment whereas prior to the present time the radiotherapist had to guess a lot. There is now some international agreement on these terms. However, Hess (1999) has argued that the combined error can only be reduced if the greatest error is reduced, and the work in his centre has shown that the clinician's interpretation of images is the main source of uncertainty in determining the PTV. The clinician uncertainty is not helped by the influence of the CT window on the choice of gross tumour volume. The addition of PET and MRI data can change the choice of target volume and image fusion (see section 5.3) should be used with caution. Hess (1999) has argued that it is useless to have very precise conformal radiotherapy if there is such a wide variation in the definition of the gross target volume.

An alternative method to cope with uncertainty in determining the PTV has been developed at DKFZ, Heidelberg. Levegrün *et al* (1997) and Levegrün and Schlegel (1998) have shown that including fuzzy logic into the determination of the PTV could lead to an improved TCP, even though the NTCPs were higher. This gave both the clinician and the patient some control over the balance of cure and side-effect. It was shown that the membership values of the fuzzy-optimized PTV decreased with increasing distance to the minimal PTV and that the extent of the crisp optimized PTV proposed to the therapist is larger towards less critical structures and smaller towards radiosensitive OARs. Remeijer *et al* (1999) have also shown that one can incorporate probabilistic information concerning the tumour volumes to create treatment margins from statistical data on geometrical accuracy.

The concept of adaptive radiation therapy in which continuous adaption of the treatment plan is made knowing the movement of the PTV over time was introduced by Yan *et al* (1997a). Yan *et al* (1999) have presented details of a method to compute the 3D shape of internal organ structures from a series of CT scans at intervals throughout the treatment. This comprises firstly registering the CT datasets using bony anatomy as landmarks. Secondly, they contoured the organs concerned (they concentrated on the prostate and rectum) and then identified fiducial points on the surface of the structures. From these points they computed a mesh of volume elements defining the volume. The volume elements must obey Newton's and Hooke's laws of motion. They constructed partial differential equations to represent the relationships between force and displacement and then solved these by finite-element analysis to determine the position of structure voxels. Then, they were able to predict the true DVH that would characterize a volume over a course of treatment including these elastic deformations.

Oldham *et al* (1999) have shown that consecutive CT images of patients indicate that target volume outlines can change over a 20 day course of radiotherapy. At the William Beaumont Hospital, Royal Oak, the PTV is therefore taken as the convex hull of the CTVs. An alternative is to minimize breathing motion by active breathing control (see chapter 4). The patient is trained for 20 min to watch the breathing trace and then the lung volume is fixed using a spirometry technique. As a result it is also possible to shrink margins.

Yan (2000) has shown how it is, in principle, possible to take MVCT images of the patient before each fraction and correct the IMRT planning for the potentially changed organ positions. Lu *et al* (2000) have shown how to combine daily fraction dose distributions applied to daily warping patient geometries. Wong *et al* (2000) have summarized the interventional and adaptive strategies.

Nuyttens *et al* (2000) have shown the movement of the small bowel during a course of IMRT for rectal cancer. They took CT images every week throughout the treatment.

5.3. MULTIMODALITY IMAGING AND THERAPY PLANNING

5.3.1. Clinical imperative and tools

Despite the commonplace availability (figure 5.5) of instruments for MRI, positron emission tomography (PET) and single photon emission computed tomography (SPECT), X-ray CT is still the workhorse for determining the PTV and the OARs for conformal planning in most clinics. It is vital that this change soon (figure 5.6).

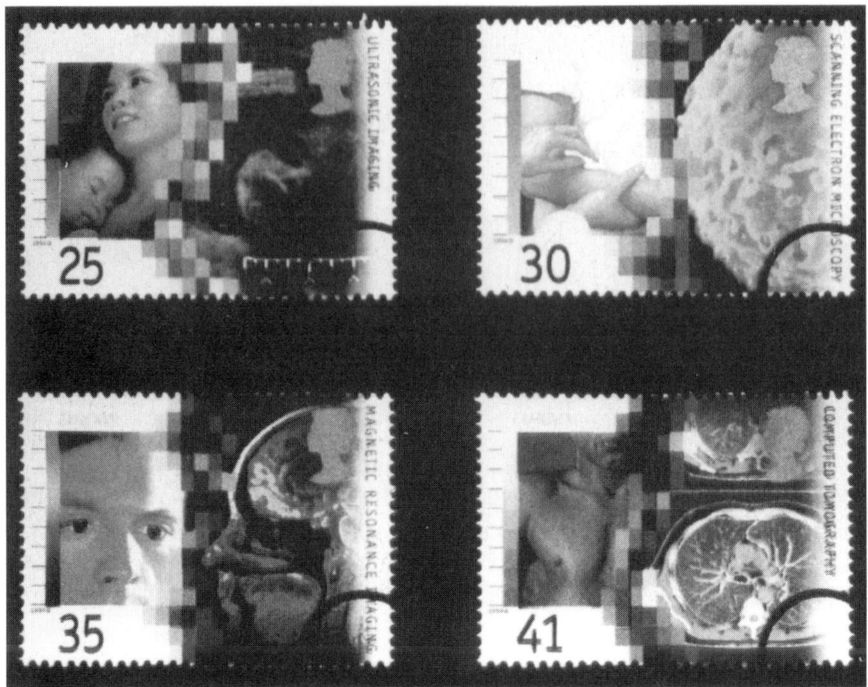

Figure 5.5. *UK postage stamps issued in September 1994 to commemorate the widespread use of medical imaging. The 25p stamp shows an ultrasound image, the 35p stamp shows a magnetic resonance image and the 41p stamp shows an X-ray CT image. Perhaps surprisingly no SPECT or PET image was used, instead the 30p stamp shows a scanning electron microscope image.*

This subject has been reviewed in depth by Khoo *et al* (1997) and commented on by Ling (2000). A simplified review of image registration applied to radiation therapy has also been written by Johnson *et al* (1999b). Hall and Jones (1999) have given examples of how MRI improves the delineation of posterior fossa tumours in children. An earlier review has been provided by Webb (1997d, chapter 7).

Van Herk *et al* (1997) have developed a method of making use of CT, MRI and SPECT data for planning with the UMPLAN system (the system emanating from the Radiation Oncology Department at Ann Arbor, University of Michigan). The system includes both chamfer matching and grey-level-based matching via the Simplex optimization algorithm. A particularly nice feature is the sliding window, two line cursors which cut the display into two or four so the matched images are shown in alternate halves or quadrants (figure 5.7). The added value of multimodality imaging was studied.

Van Herk *et al* (1997) have developed a multimodality imaging (MMI)

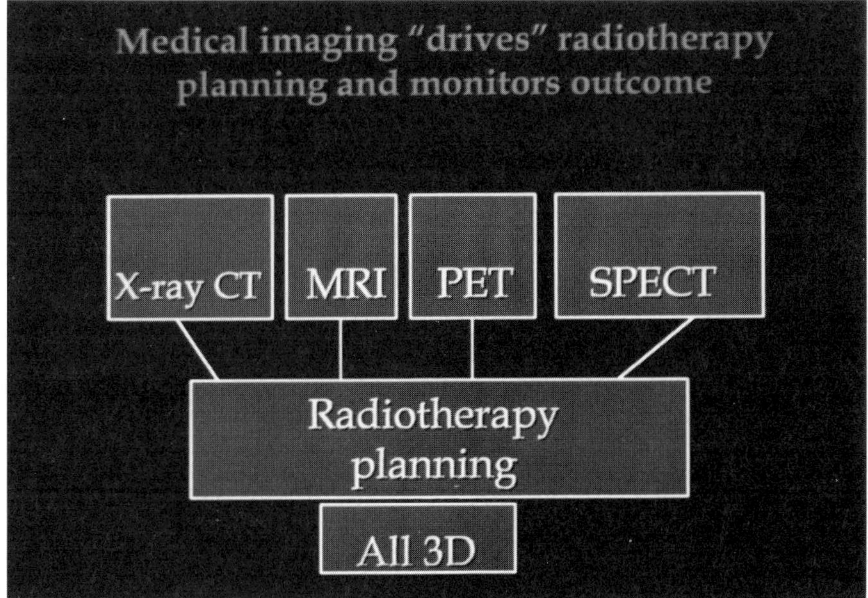

Figure 5.6. *Illustrating the need for multimodality imaging to feed into 3D conformal (intensity-modulated) planning. It is important to remember that successful CFRT and IMRT owes as much to improvements in medical imaging as to technical developments in radiation therapy.*

environment to support treatment planning for CFRT. In particular, the use of MRI, CT and SPECT data allows the visualization and quantification of organ motion. The system relies on a number of features: (i) ability to cope with imperfect image quality, specifically distortions; (ii) ability to input DICOM, interfile, ACR/NEMA and UMPLAN formats; (iii) transport and mass storage of very large data files; and (iv) development of interactive matching algorithms. These include the use of anatomical landmarks, frame-based methods, volume matching and chamfer matching. Chamfer matching minimizes a distance transform and needs to be tuned for each modality. A major advantage is the ability to remove mobile anatomy by an 'electronic eraser' (for example, the femurs which may not be in the same place for each scan and also lower jaw for head-and-neck tumours). The NKI system has quantified many uncertainties in radiotherapy.

Van Herk *et al* (1999) have developed a database to store centrally, e.g. at the trial data bureau, diagnostic images and treatment-planning information. This work was part of the EC CONQUEST project. The database was able to store scans with different modalities, e.g. CT, MRI, SPECT and PET, and to accept and merge incoming data. A hierarchical directory structure represented the data, its format and its location. CONQUEST stands for Clinical Oncology Database

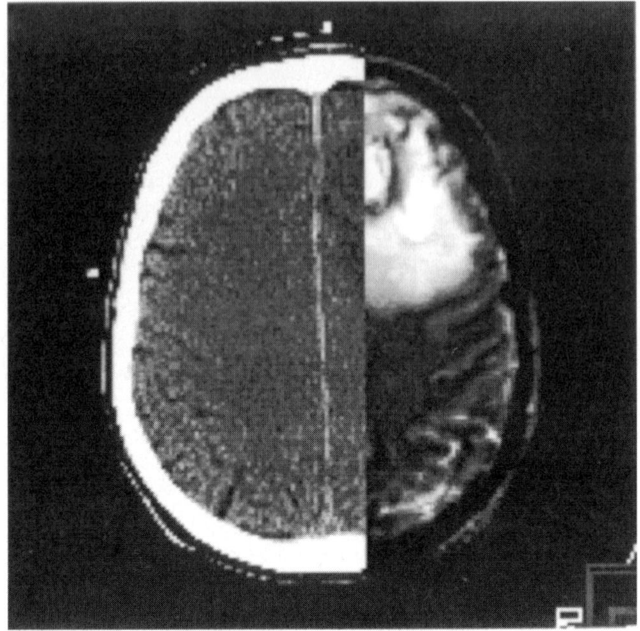

Figure 5.7. *The registration of a CT and a T_2 MRI image for a brain tumour. (From van Herk et al 1997.)*

for Quality in European Standards of Treatment. At present, the system is only available in Holland, is being used in a Dutch prostate cancer trial and has proven to be very flexible and reliable.

Davison *et al* (1997) have used chamfer matching within the ANALYZE image processing software to register MRI and X-ray CT images of malignant glioma with a spatial accuracy of 2 mm. Conversely, Gibbs *et al* (1997) and Beavis *et al* (1998a) have used MRI images alone for planning brain therapy, showing that the distortion was no more than 2 mm and that CT data were actually unnecessary.

Rosenman *et al* (1998) have reviewed 18 months of clinical CFRT practice and noted that 43% of 246 patients undergoing 3D treatment planning had multimodality image registration. Four reasons were identified. Firstly, MRI showed soft-tissue contrast better than X-ray CT. Secondly, contrast-diagnostic CT showed the tumour sometimes better than planning CT. Thirdly, the diagnostic CT or MRI images may have been pre-operative and the tumour not visible on postoperative planning CT. Fourthly, the patient may have undergone cytoreductive chemotherapy so the post-chemotherapy CT images no longer showed the original tumour volume. They concluded that the use of an additional modality changed the treatment plan by at least 1.5 cm for half of the patients and by up to 3.0 cm for one quarter of the patients. Registration was performed manually inside the University of North Carolina treatment-planning system.

Hartley-Davies (1999) has shown that magnetic resonance images are of considerable assistance in radiotherapy planning but that they are subject to geometric distortions. Open MR systems more easily reproduce the treatment geometry but these typically have restricted fields-of-view and larger distortion effects. He produced calibration schemes in which the magnetic field irregularities and gradient field non-linearities were measured and applied to correct images and also modified the imaging sequences to effectively construct a map of the magnetic field at the same time as the image is collected. Appleby (1999) has used the system to exploit the superior soft-tissue contrast of MR imaging for tumour delineation. Specifically the apex of the prostate can be determined more accurately than with CT.

Siebert *et al* (1999) have used landmark-based image fusion to integrate, in an easy and fast way, different types of imaging techniques into the clinical practice of radiotherapy.

Bisi *et al* (1999) have integrated morphological and functional information for radiotherapy treatment planning. MRI and CT images were registered using fiducial markers so obviating the need for external markers. These images were then segmented using in-house developed code. The main difficulty noted was the large number of file conversions necessary.

Gregoire (1999) has explained that the use of multimodality imaging for defining target volumes in radiotherapy is now a craft close in intention to that of a surgeon, progressing in a step-wise dissection of the normal tissues around the tumour and/or node areas, in order to determine the precise volume of the tumour and the structures at risk. By combining PET with various tracers, for example, 18-F, FDG and 11-C methionine or functional magnetic resonance imaging, (perfusion, diffusion and spectroscopy) the delineation of the clinical target volume can change.

Kessler *et al* (1999) have developed a system for registering MR, PET, SPECT and CT data which does not rely on affine transformations but instead allows the option of thin plate spline warping transformations.

One of the problems in communicating different medical imaging techniques to a radiotherapy treatment-planning system, is the non-standard format. In this sense the development of DICOM RT communication standards is a welcome improvement (Neumann 1999a). Neumann (1999b) has pointed out that one of the main issues for complementary systems is the ability to exchange and share information. In many ways the development of IMRT can be regarded as a problem in informatics.

Mongioj *et al* (1999) have used a frame-based image-registration technique to merge 3D data sets from CT, MR and SPECT. The work was carried out within the IFS image registration package available with the Nucletron PLATO TPS. The technique used the conventional method of identifying corresponding points in space, calculating the translation, rotation and scaling to align the images and subsequently reslicing the SPECT images onto the CT or MR images. Because of the low count statistics and interference with image creation in the SPECT data, taking the SPECT images with markers was made into a two stage process. Firstly, the frame and patient were imaged. Then the frame and N-shaped markers were imaged. Then the two were combined prior to image registration with other modalities. The errors in the process were assessed to have minimal clinical impact and to be comparable with the SPECT resolution and accuracy of determining the centroids of the SPECT markers. The SPECT images were of the radiopharmaceutical technetium sestaMIBI. Figure 5.8 shows that the functional tumour identified with SPECT has a smaller spatial extent than the CT image. Hence, by reducing the PTV less normal tissue damage could be achieved. The large CT extent was put down to necrosis.

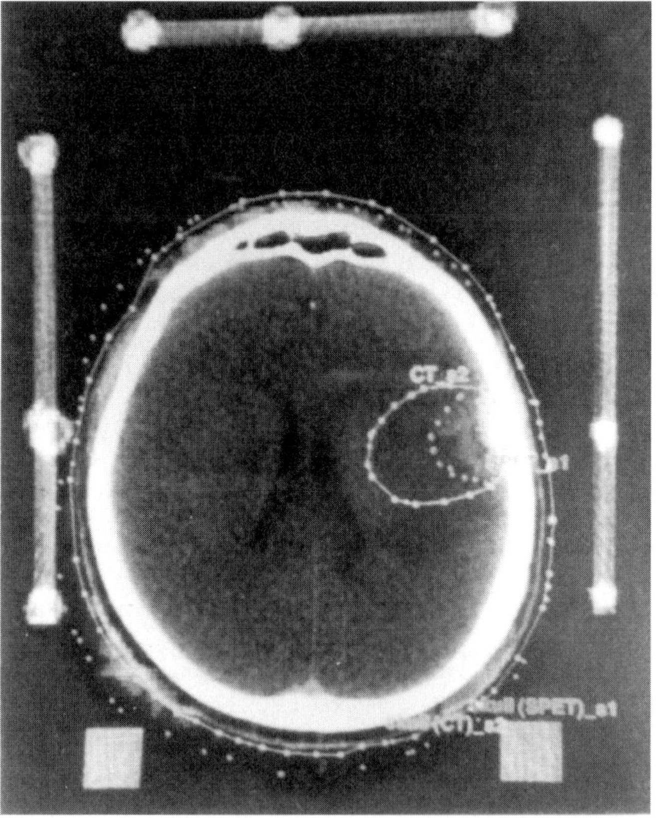

Figure 5.8. *Overlay of CT and SPECT corresponding slices after reformatting SPECT over CT image. CT lesion (delineated in dots on the full line) completely includes SPECT lesion (delineated in dots only). (From Mongioj et al 1999; reprinted with permission from Elsevier Science.)*

Sailer *et al* (1996) have concentrated on several situations where the use of multimodality imaging and image registration is essential. The first is the common case where the tumour is not visible on the treatment-planning CT slices but is visible on the MR slices. Examples are shown of planning a right tempero-parietal glioblastoma multiforme and a left fronto-temporal oligodendroglioma. A second area of concern is when the tumour has been surgically resected prior to subsequent radiation therapy. Contours defined from the pre-operative CT scans can be used to determine a PTV which is the most likely site of residual microscopic disease. A third application is when a tumour has responded to pre-radiation chemotherapy but the therapist wishes to treat the original pre-chemotherapy volume. In the latter two situations, the pre-radiotherapy images no longer display the original GTV so some form of image registration to pre-chemo or pre-surgical images is required. Sailer *et al* (1996) have implemented image registration tools within the University of North Carolina treatment-planning system. The registrations were between X-ray CT and MR and made use of a variety of matching tools. Brain provided most applications but other body sites had been planned. A particularly useful tool enabled images to 'fade' from complete 100% MR–0%CT to 0% CT–100% MR, analogous to a left–right fade in a stereophonic music system.

Pirzkall *et al* (2000) have noted the extreme difficulty in determining the contour and extent of disease of high-grade glioma. They have suggested the use of magnetic resonance spectroscopy to determine the PTV and evaluated the implications for IMRT. This paper was one of the three prize papers at the 13th ICCR, Heidelberg.

5.3.2. MMI for lung cancer planning

The planning of radiotherapy for treating lung cancer is particularly difficult because CT of the lung does not show changed lung function unless this has translated to attenuation changes. The problem has been overcome by Cai *et al* (1999) who have fused X-ray CT data sets with PET emission 3D data sets showing the distribution of fluorodeoxyglucose. This positron-emitting labelled drug has documented increased uptake in lung tumours. Cai *et al* (1999) used the chamfer matching technique. Contours were extracted from CT, from emission PET and from *transmission* PET images. The two types of transmission images were chamfer matched using the outlined lung contour and a distance metric. Then the same transformation was applied to superpose the *emission* PET data onto the CT data. The errors in the process mainly arose from the poor spatial resolution of the PET images and from the fact that the patient had arms in different positions between changing modalities. The errors were quantified using a phantom study. Image registration was performed in the context of an ADAC PINNACLE TPS. It was found that the extra information available from the functional PET images changed the PTV and was an indicator for studying the changed prognosis, although planning studies have not yet been carried out. An obvious criticism of the study was that the lung contour clearly changes with position in the breathing cycle and one wonders whether the study would have been more successful if some form of breathing control had been implemented (see chapter 4).

Levegrün and Schlegel (1998) have shown that the use of fuzzy logic can account for large inter-physician variability in the definition of PTV. The definition of PTV from CT data alone is however controversial. For example, Debois *et al* (1998) have shown, using PET FDG to determine the volume of lung tumours, that 52% of 105 patients would benefit from acquiring the additional PET scan. Of these, the PTV would increase for 18% of the patients and 34% would have a decreased PTV, in fact by as much as 47% reduction in volume.

Marks *et al* (1995) have used SPECT lung perfusion imaging to establish areas of hypoperfusion presumed to be a surrogate for non-functioning tissue. It was found that in 60% of 56 patients with lung cancer, SPECT scanning would lead to a changed PTV. Marks *et al* (1998) later showed that SPECT was a useful monitor of radiation treatment.

Ph *et al* (1999) have used dual-modality imaging to improve target definition and three-dimensional treatment planning of intrathoracic tumours. Nine patients underwent spiral CT scanning and fluorodeoxyglucose scanning using a slice thickness of 5 mm. The FDG PET scanning was performed with a dual-headed gamma camera, a PICKERPRISM XP fitted with 511 keV collimators. Matching was achieved by iteratively minimizing the cost function through three-dimensional translation and rotation for four landmarks. The registration error was assessed at better than 5 mm and malignancy was confirmed at all sites in all patients.

Uncertainty in the PTV may be a purely systematic error, constant for each patient if, for example, the patient has organ motion on the CT scanner or is set-up wrongly on the CT scanner. However, the corresponding motions on the treatment machine lead to intratreatment variations which are random. Patient movement, organ swelling between fractions is intrafraction random error.

5.3.3. MMI for prostate cancer planning

Cattaneo *et al* (1998) have shown that in contouring the prostate and seminal vesicles there is a small intra-observer variation but a large inter-observer variation in determining the PTV. (Regarding subjective measurements but in a different context, Pitkänen *et al* (1999) have studied the inter-physician (physician-to-physician) and inter-patient (patient-to-patient) variability in PTV delineation for the radiotherapy of breast cancer after conservative surgery. For one particular patient a change in PTV from 670 to 1200 cc was observed. The implication is obvious.)

For the prostate, Rasch *et al* (1998) have found that the CT-derived PTV was larger than the MRI-derived PTV and the increase was greater than inter-observer variability. As a result it is clear that multimodality definition is required.

Details of the prostate study have been given by Rasch *et al* (1999). Three independent observers measured four 3D data sets (transaxial CT, MR in three orthogonal sets of planes) for 18 patients. The data sets were registered using chamfer matching on bony landmarks and rechecked inside the UMPLAN treatment-planning system. For any given scan set, a 'scan encompassing volume' was determined being that volume encompassing all the volumes outlined by the observers and a 'common volume' determined between observers. A large ratio between encompassing and common volumes indicates a large observer variability and *vice versa*. Similarly, the same quantities were determined for a particular observer between modalities and the ratio then becomes the measure of interscan variability. The average ratio of the CT volume and the MR volume was 1.4; the axial MR-derived volume was larger than the coronal MR-derived volume by 1.1. The ratio of common volumes above was about 2.5 for the interscan variability indicating that this is the larger source of error determining prostate volume. The ratio was about 1.5 (depending on modality) *between* observers.

Uhl *et al* (1998) have performed a most interesting study to determine the accuracy of the prostate PTV. They CT imaged six patients with pathologically diagnosed prostate cancer and outlined the prostate on the NOMOS PEACOCKPLAN planning system. Then they compared the volume outlined with the true volume determined from post-operative radical prostatectomy specimens. The gross tumour volume was between 35–126% overestimated by the planning technique. The PTV with a 5 mm margin was between 233–404% overestimated. Small changes in the target volume margin obviously make big changes in these figures. The implication is that planning prepares for overtreatment.

Khoo *et al* (1999c) have compared the ability of MRI and X-ray CT to delineate the prostate and its surrounding OARs. They studied five patients, each of whom was imaged with four different MR imaging sequences and also using a conventional X-ray CT protocol. The images were viewed by three independent blinded observers who were asked to score, in a carefully quantitated way, the ability of each imaging technique to create segmented outlines of the prostatic apex, prostate, rectum, bladder and seminal vesicles. It was expected that the MRI sequences would outperform the X-ray CT technique due to the known soft-tissue contrast superiority of MRI. However, before the study, it was not known which MRI sequence would be preferable. There were significant inter-patient variabilities but almost no inter-physician variabilities.

Overall, and for each assessed pelvic structure, the four MRI sequences outperformed X-ray CT and created higher segmentation scores. A T_1-weighted 3D spoiled gradient echo (fast low angle shot—FLASH) sequence F3D provided the best segmentation mainly due to its multiplanar capability and lack of susceptibility to partial-volume effects. The imaging time was also considered feasible when compared with some other suggested sequences in the literature. In particular, the definition of the apex of the prostate was better with MRI. It has yet to be determined what the radiotherapeutic consequences of using MRI-determined volumes might be on the outcome of IMRT.

Kagawa *et al* (1997) have compared the outlining of the prostate on CT and MR scans for 22 patients. The CT and MR data were registered using selected

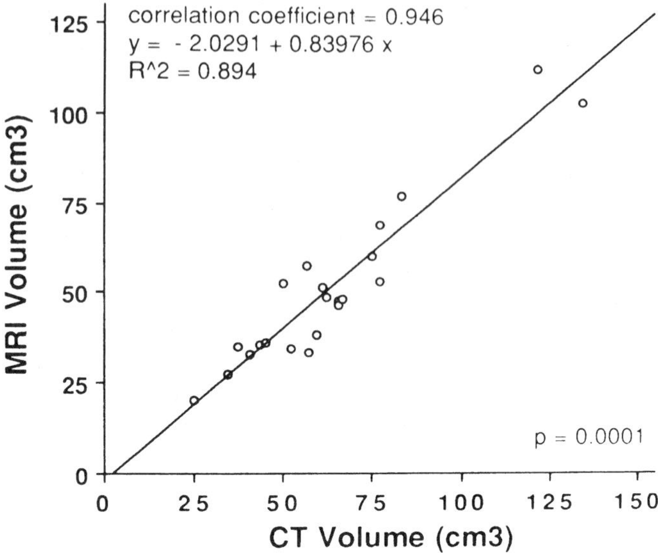

Figure 5.9. *The relationship of CT-defined prostate volume and MR-defined prostate volume. (From Kagawa et al 1997; reprinted with permission from Elsevier Science.)*

anatomical landmarks and then the MR data were resliced to the same slices as the CT data. It was determined that the registration accuracy was better than 1 mm. The MR volume was consistently lower than the CT volume. On average the MR volume was 50.9 cm^3 and the CT volume was 63.0 cm^3. Figure 5.9 shows that there is a strong linear correlation between the volume determined from the two modalities. Kagawa *et al* (1997) have also shown the correlation between the size of the prostate in three orthogonal directions estimated the two ways. This discrepancy was largest in the anterior–posterior direction. Most importantly the prostatic apex and base were likely to be misjudged from using CT alone. On the transverse MR the prostate apex was situated 7.1 mm dorsal and 15.1 mm cephalad to the tip of the urethrogram cone.

5.3.4. MMI for brain and head-and-neck cancer planning

For brain tumours, Weltens *et al* (1998) have shown that the addition of MRI to CT data did not improve inter-observer variability in PTV definition but did lead to an increase in the size of the PTV on average.

Hamilton *et al* (1997) have used SPECT in an interesting way. For a patient with an AVM, SPECT scans were performed with the patient performing certain tasks such as right-hand finger movement, right-toe movement, silent word generation etc. Then by subtracting the SPECT scans, with and without speech tasks, the area of the brain responsible for speech was identified and shielded in subsequent radiation therapy to preserve function, a very nice example of conformal avoidance. A second patient with an astrocytoma was imaged with functional MRI.

Conformal avoidance was also studied by Aldridge *et al* (1999) in the context of the Wisconsin tomotherapy apparatus. It was shown that the dose to spinal cord and parotid glands could be kept well below tolerance whilst not compromising the tumour dose. This study used a helical pitch inverse-planning technique specified by 52 rotations with a pitch of 5 mm and approximately 70 000 pencil beams.

Grosu *et al* (1998) have used ^{123}I alpha methyl tyrosin SPECT and MRI to image 60 histologically verified brain gliomas. They found the tumour volume determined from the SPECT image was larger than that from MRI in 25% of the patients and outside the enhancement region of gadolinium-weighted MRI images for 60% of the patients. As a result, fused SPECT and MRI images are now used routinely in this centre (Munich) for planning radiotherapy for these tumours.

Hosten *et al* (1998) have pointed out that T_2-weighted MRI is more capable of determining the extent of white matter oedema surrounding a glioma. This oedema is expected to contain tumour cells and so should be included in the PTV. In extra-axial brain tumours, like meningioma, the better contrast resolution achieved by MRI is of importance if the exact extent of tumour spread into the meninges is to be determined. In the pelvis, MRI multiplanar imaging capability is helpful in judging the border between rectum, prostate gland and seminal vesicles.

Khoo *et al* (1999b) studied seven patients with meningiomas involving the base of the skull using both CT and T_1-weighted MR volumetric imaging. On average, the MR appeared to define larger target volumes compared with CT. In some cases, the volume so determined was vastly different. It was advised to consider composite CT and MR volumes for treatment planning at base-of-skull meningiomas. Khoo *et al* (1999b) observed quantitative differences in the margins of the tumour as defined by MR and CT.

Adams *et al* (1999a) have shown that CT and MR data can be integrated to improve the treatment planning of stereotatic radiotherapy for paediatric patients.

5.4. PLAN IMPROVEMENT—'CONVENTIONAL' CFRT, NEURAL NETS

Read (1997) has considered the evidence for improved local control with conformal therapy, also considering the evidence for reduced metastatic spread. Considering also the dangers of CFRT with a moving patient, he has weighed the benefit with the risk. Contraindications to CFRT also include set-up errors, geographical miss and variation in individual patient sensitivity. There is clear evidence for improved response with dose escalation in the bladder, larynx, and stage-three prostate. Conformal radiotherapy also decreases the likelihood of metastatic spread and increases the time to relapse. Challenges for the future include the development of planning systems (Webb 1996) which take into account organ and patient movement (Meyer and Purdy 1996). Harris (1997) has also highlighted the movement problem.

Xing *et al* (1997c) have emphasized the continued importance of optimizing 'conventional' planning, i.e. plans with either open or wedged fields. They particularly considered the problem of determining optimum wedge angles. If there are J wedged fields then the problem reduces to determining $2J$ weights where each field is considered as a combination of open and fixed-angle wedged fields (Oldham *et al* 1996).

Xing *et al* (1997c) considered the problem as an iterative least-squares optimization of a quadratic cost function, the difference between the prescribed and calculated dose distributions. The cost function included an importance parameter to selectively emphasize the degree of conformity in different regions of space. The algorithm developed was equivalent to iterative CT computation in which the effect of updating each weight was considered in turn. Two forms of the algorithm resulted, one in which the updating took place after a complete cycle through all fields and the other, one field at a time. The former converged faster. A form of the algorithm was also presented in which the individual fields were broken down into bixels. However, the intensity of each bixel could not be individually varied since they were constrained to form the wedged fields. Two examples were shown indicating the improved dose conformality after optimum choice of wedge angles (figure 5.10).

Roach (1996) has criticized the work of Neal *et al* (1995), who studied the effect of varying the number of conventional fields, on the grounds that it is too general to state that a plan with N fields is better or worse than one with M. He argued that the field orientations and weights must be specified. Neal *et al* (1996) have countered that their simulated-annealing technique took into account the overlap between rectum and prostate PTV and also the relative importance of protecting the femoral heads. They agreed that their results were patient dependent but that the averaged results for certain numbers of fields were better than for another number. They have listed the many complicating problems in determining the optimum prostate therapy.

Optimizing oesophagus radiation therapy remains a continuing problem given the poor outcome of 'conventional therapy'. Guzel *et al* (1998) have shown that a significant improvement in sparing of the normal lung tissue results from the use of shaped conformal fields even when there is no optimization of beamweights, beam directions or wedges. The use of a two-phase treatment regime in which the fields are shaped to the BEV of the target itself improves on the use of rectangular fields and standard blocks.

Lee *et al* (1999) have conducted a comparative planning study using the DKFZ VOXELPLAN system for non-small-cell lung cancer (NSCLC). The treatment of NSCLC comprised two phases: phase one, to a large PTV comprising the mediastinal lymph nodes, both hilar nodes, the visible tumour and a margin for error and possible movement due to respiration; and phase two, to a reduced PTV including the gross tumour volume, the obviously involved lymph nodes and

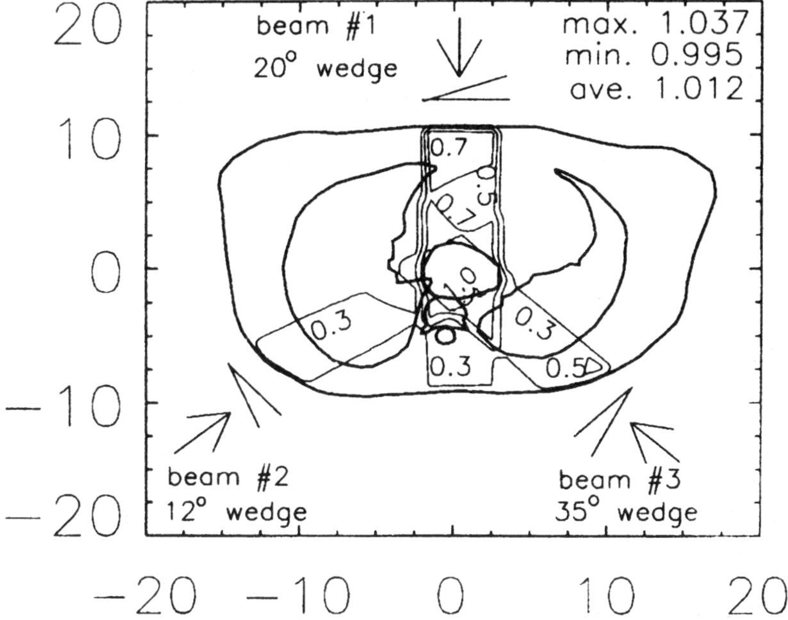

Figure 5.10. *Isodose curves for a three-field oesophageal plan with optimised wedges giving the target ten times the importance of the lung and importance factors zero for all other structures. (From Xing et al 1997c.)*

margins again. The aim was to bring the minimum dose to the first PTV to 44 Gy and then top-up the minimum in the second PTV to 60 Gy or larger. The study compared four potential X-ray beam and beamweight arrangements with a proton plan. The plans were constrained by dose-volume constraints to the normal lung and to the spinal cord.

It was found that none of the techniques could deliver a minimum dose to the second PTV to 90 Gy for all 13 patients. This *was* possible for nine out of the 13 patients studied using the proton plan. Recommendations were made on the optimum photon plan which was able to treat ten out of the 13 patients to 60 Gy and four of these to 90 Gy. The study commented on, but did not address, issues of movement, beam energy and TCP computation.

Of course, the whole issue of 'assessing' an improvement is not easy to quantify. Generally authors have stuck to the idea of showing that some 'improvement' leads to better dose statistics, e.g. a lower mean dose to the OAR for the same PTV coverage. However, clinicians often disagree on the score to be attached to a plan, i.e. its probability of acceptability. This has led to the idea of training a neural network to help assess the relative merits of competing plans. Lennernäs *et al* (1995) have trained a neural network with 18 DVHs for rectum and bladder. Clinicians scored the competing plans and the network was fed the DVH data and tuned to give the same probabilities of acceptance within limits of error. Then, the neural network was fed with nine DVHs corresponding to nine test plans (rectum and bladder DVHs). The network was shown to give acceptable probabilistic outcomes compared with those of three trained clinical judgements fed by the same DVHs. Wells and Niederer (1998) have also used a neural network to predict planning parameters based on the parameters of the contour data. Hosseini-Ashrafi *et al* (1999) have developed an artificial neural network (ANN) to determine the best plan arrangement from a pre-stored set, on which to base further optimization.

Willoughby *et al* (1996) have also trained, and then evaluated, a neural network to distinguish between the merits of different plans. Their network had five hidden nodes, and dose-volume data from competing plans were fed to 13 input nodes. Four output nodes then created a 'score' for each plan in a range of zero (useless) to four (superb). The neural network was trained with 511 plans based on data for 14 patients (plans already scored by physicians) and then used to assess a test set of 19 plans for two patients not used in the training set. It was found that the neural network could predict the scores for plans to within one point of the value scored by the physicians for 90% of the time. Given that repeat physician analyses only agreed 88% of the time, this was judged a good performance. Data concerning the construction, training, momentum, learning rate and features of the net have been given by Willoughby *et al* (1996).

Xing *et al* (1999d) have developed a knowledge-based system to predict the optimum gantry angles for IMRT beams. To do this they generated a set of template plans for geometries with known relative locations of PTV and OARs. Then, for a new real case, they used the stored knowledge to predict a best set of beams. This was compared with the brute force selection of a best set using CORVUS-based inverse planning and came up with the same angles within 3.5°.

Rowbottom *et al* (1999c) have used a neural network to predict the optimum beam orientations for treating the prostate (see also section 5.7). They trained a network using the optimum orientations found for a set of patients with prostate cancer (Rowbottom *et al* 1998a) and fed the input nodes with parameters representing the locations of the PTV and OARs. It was found that, when testing the network on a set of other patient cases, it could predict the orientations, previously found by laborious optimization techniques, to within 5–10° in most cases. Neural networks have been used in other branches of radiotherapy (Wu and Zhu 2000).

5.5. OPTIMIZATION OF IMRT

5.5.1. *Classes of optimization technique*

Bortfeld (1999b) has explained why a true inverse solution is physically impossible for IMRT. The reason is that inverse planning is a technique to *redistribute* dose rather than change the integral dose. If this were to be done optimally it would require that some negative beam intensities be generated from some beam directions. Analytic techniques which attempt to solve the inverse problem, and in turn generate negative intensities, then have to be *a posteriori* trimmed by one of two techniques. Either the negative intensities are set to zero or a dc level is added to beams in such a way that the most negative beams become zero. Either way, this disturbs the 'optimality' of the solution. As a result, direct inversion methods have been largely abandoned in favour of iterative methods.

Bortfeld (1999b) has reviewed the different techniques of optimization of IMRT. In agreement with the statements in section 5.1 he has observed that, in most cases, some parameters have to be predefined before optimizing the remaining parameters and observed that the two most commonly considered remaining free parameters are usually the beam orientations and the beam profiles. The classes of IMRT optimization algorithms are then as follows.

(1) Deterministic algorithms, for example, the gradient technique. The gradient technique is also called the steepest-descent technique and belongs to a class called downhill techniques. It is fast but may get trapped in local minima. Maximum entropy and maximum likelihood optimization are also deterministic algorithms and tend to IMRT solutions which are maximally smooth. The third type of deterministic algorithm is linear programming which has so far found application in the optimization of uniform beam therapy, not IMRT.

(2) Stochastic algorithms, conversely, are iterative optimization schemes which basically throw a dice to find the new configuration space at each iteration step. The most widely used stochastic algorithm is simulated annealing. It does not get trapped in local minima but it is slow. The two ways out of local minima are hill climbing and tunnelling. Finally, there are genetic algorithms which simulate the natural process of evolution in which the fittest solutions 'survive' (e.g. Haas 1999, Haas *et al* 1999) and iterative algebraic algorithms (Shepard *et al* 2000).

Bortfeld has argued why local minima probably present a smaller problem than anticipated. There are three possible reasons. Firstly, simple objective functions such as the least-squares objective function can be shown rigorously to have no local minima. Secondly, by selecting a good starting point in iteration space, it is possible to avoid encountering local minima. These good starting points can be achieved by using the analytical methods to provide an initial guess. Thirdly, the value of the objective function at a local minimum may not be too different from that of the global minimum, so there is no real need to escape from the local minimum. Similar observations were made by Mackie *et al* (1999b).

The advantages of gradient-descent inverse-planning techniques have been summarized by Bortfeld (1999b) who disagreed that cutting off negative values in the technique leads to suboptimal results. He had the opinion that provided the negative beam fluences are truncated to zero at each iteration step the technique can proceed.

Oelfke and Bortfeld (1999) have provided a detailed mathematical theory of inverse planning for the mathematically competent reader. They have shown for the first time that the inverse problem in rotation therapy does not have a unique solution and indeed for rotation therapy demonstrated that there are two independent solutions, i.e. fluence profiles with either positive or negative parity with respect to reflections of the gantry rotation point. Basing their work on earlier work by Cormack and colleagues they derived, for the first time, explicit analytical solutions for the rotation fluence required to deliver arbitrary dose distributions. The technique reduced to that of Brahme *et al* (1982) when such a dose distribution was circularly symmetric. No detailed examples were created.

Most of the discussion in this chapter is related to optimization of wide IMBs rotating about some isocentre. We do not specifically discuss the gamma knife but Papiez (2000) has shown that any therapy that consists of elementary concentric pencil-beam irradiations can be reproduced by radiotherapeutically equivalent rotational therapy with a wide IMB rotated about the isocentre but *not vice versa*.

5.5.2. Gradient-descent method (including KONRAD implementation) and simulated-annealing method (including CORVUS implementation) with dose-volume constraints

Bortfeld *et al* (1997a) have extended their dose-based optimization method to take into account dose-volume constraints of the form 'no more than $x\%$ of the volume may get more than y Gy'. They have shown that existing algorithms considering maximum and minimum dose constraints can simply be extended to consider dose-volume constraints. The results of tests with irradiating lung tumours close to the cord and normal lung showed that IMB optimization based on appropriately selected physical criteria involving only measurable quantities (dose and volume) can yield excellent results even in cases that were taken to be precedents for biologically-based optimization. Jones and Hoban (1999) have shown that biologically-based optimization gives very different results from physically-based optimization. Both were highly conformal but the former led to poor dose homogeneity in the PTV with a sharp fall of dose at the PTV edge. In contrast, the physically-based optimization gave a flatter PTV dose but a less sharp PTV dose contour fall off. Kåver *et al* (1999) have optimized the expectation value of the probability of uncomplicated tumour control.

The KONRAD inverse-planning system is driven by a dose-based objective function and is the commercial implementation by MRC Systems (Heidelberg) of the technique developed by Bortfeld *et al* at DKFZ. It will allow the calculation of non-coplanar beam portals. It will import contours from other treatment-planning systems. It displays a dose-volume histogram and dose distribution data updated after each iteration step and allows the user to intervene and change the controlled parameters *en route* to the solution. It will support dynamic or step-and-shoot DMLC techniques and also a compensator milling machine. Specifically, Schulze

(1997) has shown how KONRAD can integrate into the inverse-planning code the sequencer (interpreter) of the DMLC technique. The KONRAD system is also completely integrated inside the Nucletron PLATO inverse treatment-planning system such that there is no need to commission an additional treatment-planning system for IMRT. A Nucletron Consortium comprising five centres is developing inverse treatment-planning based on this PLATO ITP. The five members are: the Clatterbridge Centre for Oncology, Wirral, UK; the Childrens Hospital, Los Angeles, USA; DKFZ, Heidelberg, Germany; the University of Leuven, Belgium; and the University of Utrecht, The Netherlands. By being fully integrated into this PLATO suite of planning applications, KONRAD shares the same look and feel of all the other modules and, after optimization, the plan can be passed to the PLATO external-beam model for 3D convolution-based dose calculations using the optimized intensity distributions for each beam. Saved plans can also be exported to compensator milling machines. The KONRAD system is also marketed by Stryker Leibinger as part of VIRTUOSO and also marketed by Radionics (Jaywant *et al* 2000). Radionics link the output of KONRAD to the control of its mini MLC which is sometimes described as ConforMax (see section 3.1.3).

Papatheodorou *et al* (2000) have described the integration of the KONRAD inverse-planning system into the iSis3D treatment-planning system. This system is used for all the familiar and necessary geometric and dose constructs and KONRAD may be called within it. Also, a home-built module may be called which sequences the IMBs into the appropriate leaf movements for the DMLC technique. There is also a third home-built module which performs the dose calculation appropriate to the use of sequenced field segments (see chapter 3).

The same kind of methodology is built into the NOMOS CORVUS treatment-planning system (Carol *et al* 1997a, b, Carol 1997b) (see section 2.1) and has also been explored by Togane *et al* (1998). Bortfeld (1999a, b) has indicated how dose-volume constraints in inverse planning can improve on the application of a simple least-squares dose-based cost function. The constraints essentially squeeze the delivered dose-volume histogram into the required regions of dose-volume histogram space. When applied to OARs, these constraints can be visualized as a barrier in the dose-volume histogram with the permeability of the barrier dependent on the penalty functions assigned. When applied to the dose-volume histogram of a PTV, the application of minimal and maximal dose constraints can control the dose inhomogeneity in the target and penalty factors can be applied to permit the delivery of some dose below or above the limits.

Reinstein *et al* (1998) have also extended the Bortfeld method as implemented at Memorial Sloan Kettering Cancer Center. The extended methodology is called OPT3D and allows different critical organ limits to be assigned for the overlap and for the non-overlap portion of OARs. Modification also allowed arbitrary beam directions to avoid irradiating through the couch bars. It was found that a set of parameters could be determined that enabled this algorithm to generate IMRT plans with eight fields and for six patients treated for prostate cancer with almost complete success. The IMRT plans led to reduced dose to the femoral heads and were efficient use of planning time. Earlier work (Mohan *et al* 1994, Wang *et al* 1995) had shown that the original dose-based Bortfeld technique worked well for inverse planning for cancer of the prostate but not for lung (see review by Webb 1997d, chapter 1). This led to an exchange of correspondence between Bortfeld *et al* (1996) and Mohan and Wang (1996). Bortfeld *et al* (1996) agreed that the simple penalty-driven quadratic dose cost function would not give satisfactory results for lung. However, they pointed out that the introduction of dose-volume constraints effectively solved this problem. They argued that the use of biological indices was unfounded in hard fact. Mohan and Wang (1996) disagreed, claiming that for an arbitrary inhomogeneous dose distribution the clinical consequences depend in a complex non-linear fashion on the volumes exposed to each dose level and that TCP and NTCP are precisely the way to implement this summary of the effect of inhomogeneous dose conveniently. The conclusion of this exchange was that 'we (Mohan and Wang) agree to disagree with Bortfeld *et al*'. Progress in subsequent years has not really driven this controversy away.

Samuelsson and Johansson (1999) have studied the inverse-planning technique implemented in the commercial HELIOS inverse-dose planning option in the VARIAN/ CADPLAN treatment-planning systems. The optimization procedure is the dose-based gradient method developed at the Memorial Sloan Kettering Cancer Center. It was concluded that the choice of the number of fields and their gantry and collimator angles might be critical for defining the most optimal dose plan.

Starkschall *et al* (2000) have developed a novel way of using dose-volume constraints in inverse treatment-planning. As in the technique, for example, developed by Bortfeld, the clinician specifies hard dose-volume constraints and the inverse planning proceeds by a gradient projection technique. However, instead of penalizing computed dose-volume points which violate the dose-volume constraints, the algorithm will simply report that the constraints cannot be met and hand the problem back to the human clinician to modify the constraints, at which point the inverse planning restarts to find a feasible solution.

Arellano *et al* (2000) have criticized the use of hard dose-volume constraints in inverse treatment-planning. Instead, they have recommended that three soft constraints are applied, namely a minimum and maximum dose to the PTV and a maximum to the OAR. Their treatment-planning system then computes a whole series of feasible plans or indicates to the clinician that a solution cannot be found, and requests the clinician to make a change. This puts the control back from the computer to the clinician. It also allows a region of search space to be reached for inverse planning, that might not otherwise be reached.

5.5.3. *Projection onto convex sets, ART, SIRT, ILST and other algebraic iterative techniques*

Cho *et al* (1997, 1998c) have compared optimization by simulated annealing with 'projection onto convex sets (POCS)', finding the latter technique faster and generating smoother IMBs. POCS is a technique which does not require the specific formulation of an objective function (Lee *et al* 1997c). Cho *et al* (1998c) studied a 2D case of optimization of IMRT for the prostate and found that all the specified constraints could be met with both POCS and with the dose-volume constrained simulated-annealing technique, but that the former was some 100 times faster.

Xing *et al* (1997a, 1998b) have developed three iterative techniques for computing IMBs. These were based on the familiar imaging analogues of ART, SIRT and ILST. ILST optimization was also developed by Ahn *et al* (1997) who then compared the predicted dose distributions from IMRT with those actually delivered via a home-constructed MLC (Chang *et al* 1997). Coldham *et al* (1999) have used the method of Xing together with beam modelling for a MULTIDATA planning system to design inverse plans for delivery with a compensator.

Shepard *et al* (2000) have compared the performance of three iterative algebraic methods in which, at each iteration, beamweights were updated by either a ratio method, a least-squares method or a maximum likelihood method. Account was taken of the 'weights' of different structures, dose-volume constraints and penalties. The well-known degeneracy of the inverse-planning problem led to different dose distributions (and generating IMBs) with the same cost function (see also Webb *et al* 1998). In many circumstances the algorithms generated similar dose distributions.

5.5.4. *Gradient-descent with soft constraints*

As delivery of IMRT becomes an increasingly practical possibility, interest in optimization of IMRT (the creation of IMBs) has been renewed and there has been a proliferation of new algorithms. Spirou and Chui (1998a) have presented an algorithm which generates IMBs via a gradient inverse-planning algorithm. At the heart of the method is the use of the conjugate-gradient descent to optimize a quadratic-dose cost function. However, new features were incorporated. Firstly, 'soft constraints' were applied, which by definition are those which may be violated but with a penalty. For example, the quadratic dose difference between delivered and desired dose to OARs was included and weighted only if the dose exceeds some tolerance. A similar soft constraint could be applied to ensuring that no target points are underdosed. 'Hard constraints' are those that must never be violated, for example Spirou and Chui (1998a) forbid any negative beam intensities thus constraining the allowable search space. They argued that this is preferable to ignoring this constraint and then simply switching negative beam intensities to zero *a posteriori* since this process can seriously undo the optimization. Dose-volume constraints were also built into the algorithm.

They then applied the method to just two clinical cases, a patient with prostate cancer and one with lung cancer. Graphically presenting the outcome in terms of DVHs and dose statistics, they showed clearly that the algorithm is capable of significantly improving upon the non-IMRT plan performed by a human rather than by a computer. They cite comparable work by Oldham *et al* (1995). Beam orientations were retained for fairness of comparison. This does not represent a cohort study but it is useful data quantifying the real improvements delivered by IMRT.

There are a couple of deficiencies which they address. The first is the well-known fact that gradient-descent methods can be trapped in local minima. They offer the 'solution' that the way to overcome this is to restart the optimization from different starting configurations and observe whether the results improve. In the examples they have presented they found no improvement. This may not be entirely acceptable as an argument since stochastic iteration copes with this formally.

The second feature discussed is that the algorithm depends on the placement of the optimization points which are samples of the voxels. Different results follow from different placements. Other studies (e.g. all Webb's) have used *all* the voxels and so avoid this, albeit at the expense of increased computer time. Spirou and Chui (1998a) have also discussed the advantages and disadvantages of the quadratic cost function with respect to its control of hot and cold spots. They have presented alternative cost functions.

Finally, this work acknowledged that the resulting IMBs from inverse planning are generally very structured (see also examples in Shepard *et al* 2000) i.e. they usually have large-amplitude small-spatial scale oscillations. They have argued that this has the disadvantages that (i) such IMBs may require a long delivery time by the DMLC method, and (ii) they render the treatment very vulnerable to patient movement or any other set-up imprecision. These arguments were precisely those which underpinned the study made by Webb *et al* (1998), which by coincidence was being performed during the journal passage of this paper (see section 4.2).

5.5.5. Singular-value decomposition

Vieillevigne *et al* (1999a) have presented a novel inverse-planning algorithm for IMRT. This method performs a singular-value decomposition (SVD) on a matrix related to that linking dose to beamweight. It then selectively accepts only r highest values in the SVD thus affecting the derived weights and the objective function representing the closeness of fit of the dose distribution relative to the prescription. The behaviour with variable r is studied. The optimum solution uses all the singular-value components even if some beamweights are negative. The authors have then shown that the optimization worsens if all components are retained but when negative weights are excluded, as they have to be for physical sense. However, the optimization improves when both negative weights are removed *and* only a subset of singular values is retained. This is because the solution with truncated singular-value components has a smaller fraction of negative beamweights and thus when these are zeroed the effect is small. In fact, the solution is very close to that of using all components and allowing negative beamweights. However, the results shown for a model problem created a PTV with very inhomogeneous and probably unacceptable dose distribution. Vieillevigne *et al* (1999b) have described how the inverse SVD technique for IMRT is integrated in the DOSIGRAY 3D treatment-planning system.

Vieillevigne *et al* (1999c) have studied the influence of the beam energy on the inverse-planning process for a 2D planning technique. The methods made use of SVD. Comparisons were made between plans created at 6 and 25 MeV for a butterfly-shaped lesion. Whilst the study showed that the problem is relatively more ill-conditioned when the energy is higher, no significant influence of the energy was noted on the dose distribution for the butterfly-shaped lesion. This is in agreement with earlier work by Sternick *et al* (1997). It is worth noting that the intensity-modulated beams generated were very spiky.

Vieillevigne *et al* (2000) have presented the results of studying the condition number and SVD technique for a particular two-dimensional planning problem (the 'Brahme butterfly') as the pixel size and bixel size of the problem are varied. They have shown that for a fixed bixel size increasing the pixel size leads to an increased condition number, i.e. a more ill-conditioned problem. Correspondingly, for a fixed pixel size, decreasing the bixel size has the same effect of making the problem more ill-conditioned. They then showed that by increasing the truncation factor, i.e. the number of singular values which are rejected in the problem, as the ill-conditioning increases it turns out that the reconstructed dose distribution is remarkably similar in all cases. This proves that independent of the dose space and beam space sampling, the SVD technique has a high regularization power.

Platoni *et al* (2000) have applied the SVD technique for optimizing the inverse treatment-planning of stereotactic radiosurgery. They have shown that for choices of voxel size and stereotactic collimator size that lead to large condition numbers, i.e. ill-conditioned problems, it is necessary to truncate the small SVD contributions. There are many ways to do this and Grandjean *et al* (2000) have compared four specific regularization methods in SVD space. They found that, provided the voxel-to-pixel size remained constant, all four of the techniques investigated performed very much the same. In conclusion, it was observed that radiosurgery treatment plans can be optimized by appropriate regularizations of the ill-conditioning of the problem.

5.5.6. *Optimization of segmented fields*

In chapter 4, we saw how the group in Ghent create class solutions for treating tumours which are close to OARs by using multiple-static-fields. An example is shown in figure 5.11. Segments are created of varying field widths avoiding field-boundary matches and the planning problem is to determine the beamweights for these field segments. De Wagter *et al* (1998) have used a quadratic dose cost function for the PTV combined with a linear dose term to the OAR. This has been minimized by a constrained matrix inversion. Firstly, the dose contribution to a series of points from each beam segment was made and stored for unity beamweight. Then, the matrix inversion was performed iteratively and the weights of some 30–47 beam segments were determined. Importance factors controlling the balance of dose between PTV and OAR were adjusted during the iteration. The process needed only about 5 min on a DEC ALPHA computer and it was found that dose-escalated solutions could be obtained with escalation well beyond (120 Gy) what a human planner could determine (65–80 Gy) manually (figure 5.12) for tumours close to the spinal cord.

Woudstra and Storchi (1999) have presented a technique to optimize segmented IMBs and beam orientations. This technique can generate high-dose volumes which have a concave shape. The idea is to introduce simple beam segmentation. The method is as follows. Firstly, from each of 36 beam directions, three segments are defined. The first is a beam's-eye view of the full planning target volume projection. The second is this former beam minus the projection of any critical organ in front of the PTV and the third is the full beam's-eye view minus the projection of either a critical organ in front or beyond the PTV. The dose distribution is then computed for unit beamweight in each of these 36 $\times$ 3 fields. Then follows a process of successive weight optimization of segments. For each beam direction the beamweight segment is optimized to achieve the maximum score, where the score is determined in terms of biological control. The same process is continued for each of the three contributing field segments at each of the 36 directions, and the beam direction with the largest score is included in the treatment plan. The process then repeats cyclically to determine which beams are optimal for generating a high-dose distribution with concavity and typically of the order of ten beam directions survive, each with the three components described.

Morrill *et al* (1997) have extended the 'two-weight-per-field' method of Webb (1991a) to create a series of multiple-static MLC-shaped fields in which each field has just a few segments. The largest segment corresponded to the full beam's-eye view of the PTV and further segments corresponded to excluding specific OARs. The weight of each segment was computed by simulated annealing. For a difficult planning case involving a retroperitoneal adenopathy wrapping around the vertebral body and adjacent kidney the method gave acceptable conformation. The usual shortcoming of IMRT of a somewhat inhomogeneous dose to the PTV was however observed.

5.5.7. Simulated annealing

As we have seen, one of the earliest inverse-planning algorithms to gain wide acceptance was simulated annealing (Webb 1989, 1991a, b, 1992). When initially

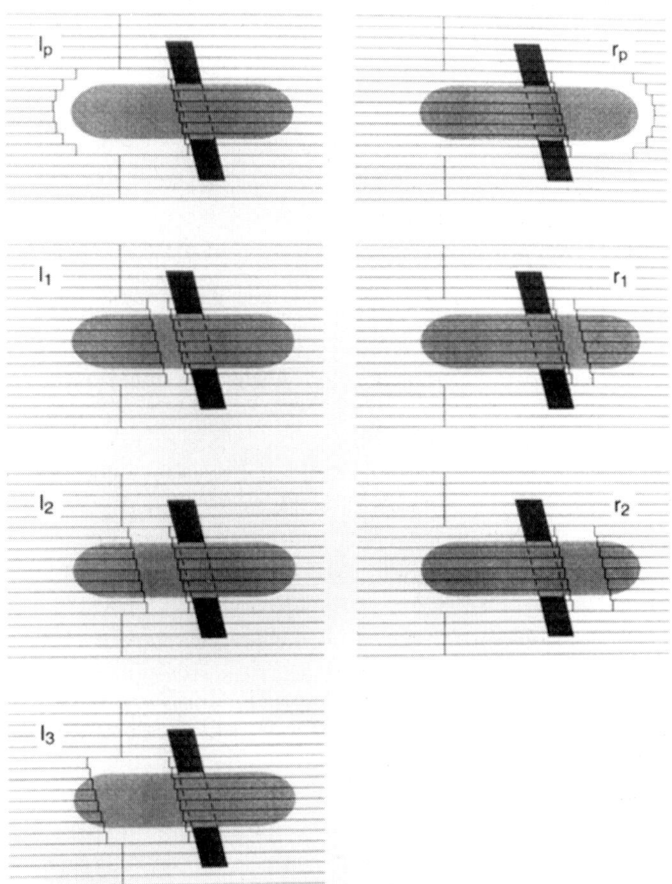

Figure 5.11. *Predefined geometry of beam segments. The spinal cord (dark grey) is surrounded by the PTV (light grey). In the beam's-eye view, the medial edges of all segments are located tangential to the spinal cord. All segments are created by an MLC. The 'parent segments' are labelled l_p and r_p and the derived segments are 1, 2 and 4 cm wide, respectively. This form of IMRT is an MSF technique. (From de Wagter et al 1998; reprinted with permission from Elsevier Science.)*

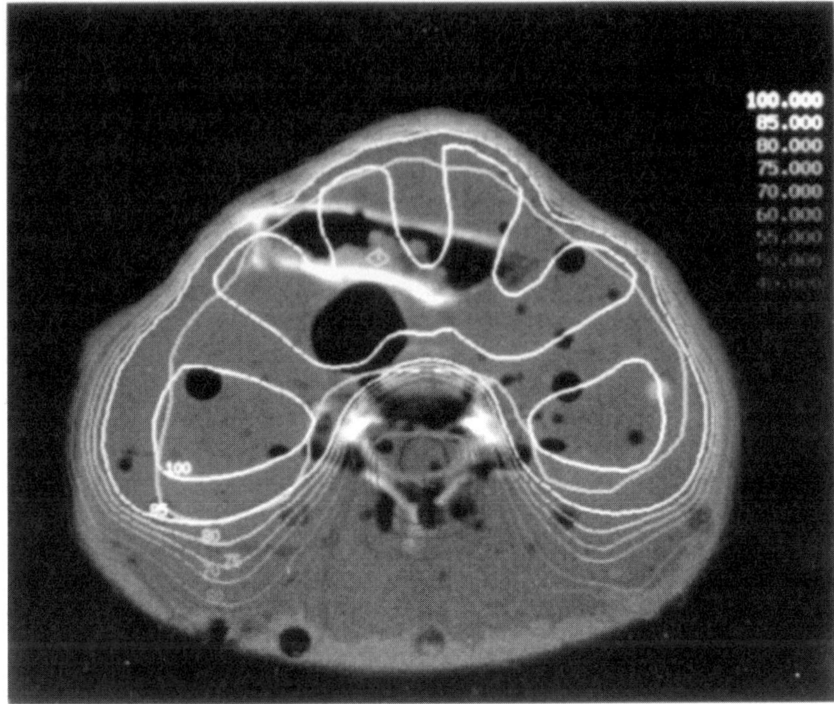

Figure 5.12. *An optimized horse-shoe shaped dose distribution for a concave target in the lower neck irradiated with 6 MV photon beam segments. The dose optimization was performed by a constrained matrix inversion. Note the crowding of the isodose lines away from the spinal cord. (From de Wagter et al 1998; reprinted with permission from Elsevier Science.)*

developed it was slow but this is not now the case and recently this technique has re-emerged. Redpath (1998) has described again most of the theory presented by Webb (1989) but in a way which emphasizes the new-found practicality. He has implemented the technique such that a million iterations can be performed in about 1 min on a Sun Sparc Ultra 170 computer. One reason that this is possible is because the beam bixels contributing to dose in the PTV at prespecified beam locations are all prestored as well as the links to dose-space voxels and their dose contribution from unit-weight bixels. A dose algorithm was implemented which catered for tissue inhomogeneities, adequately modelling the change in scattered dose. Three possible cost functions could be selected, based on importance-weighted quadratic dose or on dose-volume constraints. A particular feature was the *constraint* applied to the changes in bixel weight such that no particular bixel became vastly different from its immediate neighbours, a feature also implemented by the work of Webb *et al* (1998) (see chapter 3). An odd number of coplanar equispaced beams was used in line with the usual recommendation that beams should not be opposed and should be relatively few. The intention was to fabricate compensators or perform the DMLC technique to deliver the beam. Four clinical cases were presented, some demonstrating amazing PTV homogeneity, better than many other published methods. Possibly this is because the OAR doses were not constrained to be particularly low (e.g. 80% dose to the spinal cord for a PTV of 100% wrapping around it.)

5.5.8. IMRT planning incorporating Monte-Carlo data

Jeraj and Keall (1998) have combined the power of Monte-Carlo dose calculation with the guaranteed minimum cost of simulated annealing. They computed the dose distribution from unit fluences by Monte-Carlo methods (Mohan 1998a) and then computed the conformal dose distributions by simulated annealing. They also showed differences between dose distributions computed by this method and by convolution/superposition (Keall and Jeraj 1998 and Jeraj and Keall 2000). One of the difficulties of IMRT inverse planning is that in general it makes use of a simplified dose model for the elemental distribution for computational simplicity and speed. It is then possible to compute *a posteriori* dose distribution for the optimized fluence profiles but these may well be different from those computed with a simpler model. For example, Ma *et al* (1998c) have shown differences (greater than 5% in dose, 5 mm in isodose lines) between dose distributions computed by the EGS4/BEAM code especially in the head-and-neck, lung and breast. Pawlicki *et al* (1999a, b) have used pencil-beam kernels computed by Monte-Carlo techniques as the basis of inverse planning and have shown accuracy improvements to 3% when applied to head-and-neck treatment planning. The kernels were precomputed prior to inverse plan optimization. Walling *et al* (1998) have used the PEREGRINE Monte-Carlo code (Gibbs 1998) and Francescon *et al* (1998) the BEAM code to benchmark treatment-planning calculations for non-IMRT radiotherapy. Schulze *et al* (1997) have shown the way in which the decomposed pencil-beam algorithm in the DKFZ planning package VOXELPLAN copes with tissue inhomogeneities and have also presented results confirming the level of its accuracy.

Bogner *et al* (1999) have discussed how the inverse Monte-Carlo optimization code, IMCO, can overcome the danger of optimizing non-realistic dose distributions when large inhomogenities are present. The inverse Monte-Carlo optimization code includes the lateral scatter. The success of the optimization was demonstrated on a simple linearly-shaped target within a water lung slab phantom and also on a horseshoe-shaped target with an embedded OAR.

Laub and Nüsslin (1999) have also developed a Monte-Carlo code for implementation inside the treatment-planning system VOXELPLAN. Intensity profiles generated by the IMRT software KONRAD are used for dose calculations in VOXELPLAN. This technique links the accuracy of Monte-Carlo dose calculations with the advantages of IMRT. Laub *et al* (2000b) have extended this to include a full Monte-Carlo calculation of the dose within the stages of the inverse-planning process. They did this by first optimizing using a pencil-beam model and then recomputing with Monte-Carlo. It was shown that this could improve the conformality for plans with air structures.

Miften (2000) has studied the importance of the dose calculation model on evaluating plans created for IMRT. He used the FOCUS RTP system from Computerized Medical Systems to create three inverse treatment plans; one for the prostate, one for a head-and-neck tumour and the third for a lung tumour. The inverse treatment-planning used the model-based multi-grid superposition dose calculation algorithm which he had developed. This method correctly uses a density scaling method to scale the energy depositioned kernels which are convolved with terma. Then he evaluated the same three treatment plans using a simpler Clarkson dose calculation method. It was observed that the results in terms of dose and tumour control probability and normal tissue complication probability were very similar for the prostate case, were less similar for the head-and-neck case, and were most dissimilar for the lung case. In the latter, changes in biological outcomes of up to 14% in the NTCP and 4% in the TCP were noted. It was concluded that it is probably erroneous to use one particular simple dose model for inverse planning and then believe too strongly the biological predictions, particularly in cases where there are large tissue inhomogeneities.

5.5.9. Other optimization studies

The HELAX TMS IMRT solution also generates segmented multileaf modulation (SMLM), a form of step-and-shoot. The vendor-specific MLC restrictions are incorporated inside the treatment-planning system and, because the system uses full head-scatter modelling, the approach is more likely to generate dose distributions which are close to those measured.

Gardey *et al* (1999) have developed a virtual IMRT simulation environment for investigating the features recommended for clinical implementation.

Liu *et al* (1999) have developed an optimization routine for IMRT which makes use of a quadratic cost function. They proved that this has just a single global minimum and have constructed an algorithm, based on the theory of variation, in which at each iteration the whole profile of a 1D IMB is changed rather than its constituent discrete elements. They have planned a familiar 2D problem of a concave prostate with adjacent OARs with very simple dose model and dose prescription and no importance factors. They have shown that conformality is quite good with just five beam orientations and does not change a great deal when this is increased to seven, in line with the observations of others (see chapter 1).

Arellano *et al* (1998) have presented the dynamically penalized likelihood algorithm and coupled it to a commercial 3D planning system. They have shown the importance also of including the scattered radiation into the inverse-planning code for IMRT by using precomputed pencil-beam functions.

Mavroidis *et al* (1999) have reviewed the objectives of the dynamic radiotherapy project DYNARAD within the European community. They concentrated on cancer of the cervix, being a tumour site with concavities with involved OARs. Phantom measurements were made to confirm dose distributions created with biological and physical cost functions and it was concluded that biological objective functions allowed a more close conformation. The DYNARAD project looked forward to more novel techniques for delivering IMRT. Kåver *et al* (1999) also optimized biological functions.

Brahme and Lind (1999) have discussed the pros and cons and potential pitfalls in the use of different objective functions for dose optimization. They have pointed out that the quality of life is the important outcome, but this is a rather difficult parameter to quantify and therefore different substitutes must be used as cost functions.

Sauer *et al* (1997) and Sauer (2000) have compared objective functions for optimization using publicly available data sets specifically created for these tests (Angelos *et al* 1997). Following this, Sauer *et al* (1999) decided that simple optimization could be achieved by specifying simply a minimum target dose and then, for a single OAR, minimizing the summed dose to all voxels within it. Adding a little more complexity they considered that, when there was more than one OAR, this could be accommodated by minimizing a combination function which accounted for the relative seriality of the organs and the sensitivity to risk. They showed the results of model calculations on both a U-shaped and an L-shaped target and came to the conclusions that no more than seven beams were needed (see also section 1.2).

Keller-Reichenbecher *et al* (1997) have shown how a knowledge-based system could reduce the search space of optimization algorithms and speed treatment planning.

5.6. CHOICE OF OPTIMIZATION TECHNIQUE

An issue which is central to the question of the choice of optimization technique is the existence or otherwise of local minima in the optimization function. If there are no local minima then relatively rapid optimization schemes can be developed, such as the gradient-descent method (Rosen 1997) (see section 5.5). On the other hand, problems in which it is known that there are local minima have to be solved by stochastic search methods, such as simulated annealing, in which the ability to avoid trapping in local minima is built in (Webb 1995b, 1997c). Early optimization attempts did not always understand the need for this distinction, and certainly did not understand how to determine the presence or absence of local minima (Barrett *et al* 1983, Webb 1989).

Deasy (1997) has shown that a key determinant is the application of dose-volume constraints. He has argued that the following three types of optimization specification are all convex: (i) maximizing the TCP; (ii) maximizing the minimum target dose; and (iii) minimizing the mean-squared deviation of target dose and prescription *provided there are no dose-volume constraints*. 'Functional convexity' implies that algorithms which travel downhill will always reach the global minimum which also happens to be the only local minimum. (This is true for beamweight optimization but not for beam-direction optimization where local minima can arise even without dose-volume constraints.) This is expanded upon by Börgers (1999).

However, as soon as there are dose-volume constraints, the possibility arises of multiple local minima whatever the form of the optimization cost function. By dose-volume constraint, it is meant the specification that a certain fraction of an OAR must not receive more than a certain dose. It then arises that, in optimization space, overlapping convex sets of solutions effectively introduce non-convexities in the overall problem. The reason for this is that there are many combinations of configuration options whereby 'either this volume or that volume but not both' may be irradiated to above tolerance.

The work of Deasy (1997) is essentially a thought argument but gives insight into the effect of this DVH-guided cost on optimization strategy. Since there is a trend towards invoking dose-volume constraints in treatment-plan optimization (see e.g. section 5.5) the role of stochastic search strategies may be expected to increase. Llacer (1999) has presented a dynamically penalized likelihood method which has the property of being able to jump out of local minima and/or force a solution to be in a region where the cost function is relatively flat.

Llacer *et al* (2000) have shown that the maximum likelihood estimator in the dynamically penalized likelihood method of inverse therapy planning can be tuned using dose-volume constraints provided by the radiotherapist to create any feasible balance between homogenity in the planning target volume and dose to OARs. This technique is being incorporated into the BrainScan software package of BrainLab AG.

5.7. BEAM-ORIENTATION OPTIMIZATION

Rowbottom *et al* (1997a, b, 1998a) have studied the potential for improved TCP of the prostate by optimizing beam directions in three-field photon plans. A cost function was computed as a function of each single-beam direction for 5° intervals in gantry orientation. The function discriminated against those beams depositing large dose in OARs and included the volume effect of NTCP (figure 5.13). Then using 'gantry angle windows' which restrict the angular separation of the beams, triplets were computed which led to an improved TCP of some 5% (at NTCP = 1%) averaged over a patient population (figure 5.14). The improvement is constrained by the 'prostate–rectum overlap problem' so this 5% is a good outcome under these very constraining circumstances.

Rowbottom *et al* (2000a) have optimized the choice of beam orientations for treatment of oesophageal tumours by a different technique. For this, four beams were chosen, two fixed at 0° and 180° and the other two able to be chosen from a range of gantry orientations. Plots of the cost function components and the total

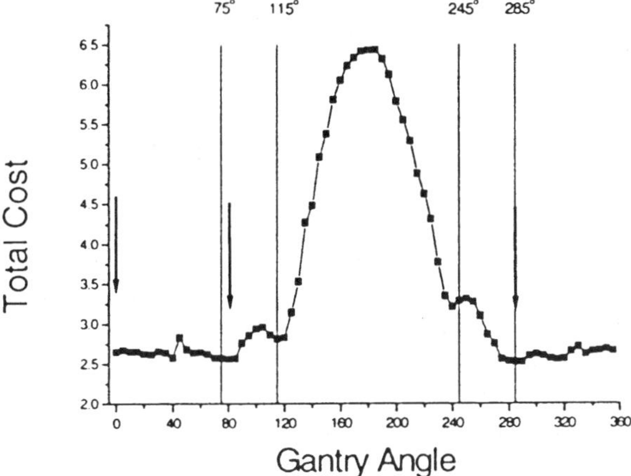

Figure 5.13. *Showing how to use a 'single-beam cost function' to help select the gantry angles for optimising non-IMRT prostate therapy. A cost function versus gantry angle plot with the allowed gantry angle windows is also displayed. The arrows show the optimal beam positions selected for this patient. (From Rowbottom et al 1997b.)*

cost (figure 5.15) were then used to select the most useful variable angles. Beam weight optimization (Oldham *et al* 1995) was applied to these choices and also to 'standard' beams, since this had already been shown to improve on the performance of a human planner. It was shown that the customized beams led to improvements in TCP and dose statistics for a cohort of eight patients with different tumour types (Rowbottom 1998).

Rowbottom *et al* (1999b) have optimized the selection of non-coplanar beams for treatment of brain tumours. This technique relied on creating a multi-beam cost function from a selection of single-beam cost functions, optimizing it and then optimizing beamweights for a standard and this customized-orientation plan. Just the lowest-cost parts of the single-beam cost function were used to reduce the search space. Beam directions were also constrained to minimize the overlap of the beams. The optimization of beam orientations improved the plan as much as did the application of IMRT (Oldham *et al* 1998c). Work at the Royal Marsden NHS Trust to optimize IMRT of the parotid by including optimization of beam directions is reported in section 4.1.12.

Pugachev *et al* (1999) have studied the inter-related problems of optimizing beam orientation and beamweights for IMRT. Gantry angles were selected by a simulated-annealing algorithm and beamlet weights subsequently computed, the most efficient way being by modified filtered backprojection. The two problems, though related, were still uncoupled. Results were an improvement compared with equispaced beams.

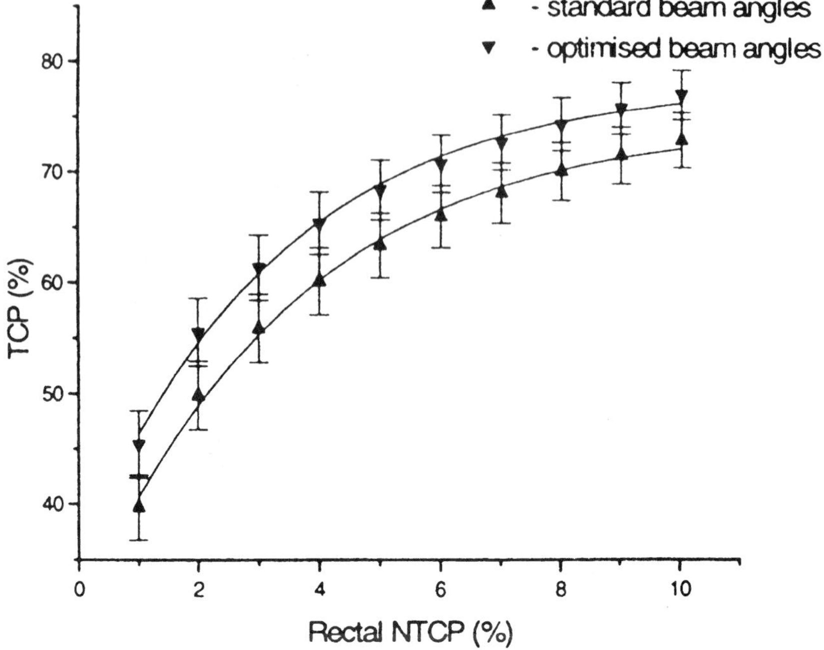

Figure 5.14. *The average TCP for twelve patients in a cohort of patients with prostate cancer as a function of NTCP. Error bars are one standard deviation of the mean. It can be seen that the optimized beam angles lead to an improved TCP compared with the values obtained for 'standard' beam orientations. (From Rowbottom et al 1997b.)*

Pugachev *et al* (2000) have compared the use of nine equispaced coplanar IMRT fields with nine coplanar IMRT fields whose orientations were optimized and thirdly with nine non-coplanar IMRT fields. They studied three planning cases, a prostate, nasopharynx and para-spine treatment. The prostate case was not greatly improved by optimizing beam directions and non-coplanarity (agreeing with earlier studies by others (see chapter 1). The other two tumour sites benefitted significantly from optimizing beam orientation and the use of non-coplanarity.

Löf *et al* (1999) have shown that by optimizing the field directions and beamweights simultaneously that P_+ increased with optimized rather than regularly spaced beams. Like the work of Webb (1999a, b) this demonstrated the diminishing returns of continuing to increase the number of beams.

Löf *et al* (2000) have shown that beam-orientation optimization can improve the performance of IMRT specifically for low numbers of treatment beams. They have developed a flexible modular planning system called ORBIT which can simultaneously optimize beam directions and beam fluence for IMRT. They

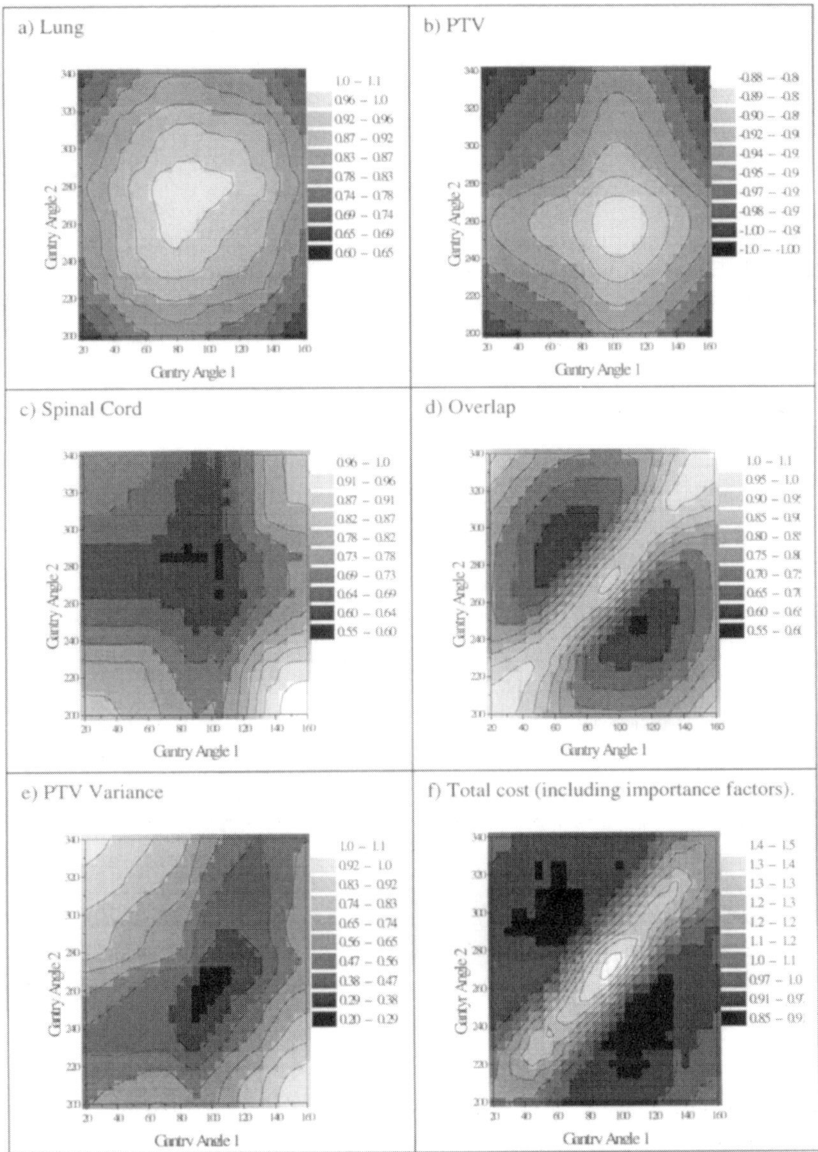

Figure 5.15. *(a)–(e) Contour-type plots for each of the terms in a single-beam cost function determining useful beam directions for treating cancer of the oesophagus, and (f) the total cost. The terms were calculated for four-beam plans with two beams fixed at 0° and 180°. The other two beams were allowed to vary in the ranges 20°–160° and 200°–340°. The concept is that the most useful beams lie in the regions of lowest cost shown darkest in these plots. (From Rowbottom 1998.)*

specifically show the need to optimize the beam orientations when there are only a few (e.g. three).

Bendl *et al* (1997) have extended the DKFZ development of the 'spherical view' to include 'beam's-eye view volumetrics'. In its original form, spherical view simply showed which beam central axes intersected OARs. The extended implementation evaluates the area of overlap between each beam and each OAR and creates a map of 'cost'. Maps may be combined for combined beams and organs. However, sometimes minima in the combined volumetry maps can lead to unacceptable hot spots and the solution is to simultaneously make use of information in the observer's view. Dickof and Ladyka (2000) have also described simple heuristics for conventional planning in which the aim is to have the computer work out orientations (and give gantry and couch angles) that minimize PTV and OAR overlap. They have commented that whilst the orientations have not been optimized they can be the basis of commencing optimization, even of IMRT.

Cho *et al* (1999) have developed a related but different concept. They have created 'target-eye view' mercator projection maps with the axes giving the angles of the line of sight of the radiation source from the target isocentre and the intensity of the map. The intensity of the map relates to scores derived from the overlap of PTV and OARs for that line of sight combined with the criticality of the overlap. This allows the user to decide whether the overlap of the BEV of the PTV with one OAR is more important than with some other OAR and to give this a higher score. Similarly, overlaps between PTV and distal OARs may be scored lower than overlaps with proximal OARs. There is also a facility to score in overlaps with margins of the beam. The extent of the overlap is not a consideration in this technique. They have proposed to develop the method of combining these data with maximal separation considerations. There is much similarity between this and some of the other concepts reviewed in this section.

Das and Marks (1997) have independently developed a method to optimize beam orientation and have applied it to the irradiation of a brain tumour. They constructed a cost function which was designed to achieve the following either independently or in combination.

(i) It opened up the beam directions maximally by maximizing

$$\Gamma = \left(\Pi_{j=1\neq i}^{N} \Pi_{i=1}^{N} \sin\left(\theta_{i,j}\right)\right)^{\frac{1}{N(N-1)}} \tag{5.1}$$

where $\theta_{i,j}$ is the angle between the central axis of the ith and jth beams. Both entrance and exit vectors were considered. The function Γ is maximized when the beams are maximally separated. In practice, the algorithm minimizes '$-\Gamma$'. Das and Marks (1997) have shown that this achieved the same result, when $N = 4$ and $N = 6$, as the method of Webb (1995a).

(ii) A 'length function' was minimized, this being either: (a) the sum of the entrance distance to the tumour isocentre for all beams; (b) the sum of the entrance and exit distance to the tumour isocentre for all beams; or (c) a length criterion representing a dose gradient along the central axis of the beams.

(iii) A function representing the amount of overlap of the beams with defined normal structures was minimized. This steers beams away from OARs. It is a 'soft' criterion because it does not exclude the possibility of beams passing through these regions.

(iv) A function forcing the choice of beam orientations to lie in the regions where no couch/gantry collisions could occur was minimized. (It is not obvious why these authors did not simply exclude these zones *a priori*.)

Das and Marks (1997) have argued that it is not possible to create a cost function optimization strategy which individually achieves all goals, since each of these elements compete. Instead they constructed an overall cost or goal function in which the individual contributions can be toggled on or off. The global minimum was found by selecting the parameters corresponding to the lowest local minimum after restarting the optimization at a number of starting points. Perhaps a simulated-annealing search strategy could have been used instead? When criteria (i) and (ii) were switched on, it was found that the OARs could be adequately protected but criterion (ii) with choice (b) forced the beams towards lying in a plane, the degree of forcing depending on the power of the criterion which was a variable. In this configuration they would give rise to a large dose gradient across the PTV. The paper did not in fact report dose-volume histograms for the PTV, which was a limitation. It was emphasized that the selected beam orientations were advisory to clinicians who could then adjust them as desired. To some extent this makes the approach less attractive since it puts back a subjective element.

Desobry *et al* (1998) have used 'utility theory' to optimize beam placement. This is essentially a modified forward-planning technique in which beams are selectively introduced observing the effects on the distribution, and accepted or rejected on the basis of improvement or otherwise.

Alber and Nüsslin (1999) have developed a technique for biologically optimizing beam directions for radiotherapy. The algorithm is predicated on the need for IMRT to optimize intensity profiles simultaneously with beam directions. The essence of the technique is that the dose is iteratively re-distributed by adding blocked or intensity-modulated fields and that the direction of each individual additional beam is optimal with respect to the running estimate of dose and beams placed up to each iterative stage. The algorithm delivers sets of beam directions which allow the same level of biological score as equi-angular directions yet with fewer beams in most cases.

Muthuswamy (1999) have compared the use of wedged plans and intensity-modulated plans for three- and seven-field irradiation of a prostate. Two kinds of wedged plans were generated. In the first, the field size was determined by the requirement of a full target coverage in the beam's-eye view, and in the second the field shape was optimized particularly at the critical organ–target overlap region. These were called field-shape optimized conventional plans (FSOC plans). For the intensity-modulated plans, simulated annealing was used to optimize the gantry angles and a conjugate-gradient technique was used to optimize the IMB fluences. Both the FSOC and the intensity-modulated plans were significantly superior to conventional plans. However, the FSOC plan was only marginally inferior to the intensity-modulated plan leading to the conclusion that IMRT for the prostate might possibly be avoided by the use of optimizing field shapes.

Bedford *et al* (1999b, 2000a) and Khoo *et al* (2000) have conducted a painstaking manual search of orientation space to optimize the placement of either three, four or six fields for improving conformal therapy of the prostate. It was shown that class solutions, somewhat different from some conventional orientations, could improve treatment.

5.8. ORGAN MOTION AND TREATMENT OUTCOME

Organ motion can affect the treatment outcome in many ways (see also section 4.2). It complicates the calculation of accurate dose distributions and so in turn affects the probabilities of biological control. Intra-treatment motion includes movement of structures in the thorax and abdomen due to breathing. In general, organ motion is asymmetric because more time is spent near the position of exhalation than inhalation (Ten Haken 1998). Balter *et al* (1998) have shown using fluoroscopy that the average patient's diaphragm remained within 25% of the range of ventilatory excursion from the average exhale position for 42% of the typical breathing cycle, but within only 25% of the range from the average inhale position for 15% of the cycle. For abdominal targets CT-based treatment planning improved using static exhale imaging. Inter-treatment changes arise for example in treating the prostate due to differential bladder and rectal filling. Ten Haken (1998) has provided an overview of the problems and some solutions.

Ten Haken *et al* (1999) have studied the effects of patient breathing on radiotherapy of the liver. Breathing was modelled using an asymmetric periodic function containing variables which allow control of the asymmetry, period and phase of the motion. Using simulated organ movement based on the use of this formula, they looked at the effects of varying the amplitude of breathing, the asymmetry and the effect of the occasional deep breath. In the worst case, a 26% overestimate of dose was made in the liver.

Yan *et al* (1997b) have taken account of the variation of the daily deformations in tissue location. By convolving the nominal spatial isodose distribution with the local probability distribution of 'tissue occupancy' they created 'adjusted dose-volume histograms' (aDVH). Jaffray and Wong (1997) have built a cone-beam CT scanner, mounted at right angles to the megavoltage source, for the purpose of creating images of the patient to determine 'target of the day'. A hundred projections are recorded over 194° of arc and a 256^3 image data set is reconstructed in some 30–50 min. This creates *a posteriori* measures of tissue occupancy, doing away with the need for sequential *a priori* measurements via conventional CT. Jaffray *et al* (1999) have shown how, by mounting two detectors on the linac, one opposite the megavoltage source and the other opposite the kilovoltage source, a dual-beam system (DBS) has been constructed (figure 5.16). They have presented a detailed study of the radiological imaging properties of both the kilovoltage and megavoltage systems for planar radiography and also for performing megavoltage computed tomography. For an equivalent contrast detail response, the kilovoltage system requires a factor of 40 lower dose at 6 MV. Whilst the megavoltage system operates close to the quantum limit, this is not the case for the kilovoltage system. Such MVCT images can be used to assess the effects of daily warping of patient geometry (Lu *et al* 2000). Jaffray *et al* (2000) have developed a cone-beam CT system on a linear accelerator using a flat-panel amorphous silicon imager.

Hesse *et al* (1997a) have a similar MVCT scanner (see section 4.4.1) which generates a single slice from 120 projections every 3°. Each projection was recorded on a 512 element detector (Wellhöfer BIS-710) and reconstruction into a 512^2 array was made by an iterative algorithm. Mubata *et al* (1997a) have also evaluated the effects of movement by randomly placing the calculated dose cube with respect to the patient instead of recomputing the dose cube. DVHs were

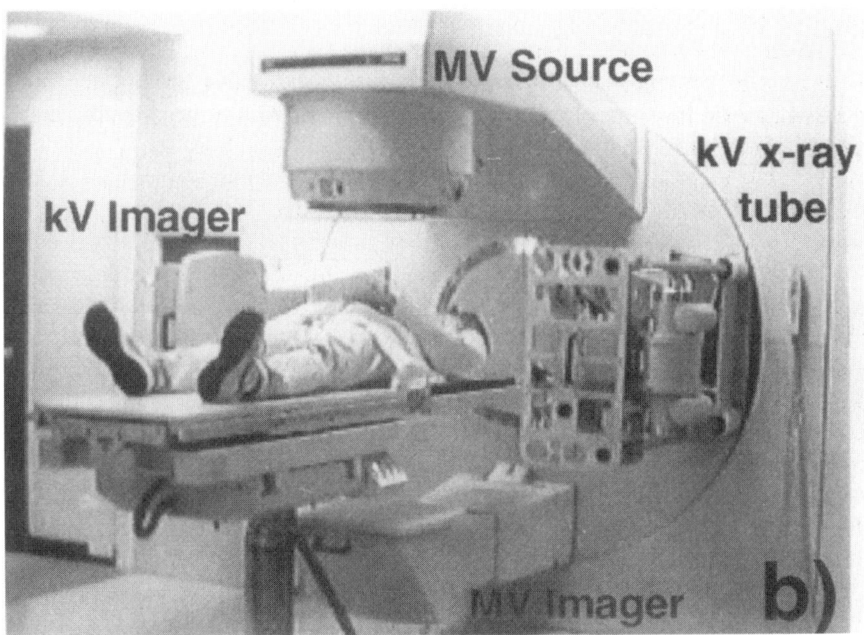

Figure 5.16. *The dual-beam system for kilovoltage and megavoltage imaging on an Elekta SL-20 linear accelerator. The kilovoltage source is mounted at a right angle to the megavoltage source and both sources are viewed by a detector system. (From Jaffray et al 1999; reprinted with permission from Elsevier Science.)*

computed for the 'with movement' situation and compared with those for the static patient.

A different approach has been adopted by Fiorino *et al* (1997). They accounted for movement by artificially widening the individual beams via a convolution with a Gaussian process, i.e. the method does not convolve the dose distribution as a whole. The beams were artificially adjusted inside the CADPLAN 3D treatment-planning system. Margins were applied, as usual, to the PTV and OAR and the size of the margins was varied, observing the effect on TCP and NTCP. For fixed margin, NTCP values turn out to decrease in correspondence to the higher values of the standard deviation of movement. This surprising result can be explained because the fraction of the rectum included in the high-dose region decreases with respect to the static situation when the standard deviation increases. The TCP also decreases with increasing standard deviation of movement for small (7.5 mm) margins. On the other hand, the TCP hardly changes with increasing standard deviation of movement when the margin was much larger (1.5 cm) whereas the NTCP value shows the same behaviour as for the smaller margin. The overall effect on uncomplicated tumour control depends on the concomitant action of the two opposed effects.

Fiorino *et al* (1999) have shown the influence of systematic errors and random errors in determining the PTV on the outcome of IMRT delivery. They studied one prostate patient with a concave-shape PTV irradiated by seven IMBs. The seven IMBs were equally spaced in angle. It was shown that the tumour control probability and fixed normal tissue complication probability actually increased when systematic errors were included.

Bakai *et al* (1999) have studied the influence of movement on IMRT of the thorax particularly when delivered using the DMLC. The effects of intrafractional motion on the tumour control probability were analysed using a multiple-convolution approach and various movement patterns were taken into account. The high gradients in individual intensity profiles can result in a decrease in tumour control probability when movement occurs and it was postulated that re-arranging intensity amongst beams can achieve smoother intensity profiles and thus reduce the influence of the inevitable volume movements on the treatment outcome. This observation is closely allied to the work of Spirou and Chui (1998a) and Webb *et al* (1998) who also observed that smoother intensity-modulated profiles have advantages for the DMLC technique. Meyer *et al* (1999a, b, 2000) have shown that smoother IMBs also have advantages for reducing the implementation error delivering IMRT via a compensator. Essentially, valleys in the IMBs provide the biggest limitations (requirement for cutters of various sizes) and so the fewer of these there are the better.

Killoran *et al* (1997) have simulated the effects of prostate movement and the effects of uncertainty in setting up treatment of the prostate using a Monte-Carlo method. They relied on the data of van Herk *et al* (1995) for prostate movement between fractions and data from Gladstone *et al* (1993) for set-up errors. Then they recomputed the dose to the PTV and OAR for each fraction taking into account sampled values of these movements. The whole treatment cycle was simulated one hundred times so that the eventual outcome was a probability map of dose. They compared four different treatment procedures subjected to this analysis. The conclusion was that the use of anisotropic margins and also some increase of fluence at the field edge was no more susceptible to movement than the use of wide margins in that the PTV coverage was maintained. However, significant reductions in the OAR dose were computed.

5.9. ANALYSIS OF TREATMENT PLANS

5.9.1. *Functional dose-volume histograms*

Lu *et al* (1997) have introduced the concept of functional dose-volume histograms for OARs. These recognize that a conventional dose-volume histogram, useful though it is, takes no account of the spatial variation in the density of functional subunits (or more generally simply function). It simply scores what fraction of the volume of a normal organ receives dose D_0 or greater. A functional dose-volume histogram, f DVH, scores, by definition, the fraction of functional subunits receiving dose D_0 or greater. To do so it requires a knowledge of the density $f(\boldsymbol{r})$. So

$$fDVH\,(D_0) = \frac{\int_{V_0} d\boldsymbol{r}\, f\,(\boldsymbol{r})\, H\,(D\,(\boldsymbol{r}) - D_0)}{\int_{V_0} d\boldsymbol{r}\, f\,(\boldsymbol{r})} \tag{5.2}$$

where the Heaviside step function $H\,(D\,(\boldsymbol{r}) - D_0)$ is unity if $D\,(\boldsymbol{r}) \geq D_0$ and zero if $D\,(\boldsymbol{r}) < D_0$. In practice, the expression can be evaluated as a summation over voxels. It is implicit that the number of FSUs is sufficiently large per voxel that the density can be well defined but the spatial variation of the density is on a larger spatial scale than the size of a voxel so that the density can be stated per voxel.

Lu *et al* (1997) have presented fDVHs for two model problems, one a representation of the treatment of a brain tumour and the other for a lung tumour (figure 5.17). In the absence of data specifying $f\,(\boldsymbol{r})$ they invented model variability. The outcome is striking in that the fDVH can be very different from the DVH. They note that the density $f\,(\boldsymbol{r})$ can be found from fMRI or SPECT and that there is clearly a need for more studies. Also it is fDVH not DVH which should form the basis of any DVH reduction to compute NTCP, and the partial volume in the Lyman formula should also be the partial functional volume.

On a related subject, Nahum (1998) has explained that when predicting complications for serial organs, such as the rectum, it may be more appropriate to compute a dose-*surface* histogram since the dose to the rectal contents included in a DVH is irrelevant. To aid the generation of DSHs surface dose maps may be generated showing the spatial location of the high dose. The rectal complications may be different for the same DSH if the *position* of the high-dose regions changes.

It has become common practice to compute TCP and NTCP through some model and to rank plans accordingly, e.g. a plan would ‘win’ over another if it

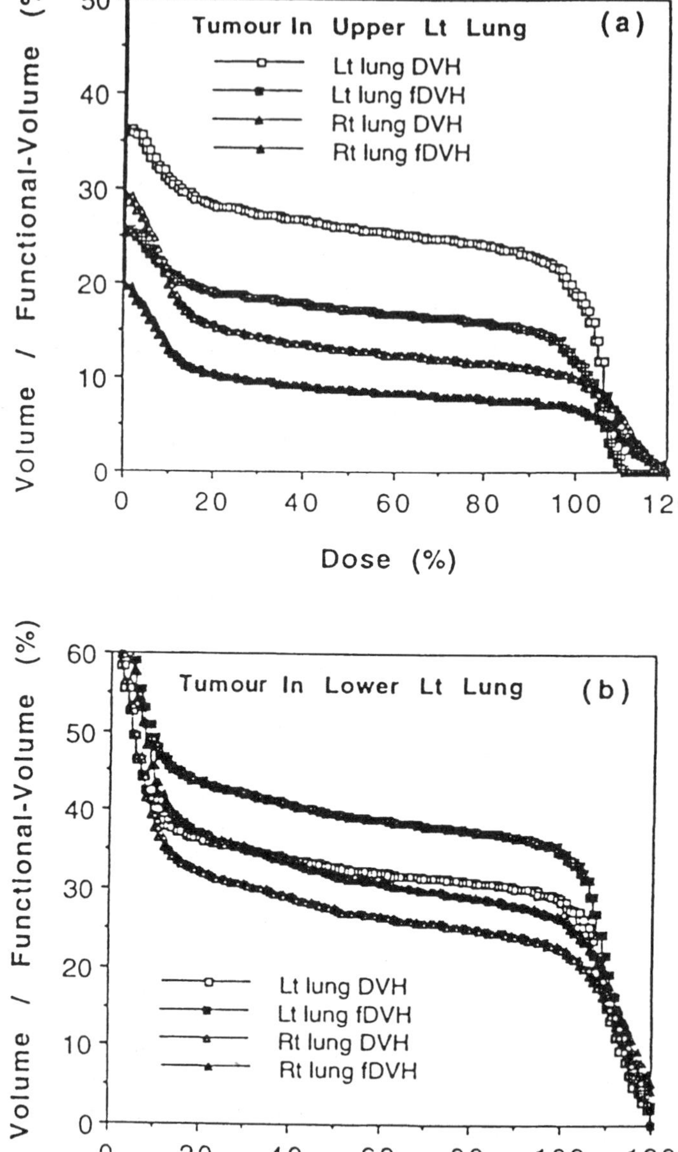

Figure 5.17. *Comparison of DVHs and fDVHs of the lung for a parallel opposed oblique plan. (a) A spherical target in the upper left lung and (b) a spherical target in the lower left lung. (From Lu et al 1997.)*

generated a higher TCP at fixed NTCP. Relative probability ranking is thus a valid method. However, it has been shown (Langer *et al* 1998) that such relative ranking does not correlate with relative P_+ ranking at all.

Dejean *et al* (2000a, b) have evaluated inverse treatment-planning for the classic Brahme dose prescription using some novel measures of conformality. The dose distribution was computed using the singular-value decomposition truncation technique and a least-squares full cost function. It was shown that the value of this cost function depended on the number of components retained in the singular-value decomposition and for this example reached a minimum when the order of the singular-value decomposition was 180. Then Dejean *et al* (2000a) went on to inspect whether the full cost function was a reasonable measure of conformality. They did this by first defining an underdosed lesion factor (ULF), which was the fractional volume of a lesion receiving dose less than some specific isodose and an overdosed healthy tissue factor (OHTF) which was the fractional volume of healthy tissue receiving a dose greater than some specified isodose. The aim is clearly to maximize (1-ULF) and minimize OHTF. Thus, by plotting vertically (1-ULF) against horizontally OHTF a receiver operating curve (ROC) results for different orders r. For this problem, it was shown that a value of $r = 180$ pushed the ROC curve up into the desirable top left position. Hence, the full cost function was seen to correlate with the receiver operating characteristic curve for the optimization of this problem. Dejean *et al* (2000a, b) have also defined a Youden index J where $J = 1-(ULF+OHTF)$. The goal is then to maximize the Youden index. Graphs can be created of the Youden index as a function of the prescribed isodose for a specified singular-value decomposition order and the peak may be then identified. By plotting this peak Youden index together with the corresponding isodose as a function of the order of the singular-value decomposition, it is possible to identify the optimum inverse-planning solution. In this case it was observed that this optimum planning solution corresponded to the minimum quadratic cost function.

5.9.2. *Equivalent uniform dose and the role of dose inhomogeneity*

Niemierko (1997) has defined for any dose distribution an equivalent uniform dose in Gy as that which, when distributed uniformly across the target volume, causes the survival of the same number of clonogens. Many formulae, of increasing complexity, are introduced taking account of an increasing number of radiobiological phenomena including: (i) absolute volume effect; (ii) non-uniform spatial distribution of clonogens; (iii) dose-per-fraction effect; (iv) proliferation effect; and (v) inhomogeneity of patient population.

Later work has shown that the equivalent uniform dose correlates better with observed treatment outcome than parameters such as mean dose or maximum dose to the target.

Niemierko (1999) has shown that the equivalent uniform dose is always bounded by the lowest dose and the mean dose for a PTV. Conversely, for an OAR, it is always bounded by the mean dose and the highest dose in the OAR. It is argued that equivalent uniform dose is a valid concept with respect to analysing plans for IMRT because of the necessarily non-uniform PTV and OAR dose distributions. However, we might note that such inhomogeneities can be largely removed by global optimization schemes as proposed by Webb (1999a, b). Ebert (2000) has argued that plans should be scored and compared using both EUD and TCP measures, the former being more appropriate if radiobiological parameters are uncertain and *vice versa*.

Wu and Zhu (2000) have used the EUD as a basis for optimizing IMRT. EUD was expressed as

$$EUD = \left[\frac{1}{N} \sum_i D_i^a \right]^{\frac{1}{a}} \tag{5.3}$$

where D_i is the dose to the ith voxel of a set of N, a is negative for tumours and positive for OARs. The objective function was

$$F = \Pi f_i \tag{5.4}$$

where the components f_i were (for tumour)

$$f_{\mathrm{T}} = \frac{1}{1 + [EUD_0/EUD]^n} \tag{5.5}$$

and for OARs

$$f_{\mathrm{OAR}} = \frac{1}{1 + [EUD/EUD_0]^n}. \tag{5.6}$$

Wu *et al* (2000a) have shown that this leads to improved IMRT dose conformity compared with the use of dose-volume constraints; the reason postulated that the latter only operate up to the specified tolerances and do not attempt to improve beyond this, unlike the EUD-based method .

Something of a myth has grown up that IMRT must necessarily produce inhomogeneous dose distributions. Whilst this is strictly true, since no algorithm can ever deposit *precisely* uniform dose to the PTV because of the laws of physics, it is often quite possible to generate a dose distribution by inverse planning which satisfies the conventional ICRU '−5%, +7%' rule. It is all a question of the importance attached to getting as uniform a dose as possible. Sometimes inhomogeneous dose is unavoidable either because the algorithm has not provided enough constraint via its cost function or because of physical difficulties, such as solving the matchline problem in slice-by-slice tomotherapy. This can sometimes lead to quoted underdoses. Goitein and Niemierko (1996) have pointed out however that the numerical value of an underdose is not enough to specify the outcome biologically. What is also needed is the volume over which such underdose occurs, in short a DVH. From this perspective, they reanalysed the matchline problem of Carol *et al* (1996) and showed the effect on TCP. Because the TCP calculation is highly non-linear, this is the only way to assess the clinical importance of underdose. They also pointed out that the location of the underdose is also important. Whereas in conventional planning it tends to be at the tumour periphery where the clonogenic cell density may be less or the margins may reduce the impact on the TCP, in IMRT this is not usually the case. Hence, they argued for a complete spatial and DVH-based analysis of the outcome of treatment planning. They also argued that dose non-uniformity, which leads to a lower dose than prescribed, should be avoided. Mohan (1999) has addressed the same issue of the need to quantify IMRT-delivered dose distributions through biological models and indeed perhaps use these to design more novel unconventional distributions.

Butler *et al* (1997) have responded on the specific issue of the matchline in slice-by-slice tomotherapy. They have argued that invasive fixing of the patient in their clinic led to a couch index accuracy of 0.01 mm. Also, because the IMRT is given in fractions, patient movement can 'feather' the matchline and distribute the inhomogeneous dose spatially. They also argued that the TCP reduction observed by Goitein and Niemierko would be less if based on prescribed rather than maximum dose.

5.10. AUTOMATIC IMPORTANCE FACTORS

It was commented above (section 5.1) that one of the limitations of 'optimization' is that the optimal solution depends on importance factors in a cost function and that, in general, the 'guessing game' on beamweight selection has been replaced by a 'guessing game' on importance factor setting. Xing *et al* (1999b) have directly commented in the same way that there can be no such thing as an optimal plan for the reason that the nominally 'optimal plan' depends on the importance factors. Xing *et al* (1999b) and Le *et al* (2000) have presented a way out of the dilemma. They have used a second stage evaluation criterion to determine the optimality of a treatment plan and then adjusted the importance parameters in the usual first stage calculation of beamweights to minimize the second stage cost function. For example, if the first stage cost function is the common quadratic function

$$F = \frac{1}{N} \sum_{n=1}^{N} r_\sigma \left[D_c(n) - D_0(n)\right]^2 \tag{5.7}$$

where D stands for dose, the subscripts 'c' and '0' stand for calculated and prescribed, and the mean square value is computed over $n = 1, 2, \ldots, N$ dose voxels and r_σ are the importance factors for each volume represented by σ. Then a second stage cost can be defined as

$$G = \sum_{D} \gamma_\sigma \left[V_\sigma(D) - V_{\sigma 0}(D)\right]^2 \tag{5.8}$$

where the terms in the square brackets represent the calculated and ideal dose-volume histogram in regions labelled by σ, and γ_σ is the relative importance of obtaining the desired DVHs. The algorithm now starts by assuming any (but sensible) set of importance factors in equation (5.7) for F and then with the importance factors *fixed* in equation (5.8) for G, computing the first value of G. Then in turn, and in a round robin, the importance factors for F are gradually varied and changes are accepted provided they lead to reductions in the cost G. It was found that the algorithm converged in about four cycles and the plans were significantly improved. Note that the spatial dose distribution is still controlled by the inner stage optimization of F and the DVHs improve by the second stage, optimizing G. Interestingly, the final set of importance factors for F was almost independent of the initial choices. The importance factors for G were set to unity for the PTV and 0.1 for the OARs. One might comment that the philosophical problem above still does not vanish because it then translates to the specification of the constant importance factors in equation (5.8) for G.

Further insight into this approach can be had by studying Xing *et al* (1999e) who have introduced a new concept into optimization. The dose distributions which are possible outcomes of IMRT inverse planning are prioritized by introducing a 'preference function' which provides a theoretical foundation for the statistical analysis of treatment plans. In the general case, empirical judgement or expert knowledge is used to construct the preference function. For example, a Gaussian preference function for the nth voxel with dose $D_c(n)$ would be specified by

$$P_n(D_c) = P_{0n} \exp\left[-\gamma_n [D_c(n) - D_0(n)]^2\right] \tag{5.9}$$

where P_{0n} is a normalization constant and γ_n represents the Gaussian parameter. The total preference function for a system of $n = 1, 2, \ldots, N$ dose voxels is then

$$P = \Pi_n P_n(D_c) = \Pi_n P_{0n} \exp\left[-\gamma_n [D_c(n) - D_0(n)]^2\right]. \tag{5.10}$$

Forming the log-preference we have

$$\begin{aligned} \ln P &= \Sigma_n \ln P_n(D_c) = \Sigma_n \ln P_{0n} \exp\left[-\gamma_n [D_c(n) - (D_0(n)]^2\right] \\ &= \ln P_{0n} - \Sigma_n \left[\gamma_n [D_c(n) - (D_0(n)]^2\right]. \end{aligned} \tag{5.11}$$

From equation (5.11) we may see that *maximizing* the total preference function P (or $\ln P$) is equivalent to *minimizing* the usual quadratic cost function between computed dose $D_c(n)$ and prescribed dose $D_0(n)$. So, the least-squares objective function, which many use, is a special case of general estimation theory. This approach gives a new insight into the options available. For example, it would be possible to skew the Gaussian so that overdose in the target is not penalized as much as underdose. This adjustment of the γ_n factors in the preference function in equation (5.10) would be equivalent to adjusting the importance factors within the PTV in the cost function in equation (5.11).

Xing *et al* (1999e) now defined a decision function in dose-volume space as

$$Q = \Pi_n Q_{\sigma 0} \exp\left[-\gamma_\sigma [V_\sigma(D) - V_{\sigma 0}(D)]^2\right]. \tag{5.12}$$

By the same argument as above, maximizing this function Q indicated that the calculated DVH matches the most desired prescription DVH, because this is equivalent to minimizing the function G in equation (5.8).

Xing *et al* (1999e) have proposed a two-step optimization procedure in which the importance factors are set almost arbitrarily in equation (5.9), the inverse planning is performed and the DVH decision function is computed. Then these parameters are adjusted in small steps, the inverse planning is re-performed and the changes accepted to maximize Q. In this way, good control is kept over the dose distributions (via equation (5.9)) whilst the distribution is also optimized in DVH space. It is important to realize that it is not adequate to work in DVH space alone because of the degeneracy of that space. It turns out that only four or five cycles of this process are needed for convergence and that the final importance factors in dose space turn out to be remarkably independent of their initial choices.

5.11. SUMMARY AND A LOOK TO THE FUTURE

In general, IMRT planning studies require that certain parameters are fixed and that the variability of outcome (resulting plan) is studied, with the optimization of those parameters remaining as variables. For example, beam directions are usually chosen before beamweights are optimized. As a result, it has been suggested that this process should be renamed 'customization' (Rowbottom *et al* 1999a) because it can be argued that, if the fixed choices were changed, a 'different optimum outcome' would result. This overcomes the difficulty of describing one optimum plan (with one set of fixed starting parameters) as 'more optimum' than another (with a different starting set), which would be an illogical use of the English language.

To date, the fixing of certain key parameters ahead of customization has led to as many unsolved questions as problems overcome. For example, one might inquire how does the outcome (resulting plan) depend on:

(i) the number of beams;
(ii) whether such beams are intensity modulated or not;
(iii) whether the beams are coplanar or not (and on the extent of non-coplanarity);
(iv) whether the beams can take any orientation in 4π or whether they are constrained to fixed orientations;
(v) whether fixed-orientation beams are equispaced or otherwise;
(vi) whether the intensity levels in the beams are allowed to be continuous or discrete and, if the latter, how many levels are set;
(vii) the choice of cost function, whether physical or biological, how importance-weighting is set, whether dose-volume constraints are applied;
(viii) whether constraints are set on the form of the beams bearing in mind the eventual practical delivery issues.

In summary, despite the large number of published studies, the ultimate question of whether the best conformality is achieved is never answered.

To a large extent, these constraints are applied for two reasons. Firstly, it is recognized that the exploration of the full set of options is impossible even with today's computer speeds. Secondly, the problem is posed with an eye to the possible methods of practical delivery. A flurry of activity in the late 1980s and into the 1990s developed inverse-planning IMRT techniques which used a large number of coplanar beams with full continuous intensity modulation, mainly because the practical delivery of such methods was yet to be developed. In the late 1990s and early 2000s the situation has changed, as we have seen. Now that we have practical delivery methods, inverse planning is being simplified to accommodate practical constraints. Fewer beam orientations are being suggested with fewer intensity quanta and planning techniques have inbuilt delivery constraints, such as requiring beams not to have sharp fluence discontinuities.

The desirable outcome of all these developments is that the clinical implementation of IMRT is rapidly becoming a practical reality. In essence a high conformality is being squeezed from the minimum departure from traditional radiotherapy. This makes sense given that the introduction of new radiotherapy techniques requires detailed quality assurance, and the involvement of a large number of professionals (physicists, engineers, clinicians, radiographers) who may at first have some resistance to change. It might also be argued that IMRT has still to win its clinical spurs, with proof positive of its clinical efficacy yet to emerge. Hence, one may expect that the majority of research should and will continue to proceed along the above lines. That is, many parameters of choice will be fixed, the inverse plan will be customized and benchmarked for improvement against some 'conventional plan', and the practical delivery method will be worked out and the treatment delivered.

However, the above approach leaves nagging doubts whether such customized plans are optimal and, if not, by how much are they suboptimal. One muses on what are the ultimate limitations on the achievement of CFRT. One questions whether in the (valid) quest to clinically introduce IMRT one will lock into suboptimal simple IMRT techniques which themselves will be the subject for rescrutiny and extension in decades to come, when perhaps the computer planning limitations or delivery limitations are less. One would like to know what is the ultimate that can be achieved using photons, even if this ultimate treatment cannot be practically delivered with today's technology. If we knew this, we would have a very real confidence in whether we are likely to improve on all the present classes of IMRT and whether creating 'the ultimate IMRT treatment' tool—the robotic linac? (see section 4.6)—is worth the effort to develop.

Bortfeld (1999a) has considered some of the reasons that hinder the practical realization of IMRT on a large scale. These include that IMRT planning still takes up too much time and that to speed up the process it would be necessary to find class solutions for a number of cases. A second limitation is that the verification of IMRT must be done for each patient and presently at DKFZ this takes between a half day and a whole day. Thirdly, the delivery of IMRT significantly takes more time than conventional treatments. This is mainly a technical problem and will be solved by the vendors in the foreseeable future. He has argued that we should aim at making IMRT not only the better technique but also the more efficient treatment technique.

De Neve *et al* (1999b) have reminisced on the origins of IMRT. Noting that the pioneering work of Brahme and co-workers in the 1980s did not find clinical application because IMRT was at that time too complicated, de Neve has considered Carol's development of the NOMOS MIMiC in 1992 to be the prime example of IMRT. Subsequent to this, there was a different speed of development of clinical implementation between very fast in the USA and very slow in Europe. The NOMOS MIMiC was very successful in the USA, but has only been applied in three European clinics. The MLC technology has also been taken up quicker in the USA than it has in mainland Europe. At the time of writing, most patients have received IMRT by means of the NOMOS MIMiC, but MLC based systems are catching up. It was observed that those university hospitals which have made most progress have certain characteristics in common. All make a large human investment in the physics of informatics, have acquired planning skills and have developed their own technology for planning data transfer and quality assurance. In general, the planning was initially case-focused and only became class solution later in the investigation. In-house developments are somewhat simpler than those facing manufacturers because equipment only has to link to the other equipment present in a particular centre, whereas a manufacturer, aiming to produce an open-ended solution, will try to tailor each piece of equipment to a range of related IMRT equipment. At the time of writing, it is not known which open-ended planning systems will survive this evolutionary process.

So what can we finally conclude about the position of photon IMRT as we have entered a new millennium. We can certainly say that there has been a decade of furious research activity since the late 1980s putting together the various techniques to plan and deliver intensity modulation. We may observe that there has been a tempering of approaches from the very exotic to quite practical methodology. We may observe some clinical implementation but not uniformly across the treatment options and certainly not in widespread use. We may conclude that there is ample paper evidence for improved dose distributions and expected outcome but little clinical proof. The future of IMRT is bright and moderately certain but there is much more work to be done.

What of *proton* IMRT? Or even proton conformal non-intensity-modulated treatments compared with photon IMRT. This is a contentious issue. It has been the subject of a 'point and counterpoint' debate between Rock Mackie and Al Smith (Hendee 1999a). Mackie has argued that:

(i) photon IMRT and proton therapy give clinically similar results;
(ii) proton therapy is ten times more expensive and therefore disadvantaged;
(iii) the proton penumbra at depth is larger than the photon penumbra causing an increase in beam area and so decreased sparing of normal structures;
(iv) photon IMRT is better at accounting for surface irregularities, tissue inhomogeneities and tissue density uncertainty.
(v) He has reminded us of the distal edge tracking proton IMRT concept he and others invented and he argued that photon and proton IMRT should join forces.

Conversely, Smith has claimed:

(i) that the basic physics of dose deposition from photon beams must give the advantage to protons whether as single or multiple beam treatments;
(ii) that proton beams can be directly aimed to avoid OARs distal to the proton range;
(iii) that treatment-planning advantages rest with proton therapy;
(iv) that Mackie has overestimated the long-term costs of proton therapy by a factor of ten.

The debate goes on. Indeed the whole subject of the role of IMRT is generally contentious and appendix 1A invites the reader to decide individually and independently.

Another recent 'point and counterpoint' argued whether the rate of evolution of radiation therapy planning and delivery systems is exceeding the evolution rate of quality assurance processes (Hendee 1999b). For the motion, S D Henderson has argued that:

(i) medical physics standards are generally created later than when the technology was initially introduced;
(ii) agreed-upon standards for QA are absent and that the very existence of symposia to discuss MU calculations and gel dosimetry indicate uncertainty in the community on what to do.

Against the motion, Biggs has countered that:

(i) the history of radiation therapy shows radiation physicists to be a remarkably conservative and careful community;
(ii) hospitals not industry always take on the required QA to assure new methods are safe;
(iii) the existence of workshops, courses, task groups, summer schools etc demonstrates that there is rapid dissemination of new knowledge and much useful interaction;
(iv) stereotactic radiosurgery—a potentially dangerous procedure if ever there was (my comment not Biggs)—has a history of careful QA and the same can be applied to IMRT.

Henderson has countered that her experience has been that, whilst some medical physicists fit Bigg's view, by no means all do and that, for example, IMRT is being introduced in clinics unused to such complexity. Biggs re-countered that most IMRT centres are large major teaching hospitals who have invested in the labour requirements to introduce IMRT.

Another 'feature' of modern scientific life is the impetus provided to scientific development through media reporting in the national press and television. Properly conducted, this informs the wider public, raises the profile of the subject and leads to tangible benefits, e.g. increased sponsorship of research and development. There

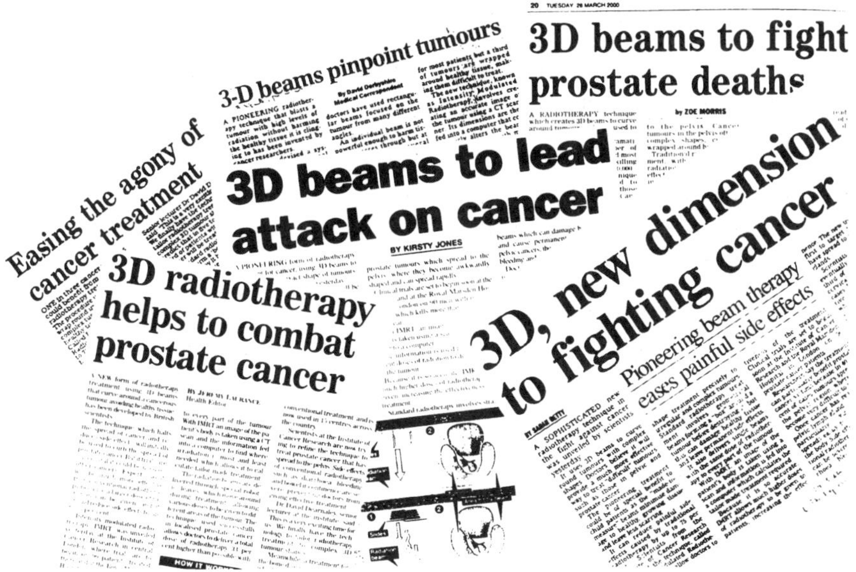
20 TUESDAY 28 MARCH 2000

3D beams to fight prostate deaths

by ZOE MORRIS

3-D beams pinpoint tumours

3D beams to lead attack on cancer

BY KIRSTY JONES

Easing the agony of cancer treatment

3D radiotherapy helps to combat prostate cancer

3D, new dimension to fighting cancer

Pioneering beam therapy eases painful side effects

HOW IT WORKS

Figure 5.18. *An example of IMRT press publicity—see text for a discussion of benefits versus downsides.*

is also no doubt it enhances the reputation of the centre and staff generating the publicity. There was a flurry of such national press activity in March 2000 reporting developments in my own centre (figure 5.18). Potential adverse consequences are a raising of public (patient) expectation that what we are engaged in developing may not be immediately available to all. Also, by definition and to some extent design, such publicity does not present the full picture and sometimes can be downright incorrect since it is generally out of the control of those who send the press releases. It is certainly not the way to present a balanced scientific overview (the role of review papers, lectures and books such as this) but it is 'out there' and has impact.

As we entered the new millennium, it was fashionable to ask distinguished scientists to predict the future development of their science. Purdy (2000a), asked to do this for radiation therapy physics, has looked at the development of the whole field of improving the physical basis of optimizing external-beam radiotherapy. In this article, were some key phrases which I quote: 'Dose delivery systems in the year 2035 will be fully automated, making use of computer-controlled IMRT techniques... current systems should be thought of only as prototypes'. Battista (2000) gave us 'Radiotherapy in the next century: ten megatrends'. One of these was the continuing evolution of IMRT and several of the others related to this. Webb (2000c) has written 'The future of photon external-beam radiotherapy: the dream and the reality' and compared what is currently expected with some more farsighted (and therefore possibly wrong) predictions. The story of the development of IMRT will go on for a long time. There are still many unanswered questions but its place in radiation therapy practice is assured.

5A. APPENDIX: 105 YEARS EARLIER...

This text has presented the state-of-the-art technology for delivering IMRT as it is in the year 2000. Some modern history and older history has been briefly mentioned. Inevitably, as major anniversaries such as the Röntgen centenary come and go, interest focuses on the origins of X-ray matters and recently there has been a reanalysis of the question 'who first performed radiotherapy?'. It would not be appropriate and is unnecessary here to tackle this again in depth since many weighty reviews exist. The most recent have been written by Kogelnik (1997) and Leszczynski and Boyko (1997) with comment by del Regato (1997).

The question is not so straightforward because the answer depends on the degree of substantiation of both the intent and outcome of the radiotherapy. Emil H Grubbé claimed to have performed the first irradiation of a patient with a breast cancer on 29 January 1896, after noticing his own skin reddening following X-ray experiments. The referral letter for the patient is in the Smithsonian Institution, and Grubbé claimed to be a manufacturer of X-ray equipment. However, the patient died a month later, Grubbé only made his claim to fame 37 years after the event and the irradiation was not documented (Grubbé 1933). It was also unlikely that Grubbé had any scientific basis for his craft. For these reasons his claim is generally discredited (Brecher and Brecher 1969, Mould 1995).

A Boston radiologist, F H Williams, treated a patient with breast cancer in November 1896 with observed pain relief. H Gocht also treated patients in November 1896 but with no lasting therapeutic benefit. Victor Despeignes was most likely the author of the first article on radiotherapy (Despeignes 1896) reporting the irradiation of a patient with a stomach cancer in July 1896 with observed 'improvement'. However, from the technical details reported, it is possible to work out that the X-ray energy was only about 20 kVp and rays could not have reached the cancer, let alone cured it. Hence, despite good intentions and scientific documentation, there could have been no scientific case for anticipated therapeutic outcome.

Leopold Freund started historic experiments in Vienna to treat a hairy nevus on a five-year old girl. The experiments were carefully carried out, documented and led to observed immediate response which lasted 80 years. The first irradiation was on 24 November 1896 and was subsequently documented (Freund 1897). Photographs prior and post irradiation exist and also records of the equipment. Freund went on to publish one of the earliest textbooks on the scientific basis of radiotherapy (Freund 1903).

There still remains contention as to who is the 'father of radiotherapy' but the two front runners are Despeignes and Freund, in that order chronologically but with more credibility to the scientific method of the latter. It is my observation that, in matters such as this, the determination of an unchallengeable claimant is a futile exercise. It all depends on one's point of view concerning what has to be achieved to establish a claim. What is however indisputable is that radiotherapy was established within a year of the discovery of the X-ray (Webb 1990).

5B. APPENDIX: POSTSCRIPT—HEIDELBERG, MAY 2000 AND BRUSSELS, JUNE 2000

This review of the field of IMRT was started in the autumn of 1996 and the last dotting of 'i's and crossing of 't's was done during a few more days in Heidelberg after the 13th ICCR, at which some 40% of the papers were on aspects of IMRT. Whilst the ICCR conferences are not the only fora for this field, many physicists regard the triennial meetings as definitive landmarks signalling what is exciting in this field. So, at the risk of overstatement, I felt the sense of time and place was significant.

The DKFZ, Heidelberg has been a pioneer of CFRT and IMRT since the 1980s and continues to be so. Some of the 'products' of its research now permeate IMRT practice throughout the world. The delegates in Heidelberg in May 2000 comprised a 'walking subset of the references in this book' (with apologies to those important contributors in the field who were not there).

What did we learn? The importance of the DMLC and the MLC MSF techniques was apparent. The technology has reached the translational research stage and many clinical applications are taking place. The vendors are fully behind IMRT. The NOMOS MIMiC still leads the league table in terms of patient numbers. The prototype Wisconsin device is nearly complete. Robotic IMRT is just beginning.

Final touches were also added just after the 1st International Workshop on IMRT in Clinical Practice in Brussels, a meeting highlighting the transition from concept to clinic.

There was a sense of optimism for IMRT despite all the well-rehearsed constraints that obstacle progress. Where will IMRT be by the time of the 14th ICCR?

REFERENCES

Aaltonen-Brahme P 1998 Nordic proposals on volume definition and margins *Radiother. Oncol.* **48** (Suppl. 1) S8

Adams E J, Cosgrove V P, Shepherd S F, Warrington A P, Bedford J L, Mubata C D, Bidmead A M and Brada M 1999b Comparison of a multi-leaf collimator with conformal blocks for the delivery of stereotactically guided conformal radiotherapy *Radiother. Oncol.* **51** 205–9

Adams E J, Suter B L, Warrington A P, Black P, Saran F, and Brada M 1999a Design and implementation of a system for treating paediatric patients with stereotactically-guided conformal radiotherapy *Radiother. Oncol.* **51** (Suppl. 1) S2

Adler J R, Chang S D, Murphy M J, Doty J, Geis P and Hancock S 1997 The Cyberknife: a frameless robotic system for radiosurgery *Stereotactic Funct. Neurosurgery* **69** 124–8

Adler J R and Cox R S 1995 Preliminary clinical experience with the CyberKnife: image guided stereotactic radiosurgery *Radiosurgery* **1** 316–26

Adler J R, Murphy M J, Chang S D and Hancock S L 1999 Image-guided robotic radiosurgery (and three post-paper published critiques) *Neurosurgery* **44** 1299–307

Adler J R, Schweikard A, Murphy M J and Hancock S 1998 Image-guided stereotactic radiosurgery: The Cyberknife *Image-guided neurosurgery: Clinical applications of surgical navigation* ed G Barnett, D Roberts and R Mancunias (St Louis, MO: Quality Medical Publishing) pp 193–204

Agazaryan N, Solberg T, Arrelano A R and Paul T 1999 Leaf sequencing for fluence modulated radiation therapy *Med. Phys.* **26** 1139

Agazaryan N, Solberg T D, Arellano A R and Paul T J 2000a Dynamic multileaf collimation and three-dimensional verification in IMRT *Proc. 13th Int. Conf. on the Use of Computers in Radiation Therapy (Heidelberg, May 2000)* ed W Schlegel and T Bortfeld (Heidelberg: Springer) pp 280–2

Agazaryan N, Solberg T D, Arellano A R and Paul T J 2000b Three-dimensional verification for dynamic multileaf collimated IMRT *Proc. 1st Int. Workshop on IMRT in Clinical Practice (IMRT2k) (Brussels, 8–9 June 2000)* ed D Verellen p 38

Ahn S D, Yi B Y, Cho B C, Kim J H, Choi E K and Chang H 1997 PC-based least square optimisation algorithm for intensity modulated fields *Proc. 12th Int. Conf. on the Use of Computers in Radiation Therapy (Salt Lake City, Utah, May 1997)* ed D D Leavitt and G Starkschall (Madison, WI: Medical Physics Publishing) pp 429–30

[1] The author recognizes that references to private communication cannot be followed up but these are inserted to indicate acknowledgement of the first knowledge of ideas. Information first received in the 'closed meetings' of the Elekta Consortium is only quoted where also presented in a later journal paper, conference report or the Elekta, publicly available journal *Wavelength*.

Ahnesjö A, Löfgren A and Saxner M 1998 Modelling of flattening filter scatter through thin and thick collimator apertures for intensity modulation using the 'step and shoot' multileaf technique *Med. Phys.* **25** A109

Alber M, Birkner M, Laub W and Nüsslin F 2000 Hyperion-an integrated IMRT planning tool *Proc. 13th Int. Conf. on the Use of Computers in Radiation Therapy (Heidelberg, May 2000)* ed W Schlegel and T Bortfeld (Heidelberg: Springer) pp 46–8

Alber M and Nüsslin F 1999 Biologically optimized beam directions for radiotherapy *Radiother. Oncol.* **51** (Suppl. 1) S6

Alber M and Nüsslin F 2000 Intensity modulated photon beams subject to a minimal surface smoothing constraint *Phys. Med. Biol.* **45** N49–52

Aldridge J S, Harari P M, Reckwerdt P J, Olivera G H, Tomé W, Fink M B and Mackie T R 1999 Development of conformal avoidance tomotherapy in the treatment of head and neck cancer *Int. J. Radiat. Oncol. Biol. Phys.* **45** (Suppl. 1) 245

Aldridge J S, Shepard D M, Mackie T R and Reckwerdt P J 1998 Conformal avoidance radiation therapy *Radiother. Oncol.* **48** (Suppl. 1) S76

Amols H I, Happersett L, Zaider M, Kutcher J, Zakian K, Hunt M, Chui C, Dyke J and Zelefsky M 1999 Brachytherapy and teletherapy intensity-modulated treatment planning for prostate using biologic imaging *Phys. Medica* **15** 231

André L, Haas B and Schär R 1997 Full size intensity modulation with the Varian dynamic multileaf collimator *Proc. Meeting: Intensity-Modulated Radiation Therapy: A Clinical Perspective (London, 26 June 1997)* (York: IPEM) pp 24–25

Angelos L, Shepard D, Sauer O and Mackie T R 1997 Data sets for studying radiotherapy optimisation *Proc. 12th Int. Conf. on the Use of Computers in Radiation Therapy (Salt Lake City, Utah, May 1997)* ed D D Leavitt and G Starkschall (Madison, WI: Medical Physics Publishing) p 480

Antonuk L E, Sprawls P and Hendee W R 1999 Point and Counterpoint: Medical physicists should seek patent protection for new ideas before publishing articles about them *Med. Phys.* **26** 2220–2

Appleby H J 1999 Use of MRI in radiotherapy treatment planning *Br. J. Radiol.* **72** (Suppl.) p 31

Arellano A, Solberg T and Llacer J 1998 Inverse treatment planning using the dynamically penalized likelihood (DPL) algorithm and a 3D pencil beam calculation engine *Med. Phys.* **25** A119

Arellano A R, Solberg T D and Llacer J 2000 A clinically oriented inverse planning implementation *Proc. 13th Int. Conf. on the Use of Computers in Radiation Therapy (Heidelberg, May 2000)* ed W Schlegel and T Bortfeld (Heidelberg: Springer) pp 532–4

Arnfield M 1998 The effect of leaf shape and transmission characteristics on the accuracy of intensity-modulated radiotherapy (IMRT) delivered with dynamic multileaf (DMLC) *Med. Phys.* **25** A202

Arnfield M R, Wu Q, Siebers J and Mohan R 1999 Dosimetric aspects of dynamic multileaf collimation for intensity modulated radiotherapy *Int. J. Radiat. Oncol. Biol. Phys.* **45** (Suppl. 1) 248

Audet C, Duzenli C, Schreiner L J, Mackay A, Harrison R, Mansour F and Peemoeller H 1997 Three dimensional MRI/Polymer gel dosimetry: Practical considerations *Proc. World Congress on Medical Physics and Biomedical Engineering and 11th Int. Conf. on Medical Physics (Nice, France, September 1997) Med. Biol. Eng. Comput.* **35** (Suppl. Part 2) 882

Augspurger M E, Teh B S, Uhl B M. Lee A G, Grant W H, Butler E B and Woo S Y 1999 Conformal intensity modulated radiation therapy for the treatment of optic sheath meningioma *Int. J. Radiat. Oncol. Biol. Phys.* **45** (Suppl. 1) 324

Bakai A, Alber M and Nüsslin F 1999 Optimisation of intensity modulated beam profiles under consideration of intrafractional organ movement *Radiother. Oncol.* **51** (Suppl. 1) S41

Baldock C and Watson S 1999 Risk assessment for the manufacture of radiation dosimetry polymer gels *Proc. 1st Int. Workshop on Radiation Therapy Gel Dosimetry (Lexington, Kentucky, 21–23 July 1999)* pp 154–6

Balog J P, Angelos L, Mackie T R and Reckwerdt P 1997 Dose computation for a one dimensional multileaf collimator *Proc. 12th Int. Conf. on the Use of Computers in Radiation Therapy (Salt Lake City, Utah, May 1997)* ed D D Leavitt and G Starkschall (Madison, WI: Medical Physics Publishing) pp 434–6

Balog J P, Mackie T and Reckwerdt P 1998 Helical versus sequential tomotherapy *Med. Phys.* **25** A203

Balog J P, Mackie T R, Reckwerdt P, Glass M and Angelos L 1999a Characterization of the output for helical delivery of intensity modulated slit beams *Med. Phys.* **26** 55–64

Balog J P, Mackie T R, Wenman D L, Glass M, Fang G and Pearson D 1999b Multileaf collimator interleaf transmission *Med. Phys.* **26** 176–86

Balter J M, Lam K L, McGinn C J, Lawrence T S and Ten Haken R K 1998 Improvement of CT-based treatment-planning models of abdominal targets using static exhale imaging *Int. J. Radiat. Oncol. Biol. Phys.* **41** 939–43

Barrett H, Barber H B, Ervin P A, Myers K J, Paxman R G, Smith W E, Wild W J and Woolfenden J M 1983 New directions in coded-aperture imaging *Information Processing in Medical Imaging* ed F Deconinck (Dordrecht: Martinus Nijhoff) pp 106–29

Barrett H H and Swindell W 1981 *Radiological Imaging: the Theory of Image Formation, Detection and Processing* (New York: Academic)

Basran P, Ansbacher W, Field C and Murray B 1998 Use of compensating filters in intensity modulated radiotherapy *Med. Phys.* **25** A207

Battista J J 2000 Radiotherapy in the next century: ten megatrends *Can. Med. Phys. Newsl.* **46** 10–46

Beavis A W, Coldham C M, Koya A and Whitton V J 1999b Film dosimetry for intensity modulated radiotherapy *Br. J. Radiol.* **72** (Suppl.) p 40

Beavis A W, Ganney P, Whitton V and Xing L 1999c Optimisation of the intensity map, in level and resolution for the delivery of IMRT *Med. Phys.* **26** 1087

Beavis, A W, Ganney P S, Whitton V J and Xing L 2000 Slide and shoot: a new method for MLC delivery of IMRT *Proc. 13th Int. Conf. on the Use of Computers in Radiation Therapy (Heidelberg, May 2000)* ed W Schlegel and T Bortfeld (Heidelberg: Springer) pp 182–4

Beavis A W, Gibbs P, Dealey R A and Whitton V J 1998a Radiotherapy treatment planning of brain tumours using MRI alone *Br. J. Radiol.* **71** 544–8

Beavis A W, Kennerley J and Whitton V J 1998b Virtual dosimetry: quality control of dynamic intensity modulated radiation therapy *Br. J. Radiol.* **71** (Suppl.) 9

Beavis A W, Whitton V I and Xing L 1999a A combination delivery mode for intensity modulated radiation therapy based on specific patient cases *Int. J. Radiat. Oncol. Biol. Phys.* **45** (Suppl. 1) 164

Beckham W A, Vincent D N and Delaney G 1997 Tangential field breast irradiation: number of planning slices versus calculated dose homogeneity *Proc. World Congress on Medical Physics and Biomedical Engineering and 11th Int. Conf. on Medical Physics (Nice, France, September 1997) Med. Biol. Eng. Comput.* **35** (Suppl. Part 2) 941

Bedford J L, Khoo V S, Dearnaley D P and Webb S 1999a Improved methods for adding transaxial 2D margins to the superior and inferior regions of a clinical target volume *Med. Dosim.* **24** 169–75

Bedford J L, Khoo V S, Dearnaley D P and Webb S 2000a Optimisation of coplanar six-field techniques for conformal radiotherapy of the prostate *Int. J. Radiat. Oncol. Biol. Phys.* **46** 231–8

Bedford J L, Khoo V S, Oldham M, Dearnaley D P and Webb S 1999b A comparison of coplanar four-field techniques for conformal radiotherapy of the prostate *Radiother. Oncol.* **51** 225–35

Bedford J L, Khoo V S, Warrington A P, Bidmead A M, Dearnaley D P and Webb S 2000c A comparison of multileaf collimation with conformal blocks for the boost phase of dose-escalated prostate radiotherapy *Radiother. Oncol.* at press

Bedford J L and Shentall G 1998 A digital method for computing target margins in radiotherapy *Med. Phys.* **25** 224–31

Bedford J L, Viviers L, Guzel Z, Childs P J, Webb S and Tait D M 2000b A quantitative treatment planning study evaluating the potential of dose escalation in conformal radiotherapy of the oesophagus *Radiother. Oncol.* at press

Bednarz G, Huq M S, Sweet J W, Hughes S, Curran W J, Anne P R and Galvin J M 1999 Forward versus inverse treatment planning for head and neck tumors that surround critical normal structures *Int. J. Radiat. Oncol. Biol. Phys.* **45** (Suppl. 1) 421–2

Bel A, Petrascu O, van de Vondel I, Coppens L, Linthout N, Verellen D and Storme G 2000 A computerized remote table control for fast on-line patient repositioning: implementation and clinical feasibility *Med. Phys.* **27** 354–8

Belanger P, Bercier Y, Parker W and Hristov D 2000 A software system for patient-specific quality control of intensity-modulated radiation treatments by Fricke-gel dosimetry *Proc. 13th Int. Conf. on the Use of Computers in Radiation Therapy (Heidelberg, May 2000)* ed W Schlegel and T Bortfeld (Heidelberg: Springer) pp 394–6

Bendl R, Keller-Reichenbecher M A, Hoess A and Schlegel W 1997 The spherical view—a tool for the interactive optimisation of irradiation directions *Proc. 12th Int. Conf. on the Use of Computers in Radiation Therapy (Salt Lake City, Utah, May 1997)* ed D D Leavitt and G Starkschall (Madison, WI: Medical Physics Publishing) pp 302–5

Benedict S H, Cardinale R M, Wu Q, Zwicker R D, Arnfield M A and Mohan R 1999 Potential for intensity modulation in stereotactic radiosurgery *Int. J. Radiat. Oncol. Biol. Phys.* **45** (Suppl. 1) 187

Bero M A, Jenneson P M, Gilboy W B, Glover P M and Doran S J 1999 Faster optical tomography with parallel-beam white light for three-dimensional dosimetry *Proc. 1st Int. Workshop on Radiation Therapy Gel Dosimetry (Lexington, Kentucky, 21–23 July 1999)* pp 136–8

Beyer D, Quiet C, Rogers K and Gurgoze E 2000 Prostate motion during a course of IMRT: the value of daily ultrasound to guide portal localization and correct for random and systematic errors *Proc. 1st Int. Workshop on IMRT in Clinical Practice (IMRT2k) (Brussels, 8–9 June 2000)* ed D Verellen p 57

Birkhoff G D 1940 On drawings composed of uniform straight lines *J. Mathematiques Pures Appliqués* **19** 221–36

Bisi F, Anselmi R and Andreucci L 1999 Determination of radiotherapeutic volumes of interest: an application of a biomedical images system in radiotherapy *Radiother. Oncol.* **51** (Suppl. 1) S7

Bleakley N, Donovan E, Eagle S, Evans P, Hansen V, Partridge M, Reise S, Symonds-Tayler R and Yarnold J 2000 *Proc. 1st Int. Workshop on IMRT in Clinical Practice (IMRT2k) (Brussels, 8–9 June 2000)* ed D Verellen p 32

Bleier A R 1998 Private communication (NOMOS e-mail 17 August 1998)

Bleier A R 2000 Private communication (NOMOS e-mail 7 January 2000)

Bleier A R, Carol M P, Curran B H, Holmes T W, Kania A A, Lalonde R J, Larson L S and Sternick E S 1997 Dose calculation for intensity-modulated radiotherapy using finite size pencil beams derived from standard measured data *Proc. 12th Int. Conf. on the Use of Computers in Radiation Therapy (Salt Lake City, Utah, May 1997)* ed D D Leavitt and G Starkschall (Madison, WI: Medical Physics Publishing) pp 437–40

Blomquist M, Sätherberg A and Karlsson M 1997 Physical features and clinical applications of scanned intensity modulated photon beams *Proc. World Congress on Medical Physics and Biomedical Engineering and 11th Int. Conf. on Medical Physics (Nice, France, September 1997) Med. Biol. Eng. Comput.* **35** (Suppl. Part 2) 1015

Blomquist M, Sätherberg A, Karlsson M and Zackrisson B 1998 Scanned intensity modulations for 50 MV photons *Phys. Med. Biol.* **43** 1185–97

Boccuzzi D E, Kim S, Pryor J, Berenstein A, Shih J A, Chiu-Tsao S T and Harrison L B 1999 A dosimetric comparison of stereotactic radiosurgery using static beams with a micro-multileaf collimator versus arcs for treatment of arteriovenous malformations *Int. J. Radiat. Oncol. Biol. Phys.* **45** (Suppl. 1) 413

Boellaard R, van Herk M and Mijnheer B J 1996 The dose response relationship of a liquid-filled electronic portal imaging device *Med. Phys.* **23** 1601–11

Boellaard R, van Herk M and Mijnheer B J 1997a A convolution model to convert transmission dose images to exit dose distributions *Med. Phys.* **24** 189–99

Boellaard R, van Herk M and Mijnheer B J 1998a A new method to obtain the midplane dose using portal in vivo dosimetry *Int. J. Radiat. Oncol. Biol. Phys.* **41** 465–74

Boellaard R, van Herk M, Uiterwaal H and Mijnheer B 1997b Two-dimensional exit dosimetry using a liquid-filled electronic portal imaging device and a convolution model *Radiother. Oncol.* **44** 149–57

Boellaard R, van Herk M, Uiterwaal H and Mijnheer B 1998b First clinical tests using a liquid-filled electronic portal imaging device and a convolution model for the verification of midplane dose *Radiother. Oncol.* **47** 303–12

Bogner L, Scherer J and Herbst M 1999 An inverse Monte Carlo optimization algorithm for conformal radiotherapy *Radiother. Oncol.* **51** (Suppl. 1) S32

Bonnett D E 1999 The measurement of dose distributions using MRI polymer gels *Radiother. Oncol.* **51** (Suppl. 1) S10

Bonnett D E, Farajollahi A R, Haas O, Mills A G, Glendinning A G, Aukett R J and Burnham K 1997 The verification of intensity-modulated conformal therapy using polymer gel dosimetry *Proc. World Congress on Medical Physics and Biomedical Engineering and 11th Int. Conf. on Medical Physics (Nice, France, September 1997) Med. Biol. Eng. Comput.* **35** (Suppl. Part 2) 1012

Börgers C 1999 The radiation therapy problem *Computational Radiology and Imaging: Therapy and Diagnostics* ed C Börgers and F Natterer *IMA Volumes in Mathematics and its Applications* **110** 1–16

Bortfeld T 1996 Methods for optimization in conformal radiotherapy *Atti del Corso: Technologie e Metodologie per la Radioterapia Conformazionale (Villa Olmo, Como, November 1996)* ed G M Cattaneo and L Conte (Como: Centro di Cultura Scientifica 'Alessandro Volta') pp 118–26

Bortfeld T 1998 Optimisation of dose distributions *Radiother. Oncol.* **48** (Suppl. 1) S57

Bortfeld T 1999a IMRT—Quo Vadis? *Radiother. Oncol.* **51** (Suppl. 1) S33

Bortfeld T 1999b Optimized planning using physical objectives and constraints *Semin. Radiat. Oncol.* **9** 20–34

Bortfeld T 1999c Inverse Treatment Planning *Course Compendium of the 1st IMRT Winter School (Heidelberg, 9–11 December 1999)* pp 117–26

Bortfeld T R, Boyer A L, Schlegel W, Kahler D L and Waldron T J 1994b Realisation and verification of three-dimensional conformal radiotherapy with modulated fields *Int. J. Radiat. Oncol. Biol. Phys.* **30** 899–908

Bortfeld T R, Kahler D, Waldron T J and Boyer A L 1994a X-ray field compensation with multileaf collimators *Int. J. Radiat. Oncol. Biol. Phys.* **28** 723–30

Bortfeld T, Levegrün S, Stein J, Zhang G, Rhein B and Schlegel W 1997b Intensity modulation with the 'step and shoot' technique using a commercial MLC *Proc. 39th ASTRO Meeting, Int. J. Radiat. Oncol. Biol. Phys.* **39** (Suppl. 2) 148

Bortfeld T and Oelfke U 1999 IMRT with mini multi-leaf collimators and compensators *Course Compendium of the 1st IMRT Winter School (Heidelberg, 9–11 December 1999)* pp 53–63

Bortfeld T, Oelfke U and Nill S 2000 What is the optimum leaf width of a multileaf collimator? *Phys. Med. Biol.* at press

Bortfeld T, Schlegel W, Dykstra C, Levegrün S and Preiser K 1996 Letter to the Editor: Physical versus biological objectives for treatment plan optimisation *Radiother. Oncology* **40** 185

Bortfeld T, Schlegel W, Höver K-H and Schulz-Ertner D 1999 Mini and micro multileaf collimators *Med. Phys.* **26** 1094

Bortfeld T, Stein J and Preiser K 1997a Clinically relevant intensity modulation optimisation using physical criteria *Proc. 12th Int. Conf. on the Use of Computers in Radiation Therapy (Salt Lake City, Utah, May 1997)* ed D D Leavitt and G Starkschall (Madison, WI: Medical Physics Publishing) pp 1–4

Bos L J, van der Horst A, Brugmans M J P, Damen E M F, Mijnheer B J and Lebesque J V 1999 Optimisation of segmented intensity modulated radiotherapy of prostate cancer *Radiother. Oncol* **51** (Suppl. 1) S6

Boston T 1997 Precision dosing—a reality with intensity modulated radiotherapy *Oncol. News Int.* **6** (May)

Boyer A L 1997 Verification and delivery of intensity-modulated treatments with dynamic multileaf collimator *Proc. Meeting: Intensity-Modulated Radiation Therapy: A Clinical Perspective (London, 26 June 1997)* (York: IPEM) pp 11–16

Boyer A 1998a MLC IV: non-conventional and future use of MLC *Med. Phys.* **25** A159

Boyer A 1998b Private communication (e-mail 12 June 1998)

Boyer A L and Bortfeld T 1995 In response to Drs Butler and Nizin *Int. J. Radiat. Oncol. Biol. Phys.* **32** 1348

Boyer A L and Bortfeld T 1997 Modulated cone beam conformal therapy *The Theory and Practice of Intensity Modulated Radiation Therapy* ed E S Sternick (Madison, WI: Advanced Medical Publishing) pp 91–106

Boyer A L, Foster K M, Geis P, Ma C-M, Mok E, Xing L, Findley D, Chang S, Lindquist E, Ahrens M and Froelich S 1998b Implementation of dynamic conformal arcs with mini-MLC *Int. J. Radiat. Oncol. Biol. Phys.* **42** 365

Boyer A L, Geis P, Grant W and Carol M 1997a Modulated beam conformal therapy for head and neck tumours *Int. J. Radiat. Oncol. Biol. Phys.* **39** 227–36

Boyer A, Ma C-M, Ma L, Kapur A and Xing L 1997b A digital quality assurance phantom for intensity-modulated conformal therapy *Proc. 39th ASTRO Meeting, Int. J. Radiat. Oncol. Biol. Phys.* **39** (Suppl. 2) 151

Boyer A L and Strait J P 1997 Delivery of intensity-modulated treatments with dynamic multileaf collimator *Proc. 12th Int. Conf. on the Use of Computers in Radiation Therapy (Salt Lake City, Utah, May 1997)* ed D D Leavitt and G Starkschall (Madison, WI: Medical Physics Publishing) pp 13–16

Boyer A, Xing L, Luxton G, Chen Y and Ma C 2000b IMRT by dynamic MLC *Proc. 13th Int. Conf. on the Use of Computers in Radiation Therapy (Heidelberg, May 2000)* ed W Schlegel and T Bortfeld (Heidelberg: Springer) pp 160–3

Boyer A, Xing L, Luxton G and Ma C-M 2000a DMLC IMRT is sufficient for all clinical purposes *Proc. 13th Int. Conf. on the Use of Computers in Radiation Therapy (Heidelberg, May 2000)* ed W Schlegel and T Bortfeld (Heidelberg: Springer) pp 7–9

Boyer A, Xing L, Ma C-M, Curran B, Hill R, Kania A and Bleier A 1999 Theoretical considerations of monitor unit calculations for intensity modulated beam treatment planning *Med. Phys.* **26** 187–95

Boyer A, Xing L, Ma L and Forster K 1998a Verification and delivery of head and neck intensity modulated radiotherapy *Med. Phys.* **25** A200

Boyer A L and Yu C X 1999 Intensity-modulated radiation therapy with dynamic multileaf collimators *Semin. Radiat. Oncol.* **9** 48–59

Brabbins D and Yan D 1999 Private communication (Elekta Oncology Systems Consortium, 27–28 May 1999)

Brahme A 1988 Optimisation of stationary and moving beam radiation therapy techniques *Radiother. Oncol.* **12** 129–40

Brahme A and Lind B 1999 Radiobiological objectives for treatment optimization *Phys. Medica* **15** 219

Brahme A, Roos J E and Lax I 1982 Solution of an integral equation encountered in rotation therapy *Phys. Med. Biol.* **27** 1221–9

Braunstein M and Levine R Y 2000 Optimum beam configurations in tomographic intensity modulated radiation therapy *Phys. Med. Biol.* **45** 305–8

Braunstein M, Levine R, Lo Y and Urie M 1999 Intensity modulated radiation therapy with tomographic reconstruction: MLC leaf width and beam number influence on dose conformation to cranial targets *Med. Phys.* **26** 1134

Brecher E and Brecher R 1969 *The Rays: a History of Radiology in the United States and Canada* (Baltimore, MD: Williams & Wilkins)

Broggi S, Fiorino C, Corletto D, Cattaneo G M and Calandrino R 1999 Improving the therapeutic ratio for prostate cancer by simple intensity-modulated beams in the treatment of concave-shaped PTVs *Radiother. Oncol.* **51** (Suppl. 1) S40

Brugmans M J P, van der Horst A, Lebesque J V and Mijnheer B J 1999 Beam intensity modulation to reduce the field sizes for conformal irradiation of lung tumours: a dosimetric study *Int. J. Radiat. Oncol. Biol. Phys.* **43** 893–904

Brugmans M, van der Horst A, Lebesque J, Mijnheer B, McShan D, Fraass B and Kessler M 1998 Planning and dosimetric verification of a computer-optimized segmented irradiation technique for the prostate *Med. Phys.* **25** A116

Budgell G J 1997 Private communication (Elekta Oncology Systems Consortium Meeting, 5–6 December 1997)

Budgell G 1998 Delivery of intensity modulated radiation therapy by dynamic multileaf collimation: resolution requirements for specifying leaf trajectories *Med. Phys.* **25** A202

Budgell G J 1999 Temporal resolution requirements for intensity modulated radiation therapy delivered by multileaf collimators *Phys. Med. Biol.* **44** 1581–96

Budgell G J, Mott J H L and Hounsell A R 2000a Clinical implementation of intensity modulated radiotherapy for customised compensation using dynamic multileaf collimation *Proc. 13th Int. Conf. on the Use of Computers in Radiation Therapy (Heidelberg, May 2000)* ed W Schlegel and T Bortfeld (Heidelberg: Springer) pp 203–5

Budgell G, Mott J and Williams P 1998a An investigation of the effects of multileaf collimator leaf position accuracy upon intensity modulated beams *Med. Phys.* **25** A177

Budgell G, Mott J, Williams P and Brown K J 2000b Requirements for leaf position accuracy for dynamic multileaf collimation *Phys. Med. Biol.* **45** 1211–27

Budgell G J, Sykes J B and Wilkinson J M 1997 Leaf collision avoidance in dynamic intensity modulation with a Philips multileaf collimator and the consequences for dose outside the field *Proc. World Congress on Medical Physics and Biomedical Engineering and 11th Int. Conf. on Medical Physics (Nice, France, September 1997) Med. Biol. Eng. Comput.* **35** (Suppl. Part 2) 1018

Budgell G J, Sykes J R and Wilkinson J M 1998b Rectangular edge synchronization for intensity modulated radiation therapy with dynamic multileaf collimation *Phys. Med. Biol.* **43** 2769–84

Bues M, Kooy H, Beatty J, Tarbell N, Chapman P and Loeffler J 1999a Feasibility of IMRT with a miniature MLC for small brain tumours *Med. Phys.* **26** 1138

Bues M, Kooy H, Hacker F L Tarbell N, Chapman P and Loeffler J 1999b Micro-IMRT for stereotactic radiotherapy of intracranial tumours *Int. J. Radiat. Oncol. Biol. Phys.* **45** (Suppl. 1) 319

Bues M, Kooy H M and Hoban P 2000 Fluence segmentation for micro intensity modulated radiotherapy with a miniature multileaf collimator *Proc. 13th Int. Conf. on the Use of Computers in Radiation Therapy (Heidelberg, May 2000)* ed W Schlegel and T Bortfeld (Heidelberg: Springer) pp 303–4

Burman C, Chui S, Kutcher G, Leibel S, Zelefsky M, LoSasso T, Spirou S, Wu Q, Yang J, Stein J, Mohan R, Fuks Z and Ling C 1997 Planning, delivery, and quality assurance of intensity modulated radiotherapy using dynamic multileaf collimator: a strategy for large-scale implementation for the treatment of carcinoma of the prostate *Int. J. Radiat. Oncol. Biol. Phys.* **39** 863–73

Burman C M, Chui C-S, LoSasso T, Kutcher G, Mohan R, Zelefsky M, Leibel S, Fuks Z and Ling C C 1999 Intensity-modulated treatment planning of the prostate

Radiotherapeutic Management of Prostate Adenocarcinoma ed A V D'Amico and G E Hanks (London: Arnold) pp 115–32

Butler E B and Grant W III 2000 Utilization of intensity modulated beam radiotherapy (IMBXRT) to diminish dose to the parotid gland in head and neck cancer *Proc. 1st Int. Workshop on IMRT in Clinical Practice (IMRT2k) (Brussels, 8–9 June 2000)* ed D Verellen p 28

Butler E B, Grant W III, Woo S, Campbell R C and Carol M P 1994 Prerequisites for clinical use of the Peacock treatment system *Radiology* **193** (Suppl.) 228

Butler E B, Teh B S, Grant W H, Uhl B M, Kuppersmith R B, Chiu J K, Donovan D T and Woo S Y 1999 SMART (Simultaneous modulated accelerated radiation therapy) boost: a new accelerated fractionation schedule for the treatment of head and neck cancer with intensity modulated radiotherapy *Int. J. Radiat. Oncol. Biol. Phys.* **45** 21–32

Butler E B, Woo S Y and Grant III W 1997 In response to Drs Goitein and Niemierko *Int. J. Radiat. Oncol. Biol. Phys.* **38** 1138

Butler E B, Woo S Y, Grant III W and Nizin P S 1995 Clinical realisation of 3D conformal intensity-modulated radiotherapy: regarding Bortfeld *et al* 1994 *IJROBP* **30** 899–908 *Int. J. Radiat. Oncol. Biol. Phys.* **32** 1547–8

Cai J, Chu J C H, Recine D, Sharma M, Nguyen C, Rodebaugh R, Saxena V A and Ali A 1999 CT and PET lung image registration and fusion in radiotherapy treatment planning using the chamfer matching method *Int. J. Radiat. Oncol. Biol. Phys.* **43** 883–91

Calandrino R, Fiorino C, Corletto D, Mangili P, Broggi S and Cattaneo G M 1998 1D non-uniform dose delivery by a single dynamic absorber: feasible applications in clinical practice *Radiother. Oncol.* **48** (Suppl. 1) S63

Cardinale R M, Benedict S H, Wu Q, Zwicker R D, Gable H E and Mohan R 1998b A comparison of three stereotactic radiotherapy techniques: arcs vs non-coplanar fixed fields vs intensity modulation *Int. J. Radiat. Oncol. Biol. Phys.* **42** 431–6

Cardinale R M, Wu Q, Benedict S H, Zwicker R and Mohan R 1998a Potential benefit of intensity modulated stereotactic radiosurgery using micro-multileaf collimation of static noncoplanar beams *Int. J. Radiat. Oncol. Biol. Phys.* **42** 359

Carrie C and Ginestet C 1997 Acute toxicity in pelvic radiotherapy: a randomised trial of conformal versus conventional treatment *Radiother. Oncol.* **44** 295

Carruthers L J, Redpath A T and Kunkler I H 1999 The use of compensators to optimise three-dimensional dose distribution in radiotherapy of the intact breast *Radiother. Oncol.* **50** 291–300

Campbell R C, Carol M, Curran B H, Huber R, Nash R, Rosen B and Richard W D 1997 The application of positional and dose distribution information in the treatment room *Proc. 12th Int. Conf. on the Use of Computers in Radiation Therapy (Salt Lake City, Utah, May 1997)* ed D D Leavitt and G Starkschall (Madison, WI: Medical Physics Publishing) pp 191–4

Carol M P 1997a IMRT: where we are today *The Theory and Practice of Intensity Modulated Radiation Therapy* ed E S Sternick (Madison, WI: Advanced Medical Publishing) pp 17–36

Carol M P 1997b Where we go from here: one person's vision *The Theory and Practice of Intensity Modulated Radiation Therapy* ed E S Sternick (Madison, WI: Advanced Medical Publishing) pp 243–52

Carol M P, Campbell R C and Huber R 1997b Inverse treatment planning for intensity modulated radiation therapy: CDVH treatment prescription with integral cost function *Proc. 39th ASTRO Meeting, Int. J. Radiat. Oncol. Biol. Phys.* **39** (Suppl. 2) 342

Carol M, Grant III W I, Bleier A R, Kania A A, Targovnik H S, Butler E B and Woo S W 1996 The field-matching problem as it applies to the Peacock three-dimensional conformal system for intensity modulation *Int. J. Radiat. Oncol. Biol. Phys.* **34** 183–7

Carol M P, Nash R V, Campbell R C, Huber R and Sternick E 1997a The development of a clinically intuitive approach to inverse treatment planning: partial volume prescription and area cost function *Proc. 12th Int. Conf. on the Use of Computers in Radiation Therapy (Salt Lake City, Utah, May 1997)* ed D D Leavitt and G Starkschall (Madison, WI: Medical Physics Publishing) pp 317–9

Carol M P, Pirzkall A and Hoess A 1999 A comparison of intensity modulated radiosurgery and beam's-eye-view stereotactic radiosurgery *Int. J. Radiat. Oncol. Biol. Phys.* **45** (Suppl. 1) 173

Carol M P, Targovnik H, Smith D and Cahill D 1992 3D planning and delivery system for optimised conformal therapy *Int. J. Radiat. Oncol. Biol. Phys.* **24** (Suppl. 1) 159

Cattaneo G M, Fiorino C, Reni M, Bolognesi A and Calandrino R 1998 Variability in contouring prostate and seminal vesicles in clinical practice: implications for conformal radiotherapy *Radiother. Oncol.* **48** (Suppl. 1) S5

Chang H S, Yi B Y, Won S C and Kim S W 1997 Development and performance test of add-on multileaf collimator for existing linear accelerator *Proc. 12th Int. Conf. on the Use of Computers in Radiation Therapy (Salt Lake City, Utah, May 1997)* ed D D Leavitt and G Starkschall (Madison, WI: Medical Physics Publishing) p 489

Chang J, Mageras G S, Chui C S, Wu Q and Lutz W 1998c Portal image verification of intensity modulated radiation therapy *Int. J. Radiat. Oncol. Biol. Phys.* **42** (Suppl.) 157

Chang S, Cullip T, Deschesne K, Chaney E, Tepper J and Siochi A 1998b Evaluation of IMRT via static MLC field sequences derived from predetermined conformal dose distributions *Int. J. Radiat. Oncol. Biol. Phys.* **42** (Suppl.) 206

Chang S, Deschesne K, Cullip T, Earnhart J, Parker S and Chaney E 1998a Contralateral breast dose from tangential breast phantom irradiation: comparison of different intensity modulation techniques *Med. Phys.* **25** A200

Chang S X, Deschesne K, Cullip T J, Parker S and Earnhart J 1999b A comparison of different intensity modulation treatment techniques for tangential breast irradiation *Int. J. Radiat. Oncol. Biol. Phys.* **45** 1305–14

Chang S D, Murphy M J, Tombropolous R and Adler J R 1999a Robotic radiosurgery *Advanced Neurological Navigation* ed E Alexander III and R J Mancunias (New York: Thieme Medical Publishers) pp 443–9

Chao C, Cheng J, Low D, Purdy J and Perez C 1999 Intensity modulated radiation therapy provides better tumour target coverage and normal tissue sparing in patients with nasopharyngeal carcinoma *Int. J. Radiat. Oncol. Biol. Phys.* **45** (Suppl. 1) 420

Chen G T Y 1997 Non-megavoltage imaging of patient treatment position *Proc. 12th Int. Conf. on the Use of Computers in Radiation Therapy (Salt Lake City, Utah, May 1997)* ed D D Leavitt and G Starkschall (Madison, WI: Medical Physics Publishing) pp 37–8

Chen Y and Ma C 1999 A leaf sequence QA tool for IMRT beam delivery *Med. Phys.* **26** 1136

Cheung K Y, Choi P H K, Chau R M C, Lee L K Y, Teo P M L and Ngar Y K 1999 The roles of multileaf collimators and micro-multileaf collimators in conformal and conventional nasopharyngeal carcinoma radiotherapy treatments *Med. Phys.* **26** 2077–85

Chen Y, Xing L, Luxton G, Li J G and Boyer A L 2000 A multi-purpose quality assurance tool for MLC-based IMRT *Proc. 13th Int. Conf. on the Use of Computers in Radiation Therapy (Heidelberg, May 2000)* ed W Schlegel and T Bortfeld (Heidelberg: Springer) pp 371–3

Cho B C J, Roa W H, Robinson D and Murray B 1999 The development of target-eye-view maps for selection of coplanar and noncoplanar beams in conformal radiotherapy treatment planning *Med. Phys.* **26** 2367–72

Cho P S, Lee S, Marks II R J, Redstone J A and Oh S 1997 Comparison of algorithms for intensity modulated beam optimisation: projection onto convex sets and simulated annealing *Proc. 12th Int. Conf. on the Use of Computers in Radiation Therapy (Salt Lake City, Utah, May 1997)* ed D D Leavitt and G Starkschall (Madison, WI: Medical Physics Publishing) pp 310–2

Cho P S, Lee S, Oh S, Sutlief S G and Phillips M H 1998c Optimzation of intensity modulated beams with volume constraints using two methods: cost function minimization and projection onto convex sets *Med. Phys.* **25** 435–43

Cho P S and Marks II R J 2000 Hardware-sensitive optimisation for intensity modulated radiotherapy *Phys. Med. Biol.* **45** 429–40

Cho P, Phillips M, Sutlief S, Gavin P and Kippenes H 1998a In vivo measurements of intensity modulated radiation therapy *Med. Phys.* **25** A201

Cho P, Sutlief S, Phillips M and Marks II R 1998b Inverse optimization of MLC-deliverable intensity modulated beams *Med. Phys.* **25** A117

Chui C 1998 The use of gravity-oriented shield in conformal and intensity-modulated irradiation therapy *Med. Phys.* **25** A149

Chui C-S 1999a The effects of intra-fraction organ motion on the delivery of intensity-modulated fields with a multileaf collimator *Med. Phys.* **26** 1087

Chui C-S 1999b Clinical implementation of IMRT *Med. Phys.* **26** 1075

Chui C-S and LoSasso T 2000 Quality assurance of IMRT *Proc. 13th Int. Conf. on the Use of Computers in Radiation Therapy (Heidelberg, May 2000)* ed W Schlegel and T Bortfeld (Heidelberg: Springer) pp 168–9

Chui C-S, Spirou S and LoSasso T 1996 Testing of dynamic multileaf collimation *Med. Phys.* **23** 635–41

Cionini L 1998 *Lecture at the 1st European Medical Physics School (Archamps, Geneva, November 1998)*

Clements R, Mayles H M O and Mayles W P M 1999 Physical compensator design for intensity modulated beams using PLATO and KONRAD *Radiother. Oncol.* **51** (Suppl. 1) S34

Coldham C M, Kennerley J, Whitton V J and Beavis A W 1999 Quality control of an inverse-planning algorithm *Proc. 5th Annu. Conf. of IPEM (Nottingham, September 1999)* p 139

Cole D J and Weatherburn H 1997 *Br. J. Radiol.* **70** (Suppl.) 13

Conte L, Stucchi P, Mordacchini C, Bianchi C, Monciardini M, Novario R, Cassani E, Mantovani R, Antognoni P and Tordiglione M 1997 The use of an electronic portal imaging device for exit dosimetry *Proc. 12th Int. Conf. on the Use of Computers in Radiation Therapy (Salt Lake City, Utah, May 1997)* ed D D Leavitt and G Starkschall (Madison, WI: Medical Physics Publishing) pp 279–81

Conway J, Robinson M H and Bragg C M 2000 A cpmparison of conventional 3D conformal radiotherapy and IMRT in the planning of radiotherapy treatment of head and neck cancer *Proc. 1st Int. Workshop on IMRT in Clinical Practice (IMRT2k) (Brussels, 8–9 June 2000)* ed D Verellen p 57

Convery D J 1997a Beam intensity modulation by dynamic multileaf collimation *Br. J. Radiol.* **70** (Suppl.) 16

Convery D J 1997b Private communication (Elekta Oncology Systems Consortium Meetings, Crawley, 8–9 May and 5–6 Decemeber 1997)

Convery D J 1998 Private communication (Elekta Oncology Systems Consortium Meetings, Crawley, 25–26 September 1998)

Convery D J, Cosgrove V P and Webb S 2000a Improving dosimetric accuracy of a dynamic MLC technique *Proc. 13th Int. Conf. on the Use of Computers in Radiation Therapy (Heidelberg, May 2000)* ed W Schlegel and T Bortfeld (Heidelberg: Springer) pp 277–9

Convery D J, Cosgrove V P and Webb S 2000b Improving dosimetric accuracy of a dynamic MLC technique *Proc. 1st Int. Workshop on IMRT in Clinical Practice (IMRT2k) (Brussels, 8–9 June 2000)* ed D Verellen p 38

Convery D and Webb S 1997 Calculation of the distribution of head-scattered radiation in dynamically-collimated MLC fields *Proc. 12th Int. Conf. on the Use of Computers in Radiation Therapy (Salt Lake City, Utah, May 1997)* ed D D Leavitt and G Starkschall (Madison, WI: Medical Physics Publishing) pp 350–3

Convery D J and Webb S 1998 Generation of discrete beam-intensity modulation by dynamic multileaf collimation under minimum leaf separation constraints *Phys. Med. Biol.* **43** 2521–38

Convery D J, Webb S and Partridge M 1998 Generation, delivery and verification of discrete beam intensity modulation using the Elekta dynamic multileaf collimator *Proc. Elekta Oncology Systems First Users Conf. (Palm Springs, CA, 1998)* pp 50–60

Corletto D, Fiorino C, Cattaneo G M, Mangili P and Calandrino R 1998 Dose calculation and dosimetry tests for clinical implementation of 1D tissue-deficit compensation by a single dynamic absorber *Radiother. Oncol.* **47** 53–62

Cosgrove V P, Convery D J, Murphy P S, Nutting C M and Webb S 2000a Dynamic MLC delivered IMRT: verification using polyacrylamide gel dosimetry *Proc. 13th Int. Conf. on the Use of Computers in Radiation Therapy (Heidelberg, May 2000)* ed W Schlegel and T Bortfeld (Heidelberg: Springer) pp 311–3

Cosgrove V P, Convery D J, Murphy P S, Nutting C M and Webb S 2000c Verifying dynamic MLC delivered IMRT with polyacrylamide gels *Proc. 1st Int. Workshop on IMRT in Clinical Practice (IMRT2k) (Brussels, 8–9 June 2000)* ed D Verellen p 38

Cosgrove V, Jahn U, Pfaender M, Bauer S, Budach V and Wurm R 1998b Clinical use of a micro multileaf collimator for conformal radiosurgery *Proc. 2nd S Takahashi Memorial Int. Workshop on Three Dimensional Conformal Radiotherapy (Nagoya, Aichi, Japan, 11–13 December 1998)* pp 44–5

Cosgrove V P, Jahn U, Pfaender M, Bauer S, Budach V and Wurm R E 1999 Commissioning of a micro multi-leaf collimator and planning system for stereotactic radiosurgery *Radiother. Oncol.* **50** 325–36

Cosgrove V P, Murphy P S, McJury M, Adams E J, Warrington A P, Leach M O and Webb S 2000b The reproducibility of polyacrylamide gel dosimetry applied to stereotactic conformal radiotherapy *Phys. Med. Biol.* **45** 1195–210

Cosgrove V, Pfaender M, Jahn U, Bauer S, Budach V and Wurm R 1998a Commissioning and clinical use of a micro multileaf collimator for conformal radiosurgery *Med. Phys.* **25** A129

Cotrutz C, Kappas C, Proimos B and Mohan R 1997 PC implementation of conformal dose distribution computation obtained with gravity oriented devices *Proc. World Congress on Medical Physics and Biomedical Engineering and 11th Int. Conf. on Medical Physics (Nice, France, September 1997) Med. Biol. Eng. Comput.* **35** (Suppl. Part 2) 937

Cotrutz C, Kappas C and Webb S 1999a Intensity modulated arc therapy (IMAT) with multi-isocentric centrally blocked rotational fields *Phys. Medica* **15** 239

Cotrutz C, Kappas C and Webb S 1999b Optimization of target dose homogeneity in rotation therapy with synchronous shielding *Radiother. Oncol.* **51** (Suppl. 1) S41

Cotrutz C, Kappas C and Webb S 2000 Intensity modulated arc therapy (IMAT) with centrally blocked beams *Phys. Med. Biol.* **45** 2185–206

Cozzi L, Fogliata A and Bolsi A 2000b Critical appraisal of different techniques for irradiating head and neck tumours: is IMRT better than conventional photons? *Proc. 1st Int. Workshop on IMRT in Clinical Practice (IMRT2k) (Brussels, 8–9 June 2000)* ed D Verellen p 58

Cozzi L, Fogliata A, Presilla S and Pueroni A 2000a Dosimetric validation of multileaf based IMRT using films, ion chambers and electronic portal imaging devices *Proc. 1st Int. Workshop on IMRT in Clinical Practice (IMRT2k) (Brussels, 8–9 June 2000)* ed D Verellen p 43

Curran B 1997 Conformal radiation therapy using a multileaf intensity modulating collimator *The Theory and Practice of Intensity Modulated Radiation Therapy* ed E S Sternick (Madison, WI: Advanced Medical Publishing) pp 75–90

Curran B, Bleier A, Carol M, Hill R, Holmes T, Kania A, Lalonde R, Larson L and Sternick E 1998 Validation of a finite size pencil beam dose model for intensity modulated radiation therapy with multileaf collimators *Med. Phys.* **25** A189

Curtin-Savard A and Podgorsak E 1998 The use of a scanning liquid ionization chamber electronic portal imaging device for verification of intensity-modulated beams *Med. Phys.* **25** A150

Curtin-Savard A J and Podgorsak E B 1999 Verification of segmented beam delivery using a commercial electronic portal imaging device *Med. Phys.* **26** 737–42

Dai J-R and Hu Y-M 1999 Intensity-modulation radiotherapy using independent collimators: an algorithm study *Med. Phys.* **26** 2562–70

Danciu C and Proimos B S 1999 Gravity oriented absorbers in conformal radiotherapy for cervix cancer *Radiother. Oncol.* **51** (Suppl. 1) S38

Das S K and Marks L B 1997 Selection of coplanar or noncoplanar beams using three-dimensional optimization based on maximum beam separation and minimized nontarget irradiation *Int. J. Radiat. Oncol. Biol. Phys.* **38** 643–55

Davison A, Miller T, Belton I P, Bolton S C and Bonnett D E 1997 An assessment of image registration in the treatment planning of tumours of the brain *Br. J. Radiol.* **70** (Suppl.) 26

Dawson D, Spies R and Carol M P 1997 A volume delivery system for intensity modulation radiation therapy *Proc. 12th Int. Conf. on the Use of Computers in Radiation Therapy (Salt Lake City, Utah, May 1997)* ed D D Leavitt and G Starkschall (Madison, WI: Medical Physics Publishing) p 481

Dawson L A, Anzai Y, Marsh L, Paulino A, Ship J and Eisbruch A 1999 Patterns of local-regional failure following parotid-sparing conformal and multisegmental intensity modulated radiotherapy for head and neck cancer *Int. J. Radiat. Oncol. Biol. Phys.* **45** (Suppl. 1) 150

Dearnaley D P 1995 Radiotherapy of prostate cancer: established results and new developments *Semin. Surg. Oncol.* **11** 50–9

Dearnaley D 1998 Prostate—UK *Proc. Conf. 'Conformal Therapy: a Clinical Perspective Meeting' (London, UK, 21 May 1998)* (York: IPEM) pp 19–20

Dearnaley D P, Khoo V S, Norman A R, Meyer L, Nahum A, Tait D, Yarnold J and Horwich A 1999 Comparison of radiation side effects of conformal and conventional radiotherapy in prostate cancer: a randomised trial *Lancet* **353** 267–72

Deasy J O 1997 Multiple local minima in radiotherapy optimisation problems with dose-volume constraints *Med. Phys.* **24** 1157–61

de Boer R 1999 Private communication (Elekta Oncology Systems Consortium, 27–28 May 1999)

Debois M, Fossard A, Vansteenkist J, de Wever W, Versscakelen J, Bogaerts J, Gatti G, Stroobants S, Dupont P, Huyskens D, Kutcher J and Vanuytsel L 1998 Impact of PET scan on the target volume in lung cancer radiotherapy planning *Radiother. Oncol.* **48** (Suppl. 1) S13

Debus J, Levegrün S, Oelfke U, Rhein B, Lohr F, Helbig A, Schultz C, Preiser K, Bortfeld T and Schlegel W 1998 Intensity modulated radiotherapy delivered by compensators: technical feasibility and clinical application *Int. J. Radiat. Oncol. Biol. Phys.* **42** 366

de Deene Y, Achten E, de Neve W and de Wagter C 1999 MRI polymer gel dosimetry in conformal radiotherapy: Quantitative comparison with film dosimetry and planning applied to a mediastinal tumor *Radiother. Oncol.* **51** (Suppl. 1) S23

de Deene Y and de Wagter C 1999 Gel dosimetry in conformal radiotherapy: validation, optimization and MR artefacts *Proc. 1st Int. Workshop on Radiation Therapy Gel Dosimetry (Lexington, Kentucky, 21–23 July 1999)* pp 75–89

de Deene Y, de Wagter C, Achten E and de Neve W 2000c On the chemical stability of 3D monomer/polymer gel dosimetry *Proc. 13th Int. Conf. on the Use of Computers in Radiation Therapy (Heidelberg, May 2000)* ed W Schlegel and T Bortfeld (Heidelberg: Springer) pp 380–2

de Deene Y, de Wagter C, de Neve W and Achten E 1998a 3D polymer gel dosimetry in conformal radiotherapy: Evaluation, artefacts and future perspectives *Radiother. Oncol.* **48** (Suppl. 1) S50

de Deene Y, de Wagter C, van Duyse B, Derycke S, Mersseman B, de Gersem W, Voet T, Achten E and de Neve 2000a Validation of MR-based polymer gel dosimetry as a preclinical three-dimensional verification tool in conformal radiotherapy *Magn. Reson. Med.* **43** 116–25

de Deene Y, de Wagter C, van Duyse B, Derycke S, de Neve W and Achten E 1998b Three-dimensional dosimetry using polymer gel and magnetic resonance imaging applied to the verification of conformal radiation therapy in head and neck cancer *Radiother. Oncol.* **48** 283–91

de Deene Y, Hanselaer P, de Wagter C, Achten E and de Neve W 2000b An investigation of the chemical stability of a monomer/polymer gel dosimeter *Phys. Med. Biol.* **45** 859–78

de Gersem W, Claus F, Vakaet L, Remouchamps V, Vanhoutte I, van Duyse B, Vermael S, de Wagter C and de Neve W 2000a General anatomy based segmentation for intensity modulated radiotherapy of head and neck cancer *Proc. 13th Int. Conf. on the Use of Computers in Radiation Therapy (Heidelberg, May 2000)* ed W Schlegel and T Bortfeld (Heidelberg: Springer) p 207

de Gersem W, Claus F, Vakaet L, Remouchamps V, Vanhoutte I, van Duyse B, Vermael S, de Wagter C and de Neve W 2000b General anatomy based segmentation for intensity modulated radiotherapy of head and neck cancer *Proc. 1st Int. Workshop on IMRT in Clinical Practice (IMRT2k) (Brussels, 8–9 June 2000)* ed D Verellen p 30

de Gersem W, de Neve W, Derycke S, de Meerleer G, van Duyse B, Vakaet L and de Wagter C 1998 Intensity modulated conformal radiotherapy (IMRT) for tumour relapses of pharyngeal cancer: interactive or automatic implementation of a class solution *Radiother. Oncol.* **48** (Suppl. 1) S148

de Gersem W R T, Derycke S, Colle C O, de Wagter C and de Neve W J 1999 Inhomogeneous target-dose distributions: a dimension more for optimisation? *Int. J. Radiat. Oncol. Biol. Phys.* **44** 461–8

de Gersem W, Derycke S, de Wagter C, Mijnheer B S and de Neve W 1997 Inhomogeneous target-dose distributions: a dimension more for optimisation? *Proc. World Congress on Medical Physics and Biomedical Engineering and 11th Int. Conf. on Medical Physics (Nice, France, September 1997) Med. Biol. Eng. Comput.* **35** (Suppl. Part 2) 920

Dejean C, Lefkopoulos D, Grandjean P, Platoni K, Vieillevigne L and Julia F 2000b Use of the receiver operating characteristics (ROC) analysis to define the conformal prescription isodose in IMRT *Proc. 1st Int. Workshop on IMRT in Clinical Practice (IMRT2k) (Brussels, 8–9 June 2000)* ed D Verellen p 21

Dejean C, Lefkopoulos D, Platoni K, Grandjean P, Vieillevigne L and Kéraudy K 2000a Investigation of ROC analysis for inverse treatment planning evaluation *Proc. 13th Int. Conf. on the Use of Computers in Radiation Therapy (Heidelberg, May 2000)* ed W Schlegel and T Bortfeld (Heidelberg: Springer) pp 20–2

Delpy D, Davies B, Fergusson-Pell M, Vadgama P, Webb S and Williams D 1998 Engineering *Clinical Futures* ed M Marinker and M Peckham (London: BMJ) pp 43–73

Del Regato J A 1997 Re: H D Kogelink's article on Leopold Freund *Radiother. Oncol.* **42** 203–11

de Meerleer G, Vakaet L, de Gersem W and de Neve W 1998 Dose escalation to the prostate while keeping the maximal rectal dose unchanged is possible with intensity-modulated radiotherapy (IMRT) in the treatment of prostate cancer *Radiother. Oncol.* **48** (Suppl. 1) S69

de Meerleer G, Vakaet L, de Gersem W and de Neve W 1999 Dose escalation to the prostate while keeping the maximal rectal dose unchanged is possible with intensity-modulated radiotherapy (IMRT) in the treatment of prostate cancer *Int. J. Radiat. Oncol. Biol. Phys.* **45** (Suppl. 1) 260

de Meerleer G, Vakaet L and de Neve W 2000 Intensity-modulated radiotherapy for prostate cancer/ three versus five beams: a planning comparison *Proc. 1st Int. Workshop on IMRT in Clinical Practice (IMRT2k) (Brussels, 8–9 June 2000)* ed D Verellen p 29

de Neve W 1999 Private communication (Elekta Oncology Systems Consortium, 27–28 May 1999)

de Neve W, Claus F, and de Wagter C 1999b Conventional RT or IMRT? Clinical considerations *Course Compendium of the 1st IMRT Winter School (Heidelberg, 9–11 December 1999)* pp 127–33 de Neve W, Claus F, de Gersem W, de Meerleer G and de Wagter C 2000a Clinical implementation of IMRT by MLC technology *Proc. 13th Int. Conf. on the Use of Computers in Radiation Therapy (Heidelberg, May 2000)* ed W Schlegel and T Bortfeld (Heidelberg: Springer) p 167

de Neve W, Claus F, de Gersem W, Vermael S and de Wagter C 2000b IMRT for head and neck cancer: clinical update *Int J. Radiat. Oncol. Biol. Phys.* **46** 698

de Neve W, de Gersem W,, Derycke S, Meerleer G D, Moerman M, Bate M-T, van Duyse B, Vakaet L, de Deene Y, Mersseman B and de Wagter C 1999a Clinical delivery of intensity modulated conformal radiotherapy for relapsed or second-primary head and neck cancer using a multileaf collimator with dynamic control *Radiother. Oncol.* **50** 301–14

de Neve W, de Gersem W, Derycke S, Vakaet L, de Meerleer B, van Duyse B and de Wagter C 1998a Class solutions to beam segmentation to step and shoot delivery: an IMRT procedure for head and neck targets *Proc. Elekta Oncology Systems First Users Conf. (Palm Springs, CA, 1998)* pp 12–23

de Neve W, Derycke S, de Gersem W, de Meerleer G, Bate M T, van Duyse B, Vakaet L, de Deene Y, Mersseman B and de Wagter C 1998b Clinical implementation of intensity modulated radiotherapy by dynamic multileaf collimation *Radiother. Oncol.* **48** (Suppl. 1) S60

de Neve W, Derycke S, de Gersem W, de Meerleer G, Bate M-T, van Duyse B, Vakaet L, de Deene Y and de Wagter C 1998c Intensity modulated radiotherapy by a dynamic multileaf collimator: class solution planning for step and shoot delivery *Proc. Conf. 'Conformal Therapy: a Clinical Perspective Meeting' (London, UK, 21 May 1998)* (York: IPEM) pp 36–7

de Neve W, de Wagter C, de Jaeger K, Thienpont M, Colle C, Derycke S and Schelfhout J 1996 Planning and delivering high doses to targets surrounding the spinal cord at the lower neck and upper mediastinal levels: static beam-segmentation technique executed by a multileaf collimator *Radiother. Oncol.* **40** 271–9

de Neve W and de Wagter C 1997 University hospital, Ghent, implements beam intensity modulation in clinical practice *Wavelength* **1** No 1 (Crawley: Elekta Oncology Systems) pp 1–7

Derycke S, de Gersem W R T, van Duyse B B R and de Neve W C J 1998 Conformal radiotherapy of stage 3 non-small-cell lung cancer: a class solution involving non-coplanar intensity-modulated beams *Int. J. Radiat. Oncol. Biol. Phys.* **41** 771–7

Derycke S, van Duyse B, de Gersem W, de Wagter C and de Neve W 1997 Non-coplanar beam intensity modulation allows large dose escalation in stage-3 lung cancer *Radiother. Oncol.* **45** 253–61

Desobry G E, Bruner D W, Movsas B and Schultheiss T E 1998 Automated segmented field selection for IMRT using utility based decision theory *Int. J. Radiat. Oncol. Biol. Phys.* **42** 206

Despeignes V 1896 Observation concernant un cas de cancer de l'estomac traité par les rayons Röntgen *Lyon. Med.* **82** 428–430, 503

de Wagter C, Colle C O, Fortan L G, van Duyse B B, van den Berge D L and de Neve W J 1998 3D conformal intensity-modulated radiotherapy planning: interactive optimisation by constrained matrix inversion *Radiother. Oncol.* **47** 69–76

de Wagter C, de Geest E, de Deene Y, Mersseman B, de Vlamynck K, Constales K and de Neve W 1997 Dosimetry of small-field photon beams using polymer-gel, film, diamond detector, ionisation chamber and TLD *Proc. World Congress on Medical*

Physics and Biomedical Engineering and 11th Int. Conf. on Medical Physics (Nice, France, September 1997) Med. Biol. Eng. Comput. **35** (Suppl. Part 2) 896

de Wagter C, Martens C, de Deene Y, de Gersem W, van Duyse B and de Neve W 1999 QA and pre-clinical dosimetric verification of IMRT *Course Compendium of the 1st IMRT Winter School (Heidelberg, 9–11 December 1999)* pp 113–26

Dickof P and Ladyka C 2000 Heuristics for conventional planning *Proc. 13th Int. Conf. on the Use of Computers in Radiation Therapy (Heidelberg, May 2000)* ed W Schlegel and T Bortfeld (Heidelberg: Springer) pp 43–5

Dirkx M L P, Essers M and de Langen M 2000 Comparison of two algorithms for leaf trajectory calculation for dynamic multileaf collimation *Proc. 1st Int. Workshop on IMRT in Clinical Practice (IMRT2k) (Brussels, 8–9 June 2000)* ed D Verellen p 38

Dirkx M L P, Essers M, van Sörnsen de Koste J R, Senan S and Heijmen B J M 1999b Beam intensity modulation for penumbra enhancement in the treatment of lung cancer *Int. J. Radiat. Oncol. Biol. Phys.* **44** 449–54

Dirkx M L P and Heijmen B J M 1999b Beam intensity modulation for penumbra enhancement in lung tissue: a dosimetric study *Int. J. Radiat. Oncol. Biol. Phys.* **45** (Suppl. 1) 407

Dirkx M L P, Heijmen B J M, Korevaar G A, van Os M J H, Stroom J C, Koper P C M and Levendag P C 1997a Field margin reduction using intensity-modulated X-ray beams formed with a multileaf collimator *Int. J. Radiat. Oncol. Biol. Phys.* **38** 1123–9

Dirkx M L P, Heijmen M J M and van Santvoort J P C 1998 Leaf trajectory calculation for dynamic multileaf collimation to realise optimized fluence profiles *Phys. Med. Biol.* **43** 1171–84

Dirkx M L P, van Ingen K M and Heijmen B J M 1999a Dosimetric verification of intensity modulated treatment plans for lung cancer patients *Radiother. Oncol.* **51** (Suppl. 1) S43

Dirkx M L P, van Santvoort J P C and Heijmen B J M 1997c Dynamic multileaf collimation to realise optimized intensity modulated beam profiles *Proc. 39th ASTRO Meeting, Int. J. Radiat. Oncol. Biol. Phys.* **39** (Suppl. 2) 150

Dirkx M L P, van Santvoort J P C, van der Kamp A and Heijmen B J M 1997b Optimisation of leaf trajectories for dynamic multileaf collimation to realise desired intensity modulated beam profiles *Proc. 12th Int. Conf. on the Use of Computers in Radiation Therapy (Salt Lake City, Utah, May 1997)* ed D D Leavitt and G Starkschall (Madison, WI: Medical Physics Publishing) pp 357–9

Djennaoui N 1997 3D dose measurements with dosimeter gels and MRI *Proc. World Congress on Medical Physics and Biomedical Engineering and 11th Int. Conf. on Medical Physics (Nice, France, September 1997) Med. Biol. Eng. Comput.* **35** (Suppl. Part 2) 882

Dobbs H J 1998 Overview of target volume definition and margins *Radiother. Oncol.* **48** (Suppl. 1) S5

Dogan H N, Leybovich L, Sethi A, Krasin M and Emami B 1999 A modified method of planning and delivery for IMRT treatments *Med. Phys.* **26** 1139

Dominiak G, Beyer D, Speiser B, Mead D and Puente F 1998 Delivery of IMRT using the NOMOS CORVUS TPS and a Siemens PRIMUS accelerator with MLC *Med. Phys.* **25** 1583

Dong L, Shiu A S, Tung S S and Hogstrom K R 1997 A pencil-beam photon algorithm for a miniature multileaf collimator *Proc. 12th Int. Conf. on the Use of Computers in*

Radiation Therapy (Salt Lake City, Utah, May 1997) ed D D Leavitt and G Starkschall (Madison, WI: Medical Physics Publishing) pp 91–4

Donovan E M, Johnson U, Shentall G, Evans P M, Neal A J and Yarnold J R 2000 Evaluation of compensators in breast radiotherapy—a planning study using multiple static fields *Int. J. Rad. Oncol. Biol. Phys.* **46** 671–9

Donovan E, Suter B and the Breast Dosimetry Group 1999 Intensity modulation in breast radiotherapy: Breast dosimetry trial ethics number 1244 *Br. J. Radiol.* **72** (Suppl.) P40

Dubal N, Chang S, Cullip T, Tracton G and Rosenman J 1998 Intensity modulation for tangential breast irradiation *Int. J. Radiat. Oncol. Biol. Phys.* **42** (Suppl.) 127

Ebert M A 2000 Viability of the EUD and TCP concepts as reliable dose indicators *Phys. Med. Biol.* **45** 441–57

Eilertsen K 1997 Automatic detection of single MLC leaf positions with corrections for penumbral effects and portal dose rate characteristics *Phys. Med. Biol.* **42** 313–34

Eisbruch A 1998 Head and neck conformal and intensity modulated irradiation *Proc. Conf. 'Conformal Therapy: a Clinical Perspective Meeting' (London, UK, 21 May 1998)* (York: IPEM) p 23

Eisbruch A, Marsh L H, Martel M K, Ship J A, Ten Haken R, Pu A T, Fraass B A and Lichter A S 1998 Comprehensive irradiation of head and neck cancer using conformal multisegmental fields: assessment of target coverage and non-involved tissue sparing *Int. J. Radiat. Oncol. Biol. Phys.* **41** 559–68

Eisbruch A, Ship J A, Martel M K, Ten Haken R K, Marsh L H, Wolf G T, Esclamado R M, Bradford C R, Terrell J E, Gebarski S S and Lichter A 1996 Parotid gland sparing in patients undergoing bilateral head and neck irradiation: techniques and early results *Int. J. Radiat. Oncol. Biol. Phys.* **36** 469–80

Eisbruch A, Ten Haken R K, Kim H M, Marsh L H and Ship J A 1999 Dose, volume, and function relationships in parotid salivary glands following conformal and intensity modulated irradiation of head and neck cancer *Int. J. Radiat. Oncol. Biol. Phys.* **45** 577–87

Elekta 1997 Not all MLCs are created equal *Wavelength* **1** No 2 (Crawley: Elekta Oncology Systems) 8–13

Elekta 1998a Company brochure 4513 370 29491 / 764 05/98

Elekta 1998b Are you ready for IMRT today? *Wavelength* **2** No 2 (Crawley: Elekta Oncology Systems) 1–9

Elekta 1998c IMRT update: William Beaumont Hospital goes clinical with IMRT *Wavelength* **2** No 4 (Crawley: Elekta Oncology Systems) 4–9

Elekta 1999a Company literature 4513 371 00421 05:99

Elekta 1999b IMRT update: Christie Hospital to implement dynamic IMRT *Wavelength* **3** No 1 (Crawley: Elekta Oncology Systems) 10–12

Elekta 1999c University Hospital, Gent, brings IMRT to everyday clinical practice *Wavelength* **3** No 2 (Crawley: Elekta Oncology Systems) 7–13

Elekta 1999d IMRT techniques enhances efficiency and homogeneity for breast treatments *Wavelength* **3** No 3 (Crawley: Elekta Oncology Systems) 1–4

Engler M J, Tsai J S, DiPetrillo T and Wazer D E 1999 The reduction of inversely planned intensity-modulated radiation tomotherapy to clinical practice *Phys. Medica* **15** 241

Ertl A, Heimberger K, Kindl P, Schöggi A, Saringer W and Lindner Ch 1997 Three-dimensional dosimetry using radiochromic film sheets and polymer gels in stereotactic gamma-knife techniques *Proc. World Congress on Medical Physics and Biomedical*

Engineering and 11th Int. Conf. on Medical Physics (Nice, France, September 1997) Med. Biol. Eng. Comput. **35** (Suppl. Part 2) 883

Essers M, Boellaard R, van Herk M, Lanson H and Mijnheer B 1996 Transmision dosimetry with a liquid-filled electronic portal imaging device *Int. J. Rad. Oncol. Biol. Phys.* **34** 931–41

Essers M, Dirkx M L P and de Langen M 2000 Clinical implementation of dynamic multileaf collimation (DMLC) on a Clinac 2300C using HELIOS and LMC *Proc. 1st Int. Workshop on IMRT in Clinical Practice (IMRT2k) (Brussels, 8–9 June 2000)* ed D Verellen p 42

Evans P M, Bleackley N, Convery D J, Donovan E M, Hansen V N, Partridge M, Reise S, Symonds-Tayler J R N and Yarnold J R 2000b The use of compensators and multiple static fields for IMRT of the breast *Proc. of the 13th Int. Conf. on the Use of Computers in Radiation Therapy (Heidelberg, May 2000)* ed W Schlegel and T Bortfeld (Heidelberg: Springer) pp 208–9

Evans P M, Donovan E M, Fenton N, Hansen V N, Moore I, Partridge M, Reise S, Suter B, Symonds-Tayler J R N and Yarnold J R 1998 Practical implementation of compensators in breast radiotherapy *Radiother. Oncol.* **49** 255–65

Evans P M, Donovan E M, Hansen V N, Hector C, Partridge M, Symonds-Tayler J R N and Webb S 2000c The design of intensity modulated breast treatments using electronic portal imaging *Proc. 16th Int. Workshop on Electronic Portal Imaging (EPI2k) (Brussels, 5–7 June 2000)* ed Jan van de Steene p 56

Evans PM, Donovan E M, Hansen V N, Partridge M and Symonds-Tayler J R N 1997b Tissue compensators combined with portal imaging *Proc. World Congress on Medical Physics and Biomedical Engineering and 11th Int. Conf. on Medical Physics (Nice, France, September 1997) Med. Biol. Eng. Comput.* **35** (Suppl. Part 2) 1070

Evans P M, Donovan E, Hector C, Partridge M, Symonds-Tayler J R N, Webb S and Yarnold J R 1999 Physical aspects of breast radiotherapy *Br. J. Radiol.* **72** (Suppl.) P57

Evans P M, Donovan E M, Partridge M, Childs P, Convery D J, Eagle S, Hansen V N, Suter B and Yarnold J R 2000a The delivery of intensity modulated radiotherapy to the breast using multiple static fields *Radiother. Oncol.* at press

Evans P M, Hansen V N, Mayles W P M, Swindell W, Torr M and Yarnold J R 1995 Design of compensators for breast radiotherapy using electronic portal imaging *Radiother. Oncol.* **37** 43–54

Evans P M, Hansen V N and Swindell W 1997a The optimum intensities for multiple static multileaf collimator field compensation *Med. Phys.* **24** 1147–56

Evans P M and Partridge M 2000 A method of improving the spatial resolution of treatments that involve a multileaf collimator *Phys. Med. Biol.* **45** 609–22

Fang G Y, Geiser B and Mackie R T 1997 Software system for the UW/GE tomotherapy prototype *Proc. 12th Int. Conf. on the Use of Computers in Radiation Therapy (Salt Lake City, Utah, May 1997)* ed D D Leavitt and G Starkschall (Madison, WI: Medical Physics Publishing) pp 332–4

Fantini M, Benassi M, Arcangeli G, Begnozzi L, Bufacchi A, Soriani A, Orvieto A and Ciuffo P 1999 Conformal irradiation by composition of narrow beams *Phys. Medica* **15** 213

Farajollahi A R, Bonnett D E, Ratcliffe A J, Aukett R J and Mills J A 1999 An investigation into the use of polymer gel dosimetry in low dose rate brachytherapy *Br. J. Radiol.* **72** 1085–92

Fiorino C, Broggi S, Corletto G M, Cattaneo R and Calandrino S 1999 Defining the PTV taking only systematic errors into account and including random errors in the treatment planning effects on TCP calculation *Radiother. Oncol.* **51** (Suppl. 1) S2

Fiorino C, Cattaneo G V, Mangili P, Ragazzi G and Calandrino R 1997 A simplified convolution method of incorporating random set-up variations in the treatment planning *Proc. 12th Int. Conf. on the Use of Computers in Radiation Therapy (Salt Lake City, Utah, May 1997)* ed D D Leavitt and G Starkschall (Madison, WI: Medical Physics Publishing) pp 178–81

Fitchard E E, Aldridge J S, Reckwerdt P J and Mackie T R 1997 Registration of tomotherapy patients using CT projection files *Proc. 12th Int. Conf. on the Use of Computers in Radiation Therapy (Salt Lake City, Utah, May 1997)* ed D D Leavitt and G Starkschall (Madison, WI: Medical Physics Publishing) pp 185–7

Fitchard E E, Aldridge J S, Reckwerdt P J and Mackie T R 1998b Registration of synthetic tomographic projection datasets using cross correlation *Phys. Med. Biol.* **43** 1645–57

Fitchard E E, Aldridge J S, Ruchala K, Fang G, Balog J, Pearson D W, Olivera G H, Schloesser E A, Wenman D, Reckwerdt P J and Mackie T R 1999 Registration using tomographic projection files *Phys. Med. Biol.* **44** 495–507

Fitchard E, Reckwerdt P, Olivera G, Shepard D and Mackie T 1998a A transformation that corrects delivery sinograms for patient misalignment *Med. Phys.* **25** A203

Fix M, Manser P, Keller H, Born E, Vetterli D, Mini R and Ruegsegger P 1999a Verification of intensity modulated beams in radiotherapy *Med. Phys.* **26** 1162

Fix M, Manser P, Keller H, Born E, Vetterli D, Mini R and Ruegsegger P 1999b Verification of intensity modulated beams in radiotherapy *Phys. Medica* **15** 178

Fogg R, Chetty I, deMarco J J, Agazaryan N and Solberg T D 2000 Modification of a virtual source model for Monte Carlo based IMRT verification *Proc. 13th Int. Conf. on the Use of Computers in Radiation Therapy (Heidelberg, May 2000)* ed W Schlegel and T Bortfeld (Heidelberg: Springer) pp 420–2

Föller M, Bortfeld T, Hädinger U, Häring P, Schultz C, Seeber S, Richter J and Schlegel W 1998 Beam-shaping with a computer controlled micromultileaf collimator *Radiother. Oncol.* **48** (Suppl. 1) S76

Followill D, Geis P and Boyer A 1997 Estimates of whole-body dose equivalent produced by beam intensity modulated conformal therapy *Int. J. Radiat. Oncol. Biol. Phys.* **38** 667–72

Fraass B A, Kessler M L, McShan D L, Marsh L H, Watson B A, Dusseau W J, Eisbruch A, Sandler H M and Lichter A S 1999a Optimization and clinical use of multisegment intensity-modulated radiation therapy for high-dose conformal therapy *Semin. Radiat. Oncol.* **9** 60–77

Fraass B A, Marsh L, Watson B, Dusseau W, Martel M, McShan D, Sandler H, Eisbruch A and Lichter A 1998 Routine clinical use of multisegment IMRT: analysis of planning strategies *Med. Phys.* **25** A109

Fraass B, McShan D and Kessler M 2000 Dose-based conformal field shaping using automated optimisation *Proc. 13th Int. Conf. on the Use of Computers in Radiation Therapy (Heidelberg, May 2000)* ed W Schlegel and T Bortfeld (Heidelberg: Springer) pp 32–4

Fraass B A, Vineberg K A, Kessler M L and McShan D L 1999b Optimization for conformal therapy using segmental IMRT *Radiother. Oncol.* **51** (Suppl. 1) S36

Fraass B A, Vineberg K A, Kim J, Kessler M L and McShan D L 1999c Conformal plan optimisation: inverse and forward planning for full and segmental IMRT *Med. Phys.* **26** 1091

Francescon P, Cavedon C and Reccanello S 1998 Comparison between commercial 2D–3D treatment planning systems and the BEAM code *Med. Phys.* **25** A129

Frencl L, Cernohuby P and Drnec I 1999 Radiation technique with gravity oriented block *Radiother. Oncol.* **51** (Suppl. 1) S56

Frenzel T, Schmidt R and Scobel W 2000 Application of modifiers for intensity modulated photon fields *Proc. 13th Int. Conf. on the Use of Computers in Radiation Therapy (Heidelberg, May 2000)* ed W Schlegel and T Bortfeld (Heidelberg: Springer) pp 188–90

Freund L 1897 Ein mit Röntgen-Strahlen behandelter Fall von Naevus pigmentosus piliferus *Wien. Med. Wsch.* **47** 428–34

Freund L 1903 *Elements of General Radio-Therapy for Practitioners* (Berlin: Ärzte, Urban und Schwarzenberg) (in German)

Fryer T D, Thomas S J, Meara S J P, Langmack K A and Dendy P P 1997 Dosimetry verification using optical tomographic imaging of gels *Proc. World Congress on Medical Physics and Biomedical Engineering and 11th Int. Conf. on Medical Physics (Nice, France, September 1997) Med. Biol. Eng. Comput.* **35** (Suppl. Part 2) 883

Fujino D, Allbright N and Wieczorek D 1999 Modelling multileaf collimators with the PEREGRINE Monte Carlo system *Med. Phys.* **26** 1085

Fuks Z 1998 Dose escalation with 3D conformal radiotherapy affects the outcome in early stage prostate cancer *Proc. 2nd S Takahashi Memorial Int. Workshop on Three Dimensional Conformal Radiotherapy (Nagoya, Aichi, Japan, 11–13 December 1998)* pp 54–5

Fuks Z 2000 Unpublished communication *13th ICCR invited lecture (Heidelberg Castle, 24 May 2000)*

Galvin J M 1999a Private communication (Elekta Oncology Systems Consortium, 27–28 May 1999); and also A comparison of forward and inverse treatment planning for IMRT *Med. Phys.* **26** 1090

Galvin J M 1999b The multileaf collimator: past, present and future *Med. Phys.* **26** 1092–3

Galvin J M 1999c Complex IMRT increases the potential for dose misadministration, and simple planning and delivery methods can achieve the same results with reduced risk of patient injury *Proc. 13th Int. Conf. on the Use of Computers in Radiation Therapy (Heidelberg, May 2000)* ed W Schlegel and T Bortfeld (Heidelberg: Springer) p 10

Galvin J M, Chen X G and Smith R M 1993 Combining multileaf fields to modulate fluence distributions *Int. J. Radiat. Oncol. Biol. Phys.* **27** 697–705

Galvin J M, Leavitt D D and Smith A A 1996 Field edge smoothing for multileaf collimators *Int. J. Radiat. Oncol. Biol. Phys.* **35** 89–94

Galvin J M, Sweet J, Bednarz G, Hughes S and Curran Jr W 1998 Optimizing 3D dose distributions without inverse treatment planning *Med. Phys.* **25** A207

Gardey K U, Mühlhaus R, Stalling D, Wust P and Felix R 1999 Linac—a research oriented platform for virtual IMRT simulation *Radiother. Oncol.* **51** (Suppl. 1) S37

Geis P, Boyer A L and Wells N H 1996 Use of a multileaf collimator as a dynamic missing-tissue compensator *Med. Phys.* **23** 1199–205

Geis P, Hill R, Kalnicki S and Wu A 1999 An analysis of speed and precision in IMRT delivery *Med. Phys.* **26** 1135

Gibbs P, Beavis A W, Dealey R A, Whitton V J and Horsman A 1997 MRI of the brain: applications in radiotherapy treatment planning *Br. J. Radiol.* **70** (Suppl.) 39

Gibbs W W 1998 Taking aim at tumours *Scientific American: in focus* (May) **598**

Gilio J, Gall K, Chang C and Garwood D 1998 Clinical results of pointing accuracy quality assurance for a robotically mounted linear accelerator used for stereotactic radiosurgery *Med. Phys.* **25** A199

Gladstone D J, Holupka E J, van Herk M, Kijewski P K, Beard C J and Svensson G K 1993 Analysis of 3D dose distribution including set-up error and stochastic target motion. Implications for set-up procedures using electronic portal imaging *Med. Phys.* **20** (3) 876

Gluckman G, Button T, Meek A, Maryanski M and Reinstein L 2000 Automated verification of 3D-Conformal and IMRT treatment planning using polymer gels *Proc. 13th Int. Conf. on the Use of Computers in Radiation Therapy (Heidelberg, May 2000)* ed W Schlegel and T Bortfeld (Heidelberg: Springer) pp 389–90

Gluckman G R, Button T M, Reinstein L E and Maryanski M J 1999 Automated verification of 3D-conformal and IMRT treatment planning using BANG-gel *Proc. 1st Int. Workshop on Radiation Therapy Gel Dosimetry (Lexington, Kentucky, 21–23 July 1999)* pp 187–9

Goitein M and Niemierko A 1996 Intensity modulated therapy and inhomogeneous dose to the tumour: a note of caution *Int. J. Rad. Oncol. Biol. Phys.* **36** 519–22

Gore J C, Ranade M, Maryanski M J and Schulz R J 1996 Radiation dose distributions in three dimensions from tomographic optical density scanning of polymer gels: 1. Development of an optical scanner *Phys. Med. Biol.* **41** 2695–704

Grandjean P, Lefkopoulos D, Dejean C, El-Balaa H, Julia F and Schlienger M 2000 Application of some regularization methods to the radiosurgery inverse problem *Proc. 13th Int. Conf. on the Use of Computers in Radiation Therapy (Heidelberg, May 2000)* ed W Schlegel and T Bortfeld (Heidelberg: Springer) pp 539–41

Grant III W and Butler E B 2000 Making IMRT a clinical routine *Proc. 1st Int. Workshop on IMRT in Clinical Practice (IMRT2k) (Brussels, 8–9 June 2000)* ed D Verellen p 32

Grant III W, Butler E B, Woo S, Targovnik H, Campbell R C and Carol M P 1994 Quality assurance of brain treatments with the Peacock treatment system *Radiology* **193** (Suppl.) 228

Grant III W and Woo S Y 1999 Clinical and financial issues for intensity-modulated radiation therapy delivery *Semin. Radiat. Oncol.* **9** 99–107

Greco S C, Uhl B M, Teh B S, Grant III W H, McGary J, Woo S Y and Butler E B 2000 Intensity-modulated radiation therapy (IMRT) for prostate cancer: analysis of prostate movement using a rectal balloon catheter *Int J. Radiat. Oncol. Biol. Phys.* **46** 749

Gregoire V 1999 Definition of the clinical target volume: a combination between physiological and imaging data *Phys. Medica* **15** 216

Groh B A, Siewerdsen J H, Drake D G, Wong J W and Jaffray D A 2000 MV and kV cone-beam CT on a medical linear accelerator *Proc. 13th Int. Conf. on the Use of Computers in Radiation Therapy (Heidelberg, May 2000)* ed W Schlegel and T Bortfeld (Heidelberg: Springer) pp 561–3

Groh B A, Spies L, Hesse B M and Bortfeld T 1999 MVCT with an α:Si flat-panel detector *Radiother. Oncol.* **51** (Suppl. 1) S25

Grosser K-H 2000 Reduction of treatment time in IMRT step and shoot irradiations by means of a changed fractionation scheme *Proc. 13th Int. Conf. on the Use of Computers in*

Radiation Therapy (Heidelberg, May 2000) ed W Schlegel and T Bortfeld (Heidelberg: Springer) pp 308–10

Grosser K-H, Bortfeld T, Rhein B, and Häring P 1999c Some remarks on IMRT step and shoot approach *Course Compendium of the 1st IMRT Winter School (Heidelberg, 9–11 December 1999)* pp 65–76

Grosser K, Bortfeld T, Rhein B and Schlegel W 1999a Elimination of underdose artifacts in intensity maps for IMRT step and shoot approach. *Int. J. Radiat. Oncol. Biol. Phys.* **45** (Suppl. 1) 247

Grosser K H, Rhein B, Häring P, Bortfeld T and Schlegel W 1999b Practical delivery of IMRT using a commercial linac with built-in MLC *Radiother. Oncol.* **51** (Suppl. 1) S36

Grosu A L, Feldmann H J, Weber W, Bartenstein P, Schwaiger M and Molls M 1998 The value of IMT-SPECT investigation in the target volume definition of brain gliomas *Int. J. Radiat. Oncol. Biol. Phys.* **42** 258

Grubbé E H 1933 Priority in the therapeutic use of X-rays *Radiology* **21** 156–62

Guan H and Zhu Y 1998 Feasibility of megavoltage portal CT using an electronic portal imaging device (EPID) and a multi-level scheme algebraic reconstruction technique (MLS-ART) *Phys. Med. Biol.* **43** 2925–37

Guerrero T M, Pawlicki T, Jiang S B, Hai J, Luxton G, Ozhasoglu C, Adler J R and Ma C-M 2000 MonteCarlo for the Cyberknife *Proc. 13th Int. Conf. on the Use of Computers in Radiation Therapy (Heidelberg, May 2000)* ed W Schlegel and T Bortfeld (Heidelberg: Springer) pp 344–6

Guerrero T M, Pawlicki T, Ma C-M, Forster K M, Xing L, Luxton G, Boyer A L, Le Q T and Goffinet D R 1999 Spinal cord dose discrepancy in IMRT treated patients at Stanford revealed by Monte Carlo dose verification: clinical summary *Int. J. Radiat. Oncol. Biol. Phys.* **45** (Suppl. 1) 411

Gustafsson A, Ahnesjö A, Löfgren A and Saxner M 1998 Intensity-modulated radiation therapy: dose-volume optimisation of segmented multileaf collimation using a gradient-based optimisation algorithm and a pencil-beam dose calculation algorithm *Med. Phys.* **25** A117

Gustafsson A, Lind B K, Svensson R and Brahme A 1995 Simultaneous optimization of dynamic multileaf collimation and scanning patterns or compensating filters using a generalized pencil beam algorithm *Med. Phys.* **22** 1141–56

Gustavsson H, Bäck S A J, Haraldsson P, Magnusson P, Love P and Olsson L E 1999 Treatment planning dose verification using MRI gel dosimetry and a Barex head and neck phantom *Proc. 1st Int. Workshop on Radiation Therapy Gel Dosimetry (Lexington, Kentucky, 21–23 July 1999)* pp 208–10

Guzel Z, Bedford J L, Childs P J, Nahum A E, Webb S, Oldham M and Tait D 1998 A comparison of conventional and conformal radiotherapy of the oesophagus: work in progress *Br. J. Radiol.* **71** 1076–82

Haas O C L 1999 *Radiotherapy treatment planning: Advances in Industrial Control Monograph* (London: Springer)

Haas O C L, Burnham K J and Mills J A 1998 Optimization of beam orientation in radiotherapy using planar geometry *Phys. Med. Biol.* **43** 2179–93

Haas O C L, Mills J A and Burnham K J 1999 Development of an intelligent oncology workstation for the 21st century *Br. J. Radiol.* **72** (Suppl.) P99

Haas O C L, Mills J A, Burnham K J, Bonnett D E and Farajollahi A R 1997a Achieving conformal dose distribution via patient specific compensators *Proc. 12th Int. Conf.*

on the Use of Computers in Radiation Therapy (Salt Lake City, Utah, May 1997) ed D D Leavitt and G Starkschall (Madison, WI: Medical Physics Publishing) p 483

Haas O C L, Mills J A, Burnham K J, Bonnett D E, Farajollahi A R, Fisher M H, Glendinning A G and Aukett R J 1997b Experimental verification of beam intensity modulated conformal radiotherapy using patient specific compensators *Br. J. Radiol.* **70** (Suppl.) 16

Hädinger U, Föller M, Neumann M, Seeber S, Richter J and Schlegel W 1998 Control system for quasi-dynamic stereotactic beam shaping with a micromultileaf collimator *Radiother. Oncol.* **48** (Suppl. 1) S101

Hädinger U, Kubesch R, Neumann M, Richter J and Schlegel W 1997 Development of a micro controller based control system for a dynamic multileaf collimator *Proc. 12th Int. Conf. on the Use of Computers in Radiation Therapy (Salt Lake City, Utah, May 1997)* ed D D Leavitt and G Starkschall (Madison, WI: Medical Physics Publishing) pp 354–6

Hall A and Jones R 1999 MRI appearances of posterior fossa tumours in children *RAD Magazine* **25** No 292, 37–8

Hamacher H W and Lenzen F 2000 A mixed integer programming approach to the multileaf collimator problem *Proc. 13th Int. Conf. on the Use of Computers in Radiation Therapy (Heidelberg, May 2000)* ed W Schlegel and T Bortfeld (Heidelberg: Springer) pp 210–2

Hamilton R J, Sweeney P J, Pelizzari C A, Yetkin F Z, Holma B L, Garada B, Weichelsbeaum R R and Chen G T Y 1997 Functional imaging in treatment planning of brain lesions *Int. J. Radiat. Oncol. Biol. Phys.* **37** 181–8

Hansen V N and Evans P M 1998b Applications of transit dosimetry *Radiother. Oncol.* **48** (Suppl. 1) S77

Hansen V N, Evans P M, Budgell G J, Mott J H L, Williams P C, Brugmans M J P, Wittkamper F W, Mijnheer B J and Brown K 1998a Quality assurance of the dose delivered by small radiation segments *Phys. Med. Biol.* **43** 2665–75

Hansen V N, Evans P M, Shentall G S, Helyer G S, Yarnold J R and Swindell W 1997 Dosimetric evaluation of compensation in radiotherapy of the breast: MLC intensity modulation and physical compensators *Radiother. Oncol.* **42** 249–56

Hansen V N, Evans P M and Swindell W 1996 The application of transit dosimetry to precision radiotherapy *Med. Phys.* **23** 713–21

Happersett L, Hunt M, Chui C, Burman C, Ling C, Zelefsky M, Leibel S and Amols H 1999 Dose painting for prostate cancer using IMRT techniques *Med. Phys.* **26** 1078

Haraldsson P, Lööf M, Magnusson P, Bäck S A J, Nystöm H, Ohihues L and Olsson L E 1998 Gel dosimetry using PoMRI: a quality assurance method for stereotactic radiosurgery *Radiother. Oncol.* **48** (Suppl. 1) S55

Harari P M, Sharda N N, Brock L K and Paliwal B R 1998 Improving dose homogeneity in routine head and neck radiotherapy with 3D custom compensation *Radiother. Oncol.* **49** 67–71

Häring P, Pirzkall A, Carol M, Runz A, Hoever K and Debus J 1999a Patient-specific phantoms for verification of intensity modulated radiotherapy *Int. J. Radiat. Oncol. Biol. Phys.* **45** (Suppl. 1) 407–8

Häring P, Rhein B and Grosser K-H 1999b A method to measure the 'real' patient plan and the results obtained from a 5-patient study *Course Compendium of the 1st IMRT Winter School (Heidelberg, 9–11 December 1999)* pp 87–92

Harris R L 1997 What to do with living, breathing, moving patients? The importance of patient positioning in radiotherapy *Br. J. Radiol.* **70** (Suppl.) 40

Harnisch G, Lalonde R, Weinhous M, Willoughby T, Mohan D and Kupelian P 1998 Clinical usability of the BAT transabdominal ultrasound system for localization of the prostate: preliminary results *Med. Phys.* **25** A204

Hartley-Davies R 1999 Removal of distortion for MRI treatment planning *Radiother. Oncol.* **51** (Suppl. 1) S6

Hector C, Evans P M and Webb S 2000b The dosimetric consequences of patient movement on three classes of intensity-modulated delivery techniques in breast radiotherapy *Proc. 13th Int. Conf. on the Use of Computers in Radiation Therapy (Heidelberg, May 2000)* ed W Schlegel and T Bortfeld (Heidelberg: Springer) pp 289–91

Hector C L, Evans P M and Webb S 2000c The dosimetric consequences of patient movement on three classes of intensity-modulated delivery techniques in breast radiotherapy *Radiother. Oncol.* at press

Hector C L, Webb S and Evans P M 2000a The dosimetric consequences of inter-fractional patient movement on conventional and intensity-modulated breast radiotherapy treatments *Radiother. Oncol.* **54** 57–64

Heijmen B J M and Dirkx M L P 1998 Beam intensity modulation with dynamic multileaf collimation *Radiother. Oncol.* **48** (Suppl. 1) S63

Heijmen B J M, Dirkx M L P and van Santvoort J P C 1997 Calculation of leaf trajectories for dynamic multileaf collimation to realise optimized intensity modulated (IM) beam profiles *Proc. World Congress on Medical Physics and Biomedical Engineering and 11th Int. Conf. on Medical Physics (Nice, France, September 1997) Med. Biol. Eng. Comput.* **35** (Suppl. Part 2) 918

Hendee W R 1999a Point and counterpoint: Intensity-modulated conformal radiation therapy and 3-dimensional treatment planning will significantly reduce the need for therapeutic approaches with particles such as protons *Med. Phys.* **26** 1185–7

Hendee W R 1999b Point and counterpoint: The rate of evolution of radiation therapy planning and delivery systems is exceeding the evolution rate of quality assurance processes *Med. Phys.* **26** 1439–41

Hepworth S J, Doran S J, Morton E J, McJury M and Oldham M 1998 Dose mapping of inhomogeneities positioned in radiosensitive polymer gels Private communication (unpublished)

Hess C F 1999 Spatial uncertainties in clinical radiotherapy: sources of error and their relative importance *Radiother. Oncol.* **51** (Suppl. 1) S1

Hesse B M, Groh B A and Spies L 1998a An on-line treatment verification system for dynamic IMRT techniques *Int. J. Radiat. Oncol. Biol. Phys.* **42** (Suppl.) 158

Hesse B M, Spies L and Groh B A 1998b Tomotherapeutic portal imaging for radiation treatment verification *Phys. Med. Biol* **43** 3607–16

Hesse B M, Spies L, Groh B, Häring P and Höver K H 1997a Treatment verification with megavoltage electronic portal imaging applied to the tomotherapy concept *Proc. 39th ASTRO Meeting, Int. J. Radiat. Oncol. Biol. Phys.* **39** (Suppl. 2) 151

Hesse B M, Spies L, Groh B, Häring P and Höver K H 1997b First results of post-treatment dose verification with mega-voltage electronic portal imaging applied to the tomotherapy concept *Proc. 12th Int. Conf. on the Use of Computers in Radiation Therapy (Salt Lake City, Utah, May 1997)* ed D D Leavitt and G Starkschall (Madison, WI: Medical Physics Publishing) pp 291–2

Hill R W, Curran B H, Strait J P and Carol M P 1997 Delivery of intensity modulated radiation therapy using computer-controlled multileaf collimators with the CORVUS inverse treatment planning system *Proc. 12th Int. Conf. on the Use of Computers in Radiation Therapy (Salt Lake City, Utah, May 1997)* ed D D Leavitt and G Starkschall (Madison, WI: Medical Physics Publishing) pp 394–7

Hoban P, Smee R and Schneider M 2000 IMRT planning at POWH: a concomitant boost treatment *Int J. Rad. Oncol. Biol. Phys.* **46** 751–2

Holmes T W, Bleier A, Carol M, Curran B, DeNisi J, Hill R, Kania A, Lalonde R, Larson L and Sternick E 1998 The CORVUS dose model revealed *Med. Phys.* **25** A144

Holmes T W, Bleier A R, Carol M P, Curran B H, Kania A A, Lalonde R J, Larson L S and Sternick E S 1997 The effect of MLC leakage on the calculation and delivery of intensity modulated radiation therapy *Proc. 12th Int. Conf. on the Use of Computers in Radiation Therapy (Salt Lake City, Utah, May 1997)* ed D D Leavitt and G Starkschall (Madison, WI: Medical Physics Publishing) pp 346–9

Holmes T W and Yu C X 1999 Direct optimisation of IMRT leaf sequences *Int. J. Radiat. Oncol. Biol. Phys.* **45** (Suppl. 1) 247–8

Hong L, Hunt M, Chui C, Spirou S, Forster K, Lee H, Yahalom J, Kutcher G J and McCormick B 1999 Intensity-modulated beam irradiation of the intact breast *Int. J. Radiat. Oncol. Biol. Phys.* **44** 1155–64

Hossain M, Huq M S and Galvin J 1999 A dosimetric issue for intensity modulated radiotherapy *Med. Phys.* **26** 1139

Hosseini-Ashrafi M E, Bagherebadian H and Yahaqi E 1999 Pre-optimisation of radiotherapy treatment planning: an artificial neural network classification aided technique *Phys. Med. Biol.* **44** 1513–28

Hosten N, Wüst P, Beier J, Lemke A-J and Felix R 1998 MRI-assisted specification/ localization of target volumes: aspects of quality control *Strahlentherapie und Oncologie* **174** (Suppl. 2) 13–8

Hounsell A R 1998a Customised compensators and class solutions using IMRT *Proc. 2nd S Takahashi Memorial Int. Workshop on Three Dimensional Conformal Radiotherapy (Nagoya, Aichi, Japan, 11–13 December 1998)* pp 38–43

Hounsell A R 1998b Monitor chamber backscatter for intensity modulated radiation therapy using multileaf collimators *Phys. Med. Biol.* **43** 445–54

Hounsell A R 1999 Dose modelling for irregularly shaped and intensity modulated radiotherapy fields *PhD thesis* University of Manchester

Hounsell A R and Jordan T J 1997 Quality control aspects of the Philips multileaf collimator *Radiother. Oncol.* **45** 225–33

Hounsell A R, Mott J H L, Budgell G J and Wilkinson J M 1999 Dose calculations in intensity modulated radiation therapy fields delivered by dynamic multileaf collimation *Radiother. Oncol.* **51** (Suppl. 1) S10

Hounsell A R and Wilkinson J M 1997b Head scatter modelling for irregular field shaping and beam intensity modulation *Phys. Med. Biol.* **42** 1737–49

Hounsell A R and Wilkinson J M 1997a Head scatter modelling for conformal radiotherapy and intensity modulated beams *Proc. 12th Int. Conf. on the Use of Computers in Radiation Therapy (Salt Lake City, Utah, May 1997)* ed D D Leavitt and G Starkschall (Madison, WI: Medical Physics Publishing) pp 216–8

Hristov D, Moftah B A, Dyke L, Souhami L, Huntzinger C and Podgorsak E B 1999 Generic inverse treatment planning for few field intensity modulated boost treatments of prostate cancer *Med. Phys.* **26** 1079

Hunt M A, Zelefsky M J, Chui C S, LoSasso T, Wolden S, Rosenzweig K E, Chong L, Amols H and Leibel S A 1999 Intensity modulated radiation therapy for primary nasopharynx cancer *Int. J. Radiat. Oncol. Biol. Phys.* **45** (Suppl. 1) 276

Huntzinger C and Hunt M 2000 A simple method to estimate IMRT monitor units *Proc. 1st Int. Workshop on IMRT in Clinical Practice (IMRT2k) (Brussels, 8–9 June 2000)* ed D Verellen p 32

Ibbott G S, Maryanski M J, Eastman P, Holcomb S D, Zhang Y, Avison R G, Sanders M and Gore J C 1997 Three-dimensional visualisation and measurement of conformal dose distributions using magnetic resonance imaging of BANG polymer gel dosimeters *Int. J. Rad. Oncol. Biol. Phys.* **38** 1097–103

Iori M, Paiusco M, Iotti C, Armaroli L and Borasi G 2000 Intensity modulated beam set-up in breast tumour treatment *Proc. 1st Int. Workshop on IMRT in Clinical Practice (IMRT2k) (Brussels, 8–9 June 2000)* ed D Verellen p 24

Jaffray D A, Drake D G, Moreau M, Martinez A A and Wong J W 1999 A radiographic and tomographic imaging system integrated into a medical linear accelerator for localisation of bone and soft-tissue targets *Int. J. Radiat. Oncol. Biol. Phys.* **45** 773–89

Jaffray D A, Siewerdsen J H, Groh B, Drake D G, Wong J W and Martinez A A 2000 Cone-beam computed tomography on a medical accelerator using a flat-panel imager *Proc. 13th Int. Conf. on the Use of Computers in Radiation Therapy (Heidelberg, May 2000)* ed W Schlegel and T Bortfeld (Heidelberg: Springer) pp 558–60

Jaffray D A and Wong J W 1997 Exploring 'target of the day' strategies for a medical linear accelerator with conebeam-CT scanning capability *Proc. 12th Int. Conf. on the Use of Computers in Radiation Therapy (Salt Lake City, Utah, May 1997)* ed D D Leavitt and G Starkschall (Madison, WI: Medical Physics Publishing) pp 172–5

James H, Atherton S, Budgell G, Kirby M and Williams P 1999 Verification of dynamic multileaf collimation using an electronic portal imaging device *Radiother. Oncol.* **51** (Suppl. 1) S25

James H, Atherton S, Budgell G, Kirby M and Williams P 2000 Verification of dynamic multileaf collimation using an electronic portal imaging device *Phys. Med. Biol.* **45** 495–509

Jaywant S M, O'Sullivan B O and Laperriere 2000 The Radionics IMRT and the Xplan MMLC *Proc. 1st Int. Workshop on IMRT in Clinical Practice (IMRT2k) (Brussels, 8–9 June 2000)* ed D Verellen p 59

Jensen J M and Hebbinghaus D 2000 Some remarks on IMRT: static versus dynamic versus step and shoot *Proc. 1st Int. Workshop on IMRT in Clinical Practice (IMRT2k) (Brussels, 8–9 June 2000)* ed D Verellen p 59

Jeraj R and Keall P 1998 Monte Carlo based inverse treatment planning *Med. Phys.* **25** A106

Jeraj R and Keall P 2000 Errors in inverse treatment planning based on inaccurate dose calculation *Proc. 13th Int. Conf. on the Use of Computers in Radiation Therapy (Heidelberg, May 2000)* ed W Schlegel and T Bortfeld (Heidelberg: Springer) pp 548–50

Jiang S B, Boyer A L and Ma C-M 1999 A hybrid system for IMRT inverse planning and dose verification *Int. J. Radiat. Oncol. Biol. Phys.* **45** (Suppl. 1) 422

Johnson T, Davison A, Bolton S C and Bonnett D E 1999b Image registration for the planning of radiotherapy of the brain. *RAD Magazine* **25** No 292, 45–6

Johnson L S, Kutchnir F, Reft C and Kung J 1999a Clinical implementation of intensity modulated radiotherapy using a dynamic multi-leaf collimator *Med. Phys.* **26** 1086

Jones L and Hoban P 1999 Should intensity modulated cost functions be biologically or physically based? *Med. Phys.* **26** 1079–80

Jones L and Hoban P 2000 The effect of motion on delivered IMRT dose distributions *Proc. 1st Int. Workshop on IMRT in Clinical Practice (IMRT2k) (Brussels, 8–9 June 2000)* ed D Verellen p 55

Jordan K 1999 Developmental issues for optical CT and gel dosimetry *Proc. 1st Int. Workshop on Radiation Therapy Gel Dosimetry (Lexington, Kentucky, 21–23 July 1999)* pp 91–7

Kachnic L, Manning M, Benedict S, Wu Q, Zwicker R and Mohan R 1998 Optimised dose conformation and normal tissue sparing for localized oesophageal cancer: a comparison of 3-dimensional conformal (3D CFRT) and intensity modulated radiotherapy (IMRT) *Int. J. Radiat. Oncol. Biol. Phys.* **42** 370

Kagawa K, Lee W R, Schultheiss T E, Hunt M A, Shaer A H and Hanks G E 1997 Initial clinical assessment of CT-MRI fusion software in localisation of the prostate for 3D conformal radiation therapy *Int. J. Radiat. Oncol. Biol. Phys.* **38** 319–25

Kania A A, Bleier A R, Carol M P, Curran B H, Holmes T W, Lalonde R J, Larson L S, Rasmussen C E and Sternick E S 1997 Computational beam model for narrow field photon radiation delivery *Proc. 12th Int. Conf. on the Use of Computers in Radiation Therapy (Salt Lake City, Utah, May 1997)* ed D D Leavitt and G Starkschall (Madison, WI: Medical Physics Publishing) pp 96–9

Kapatoes J, Olivera G, Balog J, Schloesser E, Pearson D, Fitchard E, Reckwerdt P and Mackie T 1998 Delivery verification in tomotherapy *Med. Phys.* **25** A110

Kapatoes J M, Olivera G H, Reckwerdt P J, Fitchard E E, Schloesser E A and Mackie T R 1999 Delivery verification in sequential and helical tomotherapy *Phys. Med. Biol.* **44** 1815–41

Kapatoes J M, Olivera G H, Ruchala K J, Reckwerdt P J, Smilowitz J B, Balog J P, Pearson D W and Mackie T R 2000 Database energy fluence verification and the importance of on-board CT imaging in dose reconstruction *Proc. 13th Int. Conf. on the Use of Computers in Radiation Therapy (Heidelberg, May 2000)* ed W Schlegel and T Bortfeld (Heidelberg: Springer) pp 294–6

Kapur A, Boyer A and Ma C-M 1999 Monte-Carlo studies of intensity-modulated beams in radiotherapy *Int. J. Radiat. Oncol. Biol. Phys.* **45** (Suppl. 1) 429

Kaurin D G L, Maryanski M J, Duggan D M, Morton K C and Coffey C W 1999 Use of MRI-based polymer gel dosimetry, pelvic phantom, and virtual simulation to verify set-up and calculated three-dimensional dose distribution for a prostate treatment *Proc. 1st Int. Workshop on Radiation Therapy Gel Dosimetry (Lexington, Kentucky, 21–23 July 1999)* pp 184–6

Kavanagh B D, Wu Q, Arnfield M, Segreti E M, West R J, Benedict S H, Rosenfeld E, Fulcher A S, Manning M and Mohan R 1999 Whole pelvic intensity-modulated radiotherapy (IMRT) for cervix cancer using a 'butterfly' isodose plan: in vivo dosimetric verification of reduced rectal dose and clinical feasibility evaluation *Int. J. Radiat. Oncol. Biol. Phys.* **45** (Suppl. 1) 412

Kåver G, Lind B K, Löf J, Liander A and Brahme A 1999 Stochastic optimization of intensity modulated radiotherapy to account for uncertainties in patient sensitivity *Phys. Med. Biol.* **44** 2955–69

Keall P and Baldock C 1998 Radiological properties and water equivalence of dosimetry gels *Med. Phys.* **25** A194

Keall P and Baldock C 1999 A theoretical study of the radiological properties and water equivalence of Fricke and polymer gels used for radiation dosimetry *Australas. Phys. Eng. Sci. Med.* **22** 85–91

Keall P and Jeraj R 1998 MonteCarlo vs convolution/superposition *Med. Phys.* **25** A143

Keller H, Glass M, Olivera G H and Mackie T R 2000 Monte-Carlo evaluation of a highly efficient photon detector for tomotherapy *Proc. 13th Int. Conf. on the Use of Computers in Radiation Therapy (Heidelberg, May 2000)* ed W Schlegel and T Bortfeld (Heidelberg: Springer) pp 150–2

Keller-Reichenbecher M A, Bortfeld T, Levegrün S, Stein J, Preiser K and Schlegel W 1999b Intensity modulation with the 'step and shoot' technique using a commercial MLC: a planning study *Int. J. Radiat. Oncol. Biol. Phys.* **45** 1315–24

Keller-Reichenbecher M A, Stein J, Schulze C and Bortfeld T 1999a Inverse planning for IMRT: integration of an MLC sequencer into the planning process *Radiother. Oncol.* **51** (Suppl. 1) S9

Keller-Reichenbecher M A, van Kampen M, Bendl R, Sroka-Pèrez G, Debus J and Schlegel W 1997 A knowledge-based system for rapid generation of alternative pre-optimised plans for conformal therapy *Proc. 12th Int. Conf. on the Use of Computers in Radiation Therapy (Salt Lake City, Utah, May 1997)* ed D D Leavitt and G Starkschall (Madison, WI: Medical Physics Publishing) pp 298–301

Kessen A, Grosser K-H and Bortfeld T 2000 Simplification of IMRT intensity maps by means of 1-D and 2-D median-filtering during the iterative calculation *Proc. 13th Int. Conf. on the Use of Computers in Radiation Therapy (Heidelberg, May 2000)* ed W Schlegel and T Bortfeld (Heidelberg: Springer) pp 545–7

Kessler M L, Li K and Meyer C 1999 Clinical implementation and validation of a mutual information-based automated image registration system *Phys. Medica* **15** 210

Khoo V S, Adams E J, Saran F H, Perks J, Bedford J L, Warrington A J and Brada M 1999b A comparison of clinical target volumes (CTV) determined by computed tomography (CT) versus magnetic resonance imaging (MR) for meningiomas of the base of the skull *Radiother. Oncol.* **51** (Suppl. 1) S68

Khoo V S, Bedford J L, Webb and Dearnaley D P 1998 Comparison of 2D and 3D algorithms for adding a margin to the gross tumour volume in the conformal radiotherapy planning of prostate cancer *Int. J. Radiat. Oncol. Biol. Phys.* **42** 673–9

Khoo V S, Bedford J L, Webb S and Dearnaley D P 2000 An evaluation of three-field coplanar plans for the conformal radiotherapy of prostate cancer *Radiother. Oncol.* **55** 31–40

Khoo V S, Dearnaley D P, Finnigan D J, Padhani A, Tanner S F and Leach M O 1997 Magnetic resonance imaging (MRI) considerations and applications in radiotherapy treatment planning *Radiother. Oncol.* **42** 1–15

Khoo V S, Oldham M, Adams E J, Bedford J L, Webb S and Brada M 1999a Comparison of an intensity modulated tomotherapy with stereotactically guided (conformal) radiotherapy for brain tumours *Int. J. Radiat. Oncol. Biol. and Phys.* **45** 415–25

Khoo V S, Padhani A R, Tanner S F, Finnigan D J, Leach M O and Dearnaley D P 1999c Comparison of MRI with CT for the radiotherapy planning of prostate cancer: a feasibility study *Br. J. Radiol.* **72** 590–7

Killoran J H, Kooy H M, Gladstone D J, Welte F J and Beard C J 1997 A numerical simulation of organ motion and daily set-up uncertainties: implications for radiation therapy *Int. J. Radiat. Oncol. Biol. Phys.* **37** 213–21

Kippenes H, Gavin P R and Rogers D 1998 In vivo measurements of intensity modulated beam delivery *Proc. Elekta Oncology Systems First Users Conf. (Palm Springs, CA, 1998)* P110–5

Klein E 1998 Multileaf collimation: general description, systems and technology assessment *Med. Phys.* **25** A97

Klein E E, Sohn J W, Michalski J M, Grigereit T E, Low D A, Perez C A and Purdy J A 1999b Feasibility of differential dosing of prostate and seminal vesicles using dynamic multileaf collimation *Int. J. Radiat. Oncol. Biol. Phys.* **45** (Suppl. 1) 167

Klein E E, Tepper J, Sontag M, Franklin M, Ling C and Kubo D 1999a Technology assessment of multileaf collimation: a North American users survey *Int. J. Radiat. Oncol. Biol. Phys.* **44** 705–10

Klein E, Tepper J, Sontag M, Franklin M, Ling C, Kubo H and Shiu A 1998 Technology assessment of multileaf collimator: a North American users survey *Med. Phys.* **25** A205

Knisely J P S, Liu L, Maryanski M J, Renade M, Schulz R J and Gore J C 1997 3D dosimetry for complex stereotactic radiosurgery using a tomographic optical density scanner and BANG polymer gels *Proc. 39th ASTRO Meeting, Int. J. Radiat. Oncol. Biol. Phys.* **39** (Suppl. 2) 214

Kogelnik H D 1997 Inauguration of radiotherapy as a new scientific speciality by Leopold Freund 100 years ago *Radiother. Oncol.* **42** 203–11

Kokubo M, Nagata Y, Mizowaki T, Negoro Y, Mitsumori M, Sasai K and Hiraoka M 1998 Evaluation of clinical feasibility for integrated dynamic arc conformal therapy *Int. J. Radiat. Oncol. Biol. Phys.* **42** 366

Kolmonen P, Tervo J, Jaatinen K and Lahtinen T 2000 Direct computation of the ‘step and shoot’ IMRT plan *Proc. 13th Int. Conf. on the Use of Computers in Radiation Therapy (Heidelberg, May 2000)* ed W Schlegel and T Bortfeld (Heidelberg: Springer) pp 35–6

Kolmonen P, Tervo J and Lahtinen T 1999 MLC optimization algorithm for intensity modulated radiation treatment planning *Radiother. Oncol.* **51** (Suppl. 1) S56

Korevaar E W, Heijmen B J M, Woudstra E, Huizenga H and Brahme A 1999 Mixing intensity modulated electron and photon beams: combining a steep dose fall-off at depth with sharp and depth-independent and flat beam profiles *Phys. Med. Biol.* **44** 2171–81

Kramer B A, Wafer D E, Engler M J, Tsai J and Ling M N 1998 Dosimetric comparison of stereotactic radiosurgery to intensity modulated radiotherapy *Radiat. Oncol. Investigations* **6** 18–25

Kroonwijk M, Pasma K L, Quint S, Visser A G and Heijmen B J M 1997 Development and clinical implementation of in-vivo dosimetry using an electronic portal imaging device *Proc. 12th Int. Conf. on the Use of Computers in Radiation Therapy (Salt Lake City, Utah, May 1997)* ed D D Leavitt and G Starkschall (Madison, WI: Medical Physics Publishing) pp 276–8

Kubo H 1998 Review of respiration gated radiotherapy *Med. Phys.* **25** A140

Kubo H and Hill B C 1996 Respiration gated radiotherapy treatment: a technical study *Phys. Med. Biol.* **41** 83–91

Kubo H D, Len P M, Minohara S and Mostafavi H 2000 Breathing-synchronized radiotherapy program at the University of California Davis Cancer Center *Med. Phys.* **27** 346–53

Kubo H D, Shapiro E G and Seppi E J 1999 Potential and role of a prototype amorphous silicon array electronic portal imaging device in breathing synchronized radiotherapy *Med. Phys.* **26** 2410–4

Kung J and Chen G 1998 An analysis of dose perturbations from patient misalignment in delivery of intensity modulated radiotherapy *Med. Phys.* **25** A204

Kung J H and Chen G T Y 2000 Monitor unit verification calculation in IMRT as a patient specific dosimetry QA *Proc. 13th Int. Conf. on the Use of Computers in Radiation Therapy (Heidelberg, May 2000)* ed W Schlegel and T Bortfeld (Heidelberg: Springer) pp 292–3

Kuppersmith R B, Greco S C, Teh B S, Donovan D T, Grant W, Chiu J K C, Cain R B and Butler E B 1999 Intensity-modulated radiotherapy: First results with this new technology on neoplasms of the head and neck *ENT—Ear, Nose and Throat Journal* **78** 238–51

Kuppersmith R B, Teh B S, Donovan D T, Mai W Y, Chiu J K, Grant W H, Woo S Y and Butler E B 2000 The use of intensity-modulated radiation therapy (IMRT) in the treatment of recurrent and extensive unresectable juvenile nasopharyngeal angiofibroma *Proc. 1st Int. Workshop on IMRT in Clinical Practice (IMRT2k) (Brussels, 8–9 June 2000)* ed D Verellen p 29

Kurth C, Bohsung J, Groll J, Pfaender M, Jahn U, Grebe G and Budach V 2000 Dosimetric verification of the Varian IMRT System *Proc. 1st Int. Workshop on IMRT in Clinical Practice (IMRT2k) (Brussels, 8–9 June 2000)* ed D Verellen p 42

Küster G and Bortfeld T 2000 Applicability of a multi-hole collimator for scanned photon beams: a Monte-Carlo study *Proc. 13th Int. Conf. on the Use of Computers in Radiation Therapy (Heidelberg, May 2000)* ed W Schlegel and T Bortfeld (Heidelberg: Springer) pp 179–81

Küster G, Bortfeld T and Schlegel W 1997 Monte Carlo simulations of radiation beams from radiotherapy units and beam limiting devices using the program GEANT *Proc. 12th Int. Conf. on the Use of Computers in Radiation Therapy (Salt Lake City, Utah, May 1997)* ed D D Leavitt and G Starkschall (Madison, WI: Medical Physics Publishing) pp 150–2

Kuterdem H, Cho P and Marks P 1999 Dynamic multileaf-diaphragm sequencing with adjacency gap constraints *Med. Phys.* **26** 1136

Kuterdem H G, Cho P S, Marks II R J, Phillips M H and Parsaei H 2000 Comparison of leaf sequencing techniques: dynamic vs multiple static segments *Proc. 13th Int. Conf. on the Use of Computers in Radiation Therapy (Heidelberg, May 2000)* ed W Schlegel and T Bortfeld (Heidelberg: Springer) pp 213–5

Lagendijk J J W and P Hofman 1999 The microboost concept for the improvement of local control of solid tumours *Radiother. Oncol.* **51** (Suppl. 1) S2

Lagerwaard F, Levendag P, de Pan C, van Nimwegen A and Nowak P 2000 Dosimetric comparison for boosting cancer of the nasopharynx using different techniques *Proc. 1st Int. Workshop on IMRT in Clinical Practice (IMRT2k) (Brussels, 8–9 June 2000)* ed D Verellen p 21

Lalonde R, McKee S, Campbell C, Riker R and Sternick E 1999 Quality control tests of a commercial ultrasound localisation system for image-guided radiation therapy *Med. Phys.* **26** 1119

Landberg T, Wambersie A, Akanuma A, Brahme A, Chavaudra J, Dobbs J, Gerard J P, Hanks G, Horiot J C, Johansson K A, Naudy S, Möller T, Purdy J, Suntharalingham N and Svensson H 1999 Last ICRU approaches to margin definition in external beam treatment planning *Radiother. Oncol.* **51** (Suppl. 1) S1

Langer M 1998 Application of colouring theory to reduce IMRT dose calculations *Med. Phys.* **25** A189

Langer M, Morrill S S and Lane R 1998 A test of the claim that plan rankings are determined by relative complication and tumour-control probabilities *Int. J. Radiat. Oncol. Biol. Phys.* **41** 451–7

Lappe C, Helfert S, Braun M, Pastyr O, Sturm V and Schlegel W 1997 A videogrammetric system for computer-controlled non-invasive patient positioning *Proc. 12th Int. Conf. on the Use of Computers in Radiation Therapy (Salt Lake City, Utah, May 1997)* ed D D Leavitt and G Starkschall (Madison, WI: Medical Physics Publishing) pp 168–71

Lattanzi J P, McNeeley S W, Hanlon A L, Schultheiss T E and Hanks G E 2000 Ultrasound-based stereotactic guidance of precision conformal external beam radiation therapy in clinically localized prostate cancer *Adult Urology* **55** 73–8

Lattanzi J P, McNeeley S W, Hanlon A L, Schultheiss T E, Pinover W H, Horwitz E M and Hanks G E 1999a Ultrasound-based stereotactic guidance of precision conformal external beam radiation therapy in clinically localised prostate cancer *Int. J. Radiat. Oncol. Biol. Phys.* **45** (Suppl. 1) 350–1

Lattanzi J, McNeeley S, Pinover W, Horwitz E, Das I, Schultheiss T E and Hanks G E 1999b A comparison of daily CT localization to a daily ultrasound-based system in prostate cancer *Int. J. Radiat. Oncol. Biol. Phys.* **43** 719–25

Laub W, Alber M, Birkner M and Nüsslin F 2000b IMRT with Monte Carlo dose computation: what is the benefit? *Proc. 13th Int. Conf. on the Use of Computers in Radiation Therapy (Heidelberg, May 2000)* ed W Schlegel and T Bortfeld (Heidelberg: Springer) pp 423–4

Laub W U and Nüsslin F 1999 The Monte-Carlo Code EGS4 for optimisation in inverse treatment planning *Radiother. Oncol.* **51** (Suppl. 1) S33

Laub W U, Yan D, Sharpe M, Nuyttens J, Robertson J and Wong J 2000a A comparison of IMRT planning systems in the treatment of colon-rectal cancer *Proc. 13th Int. Conf. on the Use of Computers in Radiation Therapy (Heidelberg, May 2000)* ed W Schlegel and T Bortfeld (Heidelberg: Springer) pp 529–31

Le Q T, Li J G, Pugachev A, Boyer A L and Xing L 2000 Computer-assisted selection of importance factors in inverse planning *Proc. 13th Int. Conf. on the Use of Computers in Radiation Therapy (Heidelberg, May 2000)* ed W Schlegel and T Bortfeld (Heidelberg: Springer) pp 29–31

Leadbetter S, Kane B and Logue J P 1997 Multileaf collimators in conformal radiotherapy *Rad Magazine* **23** No 267, 31–2

Lebesque J, Ten Haken R K and Wong J W 1998 Organ motion: impact on conformal therapy and methods of compensation *Int. J. Radiat. Oncol. Biol. Phys.* **42** (Suppl.) 121

Lee A G, Woo S Y, Miller N R, Safran A B, Grant W H and Butler E B 1996 Improvement in visual function in an eye with a presumed optic nerve sheath meningioma and treatment with three-dimensional conformal radiation therapy *J. Neuro Opthalmology* **16** 247–51

Lee C H M, Tait D, Nahum A E and Webb S 1999 Comparison of proton therapy and conformal X-ray therapy in non-small-cell lung cancer (NSCLC) *Br. J. Radiol.* **72** 1078–84

Lee H J, Forster K M, Sheldon J M, Wood R R, Spirou S V, Burman C S, Chui C-S, Fuks Z Y, Ling C C, Kutcher G J and Leibel S A 1997a Mixed modality intensity-modulated radiation therapy treatment planning for intracranial lesions *Proc. 39th ASTRO Meeting, Int. J. Radiat. Oncol. Biol. Phys.* **39** (Suppl. 2) 149

Lee M C, Jiang S B, Yi B-Y and Ma C-M 2000 Monte-Carlo simulations of multileaf collimated electrons *Proc. 13th Int. Conf. on the Use of Computers in Radiation Therapy (Heidelberg, May 2000)* ed W Schlegel and T Bortfeld (Heidelberg: Springer) pp 176–8

Lee S, Cho P S, Marks R J II and Oh S 1997c Conformal radiotherapy computation by the method of alternating projections onto convex sets *Phys. Med. Biol.* **42** 1065–86

Lennernäs B, Isaksson U and Nilsson S 1995 The use of artificial intelligence neural networks in the evaluation of treatment plans for external beam radiotherapy *Oncol. Rep.* **2** 863–9

Leszczynski K and Boyko S 1997 On the controversies surrounding the origins of radiation therapy *Radiother. Oncol.* **42** 213–7

Levegrün S, Hartwig K, Oelfke U, Helbig A, Schulze C, Rhein B, Debus J, Schlegel W and Bortfeld T 1998 Clinical implementation of intensity modulated treatments using compensators: dosimetric verification *Med. Phys.* **25** A150

Levegrün S and Schlegel W 1998 Improvement of target volume definition using fuzzy logic *Radiother. Oncol.* **48** (Suppl. 1) S 5

Levegrün S, van Kampen M, Washek T, Engenhart-Cabillic R and Schlegel W 1997 A knowledge-based system with fuzzy logic for the determination of the optimised extent of target volumes in conformal therapy *Proc 12th Int. Conf. on the Use of Computers in Radiation Therapy (Salt Lake City, Utah, May 1997)* ed D D Leavitt and G Starkschall (Madison, WI: Medical Physics Publishing) pp 248–51

Levendag P C, van Dieren J R, van Sörnsen de Koste J R, van der Est H, Wijers O B, Heijman B J M and Nowak J C M 1998 Beam intensity modulation using dynamic multileaf collimation in 3D CRT of primary cancers of the oropharynx and larynx, including the elective neck *Int. J. Radiat. Oncol. Biol. Phys.* **42** 222

Levendag P, Wijers O, van Sörnsen de Koste J, van der Est H and Nowak P 2000 Intensity-modulated radiation therapy: preliminary multi-institutional experience in head and neck cancer *Int J. Rad. Oncol. Biol. Phys.* **46** 709–10

Levitt S H 1999 Cost benefit for 3D conformal radiotherapy *Br. J. Radiol.* **72** (Suppl.) 39

Leybovich L B and Zakharchenko G S 1993 Device for creating prescribed dose distributions using rotational beams *USSR Patent Specification* 604210

Leybovich L B and Zakharchenko G S 1997 Another method of beam intensity modulation *Proc. World Congress on Medical Physics and Biomedical Engineering and 11th Int. Conf. on Medical Physics (Nice, France, September 1997) Med. Biol. Eng. Comput.* **35** (Suppl. Part 2) 1016

Li J G, Boyer A L and Xing L 2000 Clinical applications of combining IMRT with electron beams *Proc. 13th Int. Conf. on the Use of Computers in Radiation Therapy (Heidelberg, May 2000)* ed W Schlegel and T Bortfeld (Heidelberg: Springer) pp 526–8

Li J G, Williams S S and Boyer A L 1999a Breast-conserving radiation therapy using combined electron and IMRT technique *Int. J. Radiat. Oncol. Biol. Phys.* **45** (Suppl. 1) 312

Li J G, Xing L, Boyer A L, Hamilton R J, Spelbring D R and Turian J V 1999b Matching photon and electron fields with dynamic intensity modulation *Med. Phys.* **25** 2379–84

Libby B, Keall P, Siebers J, Zwicker R and Mohan R 1999 A Montecarlo study of dynamic IMRT dosimetry with electronic portal imaging detectors *Med. Phys.* **26** 1162

Lin Z-X, Xing L, Ma L, Le Q, Wolden S, Tate D, Donaldson S S and Boyer A L 1998 Influence of patient displacement and collimator and gantry angle misalignment on intensity modulated radiation therapy *Int. J. Radiat. Oncol. Biol. Phys.* **42** 368

Ling C C 2000 Intensity modulated radiation therapy—potential and future directions *Int J. Rad. Oncol. Biol. Phys.* **46** 698–9

Ling C C, Burman C, Chui C S, Kutcher G J, Leibel S A, LoSasso T, Mohan R, Bortfeld T, Reinstein L, Spirou S, Wang X H, Wu Q, Zelefsky M and Fuks Z 1996 Conformal radiation treatment of prostate cancer using inversely-planned intensity-modulated photon beams produced with dynamic multileaf collimation *Int. J. Rad. Oncol. Biol. Phys.* **35** 721–30

Ling C, Burman C, Chui C, Kutcher G, LoSasso T, Mageras G, Mohan R, Zelefsky M, Leibel S and Fuks Z 1998 Dose-sculpting with intensity modulated radiotherapy: the art and science of multidimensional conformal radiation treatment *Radiother. Oncol.* **48** (Suppl. 1) S79

Ling C C, Burman C, Chui C S, LoSasso T, Mohan R, Spirou S, Stein J, Wu Q, Yang J, Kutcher G J, Leibel S A, Zelefsky M and Fuks Z 1997 Implementation of photon IMRT with dynamic MLC for the treatment of prostate cancer *The Theory and Practice of Intensity Modulated Radiation Therapy* ed E S Sternick (Madison, WI: Advanced Medical Publishing) pp 219–28

Linthout N, Verellen D and Storme G 2000 The CORVUS IMRT planning system applied to the Elekta multileaf collimator: verification for the treatment of the head and neck region *Proc. 1st Int. Workshop on IMRT in Clinical Practice (IMRT2k) (Brussels, 8–9 June 2000)* ed D Verellen p 42

Linthout N and Verellen D 1999 Evaluation of the matchline region in tomotherapy *Med. Phys.* **26** 1140

Liu Y, Yin F-F and Gao Q 1999 Variation method for inverse treatment planning *Med. Phys.* **26** 356–63

Llacer J 1998 Bayesian smoothing for iterative inverse radiation treatment planning *Med. Phys.* **25** A117

Llacer J 1999 Global and local maxima in the dynamically penalised likelihood method of inverse therapy planning *Med. Phys.* **26** 1078

Llacer J, Solberg T D, Promberger C and Kuzmany A 2000 The use of the maximum likelihood estimator and the dynamically penalized likelihood methods in inverse radiation therapy planning *Proc. 13th Int. Conf. on the Use of Computers in Radiation Therapy (Heidelberg, May 2000)* ed W Schlegel and T Bortfeld (Heidelberg: Springer) pp 23–5

Lo Y, Yasuda G, Fitzgerald T, DeGaspe B and Urie M 1998 Intensity modulation for breast treatment using static multileaf collimators *Med. Phys.* **25** A209

Lo Y, Yasuda G, Fitzgerald T and Urie M 2000 Intensity modulation for breast treatment using static multileaf collimators *Int. J., Rad. Oncol. Biol. Phys.* **46** 187–94

Löf J, Liander A, Lind B K and Brahme A 1999 ORBIT: Optimization of radiation therapy beams by iterative techniques, a new universal object oriented optimization code *Phys. Medica* **15** 225

Löf J, Lind B K and Brahme A 1998 An adaptive control algorithm for optimization of intensity modulated radiotherapy considering uncertainties in beam profiles, patient set-up and internal organ motion *Phys. Med. Biol.* **43** 1605–28

Löf J, Lind B K and Brahme A 2000 ORBIT: Optimization of radiation therapy beams by iterative techniques, a new optimization code *Proc. of the 13th Int. Conf. on the Use of Computers in Radiation Therapy (Heidelberg, May 2000)* ed W Schlegel and T Bortfeld (Heidelberg: Springer) pp 49–51

Loi G, Pignoli E, Scorsetti M, Cerreta V, Somigliana A, Marchesini R, Gramaglia A, Cechiari U and Basso Ricci S 1998 Design and characterization of a dynamic multileaf collimator *Phys. Med. Biol.* **43** 3149–55

Lomax A J 2000 Heavy particles: do we need them with IM photons and electrons? *Int J. Rad. Oncol. Biol. Phys.* **46** 708

LoSasso T, Burman C, Chui C and Ling C 1998a QA and verification of intensity modulation radiotherapy *Med. Phys.* **25** A206

LoSasso T, Chui C-S, and Ling C C 1998b Physical and dosimetric aspects of a multileaf collimation system used in the dynamic mode for implementing intensity modulated radiotherapy *Med. Phys.* **25** 1919–27

Love P A, Lewis D G and Smith C W 2000 Analyzing dose distributions from a treatment planning system, Monte Carlo simulations and polymer gel measurements *Proc. 13th Int. Conf. on the Use of Computers in Radiation Therapy (Heidelberg, May 2000)* ed W Schlegel and T Bortfeld (Heidelberg: Springer) pp 386–8

Love P, Lewis D, Smith C, Haraldsson P, Magnusson P, Back S and Olsson L 1999 A three-way comparison of radiotherapy dose distributions using polymer gel dosimetry, montecarlo simulation and treatment planning systems *Med. Phys.* **26** 1085

Low D A 1999 Serial tomotherapy *Med. Phys.* **26** 1060

Low D 2000 Differences between conformal therapy and IMRT *Proc. 1st Int. Workshop on IMRT in Clinical Practice (IMRT2k) (Brussels, 8–9 June 2000)* ed D Verellen pp 36–7

Low D A, Chao K C, Gerber R L, Mutic S, Perez C A and Purdy J A 1997c Quality assurance of arc-based IMRT patient treatments: preliminary results *Proc. 39th ASTRO Meeting, Int. J. Radiat. Oncol. Biol. Phys.* **39** (Suppl. 2) 149

Low D A, Chao K C, Mutic S, Gerber R L, Perez C A and Purdy J A 1998d Quality assurance of serial tomotherapy for head and neck patient treatments *Int. J. Radiat. Oncol. Biol. Phys.* **42** 681–92

Low D A, Dempsey J F, Markman J, Mutic S, Williamson J F and Purdy J A 2000 Applicator-guided intensity modulated radiation therapy: a new approach to treating cervical cancer *Proc. 13th Int. Conf. on the Use of Computers in Radiation Therapy (Heidelberg, May 2000)* ed W Schlegel and T Bortfeld (Heidelberg: Springer) pp 185–7

Low D A, Dempsey J F, Mutic S, Markman J, Goddu S M and Purdy J A 1999d Benchmark-quality dosimetry measurements of intensity modulated radiation therapy dose distributions *Int. J. Radiat. Oncol. Biol. Phys.* **45** (Suppl. 1) 428–9

Low D A, Dempsey J F, Venkatesan R and Mutic S 1999a Evaluation of polymer gels and MRI as a 3-D dosimeter for intensity-modulated radiotherapy *Med. Phys.* **36** 1542–51

Low D A, Gerber R L, Mutic S and Purdy J A 1998c Phantoms for IMRT dose distribution measurement and treatment verification *Int. J. Radiat. Oncol. Biol. Phys.* **40** 1231–5

Low D A and Mutic S 1997 Abutment region dosimetry for sequential arc IMRT delivery *Phys. Med. Biol.* **42** 1465–70

Low D A and Mutic S 1998 A commercial IMRT treatment-planning dose-calculation algorithm *Int. J. Radiat. Oncol. Biol. Phys.* **41** 933–7

Low D A, Mutic S, Dempsey J F, Gerber R L, Bosch W R, Perez C A and Purdy J A 1999c Quantitative dosimetric verification of an IMRT planning and delivery system *Radiother. Oncol.* **49** 305–16

Low D A, Mutic S, Dempsey J F, Markman J, Goddu S M and Purdy J A 1999b Abutment region dosimetry for serial tomotherapy *Int. J. Radiat. Oncol. Biol. Phys.* **45** 193–203

Low D A, Mutic S, Gerber R L, Perez C A and Purdy J A 1998a Quantitative dosmetric verification of an IMRT planning and delivery system *Radiother. Oncol.* **49** 305–16

Low D A, Mutic S, Gerber R L and Purdy J A 1997a The quantitative verification of computer-controlled intensity-modulated radiation therapy dose planning and delivery—initial experience *Proc. 12th Int. Conf. on the Use of Computers in Radiation Therapy (Salt Lake City, Utah, May 1997)* ed D D Leavitt and G Starkschall (Madison, WI: Medical Physics Publishing) pp 410–3

Low D A, Venkatesan R, Mutic S, Goddu M, Haacke E and Purdy J 1998b Evaluation of BANG (TM) gel for intensity modulated dose distribution measurements *Med. Phys.* **25** A151

Low D A, Zhu X R and Purdy J A 1997b The influence of angular misalignment on fixed-portal intensity-modulated radiation therapy *Med. Phys.* **24** 1123–39

Lu W, Mackie T R, Keller H, Ruchala K L and Olivera G H 2000 A generalization of adaptive radiotherapy and the registration of deformable dose distributions *Proc. 13th Int. Conf. on the Use of Computers in Radiation Therapy (Heidelberg, May 2000)* ed W Schlegel and T Bortfeld (Heidelberg: Springer) pp 521–3

Lu Y, Spelbring D R and Chen G T Y 1997 Functional dose-volume histograms for functionally heterogeneous normal organs *Phys. Med. Biol.* **42** 345–56

Ma C, Ma L, Xing L, Kapur A and Boyer A 1998b Monte Carlo analysis of the response of a beam imaging phantom for IMRT *Med. Phys.* **25** A200

Ma C, Mok E, Kapur A, Brain S, Findley D and Boyer A 1998c Clinical implementation of a MonteCarlo treatment planning system *Med. Phys.* **25** A128

Ma C, Mok E, Kapur A, Pawlicki T, Findley D, Brain S and Forster K 1999b Clinical implementation of a Monte-Carlo treatment planning system *Med. Phys.* **26** 2133–43

Ma C-M, Pawlicki T, Lee M C, Jiang S B, Li J S, Deng J, Yi B, Mok E and Boyer A L 2000b Modulated electron beams for treatment of breast cancer *Proc. 13th Int. Conf. on the Use of Computers in Radiation Therapy (Heidelberg, May 2000)* ed W Schlegel and T Bortfeld (Heidelberg: Springer) pp 173–5

Ma C-M, Xing L, Mok E, Kapur A and Boyer A L 1998e Verification of IMRT dose calculations using MonteCarlo simulations *Int. J. Radiat. Oncol. Biol. Phys.* **42** (Suppl.) 207

Ma L 1998f Private communication

Ma L, Beavis A, Xing L, Ma C and Boyer A 1998d Delivery of IMRT with Ellis-type compensating filters *Med. Phys.* **25** A202

Ma L, Boyer A L, Ma C-M and Xing L 1999a Synchronizing dynamic multileaf collimators for producing two-dimensional intensity modulated fields with minimum beam delivery time *Int. J. Radiat. Oncol. Biol. Phys.* **44** 1147–54

Ma L, Boyer A L, Xing L and Ma C-M 1998a An optimised leaf-setting algorithm for beam intensity modulation using dynamic multileaf collimators *Phys. Med. Biol.* **43** 1629–43

Ma L, Geis P B and Boyer A L 1997 Quality assurance for dynamic multileaf collimator modulated fields using a fast beam imaging system *Med. Phys.* **24** 1213–20

Ma L, Haas M, Holmes T, Jacobs M, Suntharalingam M and Yu C 2000a Eliminating complex treatment set-up for intact breast or postmastectomy treatment with intensity modulated radiation therapy *Proc. 13th Int. Conf. on the Use of Computers in Radiation Therapy (Heidelberg, May 2000)* ed W Schlegel and T Bortfeld (Heidelberg: Springer) pp 200–2

Ma L, Xia P, Verhey L J and Boyer A L 1999c A dosimetric comparison of fan-beam intensity-modulated radiotherapy with gamma knife stereotactic radiosurgery for treating intermediate intracranial lesions *Int. J. Radiat. Oncol. Biol. Phys.* **45** 1325–30

MacKay R I, Graham P A, Moore C J, Logue J P and Sharrock P J 1999 Animation and radiobiological analysis of 3D motion in conformal radiotherapy *Radiother. Oncol.* **52** 43–9

Mackie T R 1997a Tomotherapy *Proc. 12th Int. Conf. on the Use of Computers in Radiation Therapy (Salt Lake City, Utah, May 1997)* ed D D Leavitt and G Starkschall (Madison, WI: Medical Physics Publishing) pp 9–11

Mackie T R 1997b Planning optimisation *Proc. Meeting: Intensity-Modulated Radiation Therapy: A Clinical Perspective (London, UK, 26 June 1997)* (York: IPEM) p 10

Mackie T R 1997c Tomotherapy *Proc. Meeting: Intensity-Modulated Radiation Therapy: A Clinical Perspective (London, UK, 26 June 1997)* (York: IPEM) p 28

Mackie T R 1997d What should the ideal planning and delivery system provide? *The Theory and Practice of Intensity Modulated Radiation Therapy* ed E S Sternick (Madison, WI: Advanced Medical Publishing) p 143–4

Mackie T R 1997e Tomotherapy: conformal radiotherapy at lower cost? *Proc. World Congress on Medical Physics and Biomedical Engineering and 11th Int. Conf. on Medical Physics (Nice, France, September 1997) Med. Biol. Eng. Comput.* **35** (Suppl. Part 2) 1015

Mackie T R 1999a Review of advanced radiotherapy delivery systems *Radiother. Oncol.* **51** (Suppl. 1) S21

Mackie T R 1999b Introduction *Semin. Radiat. Oncol.* **9** 1–3

Mackie T R, Aldridge S, Angelos L, Balog J, Coon S, Fang G, Fitchard E, Geiser B, Glass M, Iosevich S, Kapatoes J, McNutt T, Pearson D, Reckwert P, Ruchala K, Shepard D, Thompson M, Wenman D, Yang J and Zachman J 1997 Tomotherapy: rethinking the processes of radiotherapy *Proc. 12th Int. Conf. on the Use of Computers in Radiation Therapy (Salt Lake City, Utah, May 1997)* ed D D Leavitt and G Starkschall (Madison, WI: Medical Physics Publishing) pp 329–31

Mackie T R, Balog J, Ruchala K, Shepard D, Aldridge S, Fitchard E, Reckwerdt P, Oliviera G, McNutt T and Mehta M 1999b Tomotherapy *Semin. Radiat. Oncol.* **9** 108–17

Mackie T R, Olivera G H, Reckwerdt P J and Keller H 2000 Tomotherapy: optimal coplanar radiotherapy *Proc. 13th Int. Conf. on the Use of Computers in Radiation Therapy (Heidelberg, May 2000)* ed W Schlegel and T Bortfeld (Heidelberg: Springer) pp 4–6

Mackie T R, Olivera G H, Shepard D and Reckwerdt P J 1999a IMRT optimisation *Med. Phys.* **26** 1057–8

Maes A, van Esch A, Weltens C, Kutcher G, Huyskens D and van den Bogaert W 2000 Intensity-modulated conformal radiotherapy for parotid gland sparing in head and neck cancer: a dosimetric study using equiangular coplanar fields *Proc. 1st Int. Workshop on IMRT in Clinical Practice (IMRT2k) (Brussels, 8–9 June 2000)* ed D Verellen p 59

Mageras G S 1999 Respiratory motion-induced treatment uncertainties *Phys. Medica* **15** 199

Mageras G S 2000 Interventional strategies for reducing respiratory-induced motion in external beam therapy *Proc. 13th Int. Conf. on the Use of Computers in Radiation Therapy (Heidelberg, May 2000)* ed W Schlegel and T Bortfeld (Heidelberg: Springer) pp 514–6

Markman J, Low D, Halperin R, Michalski J and Purdy J 1999 In-vitro evaluation of a non-invasive ultrasound prostate localization device *Med. Phys.* **26** 1069

Marks L B, Munley M T, Sibley G, Clough R, Hollis D, Bentel G, Zhou S, Poulson J, Jaszczak R J, Coleman R E and Garypadaodlu M 1998 Pre-radiation therapy SPECT lung perfusion scans are useful in predicting changes in pulmonary function tests in patients with central mediastinal tumours *Int. J. Radiat. Oncol. Biol. Phys.* **42** 333

Marks L B, Spencer D P, Sherouse G W, Bentel G, Clough R, Vann K, Jaszczak R, Coleman R E and Prosnitz L R 1995 The role of three dimensional functional lung imaging in radiation treatment planning: the functional dose-volume histogram *Int. J. Radiat. Oncol. Biol. Phys.* **33** 65–75

Marsh L, Eisbruch A, Watson B and Martel M K 1996 Treatment planning for parotid sparing in the patient requiring bilateral neck irradiation *Med. Dosim.* **21** 7–13

Martel M K, Eisbruch A, Lawrence T S, Fraass B A, Ten Haken R K and Lichter A S 1998a Spinal cord dose from standard head and neck irradiation: implications for three-dimensional treatment planning *Radiother. Oncol.* **47** 185–9

Martel M K, Sandler H M, Cornblath W T, Marsh L H, Hazuka M B, Roa W H, Fraass B A and Lichter A S 1998b Dose-volume complication analysis for visual pathway structures of patients with advanced paranasal sinus tumours *Int. J. Radiat. Oncol. Biol. Phys.* **38** 273–84

Martens C and de Wagter C 2000 IMRT of a thyroid carcinoma: unintended dose to the medulla *Proc. 1st Int. Workshop on IMRT in Clinical Practice (IMRT2k) (Brussels, 8–9 June 2000)* ed D Verellen pp 38–9

Maryanski M J 1997 MRI-dosimetry techniques *Proc. World Congress on Medical Physics and Biomedical Engineering and 11th Int. Conf. on Medical Physics (Nice, France, September 1997) Med. Biol. Eng. Comput.* **35** (Suppl. Part 2) 873

Maryanski M J 1999 Sensitivity of BANG polymer gels to light, ultrasound and shock waves: possible new applications in medicine *Med. Phys.* **26** 1127

Maryanski M J, Audet C and Gore J C 1997 Effects of crosslinking and temperature on the dose response of a BANG polymer gel dosimeter *Phys. Med. Biol.* **42** 303–11

Maryanski M and Barry M 1998 New supersensitive polymer gel dosimeter *Med. Phys.* **25** A178

Maryanski M, Renade M, Barry M, Rangarajan A, Nath R, Hafeli U, Amols H and Wu C S 1998 High resolution 3D dosimetry for endovascular brachytherapy using optical laser CT microimaging of BANG polymer gels *Med. Phys.* **25** A107

Maryanski M, Renade M and Rangarajan A 1999 New design of optical CT microscanner for high-resolution 3D dosimetry using BANG gels *Med. Phys.* **26** 1069

Maryanski M J, Zastavker Y Z and Gore J C 1996 Radiation dose distributions in three dimensions from tomographic optical density scanning of polymer gels: B2. Optical properties of the BANG polymer gel *Phys. Med. Biol.* **41** 2705–17

Mavroidis P, Lind B K, Proimos B and Brahme A. 1999 Comparison of conformal radiation therapy techniques within the dynamic radiotherapy project DYNARAD *Phys. Medica* **15** 188

McGary J and Grant W 1998 Prostate localization using the NOMOS PEACOCK system *Med. Phys.* **25** A206

McJury M, Oldham M, Cosgrove V, Murphy P, Doran S, Leach M O and Webb S 2000 Radiotherapy dosimetry using polymer-gels: a review *Br. J. Radiol.* at press

McJury M, Oldham M, Leach M O and Webb S 1999a Dynamics of polymerisation in polyacrylamide gel (PAG) dosimeters: (1) ageing and long term stability *Phys. Med. Biol.* **44** 1863–73

McJury M, Tapper P, Cosgrove V, Murphy P S, Griffin S, Leach M O, Webb S and Oldham M 1999b Experimental 3D dosimetry around a high-dose-rate clinical Ir-192 source using a polyacrylamide-gel (PAG) dosimeter *Phys. Med. Biol.* **44** 2431–44

McNee S G, Rampling R, Dale A J and Gregor A 1998 An audit of 3D treatment planning facilities and practice in the UK *Clin. Oncol.* **10** 18–23

McNeeley S, Lattanzi J, Hanlon A, Schultheiss T and Hanks G 1999 Verification of transabdominal ultrasound (BATR) as an effective localisation modality in prostate cancer *Med. Phys.* **26** 1097

McNutt T R, Mackie T R, Reckwerdt P J and Paliwal B R 1997 Applications and implementation of the iterative convolution/superposition dose reconstruction technique *Proc. 12th Int. Conf. on the Use of Computers in Radiation Therapy (Salt Lake City, Utah, May 1997)* ed D D Leavitt and G Starkschall (Madison, WI: Medical Physics Publishing) pp 111–3

Meeks S L, Bova F J, Kim S, Tomé W A and Friedman W A 1999a Dosimetric characteristics of a double-focused miniature multileaf collimator *Med. Phys.* **25** 729–33

Meeks S L, Bova F J, Maryanski M J, Kendrick L A, Ranade M K, Buatti J M and Friedman W A 1999b Image registration of BANG gel dose maps for quantitative dosimetry verification *Int. J. Radiat. Oncol. Biol. Phys.* **43** 1135–41

Meeks S, Bova F, Maryanski M, Kendrick L, Renade M, Buatti J and Friedman W 1998a Quantitative dosimetry evaluation using BANG gel and image correlation *Med. Phys.* **25** A199

Meeks S L, Buatti J M, Bova F J, Friedman W A, Mendenhall W M and Zlotecki R A 1998b Potential clinical efficacy of intensity-modulated conformal therapy *Int. J. Radiat. Oncol. Biol. Phys.* **40** 483–95

Meyer J, Haas O C L, Mills J A, Pavin E M and Burnham K J 1999a On compensator design for intensity modulated radiation therapy (IMRT) *Br. J. Radiol.* **72** (Suppl.) p 99

Meyer J, Mills J A, Haas O C L, Parvin E M and Burnham K J 2000 Some limitations in the practical delivery of intensity modulated radiation therapy *Br. J. Radiol.* **73** 854–63

Meyer J, Mills J A, Haas, Parvin E M and Burnham K J 1999b Practical aspects for the fabrication of compensators to deliver intensity modulated radiation therapy *Proc. 5th Annu. Conf. of IPEM (Nottingham, September 1999)* p 141

Meyer J L and Purdy J A 1996 *3D Conformal Radiotherapy: a New Era in the Irradiation of Cancer* (Basel: Karger)

Michalski J M, Purdy J A, Harms W and Bosch W 1998 3D conformal radiation therapy for prostate cancer: quality assurance in a cooperative group study *Proc. Conf. 'Conformal Therapy: a Clinical Perspective Meeting' (London, UK, 21 May 1998)* (York: IPEM) p 15

Miften M 2000 The importance of the dose calculation model on plan evaluation for IMRT *Proc. 13th Int. Conf. on the Use of Computers in Radiation Therapy (Heidelberg, May 2000)* ed W Schlegel and T Bortfeld (Heidelberg: Springer) pp 234–6

Mijnheer B 1998 Conformal radiotherapy: art in science *Radiother. Oncol.* **48** (Suppl. 1) S136

Miller B, Sharpe M, Wu Y, Yan D and Wong J 1998 Monitor unit settings for intensity modulated radiation therapy performed with a multileaf collimator *Med. Phys.* **25** A109

Mohan D S, Kupelian P A and Willoughby T R 1999 Short-course intensity modulated radiotherapy for localised prostate cancer with daily transabdominal ultrasound localisation of the prostate gland: technique, acute toxicity, and early observations *Int. J. Radiat. Oncol. Biol. Phys.* **45** (Suppl. 1) 168

Mohan D S, Kupelian P A and Willoughby T R 2000 Short-course intensity-modulated radiotherapy for localized prostate cancer with daily transabdominal ultrasound localization of the prostate gland *Int. J. Radiat. Oncol. Biol. Phys.* **46** 575–80

Mohan R 1997 Intensity-modulated three-dimensional conformal radiotherapy *Proc. 39th ASTRO Meeting, Int. J. Radiat. Oncol. Biol. Phys.* **39** (Suppl. 2) 121

Mohan R 1998a Monte Carlo dose calculations for radiation treatment planning *Med. Phys.* **25** A158

Mohan R 1998b 3D conformal radiation therapy—Part 4: optimised intensity-modulated three-dimensional conformal radiation therapy *Int. J. Radiat. Oncol. Biol. Phys.* **42** (Suppl.) 113

Mohan R 1999 Intensity-modulated radiotherapy—Part 2: radiobiological aspects *Int. J. Radiat. Oncol. Biol. Phys.* **45** (Suppl. 1) 128–9

Mohan R, Cardinale R and Hagen M 1999 Comments on 'Further improvements in dose distributions are unlikely to affect cure rates' [1999 *Med. Phys.* **26** 1007–9] *Med. Phys.* **26** 2701–3

Mohan R and Wang X 1996 Letter to the Editor: Response to Bortfeld *et al* re physical versus biological objectives for treatment plan optimisation *Radiother. Oncol.* **40** 186–7

Mohan R, Wang X, Jackson A, Bortfeld T, Boyer A L, Kutcher G J, Leibel S A, Fuks Z and Ling C C 1994 The potential and limitations of the inverse radiotherapy technique *Radiother. Oncol.* **32** 232–48

Mohan R, Wu Q, Wang X and Stein J 1996 Intensity modulation optimisation, lateral transport of radiation and margins *Med. Phys.* **23** 2011–21

Mok E, Ma C and Boyer A 1999 Development of a comprehensive compensator system *Med. Phys.* **26** 1097

Mongioj V, Brusa A, Loi G, Pignoli E, Gramaglia A, Scorsetti M, Bombardieri E and Marchesini R 1999 Accuracy evaluation of fusion of CT, MR, and SPECT images using commercially available software packages (SRS PLATO and IFS) *Int. J. Radiat. Oncol. Biol. Phys.* **43** 227–34

Morrill S M, Lane R G and Langer M 1997 Coarse intensity-modulated radiation therapy: field-in-field optimisation *Proc. 12th Int. Conf. on the Use of Computers in Radiation Therapy (Salt Lake City, Utah, May 1997)* ed D D Leavitt and G Starkschall (Madison, WI: Medical Physics Publishing) pp 320–2

Mosleh-Shirazi M A, Evans P M, Swindell W, Webb S and Partridge M 1998a A cone-beam megavoltage CT scanner for treatment verification in conformal radiotherapy *Radiother. Oncol.* **48** 319–28

Mosleh-Shirazi M A, Evans P M, Swindell W, Symonds-Tayler R, Webb S and Partridge M 1998b Rapid portal imaging with a high-efficiency, large field-of-view detector *Med. Phys.* **25** 2333–46

Mott J H L 1999 Private communication (Elekta Oncology Systems Consortium, 27–28 May 1999)

Mott J H L, Ahamad A, Budgell G J and Hounsell A R 2000b Intensity modulated radiotherapy (IMRT) for tumours of the maxillary antrum *Proc. 1st Int. Workshop on IMRT in Clinical Practice (IMRT2k) (Brussels, 8–9 June 2000)* ed D Verellen p 25

Mott J H L, Budgell G J, Logue J P and Hounsell A R 2000a Clinical experience of fully dynamic IMRT in the management of bladder cancer *Proc. 1st Int. Workshop on IMRT in Clinical Practice (IMRT2k) (Brussels, 8–9 June 2000)* ed D Verellen p 32

Mott J H L, Budgell G J, Hounsell A R, Sykes J R, Wilkinson J M and Williams P C 1998a An investigative study of customised compensation using dynamic intensity modulated multileaf collimator fields *Br. J. Radiol.* **71** (Suppl.) 9

Mott J, Hounsell A, Budgell G, Wilkinson J and Williams P 1998b Customised compensation by dynamic multileaf collimation *Med. Phys.* **25** 1581

Mott J H L, Hounsell A, Budgell G J, Wilkinson J M and Williams P C 1998c Customised compensation using intensity modulated beams *Radiother. Oncol.* **48** (Suppl. 1) S70

Mott J H L, Hounsell A R, Budgell G J, Wilkinson J W and Williams P C 1999 Customised compensation using intensity modulated beams delivered by dynamic collimation *Radiother. Oncol.* **53** 59–65

Mould R F 1995 Invited review: The early years of radiotherapy with emphasis on X-ray and radium apparatus *Br. J. Radiol.* **68** 567–82

Mubata C D, Bidmead A M, Ellingham L M and Dearnaley D P 1997a Incorporating patient set-up errors in 3D treatment planning *Proc 12th Int. Conf. on the Use of Computers in Radiation Therapy (Salt Lake City, Utah, May 1997)* ed D D Leavitt and G Starkschall (Madison, WI: Medical Physics Publishing) pp 188–90

Mubata C D, Bidmead A M, Ellingham L M, Thompson V and Dearnaley D P 1998 Portal imaging protocol for radical dose-escalated radiotherapy treatment of prostate cancer *Int. J. Radiat. Oncol. Biol. Phys.* **40** 221–31

Mubata C D, Childs P and Bidmead A M 1997b A quality assurance procedure for the Varian multi-leaf collimator *Phys. Med. Biol.* **42** 423–31

Mueller M 1996 Morris Wizenbergh MD, Director of Radiation Oncology, The Cancer Centre at Mercy—Using the Peacock system to plan and deliver true 3D conformal radiation therapy *MD News Central Oklahoma edn* July

Murmann A, Lassen S and Nyström H 1999 Verification of intensity modulated radiation therapy with electronic portal imaging *Radiother. Oncol.* **51** (Suppl. 1) S42

Murphy M J and Cox R S 1996 The accuracy of dose localisation for an image-guided frameless radiosurgery system *Med. Phys.* **23** 2043–9

Murphy M J, Le Q, Martin D, Poen J, Adler J R, Forster K, Hai J, Luxton G, Bodduluri M and Dooley J 2000a Image-guided radiosurgery in the thorax and abdomen *Comput. Aided Surgery* at press

Murphy P S, Cosgrove V, Leach M O and Webb S 2000c A modified BANG-type gel for radiotherapy dosimetry: assessment by MRI and MRS *Phys. Med. Biol.* at press

Murphy P S, Cosgrove V P, Schwarz A, Webb S and Leach M O 2000b Proton spectroscopic imaging of polyacrylamide gel dosimeters for absolute radiation dosimetry *Phys. Med. Biol.* at press

Muthuswamy M 1999 A study of the potential impact of field shape optimization *Radiother. Oncol.* **51** (Suppl. 1) S41

Mutic S and Low D 1998a Superficial doses from serial tomotherapy delivery *Med. Phys.* **25** A110

Mutic S and Low D 2000 Superficial doses from serial tomotherapy delivery *Med. Phys.* **27** 163–5

Mutic S and Low D 1998b Whole-body dose from tomotherapy delivery *Int. J. Radiat. Oncol. Biol. Phys.* **42** 229–32

Mutic S, Low D, Klein E and Purdy J 1999 Room shielding requirements for IMRT *Med. Phys.* **26** 1086

Nahum A 1998 Can dose volume information predict NTCP? *Proc. Conf. 'Conformal Therapy: a Clinical Perspective Meeting' (London, UK, 21 May 1998)* (York: IPEM) pp 12–14

Nakagawa K 1998 Technological and clinical experiences on megavoltage CT-assisted 3D-CRT in extracranial lesions: radiosurgery for the chest metastases *Proc. 2nd S Takahashi Memorial Int. Workshop on Three Dimensional Conformal Radiotherapy (Nagoya, Aichi, Japan, 11–13 December 1998)* pp 56–61

Nash R V, Carol M P, Huber R E and Campbell R C 1997 Incremental dose calculation optimisation of inverse treatment planning for intensity modulated radiation therapy *Proc. 12th Int. Conf. on the Use of Computers in Radiation Therapy (Salt Lake City, Utah, May 1997)* ed D D Leavitt and G Starkschall (Madison, WI: Medical Physics Publishing) pp 431–3

Neal A, Oldham M and Dearnaley D 1995 Comparison of treatment techniques for conformal radiotherapy of the prostate using dose-volume histograms and normal tissue complication probabilities *Radiother. Oncol.* **37** 29–34

Neal A, Oldham M and Dearnaley D 1996 Letter to the Editor *Radiother. Oncol.* **40** 86–7

Nedzi L A, Kooy H M, Alexander E III, Svensson G K and Loeffler J S 1993 Dynamic field shaping for stereotactic radiosurgery: a modelling study *Int. J. Radiat. Oncol. Biol. Phys.* **25** 859–69

Neumann M 1999a The impact of Dicom in radiotherapy. Is Dicom ready for radiotherapy? *Radiother. Oncol.* **51** (Suppl. 1) S1

Neumann M 1999b QA for IMRT: information systems *Course Compendium of the 1st IMRT Winter School (Heidelberg, 9–11 December 1999)* pp 99–112

Newton C M 1970 What next in radiation treatment optimisation? *Computers in Radiotherapy: Proc. 3rd Int. Conf. on Computers in Radiotherapy (Glasgow) Br. J. Radiol.* Special Report 5, pp 83–9

Niemierko A 1997 Reporting and analysing dose distributions: a concept of equivalent uniform dose *Med. Phys.* **24** 103–10

Niemierko A 1999 Reporting and analysing dose for IMRT *Med. Phys.* **26** 1070–1

Nowak P, Levendag P, van Sörnsen de Koste J, van der Est H and Wijers O 2000 Intensity-modulated radiation therapy: preliminary multi-institutional experience in head and neck cancer *Proc. 1st Int. Workshop on IMRT in Clinical Practice (IMRT2k) (Brussels, 8–9 June 2000)* ed D Verellen p 24

Nutting C M 2000 3DCRT vs IMRT: Progress and problems *Proc. 1st Int. Workshop on IMRT in Clinical Practice (IMRT2k) (Brussels, 8–9 June 2000)* ed D Verellen p 20

Nutting C M, Bedford J L, Cosgrove V P, Tait D M, Dearnaley D P and Webb S 2000a Radiotherapy for carcinoma of the oesophagus: a comparison of 3D conformal and intensity-modulated techniques, at press

Nutting C M, Convery D J, Cosgrove V P, Rowbottom C, Padhani A R, Webb S and Dearnaley D P 2000c Reduction of small and large bowel irradiation using an optimised intensity-modulated pelvic radiotherapy technique (IMRT) in patients with prostate cancer *Radiother. Oncol.* at press

Nutting C M, Convery D J, Cosgrove V P, Rowbottom C, Padhani A R, Webb S and Dearnaley D P 2000e An evaluation of intensity modulation in pelvic radiotherapy for prostate cancer *Proc. 1st Int. Workshop on IMRT in Clinical Practice (IMRT2k) (Brussels, 8–9 June 2000)* ed D Verellen p 24

Nutting C M, Convery D J, Cosgrove V P, Rowbottom C, Vini L, Harmer C L, Dearnaley D P and Webb S 2000d An evaluation of the role of intensity-modulated radiotherapy (IMRT) in the treatment of carcinoma of the thyroid *Radiother. Oncol.* at press

Nutting C, Dearnaley D P and Webb S 2000b Intensity modulated radiation therapy: a clinical review *Br. J. Radiol.* **73** 459–69

Nutting C, Rowbottom C, Convery D, Cosgrove V, Dearnaley D P, Harmer C, Henk J M and Webb S 1999 Intensity modulated radiotherapy (IMRT) in tumours of the head and neck *Br. J. Radiol.* **72** (Suppl.) P98

Nuyttens *et al* 2000 Intensity modulated radiation therapy (IMRT) for rectal cancer: the effect of small bowel mobility *Proc. 1st Int. Workshop on IMRT in Clinical Practice (IMRT2k) (Brussels, 8–9 June 2000)* ed D Verellen p 55

Oelfke U and Bortfeld T 1999 Inverse planning for X-ray rotation therapy: a general solution of the inverse problem *Phys. Med. Biol.* **44** 1089–104

Ohara K, Okomura T, Akisada M, Inada T, Mori T, Yukota H and Calaguas M J B 1989 Irradiation synchronized with respiration gate *Int. J. Rad. Oncol. Biol. Phys.* **17** 853–7

Olch A J and Lavey R S 2000 Applications of IMRT in pediatric radiation oncology *Proc. 1st Int. Workshop on IMRT in Clinical Practice (IMRT2k) (Brussels, 8–9 June 2000)* ed D Verellen p 25

Oldham M 1997 The impact of computer optimisation on treatment planning *Proc. World Congress on Medical Physics and Biomedical Engineering and 11th Int. Conf. on Medical Physics (Nice, France, September 1997) Med. Biol. Eng. Comput.* **35** (Suppl. Part 2) 919

Oldham M, Baustert I, Lord C, Smith TAD, McJury M, Warrington J, Leach M O and Webb S 1998a An investigation into the dosimetry of a nine-field tomotherapy irradiation using BANG-gel dosimetry *Phys. Med. Biol.* **43** 1113–32

Oldham M, Baustert I, Smith T, Webb S and Leach M O 1997a BANG gel dosimetry and the verification of intensity modulated radiotherapy plans *Br. J. Radiol.* **70** (Suppl.) 99

Oldham M, Baustert I, Smith T, Webb S and Leach M O 1997b BANG gel dosimetry applied to the verification of static tomotherapy *Proc. 12th Int. Conf. on the Use of Computers in Radiation Therapy (Salt Lake City, Utah, May 1997)* ed D D Leavitt and G Starkschall (Madison, WI: Medical Physics Publishing) p 480

Oldham M, Khoo V, Rowbottom C, Bedford J and Webb S 1998c A case study comparing the relative benefit of optimising beamweights, wedge angles, beam orientation, and tomotherapy in stereotactic radiotherapy of the brain *Phys. Med. Biol.* **43** 2123–46

Oldham M, McJury M, Baustert I B, Webb S and Leach M O 1998b Improving calibration accuracy in gel dosimetry *Phys. Med. Biol.* **43** 2709–20

Oldham M, Neal A and Webb S 1995 A comparison of conventional forward planning with inverse planning for 3D conformal radiotherapy of the prostate *Radiother. Oncol.* **35** 248–62; and **37** 171–2

Oldham M, Neal A and Webb S 1996 The optimisation of wedge filters in radiotherapy of the prostate *Radiother. Oncol.* **37** 209–20

Oldham M, Sapareto S A, Yan D, Jaffray D A, Sharpe M B and Wong J 1999 Optimising clinical outcome in 3D treatment planning *Radiother. Oncol.* **51** (Suppl. 1) S45

Oldham M, Siewerdsen J H and Jaffray D A 2000 An initial investigation of optical CT and MR scanning of gel dosimeters *Proc. 13th Int. Conf. on the Use of Computers in*

Radiation Therapy (Heidelberg, May 2000) ed W Schlegel and T Bortfeld (Heidelberg: Springer) pp 383–5

Oldham M and Webb S 1997a Film verification study of a nine field intensity modulated plan *Br. J. Radiol.* **70** (Suppl.) 16

Oldham M and Webb S 1997b A nine field static tomotherapy and delivery study *Proc. 12th Int. Conf. on the Use of Computers in Radiation Therapy (Salt Lake City, Utah, May 1997)* ed D D Leavitt and G Starkschall (Madison, WI: Medical Physics Publishing) pp 398–401

Oldham M and Webb S 1997c IMRT delivery with the NOMOS MIMiC: Part 2 Verification using film and gel dosimetry *Proc. Meeting: Intensity-Modulated Radiation Therapy: A Clinical Perspective (London, UK 26 June 1997)* (York: IPEM) pp 19–22

Oldham M and Webb S 1997d Intensity-modulated radiotherapy by means of static tomotherapy: a planning and verification study *Med. Phys.* **24** 827–36

Olivera G H, Fitchard E E, Reckwerdt P J, Ruchala K and Mackie T R 2000 Delivery modification as an alternative to patient repositioning in tomotherapy *Proc. 13th Int. Conf. on the Use of Computers in Radiation Therapy (Heidelberg, May 2000)* ed W Schlegel and T Bortfeld (Heidelberg: Springer) pp 297–9

Olivera G H, Shepard D M, Reckwerdt P J, Ruchala K, Zachman J, Fichard E E and Mackie T R 1998 Maximum likelihood as a common computational framework in tomotherapy *Phys. Med. Biol.* **43** 3277–94

Pagnini P G, Wazer D E, Engler M J and Tsai J S 1999 A dosimetric analysis of multiple couch angle optimised dual-dynamic intensity modulated radiotherapy plans for the treatment of intracranial lesions *Int. J. Radiat. Oncol. Biol. Phys.* **45** (Suppl. 1) 246

Palta J 1998 Multileaf collimator dosimetry *Med. Phys.* **25** A139

Papatheodorou S, Diot L, Schultz C. Gaboriaud G, Pontvert D and Rosenwald J-C 2000 Integration of the KONRAD inverse planning tool into the iSis3D treatment planning system *Proc. 13th Int. Conf. on the Use of Computers in Radiation Therapy (Heidelberg, May 2000)* ed W Schlegel and T Bortfeld (Heidelberg: Springer) pp 54–6

Papatheodorou S, Murillo M C, Zefkili, Gaboriaud G and Rosenwald J C 1999 Extension of primary-scatter separation dose model to account for intensity modulation generated by dynamic MLC *Radiother. Oncol.* **51** (Suppl. 1) S9

Papiez L 2000 On the equivalence of rotational and concentric therapy *Phys. Med. Biol.* **45** 399–409

Pappas E, Maris T, Angelopoulos A, Paparigopoulou M, Sakelliou L, Sandilos P, Voyiiatzi S and Vlachos L 1999 A new polymer gel for magnetic resonance imaging (MRI) radiation dosimetry *Phys. Med. Biol.* **44** 2677–84

Park J, Chen C, Liu C, Anghaie S and Palta J R 1999 Use of dynamic multileaf collimator as a dose compensator *Med. Phys.* **26** 1139

Parker W, Tsien C I, Gingras C and Freeman C R 1999 3D treatment planning for pediatric whole CNS irradiation using CT-simulation and intensity modulated radiation therapy *Int. J. Radiat. Oncol. Biol. Phys.* **45** (Suppl. 1) 427

Parsaei H, Phillips M H, Cho P S, Kuterdem G H, Kippenes H and Gavin P 2000 Delivery and dosimetry verification in IMRT *Proc. 13th Int. Conf. on the Use of Computers in Radiation Therapy (Heidelberg, May 2000)* ed W Schlegel and T Bortfeld (Heidelberg: Springer) pp 283–5

Partridge M, Donovan E, Fenton N, Reise S and Blane S 1999 Clinical implementation of a computer controlled milling machine for compensating filter production. *Br. J. Radiol.* **72** 1099–103

Partridge M, Evans P M and Mosleh-Shirazi M A 1998a Linear accelerator output variations and their consequences for megavoltage imaging *Med. Phys.* **25** 1443–61

Partridge M, Evans P M and Mosleh-Shirazi M A 1997a Accelerator output fluctuation: the effect on megavoltage CT and portal image quality *Proc. 12th Int. Conf. on the Use of Computers in Radiation Therapy (Salt Lake City, Utah, May 1997)* ed D D Leavitt and G Starkschall (Madison, WI: Medical Physics Publishing) pp 288–90

Partridge M, Evans P M, Mosleh-Shirazi M A and Convery D 1998b Independent verification using portal imaging of intensity-modulated beam delivery by the dynamic MLC technique *Med. Phys.* **25** 1872–9

Partridge M, Mosleh-Shirazi M A and Evans P 1997b Accelerator output fluctuation: the effect on megavoltage CT and portal image quality *Br. J. Radiol.* **70** (Suppl.) 39

Partridge M, Symonds-Tayler J R N and Evans P M 2000a Verification of dynamic MLC beam delivery using electronic portal imaging *Proc. 13th Int. Conf. on the Use of Computers in Radiation Therapy (Heidelberg, May 2000)* ed W Schlegel and T Bortfeld (Heidelberg: Springer) pp 556–7

Partridge M, Symonds-Tayler J R N and Evans P M 2000b The use of electronic portal imaging for verification of dynamic MLC beam delivery *Proc. 16th Int. Workshop on Electronic Portal Imaging (EPI2k) (Brussels, 5–7 June 2000)* ed Jan van de Steene p 58

Pasma K, Dirkx M, Kroonwijk M, Visser A and Heijmen B 1999a Dosimetric verification of intensity modulated fields produced with dynamic multileaf collimation using an electronic portal imaging device (EPID) *Radiother. Oncol.* **51** (Suppl. 1) S33

Pasma K L, Dirkx M L P, Kroonwijk M, Visser A G and Heijmen B J M 1999e Dosimetric verification of intensity modulated fields produced with dynamic multileaf collimation using an electronic portal imaging device *Med. Phys.* **26** 2373–8

Pasma K, Dirkx M, Kroonwijk M, Visser A and Heijmen B 1999d Dosimetric verification of intensity modulated fields produced with dynamic multileaf collimation using an electronic portal imaging device *Int. J. Radiat. Oncol. Biol. Phys.* **45** (Suppl. 1) 206

Pasma K L, Kroonwijk M, de Boer J C J, Visser A G and Heijmen B J M 1998 Accurate portal dose measurement with a fluoroscopic electronic portal imaging device (EPID) for open and wedged beams and dynamic multileaf collimation *Phys. Med. Biol.* **43** 2047–60

Pasma K L, Kroonwijk M, van Dieren E B and Visser A G 1999c Verification of compensator thicknesses using a fluoroscopic electronic portal imaging device *Med. Phys.* **36** 1524–9

Pasma K, Kroonwijk M, van Dieren E, Visser A and Heijmen B 1999b Simple and accurate verification of compensator thicknesses using an electronic portal imaging device (EPID) *Radiother. Oncol.* **51** (Suppl. 1) S25

Pasma K L, Kroonwijk M, Visser A G and Heijmen B J M 1997 Portal dose measurements with a video-based electronic portal imaging device using a deconvolution algorithm *Proc. 12th Int. Conf. on the Use of Computers in Radiation Therapy (Salt Lake City, Utah, May 1997)* ed D D Leavitt and G Starkschall (Madison, WI: Medical Physics Publishing) pp 282–4

Paul T, Leu M, Rosemark P, Solberg T and Smathers J 1998 Characteristics of an a-Si-H (amorphous silicon) array detector for real-time dosimetric verification in dynamically shaped radiotherapy *Med. Phys.* **25** A151

Paul T, Solberg T, Leu M, Rosemark P, Agazaryan N and Smathers J 2000a Feedback and interlock triggering for real-time IMRT verification *Proc. 13th Int. Conf. on the Use of Computers in Radiation Therapy (Heidelberg, May 2000)* ed W Schlegel and T Bortfeld (Heidelberg: Springer) pp 145–6

Paul T, Solberg T, Leu M, Rosemark P, Mollenhauer U, Agazaryan N and Smathers J 2000b Real-time verification and feedback for clinically deliverable IMRT plans *Proc. 1st Int. Workshop on IMRT in Clinical Practice (IMRT2k) (Brussels, 8–9 June 2000)* ed D Verellen p 43

Paul T, Solberg T, Leu M, Rosemark P, Smathers J, Mollenhauer U and Agazaryan N 1999 Independent, real-time verification of dynamically shaped intensity modulated radiotherapy (IMRT) using an amorphous silicon (a-Si:H) detector array *Med. Phys.* **26** 1065

Pawlicki T, Guerrero T, Jiang S B, Deng J, Li J, Boyer A L and Ma C M 1999a Investigation of the dose perturbations in IMRT treatment plans caused by patient heterogeneities *Int. J. Radiat. Oncol. Biol. Phys.* **45** (Suppl. 1) 164–5

Pawlicki T, Jiang S, Deng J, Li J and Ma C 1999b Monte Carlo calculated beamlets for photon beam inverse planning *Med. Phys.* **26** 1064–5

Peacock J H, Nahum A E and Steel G G 1998 Normal-tissue effects in radiotherapy: physics meets biology. Report on a workshop held at Hartsfield Manor, Betchworth, Surrey, UK, 14–16 July 1997 *Int. J. Radiat. Biol.* **73** 341–4

Pfaender M, Grebe G, Budach V and Würm R 1999 Dosimetry with BANG-dosimeters regarding slim shaped parts of lesions for stereotactic radiation with a linac and a micro-multi-leaf-collimator *Proc. 1st Int. Workshop on Radiation Therapy Gel Dosimetry (Lexington, Kentucky, 21–23 July 1999)* pp 190–2

Ph G, Grahek D, Montravers F, Caselles O, Courbon F, Dupont S, Rosenwald J C, Housset M and Talbot J N 1999 Does CT and [18F]-FDG fusion improve the radiotherapy treatment planning in non-small cell lung cancer? *Phys. Medica* **15** 234

Phillips M H 1997 Private communication (Elekta Oncology Systems Consortium Meetings, Crawley, 8–9 May and 5–6 December 1997)

Phillips M H 1999 Private communication (Elekta Oncology Systems Consortium, 27–28 May 1999)

Phillips M H, Cho P S and Sutleif S 1998 Implementation issues for IMRT: inverse planning to MLC prescription files *Proc. Elekta Oncology Systems First Users Conf. (Palm Springs, CA)* P28–37

Phillips M H, Parsaei H, Cho P S, Kuterdem G K, Kippenes H and Gavin P 2000 Image correlation and in vivo dosimetry in IMRT *Proc. 13th Int. Conf. on the Use of Computers in Radiation Therapy (Heidelberg, May 2000)* ed W Schlegel and T Bortfeld (Heidelberg: Springer) pp 377–9

Phillips M H, Singer K M and Hounsell A R 1999a The macro pencil beam model: clinical implications for conformal and intensity modulated radiation therapy *Phys. Med. Biol.* **44** 1067–88

Phillips M H, Singer K M and Miller E 1999b Adaptation of an optically-guided stereotactic localization system for clinical use in conformal therapy and IMRT *Med. Phys.* **26** 1096–7

Pickett B, Verhey L and Roach III M 1997 Fixed field intensity modulation to treat a dominant intra-prostatic lobe (DIL) to greater than 90 Gy compared to seven field three-dimensional non-coplanar radiotherapy *Proc. 39th ASTRO Meeting, Int. J. Radiat. Oncol. Biol. Phys.* **39** (Suppl. 2) 194

Pickett B, Vigneault E, Kurhanewicz J, Verhey L and Roach M 1999 Static field intensity modulation to treat a dominant intra-prostatic lesion to 90 Gy compared to seven field 3-dimensional radiotherapy *Int. J. Radiat. Oncol. Biol. Phys.* **43** 921–9

Piermattei A, Delle Canne S, Azario L, Fidanzio A, Soriani A, Orvieto A and Fantini M 1999 Linac Novac7 electron beam calibration using GAF-Chromic film *Phys. Medica* **15** 277–83

Pignoli E, Gramaglia A, Cerchiari U, Somigliana A, Loi G, Zonca G, Marchesini R, Scorsetti M, Casolino D, Luzzara M, Ongania E and Lombardi F 1997 Characteristics of a prototype of a focalized multileaf collimator for conformal radiotherapy *Proc. World Congress on Medical Physics and Biomedical Engineering and 11th Int. Conf. on Medical Physics (Nice, France, September 1997) Med. Biol. Eng. Comput.* **35** (Suppl. Part 2) 1017

Pirzkall A, Graves E, McKnight T, Larson D A, Sneed P K, Wara W M, Nelson S J and Verhey L J 2000 Magnetic resonance spectroscopy guided integrated boost irradiation for high grade gliomas using IMRT *Proc. 13th Int. Conf. on the Use of Computers in Radiation Therapy (Heidelberg, May 2000)* ed W Schlegel and T Bortfeld (Heidelberg: Springer) pp 156–7

Pitkänen M, Ojala A, Holli K and Laippala P 1999 Inter-physician and inter-patient variability in PTV delineation for the radiotherapy of breast cancer after conservative surgery *Radiother. Oncol.* **51** (Suppl. 1) S13

Platoni, K, Lefkopoulos D, Grandjean P and Keraudy K 2000 The influence of dose matrix and voxel dimensions for the dose calculation on inverse treatment planning for stereotactic radiosurgery *Proc. 13th Int. Conf. on the Use of Computers in Radiation Therapy (Heidelberg, May 2000)* ed W Schlegel and T Bortfeld (Heidelberg: Springer) pp 57–9

Ploeger L S, Monique H P, Smitsmans M H P, Gilhuijs K G A and van Herk M 2000 Accurate measurement of the dynamic response of a scanning EPID *Proc. 16th Int. Workshop on Electronic Portal Imaging (EPI2k) (Brussels, 5–7 June 2000)* ed Jan van de Steene p 59

Pollack A 1999 What's your position? We have a need to know *Int. J. Rad. Oncol. Biol. Phys.* **43** 705–6

Popple R and Rosen I 2000 Delivery of multiple IMRT fields using a single physical attenuator *Proc. 13th Int. Conf. on the Use of Computers in Radiation Therapy (Heidelberg, May 2000)* ed W Schlegel and T Bortfeld (Heidelberg: Springer) pp 191–3

Portelance L, Chao C, Grigsby P W and Bennet H 1999 Intensity modulated radiation therapy (IMRT) may reduce small bowel, rectum and bladder complications in patients with cervical cancer treated to the pelvis and paraaortic nodes *Int. J. Radiat. Oncol. Biol. Phys.* **45** (Suppl. 1) 211

Preiser K, Bortfeld T, Hartwig K, Schlegel W and Stein J 1997 A new program for inverse radiotherapy planning *Proc. 12th Int. Conf. on the Use of Computers in Radiation Therapy (Salt Lake City, Utah, May 1997)* ed D D Leavitt and G Starkschall (Madison, WI: Medical Physics Publishing) pp 425–8

Proimos B S and Danciu C 1997a 'Photomodelling': a simple conformal radiotherapy technique modelling gravity oriented protectors and photocopying them in the patient *Proc. World Congress on Medical Physics and Biomedical Engineering and 11th Int. Conf. on Medical Physics (Nice, France, September 1997) Med. Biol. Eng. Comput.* **35** (Suppl. Part 2) 936

Proimos B S and Danciu C 1997b THERA by GOD: THERApy art by a gravity oriented device *Proc. World Congress on Medical Physics and Biomedical Engineering and 11th Int. Conf. on Medical Physics (Nice, France, September 1997) Med. Biol. Eng. Comput.* **35** (Suppl. Part 2) 1029

Pugachev A, Li J G, Boyer A L and Xing L 2000 Beam orientations in IMRT: to optimise or not to optimise *Proc. 13th Int. Conf. on the Use of Computers in Radiation Therapy (Heidelberg, May 2000)* ed W Schlegel and T Bortfeld (Heidelberg: Springer) pp 37–9

Pugachev A, Xing L and Boyer A 1999 Beam orientation optimisation in intensity modulated radiotherapy *Med. Phys.* **26** 1065

Purdy J A 1996 Intensity-modulated radiation therapy *Int. J. Radiat. Oncol. Biol. Phys.* **35** 845–6

Purdy J A 1997 The development of intensity modulated radiation therapy *The Theory and Practice of Intensity Modulated Radiation Therapy* ed E S Sternick (Madison, WI: Advanced Medical Publishing) pp 1–15

Purdy J A 1999 ICRU report 62: a review of the new supplement to ICRU report 50 *Med. Phys.* **26** 1070

Purdy J A 2000a Future directions in 3-D treatment planning and delivery: a physicist's perspective *Int. J. Radiat. Oncol. Biol. Phys.* **46** 3–6

Purdy J A 2000b Technical and clinical requirements for 3D CRT *Int. J. Radiat. Oncol. Biol. Phys.* **46** 694

Purdy J A 2000c Intensity modulated radiation therapy: what is it and how is it delivered? *Int J. Rad. Oncol. Biol. Phys.* **46** 697

Que W 1999a Comparison of algorithms for multileaf collimator field segmentation *Med. Phys.* **26** 1134

Que W 1999b Comparison of algorithms for multileaf collimator field segmentation *Med. Phys.* **26** 2390–6

Rad 1999 IMRT consortium now focuses attention on clinical requirements RAD Magazine **25** No 292, p 46

Ramsey C R, Cordrey I L and Oliver A L 1999 A comparison of beam characteristics for gated and nongated clinical X-ray beams *Med. Phys.* **26** 2086–91

Ranade M, d'Errico F, Barry M and Maryanski M 1998 Clinical 3D dosimetry by optical laser CT scanning of BANG polymer gels *Med. Phys.* **25** A192

Rasch C W 1999 Private communication (Elekta Oncology Systems Consortium, 27–28 May 1999)

Rasch C, Barillot I, Remeijer P, Touw A, van Herk M and Lebesque J V 1999 Definition of the prostate in CT and MRI: a multi-observer study *Int. J. Radiat. Oncol. Biol. Phys.* **43** 57–66

Rasch C, Remeijer P, Barillot I, van Herk M and Lebesque J V 1998 Definition of the prostate in CT and MRI: a multi-observer study *Radiother. Oncol.* **48** (Suppl. 1) S14

Rasmussen C E, Carol M P, Curran B H, Holmes T W and Kania A A 1997 A comparison of the performance of genetic algorithms with simulated annealing for the optimisation of beam weights in intensity modulated radiation therapy *Proc. 12th Int. Conf. on the Use of Computers in Radiation Therapy (Salt Lake City, Utah, May 1997)* ed D D Leavitt and G Starkschall (Madison, WI: Medical Physics Publishing) p 478

Read G 1997 Conformal radiotherapy *Br. J. Radiol.* **70** (Suppl.) 5

Reckwerdt P J, Mackie T R, Balog J and McNutt T R 1997 Three dimensional inverse treatment optimisation for tomotherapy *Proc. 12th Int. Conf. on the Use of Computers in Radiation Therapy (Salt Lake City, Utah, May 1997)* ed D D Leavitt and G Starkschall (Madison, WI: Medical Physics Publishing) pp 420–2

Reckwerdt P J, Olivera G H, Shepard D M and Mackie T R 1998 Helical intensity modulation optimization as used in tomotherapy *Med. Phys.* **25** A117

Reckwerdt P J, Olivera G H, Shepard D M and Mackie T R 2000 Case studies in tomotherapy optimization: breast, prostate, mesothelioma and nasopharyngeal treatments *Proc. 13th Int. Conf. on the Use of Computers in Radiation Therapy (Heidelberg, May 2000)* ed W Schlegel and T Bortfeld (Heidelberg: Springer) pp 60–2

Redpath A T 1998 Planning of beam intensity modulation using an advanced 3D dose calculation algorithm and a simulated annealing method *Radiother. Oncol.* **49** 295–304

Reinstein L E, Wang X-H, Burman C M, Chen Z, Mohan R, Kutcher G, Leibel S A and Fuks Z 1998 A feasibility study of automated inverse treatment planning for cancer of the prostate *Int. J. Radiat. Oncol. Biol. Phys.* **40** 207–14

Reise S, Bleackley N, Donovan E, Evans P, Hansen V, Partridge M and Yarnold J 2000 IMRT for breast radiotherapy using physical compensators and MLC *Int J. Rad. Oncol. Biol. Phys.* **46** 748

Remeijer P, van Herk M, Rasch C and Lebesque J V 1999 A probabilistic recipe to create treatment margins from statisical data on geometrical accuracy *Radiother. Oncol.* **51** (Suppl. 1) S2

Rhein B, Grosser K.H, Häring P and Bortfeld T 1999a The dose verification concept in intensity modulated radiation therapy (IMRT) at the German Cancer Research Centre *Radiother. Oncol.* **51** (Suppl. 1) S44

Rhein B, Grosser K.H, Häring P and Bortfeld T 1999b Experience with the dosimetric verification in intensity modulated radiotherapy (IMRT) at the German Cancer Research Centre *Int. J. Radiat. Oncol. Biol. Phys.* **45** (Suppl. 1) 408

Rhein B, Grosser K-H, Häring P, Bortfeld T and Debus J 1999c The dosimetric verification in intensity modulated radiotherapy at the German Cancer Research Centre *Course Compendium of the 1st IMRT Winter School (Heidelberg, 9–11 December 1999)* pp 77–86

Rhein B and Häring P 2000 The dosimetric verification in intensity modulated radiotherapy at the German Cancer Research Centre *Course Compendium of the 2nd IMRT Winter School (Heidelberg, 19–20 May 2000)* pp 77–86

Richter J, Hädinger U, Kölbl O, Neumann M and Flentje M 2000 Realisation of IMRT (step and shoot technique) and MLC rotation technique using the linac PRIMUS (Siemens) *Proc. 1st Int. Workshop on IMRT in Clinical Practice (IMRT2k) (Brussels, 8–9 June 2000)* ed D Verellen p 59

Richter J, Neumann M and Bortfeld T 1997 Dynamic multileaf collimator rotation techniques versus intensity modulated fixed fields *Proc. 12th Int. Conf. on the Use of Computers in Radiation Therapy (Salt Lake City, Utah, May 1997)* ed D D Leavitt and G Starkschall (Madison, WI: Medical Physics Publishing) pp 335–7

Riker R J, Carol M P, DeNisi J A and Fang W 1997 Human factors problems in optimisation and evaluation of RT image coregistration *Proc. 12th Int. Conf. on the Use of*

Computers in Radiation Therapy (Salt Lake City, Utah, May 1997) ed D D Leavitt and G Starkschall (Madison, WI: Medical Physics Publishing) pp 241–2

Roach III M 1996 Letter to the Editor *Radiother. Oncol.* **40** 85–7

Roberts S A and Hendry J H 1998 A realistic closed-form radiobiological model of clinical tumour control data incorporating intertumour heterogeneity *Int. J. Radiat. Oncol. Biol. Phys.* **41** 689–99

Robertson J M, Yan D, Girimonte P E and Kota K 1999 The potential benefit of intensity modulated radiation therapy (IMRT) for rectal cancer *Int. J. Radiat. Oncol. Biol. Phys.* **45** (Suppl. 1) 248–9

Rosen I I 1997 Treatment planning for IMRT *The Theory and Practice of Intensity Modulated Radiation Therapy* ed E S Sternick (Madison, WI: Advanced Medical Publishing) pp 37–49

Rosenman J G, Miller E P, Tracton G and Cullip T J 1998 Image registration: an essential part of radiation therapy treatment planning *Int. J. Radiat. Oncol. Biol. Phys.* **40** 197–205

Rowbottom C G 1998 Optimisation of beam orientations in conformal radiotherapy treatment planning *PhD Thesis* London University

Rowbottom C G, Nutting C M and Webb S 2000c Beam-orientation optimisation of IMRT: clinical application to parotid gland tumours *Proc. 1st Int. Workshop on IMRT in Clinical Practice (IMRT2k) (Brussels, 8–9 June 2000)* ed D Verellen p 30

Rowbottom C G, Nutting C M and Webb S 2000b Beam-orientation optimisation of IMRT: clinical application to parotid gland tumours *Radiother. Oncol.* at press

Rowbottom C G, Oldham M and Webb S 1997a Optimisation of beam orientations in radiotherapy treatment planning *Br. J. Radiol.* **70** (Suppl.) 6

Rowbottom C G, Oldham M and Webb S 1999b Constrained customisation of non-coplanar beam-orientations in radiotherapy of brain tumours *Phys. Med. Biol.* **44** 383–99

Rowbottom C G, Webb S and Oldham M 1997b Determination of the optimum beam directions in conformal radiotherapy planning *Proc. 12th Int. Conf. on the Use of Computers in Radiation Therapy (Salt Lake City, Utah, May 1997)* ed D D Leavitt and G Starkschall (Madison, WI: Medical Physics Publishing) 306–9

Rowbottom C G, Webb S and Oldham M 1998a Improvements in prostate radiotherapy from the customisation of beam directions *Med. Phys.* **25** 1171–9

Rowbottom C G, Webb S and Oldham M 1999a Is it possible to optimise a radiotherapy treatment plan? *Int. J. Radiat. Oncol. Biol. Phys.* **43** 698–9

Rowbottom C G, Webb S and Oldham M 1999c Beam-orientation customisation using an artificial neural network *Phys. Med. Biol.* **44** 2251–62

Rowbottom C G, Webb S and Oldham M 2000a Customisation of beam orientations using a multiple-beam cost function: application to patients with cancer of the oesophagus. Private communication (unpublished)

Ruchala K J, Olivera G H, Kapatoes J M, Smilowitz J B, Schloesser E A, Pearson D W, Balog J P, Reckwerdt P J and Mackie T R 2000 Tomographic verification of tomotherapy—before, during, and after treatment *Proc. 13th Int. Conf. on the Use of Computers in Radiation Therapy (Heidelberg, May 2000)* ed W Schlegel and T Bortfeld (Heidelberg: Springer) pp 314–6

Ruchala K, Olivera G, Mackie T 1999b Can MLC leakage and transmission be used to obtain CT images during treatment? *Med. Phys.* **26** 1086

Ruchala K, Olivera G, Mackie T 1999a Megavoltage CT imaging using the incomplete data collected during modulated treatment delivery *Med. Phys.* **26** 1064

Ruchala K, Olivera G, Mackie T and Reckwerdt P 1998 Megavoltage CT on a tomotherapy system: a comparison of maximum likelihood vs filtered back-projection reconstruction algorithms *Med. Phys.* **25** A197

Ruchala K J, Olivera G H, Schloesser E A and Mackie T R 1999c Megavoltage CT on a tomotherapy system *Phys. Med. Biol.* **44** 2597–621

Ruchala K J, Olivera G H, Schloesser E A, Hinderer R and Mackie T R 2000 Calibration of a tomotherapeutic MVCT system *Phys. Med. Biol.* **45** N27–36

Sailer S L, Rosenman J G, Soltys M, Cullip T J and Chen J 1996 Improving treatment planning accuracy through multimodality imaging *Int. J. Radiat. Oncol. Biol. Phys.* **35** 117–24

Salter B and Hevezi J 1999 Non-parallel plane intensity modulated radiosurgery (IMRS) *Med. Phys.* **26** 1137

Salz H, Wiezorek T, Scheithauer M, Müller, R and Wendt T 2000 Evaluation and quality control of a compensation system for intensity-modulated radiotherapy *Proc. 13th Int. Conf. on the Use of Computers in Radiation Therapy (Heidelberg, May 2000)* ed W Schlegel and T Bortfeld (Heidelberg: Springer) pp 216–7

Samuelsson A and Johansson K-A 1999 Investigation of a commercial inverse dose planning system *Radiother. Oncol.* **51** (Suppl. 1) S41

Samuelsson A, Johansson K-A and Mercke C 2000 Method of normalising and sensitivity to set-up variations of IMRT treatment plans *Proc. 13th Int. Conf. on the Use of Computers in Radiation Therapy (Heidelberg, May 2000)* ed W Schlegel and T Bortfeld (Heidelberg: Springer) pp 63–4

Sauer O A 2000 Influence of the dose response model for normal tissue on the dose distribution in IMRT *Proc. 13th Int. Conf. on the Use of Computers in Radiation Therapy (Heidelberg, May 2000)* ed W Schlegel and T Bortfeld (Heidelberg: Springer) pp 260–2

Sauer O A and Bratengaier K 1999 Use of constrained optimization for radiotherapy treatment planning *Radiother. Oncol.* **51** (Suppl. 1) S42

Sauer O A, Shepard D M, Angelos L and Mackie T R 1997 A comparison of objective functions for use in radiotherapy optimisation *Proc. 12th Int. Conf. on the Use of Computers in Radiation Therapy (Salt Lake City, Utah, May 1997)* pp 313–6

Sauer O A, Shepard D M and Mackie T R 1999 Application of constrained optimization to radiotherapy planning *Med. Phys.* **26** 2359–66

Saxner M, Löfgren A and Ahnesjö 1997 Integration of head scatter fluence calculations in treatment planning *Proc. 12th Int. Conf. on the Use of Computers in Radiation Therapy (Salt Lake City, Utah, May 1997)* ed D D Leavitt and G Starkschall (Madison, WI: Medical Physics Publishing) pp 213-5

Scherer M, Grosser K H and Schlegel W 1999 Planning of IMRT, using a micro-multileaf-collimator, in comparison with other IMRT delivery techniques *Radiother. Oncol.* **51** (Suppl. 1) S66

Schlegel W 1999 3D Treatment Planning and Optimisation *Radiother. Oncol.* **51** (Suppl. 1) S1 79–82

Schlegel W, Pastyr O, Herfarth K, Kuhr G, Lappe C, Echner G and Schneberger M 2000 Advances in patient immobilisation and positioning techniques *Proc. 13th Int. Conf. on the Use of Computers in Radiation Therapy (Heidelberg, May 2000)* ed W Schlegel and T Bortfeld (Heidelberg: Springer) pp 508–10

Schlegel W, Pastyr O and Höver K-H 1997b Multileaf-collimators in stereotactic conformal irradiation *Proc. World Congress on Medical Physics and Biomedical Engineering and*

11th Int. Conf. on Medical Physics (Nice, France, September 1997) Med. Biol. Eng. Comput. **35** (Suppl. Part 2) 1044

Schlegel W, Pastyr O, Kubesch R, Diemer T, Küster G, Rhein R and Höver K-H 1997c A computer controlled high-resolution micro-multileaf collimator for stereotactic conformal radiotherapy *Proc. 39th ASTRO Meeting, Int. J. Radiat. Oncol. Biol. Phys.* **39** (Suppl. 2) 183

Schlegel W, Pastyr O, Kubesch R, Stein J, Diemer T, Höver K-H and Rhein R 1997a A computer controlled micro-multileaf collimator for stereotactic conformal radiotherapy *Proc. 12th Int. Conf. on the Use of Computers in Radiation Therapy (Salt Lake City, Utah, May 1997)* ed D D Leavitt and G Starkschall (Madison, WI: Medical Physics Publishing) pp 79–82

Schmidt W F O, Nespor W, Wirth P, Büchler F and Hawliczek R 2000 IMRT in routine—experiences from a feasibility study *Proc. 13th Int. Conf. on the Use of Computers in Radiation Therapy (Heidelberg, May 2000)* ed W Schlegel and T Bortfeld (Heidelberg: Springer) pp 218

Schreiner L, Gallant G, Salomons G, Cartile and Kerr A 1998 Viability of a cobalt-60 tomotherapy unit *Med. Phys.* **25** 1584

Schulz R J 1999a Further improvements in dose distributions are unlikely to affect cure rates *Med. Phys.* **26** 1007–9

Schulz R J 1999b Response to 'Comments on "Further improvements in dose distributions are unlikely to affect cure rates [1999 *Med. Phys.* **26** 1007–9]"' *Med. Phys.* **26** 2704–5

Schulze C, Pijpelink J, Linton N, Bortfeld T and Schlegel W 1997 3D photon dose calculation: from a scientific tool to routine planning *Proc. 12th Int. Conf. on the Use of Computers in Radiation Therapy (Salt Lake City, Utah, May 1997)* ed D D Leavitt and G Starkschall (Madison, WI: Medical Physics Publishing) pp 43–5

Schweikard A, Bodduluri M, Tombropoulos R and Adler J R 1994a Planning, calibration, and collision avoidance for image-guided robotic radiosurgery *Proc. Int. IEEE Workshop on Intelligent Robots and Systems (IROS)* pp 854–61

Schweikard A, Bodduluri M, Tombropoulos R and Adler J R 1996 Planning, calibration, and collision avoidance for image-guided radiosurgery *Proc. Symp. 'Principles and Practice of 3-D Radiation Treatment Planning' (Munich, March 1996)* (Munich: Klinikum Rechts der Isar, Technische Universität)

Schweikard A, Tombropoulos R Z, Kavraki L E, Adler J R and Latcombe J-C 1994b Treatment planning for a radiosurgical system with general kinematics *Proc. IEEE Int. Conf. on Robotics and Automation* pp 1720–7

Schweikard A, Tombropoulos R and Adler J R 1995 Robotic radiosurgery with beams of adaptable shapes *Proc. 1st Int. Conf. on Computer Vision, Virtual Reality and Robotics in Medicine (CVRMed'95): Computer Vision and Robotics in Medicine (Lecture Notes in Computer Science (LNCS) vol 905* ed N Ayache (New York: Springer) pp 138–49

Seco J, Evans P M and Webb S 2000 Modelling the effects of IMRT delivery: constraints and incorporation of beam smoothing into inverse planning *Proc. 13th Int. Conf. on the Use of Computers in Radiation Therapy (Heidelberg, May 2000)* ed W Schlegel and T Bortfeld (Heidelberg: Springer) pp 542–4

Shahnazi K, Yue J, Bongiorni P, Nath R and Maryanski M 1999 Absence of doserate effects on optical response of BANG polymer gels at low and intermediate doserates *Med. Phys.* **26** 1126

Sharif A, Katsoula M, Welch M E and Procter E K 1999 A comparison between single and phased field techniques using the Elekta multi-leaf collimator (MLC) for pituitary treatments on a Rando phantom *Phys. Medica* **15** 224

Sharma K, Newman W S, Weinhous M S, Crownover R L and Macklis R M 1999 Radiation therapy of a moving target: traditional versus dynamically-gated treatment *Int. J. Radiat. Oncol. Biol. Phys.* **45** (Suppl. 1) 426

Sharpe M B 1997 Private communication (Elekta Oncology Systems Consortium Meetings, Crawley, 8–9 May and 5–6 December 1997)

Sharpe M B, Frazier R, Kestin L, Yan D, Jaffray D A, Crilly R J, Wong J W, Martinez A A and Vicini F A 2000 Treatment of breast cancer with intensity modulated radiation therapy *Proc. 1st Int. Workshop on IMRT in Clinical Practice (IMRT2k) (Brussels, 8–9 June 2000)* ed D Verellen p 55

Sharpe M B, Kim L, Musselwhite J and Wong J 1999 An improved active breathing control (ABC) apparatus to immobilise breathing motion *Med. Phys.* **26** 1106

Sharpe M B, Yan D, Jaffray D A, Wu Y, Miller B M, Brabbins D S, Robertson J M, Martinez A A and Wong J W 1998 A comprehensive process for the clinical implementation of intensity modulated radiotherapy *Int. J. Radiat. Oncol. Biol. Phys.* **42** 365

Sharpe M B, Miller B M and Wong J W 2000 Compensation of X-ray beam penumbra in conformal radiotherapy *Med. Phys.* at press

Shepard D M, Angelos L, Sauer O A and Mackie T R 1997 A simple model for examining radiotherapy optimisation *Proc. 12th Int. Conf. on the Use of Computers in Radiation Therapy (Salt Lake City, Utah, May 1997)* ed D D Leavitt and G Starkschall (Madison, WI: Medical Physics Publishing) pp 323–6

Shepard D M, Olivera G, Angelos L, Sauer O, Reckwerdt P and Mackie T R 1999 A simple model for examining issues in radiotherapy optimization *Med. Phys.* **26** 1212–21

Shepard D M, Olivera G, Reckwerdt P J and Mackie T R 2000 Iterative approaches to dose optimisation *Phys. Med. Biol.* **45** 69–90

Sherouse G W 1997 Conformal radiotherapy without multileaf collimators *Unpublished delegate's notes for ESTRO Pre-Meeting Workshop on Challenges in Conformal Radiotherapy prior to World Congress on Medical Physics and Biomedical Engineering and 11th Int. Conf. on Medical Physics (Nice, France, September 1997)*

Shiomi H, Inque T, Yamazaki H, Tanaka E, Yoshida K, Imai A, Yoshioka Y and Inque T 1998 Preliminary report of stereotactic irradiation using cyberknife *Proc. 2nd S Takahashi Memorial Int. Workshop on Three Dimensional Conformal Radiotherapy (Nagoya, Aichi, Japan, 11–13 December 1998)* p 98

Shirato H, Shimizu T, Akita H, Kurauchi N, Shinohara N, Ogura S, Harabayashi T, Aoyama H and Miyasaka K 1999 Fluoroscopic real-time tumour tracking radiotherapy *Int. J. Radiat. Oncol. Biol. Phys.* **45** (Suppl. 1) 205

Shiu A S, Dong L and Maor M 1998 Clinical implementation of a miniature multileaf collimator for conformal stereotactic radiotherapy *Med. Phys.* **25** A129

Shiu A S, Kooy H M, Ewton J R, Tung S S, Wong J, Antes K and Maor M H 1997b Comparison of miniature multileaf collimation (MMLC) with circular collimation for stereotactic treatment *Int. J. Radiat. Oncol. Biol. Phys.* **37** 679–88

Shiu A S, Wong J, Ewton J E, Rittichier H E, Tung S S and Dong L 1997a Computer-controlled miniature multileaf collimator *Proc. 12th Int. Conf. on the Use of Computers in Radiation Therapy (Salt Lake City, Utah, May 1997)* ed D D Leavitt and G Starkschall (Madison, WI: Medical Physics Publishing) pp 83–5

Siebers J V, Keall P J, Arnfield M, Kim J O and Mohan R 2000 Dynamic-MLC modeling for Monte Carlo dose calculations *Proc. 13th Int. Conf. on the Use of Computers in Radiation Therapy (Heidelberg, May 2000)* ed W Schlegel and T Bortfeld (Heidelberg: Springer) pp 455–7

Siebert F A, Schultze J and Kimmig B 1999 Landmarked based image fusion in radiotherapy *Radiother. Oncol.* **51** (Suppl. 1) S7

Siemens 1998 Company literature. See also Siochi R A C Segmentation optimization in IMFAST. Siemens Medical Systems Oncology Care Systems Document Number 90.000974 (9 July 1998)

Sixel K E 1999 Dosimetric evaluation of a 3D treatment planning system for forward planning of intensity modulated fields *Radiother. Oncol.* **51** (Suppl. 1) S9

Siochi R A C 1997 Optimisation of an intensity modulated field *US Patent Specification* 5663999

Siochi R 1998a Leaf sequencing optimisation for static IMRT *Med. Phys.* **25** A200

Siochi R 1998b Fluence error reduction for static IMRT by segment weighting *Med. Phys.* **25** 1584

Siochi R 1998c A fast head scatter calculation algorithm for static IMRT *Med. Phys.* **25** 1576

Siochi R A C 1999a Accuracy and time considerations in intensity modulated treatment planning *Med. Phys.* **26** 1086

Siochi R A C 1999b Virtual micro-IMRT *Med. Phys.* **26** 1096

Siochi R A C 1999c Minimising static intensity modulation delivery time using an intensity solid paradigm *Int. J. Radiat. Oncol. Biol. Phys.* **43** 671–80

Siochi R A C 1999d Penumbra considerations in fluence calculation and verification of intensity modulated fields *Int. J. Radiat. Oncol. Biol. Phys.* **45** (Suppl. 1) 414

Siochi R A C 2000 Optimized decomposition of virtual micro intensity maps *Proc. 13th Int. Conf. on the Use of Computers in Radiation Therapy (Heidelberg, May 2000)* ed W Schlegel and T Bortfeld (Heidelberg: Springer) pp 300–2

Smith A R 1996 Regarding Smith and Ling 1995 IJROBP **33** 977 *Int. J. Rad. Oncol. Biol. Phys.* **35** 635–6

Smith V, Xia P and Verhey L 1998 Comparison of multi-field conformal versus intensity-modulated treatment plans based on complication and control probabilities *Med. Phys.* **25** A202

Smitsmans M H P, Ploeger L S, Gilhuijs K G A and van Herk M 2000 A system for field shape verification during dynamic IMRT *Proc. 16th Int. Workshop on Electronic Portal Imaging (EPI2k) (Brussels, 5–7 June 2000)* ed Jan van de Steene p 60

Söderström S and Brahme A 1993 Optimisation of the dose delivery in a few field techniques using radiobiological objective functions *Med. Phys.* **20** 1201–9

Söderström S and Brahme A 1995 Which is the most suitable number of photon beams in coplanar radiation therapy? *Int. J. Radiat. Oncol. Biol. Phys.* **33** 151–9

Solberg T D, Paul T J, Agazaryan N N, Urmanita T, Arellano A R, Llacer J, Boone R A, Fogg R, DeMarco J J, Chetty I and Smathers J B 2000 Dosimetry of gated intensity modulated radiotherapy *Proc. 13th Int. Conf. on the Use of Computers in Radiation Therapy (Heidelberg, May 2000)* ed W Schlegel and T Bortfeld (Heidelberg: Springer) pp 286–8

Sontag M 1999 Respiratory gated radiotherapy *Med. Phys.* **26** 1112

Sontag M and Burnham B 1998 Design and clinical implementation of a system for respiratory gated radiotherapy *Med. Phys.* **25** A126

Sontag M, Burnham B, Guber M and Patra P 1998 Linear accelerator beam characteristics and dosimetry for low monitor unit treatment delivery *Med. Phys.* **25** 1583

Spies L, Ebert M, Groh B A, Hesse B M and Bortfeld T 2000b Quantitative cone beam megavoltage CT *Proc. 13th Int. Conf. on the Use of Computers in Radiation Therapy (Heidelberg, May 2000)* ed W Schlegel and T Bortfeld (Heidelberg: Springer) pp 564–6

Spies L, Evans P M, Partridge M, Hansen V N and Bortfeld T 2000a Direct measurement and analytical modeling of scatter in portal imaging *Med. Phys.* **27** 462–71

Spies L, Hesse B M and Groh B A 1998 Online dose reconstruction in tomotherapy *Radiother. Oncol.* **48** (Suppl. 1) S206

Spirou S V and Chui C S 1994 Generation of arbitrary intensity profiles by dynamic jaws or multileaf collimators *Med. Phys.* **21** 1031–41

Spirou S V and Chui C-S 1998a A gradient inverse planning algorithm with dose-volume constraints *Med. Phys.* **25** 321–33

Spirou S V and Chui C S 1998b The influence of head scatter and leaf edge effects on dynamic multileaf collimation *Med. Phys.* **25** A109

Spirou S V, Chui C S and LoSasso T 1999 A method to generate intensity-modulated fields that exceed the maximum field size limitations imposed by multileaf collimator designs *Phys. Medica* **15** 210

Starkschall G, Pollack A and Stevens C W 2000 Using a dose-volume feasibility search algorithm for radiation treatment planning *Proc. of the 13th Int. Conf. on the Use of Computers in Radiation Therapy (Heidelberg, May 2000)* ed W Schlegel and T Bortfeld (Heidelberg: Springer) pp 11–13

Stein J 1997 Investigations on intensity modulated treatment techniques in conformal radiotherapy *PhD Thesis* University of Heidelberg

Stein J 1999 Inverse Planning: control data generation *Course Compendium of the 1st IMRT Winter School (Heidelberg, 9–11 December 1999)* pp 27–38

Stein J, Bortfeld T, Dörschel B and Schlegel W 1994 Dynamic X-ray compensation for conformal radiotherapy by means of multileaf collimation *Radiother. Oncol.* **32** 163–73

Stein J, Bortfeld T, Preiser K, Keller-Reichenbacher M A, Levegrün S, Schultz C and Schlegel W 1998 Inverse planning: considerations of limitations of the delivery of intensity modulated fields *Radiother. Oncol.* **48** (Suppl. 1) S69

Stein J, Hartwig K, Levegrün S, Zhang G, Preiser K, Rhein B, Debus J and Bortfeld T 1997a Intensity-modulated treatments: compensators vs multileaf modulation *Proc. 12th Int. Conf. on the Use of Computers in Radiation Therapy (Salt Lake City, Utah, May 1997)* ed D D Leavitt and G Starkschall (Madison, WI: Medical Physics Publishing) pp 338–41

Stein J, Mohan R, Wang X-H, Bortfeld T, Wu Q, Preiser K, Ling C C and Schlegel W 1997b Number and orientation of beams in intensity-modulated radiation treatments *Med. Phys.* **24** 149–60

Steinberg T 1999 Improved beam edge conformity using an MLC shift technique *Med. Phys.* **26** 1098

Sternick E S 1997a Oral comment *London IMRT Conf. (London, UK, 26 June 1997)*

Sternick E S 1997b Acknowledgement *The Theory and Practice of Intensity Modulated Radiation Therapy* ed E S Sternick (Madison, WI: Advanced Medical Publishing)

Sternick E S, Bleier A R, Carol M P, Curran B H, Holmes T W, Kania A A, Lalonde R and Larson L S 1997 Intensity modulated radiation therapy: what photon energy is best?

Proc. 12th Int. Conf. on the Use of Computers in Radiation Therapy (Salt Lake City, Utah, May 1997) ed D D Leavitt and G Starkschall (Madison, WI: Medical Physics Publishing) pp 418–9

Sternick E S and Carol M P 1997 An integrated planning/delivery system for intensity modulated radiation therapy (IMRT) *Proc. World Congress on Medical Physics and Biomedical Engineering and 11th Int. Conf. on Medical Physics (Nice, France, September 1997) Med. Biol. Eng. Comput.* **35** (Suppl. Part 2) 938

Sternick E S, Carol M P and Grant III W 1998 Intensity modulated radiotherapy *Treatment Planning in Radiation Oncology* Ch 9, (Baltimore, MD: Williams & Wilkins) pp 187–213

Sternick E S and Sussman M L 1997 Safety management of IMRT *The Theory and Practice of Intensity Modulated Radiation Therapy* ed E S Sternick (Madison, WI: Advanced Medical Publishing) pp 145–59

Sultanem K, Shu H K, Xia P, Akazawa C, Quivey J M, Verhey L J and Fu K K 1999 3-D intensity modulated radiotherapy (IMRT) in the treatment of nasopharyngeal carcinoma: the UCSF experience *Int. J. Radiat. Oncol. Biol. Phys.* **45** (Suppl. 1) 151

Sun A Y and Meng J S 1997 Virtual 5mm-width multileaf collimation *Proc. 39th ASTRO Meeting, Int. J. Radiat. Oncol. Biol. Phys.* **39** (Suppl. 2) 166

Sutlief S G, Phillips M H, Kalet I J and Cho P S 1997 Integration and verification of intensity modulated radiation therapy into a computer treatment planning program (PRISM) *Proc. 12th Int. Conf. on the Use of Computers in Radiation Therapy (Salt Lake City, Utah, May 1997)* ed D D Leavitt and G Starkschall (Madison, WI: Medical Physics Publishing) pp 441–4

Svatos M, Verhey L and Steinberg T 1999 The use of multiple static fields to smooth MLC field edges *Med. Phys.* **26** 1152

Svatos M, Xia P, Rosenman J, Chang S and Verhey L 2000 Comparison of two clinical inverse planning engines *Proc. 1st Int. Workshop on IMRT in Clinical Practice (IMRT2k) (Brussels, 8–9 June 2000)* ed D Verellen p 21

Svensson R, Källman P and Brahme 1994 Analytic solution for the dynamic control of multileaf collimators *Phys. Med. Biol.* **39** 37–61

Svensson R, Lind B K and Brahme 1997 Biologically optimized dose delivery with narrow scanned photon beams—can the expected clinical outcome be improved by shortening the SSD *Proc. 12th Int. Conf. on the Use of Computers in Radiation Therapy (Salt Lake City, Utah, May 1997)* ed D D Leavitt and G Starkschall (Madison, WI: Medical Physics Publishing) pp 402–5

Svensson R, Lind B and Brahme A 1999 A new compact treatment unit design combining narrow pencil beam scanning and segmental multileaf collimation *Radiother. Oncol.* **51** (Suppl. 1) S21

Swindell W and Evans P M 1996 Scattered radiation in portal images: a Monte-Carlo simulation and an empirical model *Med. Phys.* **23** 63–74

Sykes J R and Williams P C 1998 An experimental investigation of the tongue and groove effect for the Philips multileaf collimator *Phys. Med. Biol.* **43** 3157–65

Symonds-Tayler J R N and Webb S 1998 Gap-stepped MLC leaves with filler blades can eliminate tongue-and-groove underdoses for delivering IMRT with maximum efficiency *Phys. Med. Biol.* **43** 2393–5

Tait D M and Nahum A E 1997 Reply to Acute toxicity in pelvic radiotherapy: a randomised trial of conformal versus conventional treatment *Radiother. Oncol.* **44** 295–6

Tait D M, Nahum A E, Meyer L C, Law M, Dearnaley D P, Horwich A, Mayles W P and Yarnold J R 1997 Acute toxicity in pelvic radiotherapy: a randomised trial of conformal versus conventional treatment *Radiother. Oncol.* **42** 121–36

Tate D, Le Q-T, Xing L, Goffinet D, Poen J, Wolden S and Boyer A 1998 Intensity modulated irradiation of nasopharynx carcinoma with inclusion of cervical and supraclavicular lymph nodes *Int. J. Rad. Oncol. Biol. Phys.* **42** 223

Teh B S, Lu H H, Sobremonte S, Bellezza D, Nizin P, Grant W III, Carpenter S and Butler E B 2000b The potential use of intensity-modulated radiotherapy (IMRT) in women with pectus excavatum desiring breast conserving therapy *Proc. 1st Int. Workshop on IMRT in Clinical Practice (IMRT2k) (Brussels, 8–9 June 2000)* ed D Verellen p 24

Teh B S, Mai W Y, Uhl B M, Augspurger M E, Grant W H and Butler 2000e Intensity modulated radiation therapy (IMRT) following radical prostatectomy: lower acute genitourinary (GU) toxicity profile compared to primary IMRT for prostate cancer *Proc. 1st Int. Workshop on IMRT in Clinical Practice (IMRT2k) (Brussels, 8–9 June 2000)* ed D Verellen p 29

Teh B S, Mai W Y, Uhl B M, Augspurger M E, Grant W H and Butler 2000d Intensity modulated radiation therapy (IMRT) for prostate cancer: a study of dose volume relationship with clinical toxicity *Proc. 1st Int. Workshop on IMRT in Clinical Practice (IMRT2k) (Brussels, 8–9 June 2000)* ed D Verellen p 28

Teh B S, Mai W Y, Uhl B M, Augspurger M E, Grant W H, McGary J, Herman J R, Nizin P and Butler E B 2000f Intensity modulated radiotherapy (IMRT) for localized prostate cancer: preliminary results of acute toxicity compared to conventional and six-field conformal approach *Proc. of the 1st Int. Workshop on IMRT in Clinical Practice (IMRT2k) (Brussels, 8–9 June 2000)* ed D Verellen p 29

Teh B S, Uhl B M, Augspurger M E, Grant W H, McGary J, Herman J R, Nizin P, Woo S Y, Butler E B 1998 Intensity modulated radiotherapy (IMRT) for localized prostate cancer: preliminary results of acute toxicity compared to conventional and six-field conformal approach *Int. J. Radiat. Oncol. Biol. Phys.* **42** 219

Teh B S, Uhl B M, Mai W Y, Berner B M, Augspurger M E, Grant W H, Butler E B 2000c Combined intensity modulated radiation therapy (IMRT) and radioactive gold seed implant (RGSI) in the treatment of prostate cancer *Proc. 1st Int. Workshop on IMRT in Clinical Practice (IMRT2k) (Brussels, 8–9 June 2000)* ed D Verellen p 28

Teh B S, Woo S Y, Mai W H, Uhl B M, Augspurger M E and Butler E B 2000a Intensity modulated radiation therapy (IMRT) following radical prostatectomy: lower acute genitourinary (GU) toxicity profile compared to primary IMRT for prostate cancer *Int J. Rad. Oncol. Biol. Phys.* **46** 749–50

Ten Haken R K 1998 Incorporation of patient set-up errors and organ motion in conformal radiotherapy *Proc. Conf. 'Conformal Therapy: a Clinical Perspective Meeting' (London, UK, 21 May 1998)* (York: IPEM) pp 7–10

Ten Haken R K, Lujan A E, Balter J M and Larsen E W 1999 Toward dose calculations that include the effects of patient breathing *Phys. Medica* **15** 205

Tepper J E 1998 Outcomes assessment of technology *Int. J. Radiat. Oncol. Biol. Phys.* **42** (Suppl.) 120

Thompson H, Evans M, Parker W and Fallone B 1998 Study of numerically produced compensators for conventional and intensity modulated beam therapy *Med. Phys.* **25** A204

Ting J, Butker E, Elder E, Mitchum M, Ghavidel S and Davis L 1999 A leaf sequencing algorithm for dynamic multileaf collimators *Med. Phys.* **26** 1137

Togane D, Hamilton R, Boyer A and Xing L 1998 Dose volume histogram based optimisation for intensity modulated radiation therapy *Med. Phys.* **25** A118

Tombropoulos R Z, Adler J R and Latombe J-C 1999 CARABEAMER: a treatment planner for a robotic radiosurgical system with general kinematics *Med. Image Anal.* **3** 237–64

Tombropoulos R Z, Latcombe J-C and Adler J R 1998 Inverse treatment planning for the CyberKnife *Radiosurgery* **2** 236–50 (Basel: Karger)

Tombropoulos R Z, Latcombe J-C and Adler J R 1996 A general algorithm for beam selection in radiosurgery *Preprints of the IARP Workshop on Medical Robotics* pp 91–8

Tombropoulos R Z, Schweikard A, Latcombe J-C and Adler J R 1995 Treatment planning for image-guided robotics radiosurgery *Proc. 1st Int. Conf. on Computer Vision, Virtual Reality and Robotics in Medicine (CVRMed'95): Computer Vision and Robotics in Medicine (Lecture Notes in Computer Science (LNCS) vol 905* ed N Ayache (New York: Springer) pp 131–7

Tsai J and Engler M 1999 Dependence of dose output on the switch rate of an intensity modulated radiation therapy (IMRT) collimator *Med. Phys.* **26** 1086

Tsai J S, Wazer D E, Ling M N, Wu J K, Fagundes M, DiPetrillo T, Kramer B, Koistinen M and Engler M J 1998 Dosimetric verification of the dynamic intensity-modulated radiation therapy of 92 patients *Int. J. Radiat. Oncol. Biol. Phys.* **40** 1213–30

Tyburski L L, Kestin L, Yan D, Vicini F, Sharpe M and Wong J 1999 An efficient method to improve dose uniformity for tangential breast radiotherapy using multiple MLC segments *Int. J. Radiat. Oncol. Biol. Phys.* **45** (Suppl. 1) 423–4

Twining E W 1937 Tomography, by means of a simple attachment to the Potter–Bucky couch *Br. J. Radiol.* **10** 332–47

Uhl B M, Dong L, Teh B S, Augspurger M E, Grant W H and Butler E B 1999 Intensity modulated radiotherapy (IMRT) for prostate cancer: rectal mucosal sparing by dosimetric effect of a rectal balloon *Int. J. Radiat. Oncol. Biol. Phys.* **45** (Suppl. 1) 418–9

Uhl B M, Teh B S, Wheeler T, Scardino P T, Woo S Y and Butler E B 1998 Intensity-modulated radiation therapy (IMRT) for localised prostate cancer: a comparison of Peacock conformal treatment volumes and pathologic radical prostatectomy specimens *Int. J. Radiat. Oncol. Biol. Phys.* **42** 292

Vakaet L, de Gersem W, de Meerleer G and de Neve W 1998 PANIC (Prevent ANy Impending Collision): a 3D vizualization and collision prevention software tool for complex isocentric beam set-ups *Radiother. Oncol.* **48** (Suppl. 1) S148

van Asselen B, Raaijmakers C P J, Hofman P and Lagendijk J J W 2000 An improved technique for tangential breast irradiation applying MLC field shaping and inverse planning *Proc. 13th Int. Conf. on the Use of Computers in Radiation Therapy (Heidelberg, May 2000)* ed W Schlegel and T Bortfeld (Heidelberg: Springer) pp 197–9

van der Heide U A, Nederveen A J, Hofman P, Welleweerd J J and Lagendijk J J W 2000 Microboost treatment of prostate tumours *Proc. 13th Int. Conf. on the Use of Computers in Radiation Therapy (Heidelberg, May 2000)* ed W Schlegel and T Bortfeld (Heidelberg: Springer) pp 537–8

van Dieren E, Nowak P, van Sornsen de Koste J, van der Est H, Wijers O, Heijmen B and Levendag P 1998 Beam intensity modulation using dynamic MLC in 3D CRT of primary cancers in the oropharynx or larynx, including the elective neck *Med. Phys.* **25** A150 van Esche A, Sommer P, Bogaerts R, Rijnders A, Reymen R, Kutcher G J and Huyskens D 2000 Dosimetric treatment verification for IMRT: towards clinical implementation *Proc. 16th Int. Workshop on Electronic Portal Imaging (EPI2k) (Brussels, 5–7 June 2000)* ed Jan van de Steene p 57

van Herk M, Bruce A, Kroes G, Shouman T, Touw A and Lebesque J V 1995 Quantification of organ motion during conformal radiotherapy of the prostate by three dimensional (3D) image registration *Int. J. Radiat. Oncol. Biol. Phys.* **33** 1311–20

van Herk M, de Munck J C and Peters A R 1997 A comprehensive system for three-dimensional image registration in radiotherapy *Proc. 12th Int. Conf. on the Use of Computers in Radiation Therapy (Salt Lake City, Utah, May 1997)* ed D D Leavitt and G Starkschall (Madison, WI: Medical Physics Publishing) pp 237–40

van Herk M, Meinders J, Zijp L, Hoogeman M S and Lebesque J V 1999 The Conquest multimodality image database for clinical trials *Radiother. Oncol.* **51** (Suppl. 1) S6

van Santvoort J P C and Heijmen B J M 1996 Dynamic multileaf collimation without 'tongue-and-groove' underdosage effects *Phys. Med. Biol.* **41** 2091–105

van Santvoort J, Olofsen M, Quint S, van den Berg H and Verstraate M 1999 Inverse treatment planning, using dose-volume constraints, applied to cervical tumours: a pilot study *Radiother. Oncol.* **51** (Suppl. 1) S9

Varian 1997 Product literature RAD2080 10/97

Verellen D and Linthout N 1999 Whole body equivalent dose estimates in tomotherapy *Med. Phys.* **26** 1137

Verellen D, Linthout N, Bel A, van den Berge D, Bel A and Storme G 1997a Initial experience with intensity-modulated conformal radiation therapy for treatment of the head and neck region *Int. J. Radiat. Oncol. Biol. Phys.* **39** 99–114

Verellen D, Linthout N, Bel A, van den Berge D and Storme G 1997b Intensity-modulated conformal radiotherapy for treatment of the head and neck region: the AZ VUB experience *Proc. 2nd Int. Symposium on Special Aspects of Radiotherapy (Berlin, 1–3 May 1997)*

Verellen D, Linthout N and Storme G 1998 Target localization and treatment verification for intensity modulated conformal radiation therapy of the head and neck region *Strahlentherapie und Oncologie* **174** (Suppl. 2) 19–27

Verellen D and Vanhavere F 1999 Risk assessment of radiation-induced malignancies based on whole-body equivalent dose estimates for IMRT treatment in the head and neck region *Radiother. Oncol.* **53** 199–203

Verhey L J 1997 Private communication (Lecture at the Royal Marsden NHS Trust, 10 October 1997)

Verhey L J 1998 Physical considerations in the use of intensity modulated radiotherapy to produce 3D conformal dose distributions *Proc. 2nd S Takahashi Memorial Int. Workshop on Three Dimensional Conformal Radiotherapy (Nagoya, Aichi, Japan, 11–13 December 1998)* pp 24–31

Verhey L J 1999a Comparison of three-dimensional conformal radiation therapy and intensity-modulated radiation therapy systems *Semin. Radiat. Oncol.* **9** 78–98

Verhey L J 1999b IMRT with conventional MLC's *Med. Phys.* **26** 1072–3

Verhey L J, Xia P and Akazawa P 1997a Intensity modulation for head and neck lesions using multiple static multileaf collimator (MLC) fields *Proc. 12th Int. Conf. on the Use of Computers in Radiation Therapy (Salt Lake City, Utah, May 1997)* ed D D Leavitt and G Starkschall (Madison, WI: Medical Physics Publishing) pp 327–8

Verhey L J, Xia P and Akazawa P 1997b Clinically practicable intensity modulation for complex head and neck lesions using multiple static MLC fields *Proc. 39th ASTRO Meeting, Int. J. Radiat. Oncol. Biol. Phys.* **39** (Suppl. 2) 237

Verhey L J, Xia P and Akazawa P 1998 Clinical implementation of intensity modulation using static sequential MLC fields *Med. Phys.* **25** A109–10

Vetterli D, Born E J and Mini R 2000 Verification of intensity modulated beams in radiotherapy using electronic portal imaging *Proc. 16th Int. Workshop on Electronic Portal Imaging (EPI2k) (Brussels, 5–7 June 2000)* ed Jan van de Steene p 61

Viellevigne L, Berre F, Lefkopoulos D, Grandjean P, Kafrouni H, Diaz J C and Noël J N 1999b A 3D inverse planning based on the singular value decomposition: Preliminary results *Radiother. Oncol.* **51** (Suppl. 1) S37

Vieillevigne L, Lefkopoulos D, Grandjean P, Dejean C and Julia F 2000 Influence of the sampling dose and beams spaces on the ill conditioning and on quality of inverse dose distribution for IMRT *Proc. 13th Int. Conf. on the Use of Computers in Radiation Therapy (Heidelberg, May 2000)* ed W Schlegel and T Bortfeld (Heidelberg: Springer) pp 65–7

Vieillevigne L, Lefkopoulos D, Grandjean P, Mazurier J and Keraudy K 1999c Influence of the energy in the inverse process for a 2D planning technique *Radiother. Oncol.* **51** (Suppl. 1) S10

Vieillevigne L, Lefkopoulos D, Grandjean P, Touboul E and Housset M 1999a Intensity modulation optimization by inverse planning into the singular value decomposion space: a preliminary study *Phys. Medica* **15** 11–20

Vineberg K A, Martel M K, Kessler M L, McShan D L, Kim J J, Sandler H M and Fraass B A 1999 Dose escalation of brain tumours to 100+ Gy using automated IMRT optimisation *Int. J. Radiat. Oncol. Biol. Phys.* **45** (Suppl. 1) 270

Walling R, Hartmann Siantar C, Albright N, Wieczorek D, Knapp D, Verhey L, May S and Moses E 1998 Clinical validation of the PEREGRINE MonteCarlo dose calculation system for photon beam teletherapy *Med. Phys.* **25** A128

Wambersie A 1999 Prescribing and reporting photon beam therapy: the problem of margins (the recent ICRU recommendations, Report number 62, 1999) *Phys. Medica* **15** 216

Wang X-H, Mohan R, Jackson A, Leibel S A, Fuks Z and Ling C C 1995 Optimisation of intensity modulated 3D conformal treatment plans based on biological indices *Radiother. Oncol.* **37** 140–52

Wang X, Spirou S, LoSasso T, Chui C S and Mohan R 1996 Dosimetric verification of an intensity modulated treatment *Med. Phys.* **23** 317–27

Watson G A, Leavitt D D, Tobler M, Gaffney D K and Gibbs F A 2000 Application of enhanced dynamic wedge to stereotactic radiotherapy *Proc. 13th Int. Conf. on the Use of Computers in Radiation Therapy (Heidelberg, May 2000)* ed W Schlegel and T Bortfeld (Heidelberg: Springer) pp 194–6

Wazer D E, Ling M, Kramer B, Wu J, Engler M, Tsai J-S and Fagundes M 1997 Clinical applications of IMRT for malignant tumours of the CNS *The Theory and Practice of Intensity Modulated Radiation Therapy* ed E S Sternick (Madison, WI: Advanced Medical Publishing) pp 177–93

Webb S 1989 Optimisation of conformal radiotherapy dose distributions by simulated annealing *Phys. Med. Biol.* **34** 1349–69

Webb S 1990 *From the Watching of Shadows: the Origins of Radiological Tomography* (Bristol: Institute of Physics Publishing)

Webb S 1991a Optimisation by simulated annealing of three-dimensional conformal treatment planning for radiation fields defined by a multileaf collimator *Phys. Med. Biol.* **36** 1201–26

Webb S 1991b Optimisation of conformal radiotherapy dose distributions by simulated annealing: 2. Inclusion of scatter in the 2D technique *Phys. Med. Biol.* **36** 1227–37

Webb S 1992 Optimisation by simulated annealing of three-dimensional, conformal treatment planning for radiation fields defined by a multileaf collimator: 2. Inclusion of two-dimensional modulation of the X-ray intensity *Phys. Med. Biol.* **37** 1689–1704

Webb S 1993 *The Physics of Three-Dimensional Radiation Therapy: Conformal Therapy, Radiosurgery and Treatment Planning* (Bristol: Institute of Physics Publishing)

Webb S 1994a Tomotherapy and beamweight stratification *The Use of Computers in Radiation Therapy: Proc. 11th Conf.* ed A R Hounsell *et al* (Manchester) pp 58–9

Webb S 1994b Optimising the planning of intensity-modulated radiotherapy *Phys. Med. Biol.* **39** 2229–46

Webb S 1995a The problem of isotropically orienting *N* converging vectors in space with application to radiotherapy planning *Phys. Med. Biol.* **40** 945–54

Webb S 1995b Optimising radiation therapy inverse treatment planning using the simulated annealing technique. Special Issue of *Intern. J. Imaging Systems* Optimisation of the three dimensional dose delivery and tomotherapy **6** 71–9

Webb S 1996 Technical characteristics of a 3D treatment planning system for an effective clinical use *Evaluation of 3D Treament Planning Systems for Clinical Use in Radiotherapy* ed L Andreucci (Pisa: Giardini) pp 90–106

Webb S 1997a Dynamic modulation with a prototype NOMOS multileaf intensity-modulating collimator apparatus *Br. J. Radiol.* **70** (Suppl.) 16

Webb S 1997b Historic overview of IMRT *Proc. Meeting: Intensity-Modulated Radiation Therapy: A Clinical Perspective (London, UK, 26 June 1997)* (York: IPEM) pp 4–5

Webb S 1997c Inverse planning for IMRT: the role of simulated annealing *The Theory and Practice of Intensity Modulated Radiation Therapy* ed E S Sternick (Madison, WI: Advanced Medical Publishing) pp 51–73

Webb S 1997d *The Physics of Conformal Radiotherapy: Advances in Technology* (Bristol: Institute of Physics Publishing)

Webb S 1998a Configuration options for intensity-modulated radiation therapy using multiple-static fields shaped by a multileaf collimator *Phys. Med. Biol.* **43** 241–60

Webb S 1998b Configuration options for intensity-modulated radiation therapy using multiple-static fields shaped by a multileaf collimator: 2. Constraints and limitations on 2D modulation *Phys. Med. Biol.* **43** 1481–95

Webb S 1998c The physics of radiation treatment *Phys. World* November 39–43

Webb S 1998d Advances in treatment with intensity-modulated conformal radiotherapy *Tumori* **84** 112–26

Webb S 1998e Intensity-modulated radiation therapy: dynamic MLC (DMLC) therapy, multisegment therapy and tomotherapy: an example of QA in DMLC therapy *Strahlentherapie und Oncologie* **174** (Suppl. 2) 8–12

Webb S 1999a Conformal intensity-modulated radiotherapy (IMRT) delivered by robotic linac—testing IMRT to the limit? *Phys. Med. Biol.* **44** 1639–54

Webb S 1999b Conformal intensity-modulated radiotherapy (IMRT) delivered by robotic linac—testing IMRT to the limit? *Phys. Medica* **15** 161

Webb S 1999c Delivery of IMRT with a MLC *Phys. Medica* **15** 160

Webb S 1999d IMRT Delivery Techniques: general concepts *Course Compendium of the 1st IMRT Winter School (Heidelberg, 9–11 December 1999)* pp 39–52

Webb S 2000a Inverse planning for IMRT with a robotically-mounted linac: parameterisation of this ultimate (?) IMRT technique *Proc. 13th Int. Conf. on the Use of Computers in Radiation Therapy (Heidelberg, May 2000)* ed W Schlegel and T Bortfeld (Heidelberg: Springer) pp 1–3

Webb S 2000b Conformal intensity-modulated radiotherapy (IMRT) delivered by robotic linac—conformality versus efficiency of dose delivery *Phys. Med. Biol.* **45** 1715–30

Webb S 2000c The future of photon external-beam radiotherapy: the dream and the reality *Boston lecture, May 2000 Int. J. Rad. Oncol. Biol. Phys.* at press

Webb S 2000d Advances in three-dimensional conformal radiation therapy with intensity modulation *Lancet Oncol.* **1** 30–6

Webb S, Bortfeld T, Stein J and Convery D J 1997a The effect of stair-step transmission on the 'tongue-and-groove problem' in dynamic radiotherapy with a multileaf collimator *Phys. Med. Biol.* **42** 595–602

Webb S, Convery D J, Bortfeld T and Stein J 1997b A general analysis of the 'tongue-and-groove' effect in dynamic MLC therapy *Proc. 12th Int. Conf. on the Use of Computers in Radiation Therapy (Salt Lake City, Utah, May 1997)* ed D D Leavitt and G Starkschall (Madison, WI: Medical Physics Publishing) pp 342–5

Webb S, Convery D J and Evans P M 1998 Inverse planning with constraints to generate smoothed intensity-modulated beams *Phys. Med. Biol.* **43** 2785–94

Webb S, Convery D J and Evans P M 1999 Inverse planning with constraints to generate smoothed intensity-modulated beams *Radiother. Oncol.* **51** (Suppl. 1) S8

Webb S and Oldham M 1996 A method to study the characteristics of 3D dose distributions created by superposition of many intensity-modulated beams delivered via a slit aperture with multiple absorbing vanes *Phys. Med. Biol.* **41** 2135–53

Webb S and Oldham M 1997a IMRT delivery with the NOMOS MIMiC: Part 1. Modelling treatment delivery *Proc. Meeting: Intensity-Modulated Radiation Therapy: A Clinical Perspective (London, UK, 26 June 1997)* (York: IPEM) pp 17–18

Webb S and Oldham M 1997b Advanced methods for delivering conformal therapy and dose planning techniques *Phys. Medica* **13** (Suppl. 1) 39–44

Webb S and Oldham M 1998 The NOMOS MIMiC for IMRT *Radiother. Oncol.* **48** (Suppl. 1) S64

Welch M 1999a The application of simple techniques makes the Elekta 40 by 40 cm multileaf collimator as accurate but more versatile than a mini multileaf collimator *Med. Phys.* **26** 1152

Welch M E 1999b Simple techniques applied to an Elekta 40 × 40 cm2 multileaf collimator (MLC) equals the accuracy but exceeds the versatility of a mini multileaf collimator *Phys. Medica* **15** 209

Wells D M and Niederer J 1998 A medical expert system approach using artificial neural networks for standardized treatment planning *Int. J. Radiat. Oncol. Biol. Phys.* **41** 173–82

Weltens C, Menten J, Feron M, Bellon E, Demaerel P, Maes F and van der Schuren E 1998 Interobserver variations in gross tumour volume delineation of brain tumours on CT and impact of MRI *Radiother. Oncol.* **48** (Suppl. 1) S14

Wijers G, Levendag P, van Dieren E, van Sörnsen de Koste J, van der Est H, Heijmen B and Nowak P 1998 Beam intensity modulation using dynamic MLC in 3-DCRT of primary cancers of the oropharynx and larynx, including the elective neck *Radiother. Oncol.* **48** (Suppl. 1) S17

Wilkinson J, Williams P, Hounsell A, Budgell G and Mott J 1998 Customised compensation and class solutions *Proc. Elekta Oncology Systems First Users Conf. (Palm Springs, CA, 1998)* pp 66–74

Williams P C 1998 Conformal therapy—where can it be used and is it cost beneficial? *Radiother. Oncol.* **48** (Suppl. 1) S60

Williams P C 1999 Private communication (Elekta Oncology Systems Consortium, 27–28 May 1999) and A high resolution collimating device—an affordable alternative to a micro-multileaf collimator *Med. Phys.* **26** 1096

Williams P, Budgell G, Mott J, Perrin B, James H, Atherton S and Hounsell A 1999a In vivo verification of clinical intensity-modulated radiotherapy delivered via dynamic multileaf collimation *Med. Phys.* **26** 1085

Williams S S, Xing L, Boyer A and Goffinet D R 1998 Intensity modulated treatment of breast cancer with inclusion of supraclavicular and internal mammary lymph nodes *Int. J. Radiat. Oncol. Biol. Phys.* **42** 370

Willoughby T, Kupelian P and Weinhouse M 2000 Ultrasound for patient positioning in radiation therapy *Proc. 13th Int. Conf. on the Use of Computers in Radiation Therapy (Heidelberg, May 2000)* ed W Schlegel and T Bortfeld (Heidelberg: Springer) p 517

Willoughby T R, Starkschall G, Janian N A and Rosen I I 1996 Evaluation and scoring of radiotherapy treatment plans using an artificial neural network *Int. J. Radiat. Oncol. Biol. Phys.* **34** 923–30

Wittkämper F W, Brugmans M J P, Lebesque J V, van der Horst A and Mijnheer R J 1998 Clinical implementation of IMRT in the Netherlands Cancer Institute *Proc. Elekta Oncology Systems First Users Conf. (Palm Springs, CA, 1998)* pp 76–83

Wong J 1999 Private communication (Elekta Oncology Systems Consortium, 27–28 May 1999)

Wong J 2000 Private communication (Heidelberg, May 2000)

Wong J, Kini V, Sharpe M, Jaffray D, Yan D, Stromberg J and Robertson J 1998a Treatment with active breathing control (ABC) in the thoracic and upper abdominal regions *Med. Phys.* **25** A127

Wong J, Sharpe M and Jaffray D 1997a The use of active breathing control (ABC) to minimise breathing motion in conformal therapy *Proc. 12th Int. Conf. on the Use of Computers in Radiation Therapy (Salt Lake City, Utah, May 1997)* ed D D Leavitt and G Starkschall (Madison, WI: Medical Physics Publishing) pp 220–2

Wong J W, Sharpe M B, Jaffray D A, Kini V R, Robertson J M, Stromberg J S and Martinez A A 1999 The use of active breathing control (ABC) to reduce margin for breathing motion *Int. J. Radiat. Oncol. Biol. Phys.* **44** 911–9

Wong J, Sharpe M, Jaffray D, Robertson J M, Stromberg J S, Kini V R and Martinez A A 1997b The use of active breathing control (ABC) to minimise breathing motion in conformal therapy *Proc. 39th ASTRO Meeting, Int. J. Radiat. Oncol. Biol. Phys.* **39** (Suppl. 2) 164

Wong J, Sharpe M, Yan D and Jaffray D 1998b A comprehensive intensity modulated radiation therapy (IMRT) process *Radiother. Oncol.* **48** (Suppl. 1) S64

Wong J, Yan D, Jaffray D A, Edmundsen G and Martinez A A 2000 Interventional strategies to counter the effects of inter-fraction treatment variation *Proc. 13th Int. Conf. on the Use of Computers in Radiation Therapy (Heidelberg, May 2000)* ed W Schlegel and T Bortfeld (Heidelberg: Springer) pp 511–3

Woo S Y, Butler B and Grant W H 1997 Clinical experience: benign tumours of the CNS and head and neck tumours *The Theory and Practice of Intensity Modulated Radiation Therapy* ed E S Sternick (Madison, WI: Advanced Medical Publishing) pp 195–8

Woo S Y, Grant III W, Bellezza D, Grossman R, Gildenberg P, Carpenter L S, Carol M and Butler E B 1996 A comparison of intensity modulated conformal therapy with a conventional external beam stereotactic radiosurgery system for the treatment of single and multiple intracranial legions *Int. J. Radiat. Oncol. Biol. Phys.* **35** 593–7

Woo S Y, Teh B S, Horowitz M, Dauser R, Strother D, McClain K, Chiu J K and Butler E B 1998 Initial experience of intensity modulated radiotherapy (IMRT) for children with brain or head and neck tumours *Int. J. Radiat. Oncol. Biol. Phys.* **42** 358

Woo S Y, Teh B S, Horowitz M, Dauser R, Strother D, McClain K, Chiu J K and Butler E B 2000 Initial experience of intensity-modulated radiotherapy (IMRT) for pediatric brain tumours *Proc. 1st Int. Workshop on IMRT in Clinical Practice (IMRT2k) (Brussels, 8–9 June 2000)* ed D Verellen p 28

Woudstra E and Storchi P R M 1999 Simultaneous optimisation of segmented intensity modulated beams and beam orientations *Radiother. Oncol.* **51** (Suppl. 1) S45

Wu Q, Manning M, Schmidt-Ullrich R and Mohan R 2000b The potential for sparing of parotids and escalation of biologically effective dose with intensity-modulated radiation treatments of head and neck cancers: a treatment design study *Int. J. Radiat. Oncol. Biol. Phys.* **46** 195–205

Wu Q, Mohan R and Niemierko A 2000a IMRT optimisation based on the generalised equivalent uniform dose (EUD) *Proc. 13th Int. Conf. on the Use of Computers in Radiation Therapy (Heidelberg, May 2000)* ed W Schlegel and T Bortfeld (Heidelberg: Springer) pp 17–19

Wu Q, Mohan R and Schmidt-Ullrich R 1999c Designing IMRT plans for head and neck cancers *Med. Phys.* **26** 1079

Wu Q J, Xing L and Sibata C H 1999b Intensity modulated stereotactic radiosurgery using a micro-multileaf compared with gamma knife and linac arc radiosurgery *Int. J. Radiat. Oncol. Biol. Phys.* **45** (Suppl. 1) 167

Wu X and Zhu Y 2000 A neural network regression model for relative dose computation *Phys. Med. Biol.* **45** 913–22

Wu Y, Yan D, Sharpe M, Miller B and Wong J 1999a A method of generating multiple static fields and delivering intensity modulated radiotherapy *Med. Phys.* **26** 1137

Xia P, Chaung C, Nguyen-Tan F, Fu K K and Verhey L J 2000 Computer simulated patient motion and set-up uncertainties in intensity modulated radiotherapy *Proc. 13th Int. Conf. on the Use of Computers in Radiation Therapy (Heidelberg, May 2000)* ed W Schlegel and T Bortfeld (Heidelberg: Springer) pp 317–9

Xia P, Fu K, Akazawa C and Verhey L 1998 Combination of intensity modulated radiotherapy and 3D conformal radiotherapy for head and neck tumours *Med. Phys.* **25** A149

Xia P, Geis P, Xing L, Ma C, Findley D, Forster K and Boyer A 1999a Physical characteristics of a miniature multileaf collimator *Med. Phys.* **26** 65–70

Xia P and Verhey L 1998 Multileaf collimator leaf-sequencing algorithm for intensity modulated beams with multiple static segments *Med. Phys.* **25** 1424–34

Xia P, Wong G, Curran B and Verhey L 1999b Dosimetric aspects of intensity modulation in serial tomotherapy *Med. Phys.* **26** 1140

Xiao Y 1999a Implementation of a fast algorithm for 3D inverse treatment planning for IMRT *Med. Phys.* **26** 1079

Xiao Y 1999b An optimized forward planning technique for intensity modulated treatment of prostate cancer *Int. J. Radiat. Oncol. Biol. Phys.* **45** (Suppl. 1) 415–6

Xiao Y, Galvin J and Valicenti R 1999 Optimization of weightings for the beams selected through forward planning process *Med. Phys.* **26** 1079

Xing L, Boyer A L, Kapp D S and Hoppe R T 1997b Improving the match of abutting photon fields by dynamically modulating photon beams *Proc. World Congress on Medical Physics and Biomedical Engineering and 11th Int. Conf. on Medical Physics (Nice, France, September 1997) Med. Biol. Eng. Comput.* **35** (Suppl. Part 2) 921

Xing L, Chen Y, Luxton G, Li J G and Boyer A L 2000c Monitor unit calculation for intensity modulated photon beam *Proc. 1st Int. Workshop on IMRT in Clinical Practice (IMRT2k) (Brussels, 8–9 June 2000,)* ed D Verellen pp 32–3

Xing L, Chen Y, Luxton G, Li J G and Boyer A L 2000a Monitor unit calculation for an intensity modulated photon field by a simple scatter-summation algorithm *Phys. Med. Biol.* **45** N1–N7

Xing L, Chen Y, Luxton G, Li J G and Boyer A L 2000b Monitor unit calculation for intensity modulated photon field *Proc. 13th Int. Conf. on the Use of Computers in Radiation Therapy (Heidelberg, May 2000)* ed W Schlegel and T Bortfeld (Heidelberg: Springer) pp 374–6

Xing L, Curran B, Hill R, Holmes T, Ma L and Boyer A 1998a Commissioning and testing of an inverse treatment planning system *Med. Phys.* **25** A150

Xing L, Curran B, Hill R, Holmes T, Ma L, Forster K M and Boyer A 1999c Dosimetric verification of a commercial inverse treatment planning system *Phys. Med. Biol.* **44** 463–78

Xing L, Li J and Boyer A 1999a Optimisation of importance factors in inverse planning *Med. Phys.* **26** 1157

Xing L, Li J G, Donaldson S, Le Q T and Boyer A L 1999b Optimisation of importance factors in inverse planning *Phys. Med. Biol.* **44** 2525–36

Xing L, Li J G, Pugachev A, Le Q T and Boyer A L 1999e Estimation theory and model parameter selection for therapeutic treatment plan optimization *Med. Phys.* **26** 2348–58

Xing L, Hamilton R J, Pelizzari C A and Chen G T Y 1997a Iterative algorithms for three-dimensional inverse treatment planning *Proc. 12th Int. Conf. on the Use of Computers in Radiation Therapy (Salt Lake City, Utah, May 1997)* ed D D Leavitt and G Starkschall (Madison, WI: Medical Physics Publishing) pp 423–4

Xing L, Hamilton R J, Spelbring D, Pelizzari C A, Chen G T Y and Boyer A L 1998b Fast iterative algorithms for three-dimensional inverse treatment planning *Med. Phys.* **25** 1845–9

Xing L, Lin Z-X, Donaldson S S, Le Q T, Tate D, Goffinet D R, Wolden S, Ma L and Boyer A L 2000d Dosimetric effects of patient displacement and collimator and gantry angle misplacement on intensity modulated radiation therapy *Radiother. Oncol.* **56** 97–108

Xing L, Pelizzari C, Kuchnir F T and Chen G T Y 1997c Optimisation of relative weights and wedge angles in treatment planning *Med. Phys.* **24** 215–21

Xing L, Pugachev A, Li J, Le Q, Donaldson S, Goffinet D, Hancock S and Boyer A 1999d A medical knowledge based system for the selection of beam orientations in intensity-modulated radiation therapy (IMRT) *Int. J. Radiat. Oncol. Biol. Phys.* **45** (Suppl. 1) 246

Yan D 2000 Treatment strategy for daily image feedback adaptive radiotherapy *Proc. 13th Int. Conf. on the Use of Computers in Radiation Therapy (Heidelberg, May 2000)* ed W Schlegel and T Bortfeld (Heidelberg: Springer) pp 518–20

Yan D, Jaffray D and Wong J 1997b Accounting for deformation of organs in dose/volume evaluation *Proc. 12th Int. Conf. on the Use of Computers in Radiation Therapy (Salt Lake City, Utah, May 1997)* ed D D Leavitt and G Starkschall (Madison, WI: Medical Physics Publishing) pp 166–7

Yan D, Jaffray D A and Wong J W 1999 A model to accumulate fractionated dose in a deforming organ *ASTRO 1997 presentation*

Yan D, Vicini F, Wong J and Martinez A 1997a Adaptive radiation therapy *Phys. Med. Biol.* **42** 123–32

Yang J N, Mackie T R, Reckwerdt P, Deasy J O and Thomadsen B R 1997 An investigation of tomotherapy beam delivery *Med. Phys.* **24** 425–35

Yi B, Chen Y and Boyer A 2000 The effect of beamlet size on IMRT optimization *Proc. 13th Int. Conf. on the Use of Computers in Radiation Therapy (Heidelberg, May 2000)* ed W Schlegel and T Bortfeld (Heidelberg: Springer) pp 305–7

Yi B, Forster K and Boyer A 1999a The effects of beamlet sizes on IMRT optimisation *Med. Phys.* **26** 1139

Yi B, Mok E, Xing L and Boyer A L 1999b A comparison of 3-dimensional treatment planning: conventional conformal therapy, static gantry IMRT and dynamic arc IMRT *Int. J. Radiat. Oncol. Biol. Phys.* **45** (Suppl. 1) 165

Yin F-F, Guan H and Kim J H 2000 A full 3-D verification technique for static field IMRT *Proc. 13th Int. Conf. on the Use of Computers in Radiation Therapy (Heidelberg, May 2000)* ed W Schlegel and T Bortfeld (Heidelberg: Springer) pp 158-9

Yu C 1998a Design considerations for the sides of multileaf collimator leaves *Phys. Med. Biol.* **43** 1335–42

Yu C 1998b Effects of leakage, tongue-and-groove leaf design and output variations on intensity modulation with multileaf collimator *Med. Phys.* **25** A205

Yu C, Li A, Ma L, Shepard D, Sarfaraz M, Holmes T, Suntharalingham M and Mansfield C 2000a Clinical implementation of intensity-modulated arc therapy *Elekta Oncology Symposium on IMRT (Thomas Jefferson Univerity, PA, USA, 17 March 2000)*

Yu C, Chen D-J, Li A, Ma L, Shepard D and Sarfaraz M 2000b Intensity-modulated arc therapy: clinical implementation and experience *Proc. 13th Int. Conf. on the Use of Computers in Radiation Therapy (Heidelberg, May 2000)* ed W Schlegel and T Bortfeld (Heidelberg: Springer) pp 164–6

Yu C, Jaffray D A, Martinez A A and Wong J W 1997b Predicting the effects of organ motion on the dose delivered by dynamic intensity modulation. *Proc. 39th ASTRO Meeting. Int. J. Radiat. Oncol. Biol. Phys.* **39** (Suppl. 2) 164

Yu C, Jaffray D A and Wong J W 1997a Calculating the effects of intra-treatment organ motion on dynamic intensity modulation *Proc. 12th Int. Conf. on the Use of Computers in Radiation Therapy (Salt Lake City, Utah, May 1997)* ed D D Leavitt and G Starkschall (Madison, WI: Medical Physics Publishing) pp 231–3

Yu C, Jaffray D A and Wong J W 1998 The effects of intra-fraction organ motion on the delivery of dynamic intensity modulation *Phys. Med. Biol.* **43** 91–104

Yu C and Sarfarez M 1998a Effects of leakage, tongue and groove leaf design and output variations on intensity modulated radiation therapy *Proc. Elekta Oncology Systems First Users Conf. (Palm Springs, CA)* P24–6

Yu C and Sarfarez M 1998b Effects of leakage, tongue and groove leaf design and output variations on intensity modulated radiation therapy *Int. J. Radiat. Oncol. Biol. Phys.* **42** (Suppl.) 205

Yu C X, Symons M J, Du M N, Martinez A A and Wong J W 1995 A method for implementing dynamic photon beam intensity modulation using independent jaws and a multileaf collimator *Phys. Med. Biol.* **40** 769–87

Zakharchenko G S 1997 Technology of rapid programmated changing of densitive structure of material medium in zone of irradiation of rotation therapy unit from treatment to treatment *Proc. World Congress on Medical Physics and Biomedical Engineering and 11th Int. Conf. on Medical Physics (Nice, France, September 1997) Med. Biol. Eng. Comput.* **35** (Suppl. Part 2) 1029

Zelefsky M J, Fuks Z, Happersett L, Lee H J, Ling C C, Burman C M, Hunt M, Venkatraman E S, Jackson A A and Leibel S A 1999 Improved conformality and reduced toxicity with high-dose intensity modulated radiation therapy (IMRT) for patients with prostate cancer *Int. J. Radiat. Oncol. Biol. Phys.* **45** (Suppl. 1) 170

INDEX

[1] Numbers in **bold** refer to **figures**.